Dynamic Prediction in Clinical Survival Analysis

MONOGRAPHS ON STATISTICS AND APPLIED PROBABILITY

General Editors

1 Stochastic Population Models in Ecology and Epidemiology *M.S. Barlett* (1960)
2 Queues *D.R. Cox and W.L. Smith* (1961)
3 Monte Carlo Methods *J.M. Hammersley and D.C. Handscomb* (1964)
4 The Statistical Analysis of Series of Events *D.R. Cox and P.A.W. Lewis* (1966)
5 Population Genetics *W.J. Ewens* (1969)
6 Probability, Statistics and Time *M.S. Barlett* (1975)
7 Statistical Inference *S.D. Silvey* (1975)
8 The Analysis of Contingency Tables *B.S. Everitt* (1977)
9 Multivariate Analysis in Behavioural Research *A.E. Maxwell* (1977)
10 Stochastic Abundance Models *S. Engen* (1978)
11 Some Basic Theory for Statistical Inference *E.J.G. Pitman* (1979)
12 Point Processes *D.R. Cox and V. Isham* (1980)
13 Identification of Outliers *D.M. Hawkins* (1980)
14 Optimal Design *S.D. Silvey* (1980)
15 Finite Mixture Distributions *B.S. Everitt and D.J. Hand* (1981)
16 Classification *A.D. Gordon* (1981)
17 Distribution-Free Statistical Methods, 2nd edition *J.S. Maritz* (1995)
18 Residuals and Influence in Regression *R.D. Cook and S. Weisberg* (1982)
19 Applications of Queueing Theory, 2nd edition *G.F. Newell* (1982)
20 Risk Theory, 3rd edition *R.E. Beard, T. Pentikäinen and E. Pesonen* (1984)
21 Analysis of Survival Data *D.R. Cox and D. Oakes* (1984)
22 An Introduction to Latent Variable Models *B.S. Everitt* (1984)
23 Bandit Problems *D.A. Berry and B. Fristedt* (1985)
24 Stochastic Modelling and Control *M.H.A. Davis and R. Vinter* (1985)
25 The Statistical Analysis of Composition Data *J. Aitchison* (1986)
26 Density Estimation for Statistics and Data Analysis *B.W. Silverman* (1986)
27 Regression Analysis with Applications *G.B. Wetherill* (1986)
28 Sequential Methods in Statistics, 3rd edition
G.B. Wetherill and K.D. Glazebrook (1986)
29 Tensor Methods in Statistics *P. McCullagh* (1987)
30 Transformation and Weighting in Regression
R.J. Carroll and D. Ruppert (1988)
31 Asymptotic Techniques for Use in Statistics
O.E. Bandorff-Nielsen and D.R. Cox (1989)
32 Analysis of Binary Data, 2nd edition *D.R. Cox and E.J. Snell* (1989)
33 Analysis of Infectious Disease Data *N.G. Becker* (1989)
34 Design and Analysis of Cross-Over Trials *B. Jones and M.G. Kenward* (1989)
35 Empirical Bayes Methods, 2nd edition *J.S. Maritz and T. Lwin* (1989)
36 Symmetric Multivariate and Related Distributions
K.T. Fang, S. Kotz and K.W. Ng (1990)
37 Generalized Linear Models, 2nd edition *P. McCullagh and J.A. Nelder* (1989)
38 Cyclic and Computer Generated Designs, 2nd edition
J.A. John and E.R. Williams (1995)
39 Analog Estimation Methods in Econometrics *C.F. Manski* (1988)
40 Subset Selection in Regression *A.J. Miller* (1990)
41 Analysis of Repeated Measures *M.J. Crowder and D.J. Hand* (1990)
42 Statistical Reasoning with Imprecise Probabilities *P. Walley* (1991)
43 Generalized Additive Models *T.J. Hastie and R.J. Tibshirani* (1990)
44 Inspection Errors for Attributes in Quality Control
N.L. Johnson, S. Kotz and X. Wu (1991)
45 The Analysis of Contingency Tables, 2nd edition *B.S. Everitt* (1992)

46 The Analysis of Quantal Response Data *B.J.T. Morgan* (1992)
47 Longitudinal Data with Serial Correlation—A State-Space Approach *R.H. Jones* (1993)
48 Differential Geometry and Statistics *M.K. Murray and J.W. Rice* (1993)
49 Markov Models and Optimization *M.H.A. Davis* (1993)
50 Networks and Chaos—Statistical and Probabilistic Aspects *O.E. Barndorff-Nielsen, J.L. Jensen and W.S. Kendall* (1993)
51 Number-Theoretic Methods in Statistics *K.-T. Fang and Y. Wang* (1994)
52 Inference and Asymptotics *O.E. Barndorff-Nielsen and D.R. Cox* (1994)
53 Practical Risk Theory for Actuaries *C.D. Daykin, T. Pentikäinen and M. Pesonen* (1994)
54 Biplots *J.C. Gower and D.J. Hand* (1996)
55 Predictive Inference—An Introduction *S. Geisser* (1993)
56 Model-Free Curve Estimation *M.E. Tarter and M.D. Lock* (1993)
57 An Introduction to the Bootstrap *B. Efron and R.J. Tibshirani* (1993)
58 Nonparametric Regression and Generalized Linear Models *P.J. Green and B.W. Silverman* (1994)
59 Multidimensional Scaling *T.F. Cox and M.A.A. Cox* (1994)
60 Kernel Smoothing *M.P. Wand and M.C. Jones* (1995)
61 Statistics for Long Memory Processes *J. Beran* (1995)
62 Nonlinear Models for Repeated Measurement Data *M. Davidian and D.M. Giltinan* (1995)
63 Measurement Error in Nonlinear Models *R.J. Carroll, D. Rupert and L.A. Stefanski* (1995)
64 Analyzing and Modeling Rank Data *J.J. Marden* (1995)
65 Time Series Models—In Econometrics, Finance and Other Fields *D.R. Cox, D.V. Hinkley and O.E. Barndorff-Nielsen* (1996)
66 Local Polynomial Modeling and its Applications *J. Fan and I. Gijbels* (1996)
67 Multivariate Dependencies—Models, Analysis and Interpretation *D.R. Cox and N. Wermuth* (1996)
68 Statistical Inference—Based on the Likelihood *A. Azzalini* (1996)
69 Bayes and Empirical Bayes Methods for Data Analysis *B.P. Carlin and T.A Louis* (1996)
70 Hidden Markov and Other Models for Discrete-Valued Time Series *I.L. MacDonald and W. Zucchini* (1997)
71 Statistical Evidence—A Likelihood Paradigm *R. Royall* (1997)
72 Analysis of Incomplete Multivariate Data *J.L. Schafer* (1997)
73 Multivariate Models and Dependence Concepts *H. Joe* (1997)
74 Theory of Sample Surveys *M.E. Thompson* (1997)
75 Retrial Queues *G. Falin and J.G.C. Templeton* (1997)
76 Theory of Dispersion Models *B. Jørgensen* (1997)
77 Mixed Poisson Processes *J. Grandell* (1997)
78 Variance Components Estimation—Mixed Models, Methodologies and Applications *P.S.R.S. Rao* (1997)
79 Bayesian Methods for Finite Population Sampling *G. Meeden and M. Ghosh* (1997)
80 Stochastic Geometry—Likelihood and computation *O.E. Barndorff-Nielsen, W.S. Kendall and M.N.M. van Lieshout* (1998)
81 Computer-Assisted Analysis of Mixtures and Applications—Meta-Analysis, Disease Mapping and Others *D. Böhning* (1999)
82 Classification, 2nd edition *A.D. Gordon* (1999)
83 Semimartingales and their Statistical Inference *B.L.S. Prakasa Rao* (1999)
84 Statistical Aspects of BSE and vCJD—Models for Epidemics *C.A. Donnelly and N.M. Ferguson* (1999)
85 Set-Indexed Martingales *G. Ivanoff and E. Merzbach* (2000)
86 The Theory of the Design of Experiments *D.R. Cox and N. Reid* (2000)
87 Complex Stochastic Systems *O.E. Barndorff-Nielsen, D.R. Cox and C. Klüppelberg* (2001)
88 Multidimensional Scaling, 2nd edition *T.F. Cox and M.A.A. Cox* (2001)

89 Algebraic Statistics—Computational Commutative Algebra in Statistics
G. Pistone, E. Riccomagno and H.P. Wynn (2001)

90 Analysis of Time Series Structure—SSA and Related Techniques
N. Golyandina, V. Nekrutkin and A.A. Zhigljavsky (2001)

91 Subjective Probability Models for Lifetimes
Fabio Spizzichino (2001)

92 Empirical Likelihood *Art B. Owen* (2001)

93 Statistics in the 21st Century
Adrian E. Raftery, Martin A. Tanner, and Martin T. Wells (2001)

94 Accelerated Life Models: Modeling and Statistical Analysis
Vilijandas Bagdonavicius and Mikhail Nikulin (2001)

95 Subset Selection in Regression, Second Edition *Alan Miller* (2002)

96 Topics in Modelling of Clustered Data
Marc Aerts, Helena Geys, Geert Molenberghs, and Louise M. Ryan (2002)

97 Components of Variance *D.R. Cox and P.J. Solomon* (2002)

98 Design and Analysis of Cross-Over Trials, 2nd Edition
Byron Jones and Michael G. Kenward (2003)

99 Extreme Values in Finance, Telecommunications, and the Environment
Bärbel Finkenstädt and Holger Rootzén (2003)

100 Statistical Inference and Simulation for Spatial Point Processes
Jesper Møller and Rasmus Plenge Waagepetersen (2004)

101 Hierarchical Modeling and Analysis for Spatial Data
Sudipto Banerjee, Bradley P. Carlin, and Alan E. Gelfand (2004)

102 Diagnostic Checks in Time Series *Wai Keung Li* (2004)

103 Stereology for Statisticians *Adrian Baddeley and Eva B. Vedel Jensen* (2004)

104 Gaussian Markov Random Fields: Theory and Applications
Håvard Rue and Leonhard Held (2005)

105 Measurement Error in Nonlinear Models: A Modern Perspective, Second Edition
Raymond J. Carroll, David Ruppert, Leonard A. Stefanski, and Ciprian M. Crainiceanu (2006)

106 Generalized Linear Models with Random Effects: Unified Analysis via H-likelihood
Youngjo Lee, John A. Nelder, and Yudi Pawitan (2006)

107 Statistical Methods for Spatio-Temporal Systems
Bärbel Finkenstädt, Leonhard Held, and Valerie Isham (2007)

108 Nonlinear Time Series: Semiparametric and Nonparametric Methods
Jiti Gao (2007)

109 Missing Data in Longitudinal Studies: Strategies for Bayesian Modeling and Sensitivity Analysis
Michael J. Daniels and Joseph W. Hogan (2008)

110 Hidden Markov Models for Time Series: An Introduction Using R
Walter Zucchini and Iain L. MacDonald (2009)

111 ROC Curves for Continuous Data
Wojtek J. Krzanowski and David J. Hand (2009)

112 Antedependence Models for Longitudinal Data
Dale L. Zimmerman and Vicente A. Núñez-Antón (2009)

113 Mixed Effects Models for Complex Data
Lang Wu (2010)

114 Intoduction to Time Series Modeling
Genshiro Kitagawa (2010)

115 Expansions and Asymptotics for Statistics
Christopher G. Small (2010)

116 Statistical Inference: An Integrated Bayesian/Likelihood Approach
Murray Aitkin (2010)

117 Circular and Linear Regression: Fitting Circles and Lines by Least Squares
Nikolai Chernov (2010)

118 Simultaneous Inference in Regression *Wei Liu* (2010)

119 Robust Nonparametric Statistical Methods, Second Edition *Thomas P. Hettmansperger and Joseph W. McKean* (2011)

120 Statistical Inference: The Minimum Distance Approach
Ayanendranath Basu, Hiroyuki Shioya, and Chanseok Park (2011)

121 Smoothing Splines: Methods and Applications *Yuedong Wang* (2011)

122 Extreme Value Methods with Applications to Finance *Serguei Y. Novak* (2012)

123 Dynamic Prediction in Clinical Survival Analysis *Hans C. van Houwelingen and Hein Putter* (2012)

Monographs on Statistics and Applied Probability 123

Dynamic Prediction in Clinical Survival Analysis

Hans C. van Houwelingen
Hein Putter

CRC Press
Taylor & Francis Group
Boca Raton London New York

CRC Press is an imprint of the
Taylor & Francis Group, an **informa** business
A CHAPMAN & HALL BOOK

CRC Press
Taylor & Francis Group
6000 Broken Sound Parkway NW, Suite 300
Boca Raton, FL 33487-2742

Version Date: 20111005

International Standard Book Number: 978-1-4398-3533-3 (Hardback)

Visit the Taylor & Francis Web site at
http://www.taylorandfrancis.com

and the CRC Press Web site at
http://www.crcpress.com

Contents

Preface

The inspiration for this book stems from the long-lasting involvement of our department (Department of Medical Statistics of Leiden University Medical Center, the Netherlands) in clinical research, especially in clinical trials for different types of cancer. In cancer trials the usual endpoint to assess the success of a treatment is "survival". The question that is often asked by a patient after a diagnosis of cancer is, "How long will I live?" This question about the patient's prognosis is mostly answered by providing estimates of the probability of mortality "x" years after diagnosis/start of treatment. The value of "x" used depends on the severity of the cancer. For severe cancers, $x = 1$ or $x = 2$ is relevant, while for milder cancers, $x = 5$ or even $x = 10$ is more realistic.

In the last thirty years, statisticians have developed many models and techniques for the analysis of data arising in clinical trials or from cancer registries. Nowadays, the Kaplan-Meier survival curve (Kaplan & Meier 1958) and the Cox regression model for survival data (Cox 1972) are standard elements in the training of medical doctors, and the papers describing these statistical techniques are among the most frequently cited scientific papers. In Web of Science, a search on May 12, 2011 for the papers by Kaplan & Meier and by Cox resulted in 34,946 and 25,149 hits. The number of citations in 2010 was 681 and 431, respectively, even after many decades.

Over the years our department has contributed significantly to the development of statistical models for survival data. This book summarizes part of this research, both published and ongoing.

The emphasis is on the dynamic use of predictive models. The question, "How long will I live?" is not only asked at the start of the treatment, but could be (and will be) asked at follow-up visits. Gradually, the question may change to, "Am I cured?" It will be discussed how these questions can be answered using traditional models, and new approaches will be presented that have been developed in the last few years.

Our ideas will be exemplified by application on six data sets that arise from our clinical cooperations. The data sets are described in Appendix A. The software used in the data analysis will be detailed in Appendix B. Software and data are also available on the book's website, `www.msbi.nl\DynamicPrediction`. This will allow the reader to "replay" our analyses and apply them to his or her own data.

Content

The book consists of four parts. Part I deals with prognostic models for survival data using (clinical) information available at baseline, based on the Cox model. It reviews the Cox model and discusses its use in dynamic prediction. Moreover, it discusses issues like measurement of prediction error, validation and calibration.

Part II is about prognostic models for survival data using (clinical) information available at baseline, when the proportional hazards assumption of the Cox model is violated. Mechanisms are discussed that can lead to violation of the Cox model, specifically the proportional hazards assumption. The dominant mechanism is the presence of heterogeneity in survival, often described by the frailty concept. Different extensions of the Cox model will be presented, most of them inspired by the frailty concept. Finally, a different paradigm, *landmarking*, is introduced that aims at direct estimation of the predictive probabilities of interest.

Part III is dedicated to the use of time-dependent information in dynamic prediction. This information could consist of biomarkers like prostate cancer antigen, or the occurrence of intermediate events. The landmarking approach is very effective in using such information. An elaborate analysis of a data set of breast cancer patients with long-term follow-up and information on relapse and metastasis shows how a prudent answer can be given to the "Am I cured?" question.

In Part IV the (im)possibilities of genomic data in the prediction of survival are explored. The use of gene expression measured in tumor tissue, described as a *gene expression signature* raised a lot of interest in the early 2000s. The well-known breast cancer data of the Netherlands Cancer Institute (van't Veer et al. 2002, van de Vijver et al. 2002) are re-analyzed. The long term effect of such information is assessed as well as the improvement over predictors based on clinical information.

The analyses of the different data sets throughout the book are meant to demonstrate how predictive models can be obtained from proper data sets. They do not pretend to give the *final* predictive model for patients similar to those in these data sets. This is not a textbook on clinical epidemiology but on statistical methodology.

Philosophy

The main purpose of this book is to show how dynamic predictions of survival can be obtained that allow the use of dynamic information without creating very complicated models and procedures.

Such models cannot be defined without proper exploratory analysis of the data on which they should be based. Traditional survival analyses by the Cox model or similar models are a prerequisite to gain insight into the relevance of covariates and the way they are related to survival. However, statistical modeling requires some prudence. Trying to get too much information out of a single data set may lead to overfitting and loss of predictive potential. Therefore, there is less emphasis on variable selection and model building as such.

It will be stressed that predictive models should not be sensitive to model assumptions like the proportional hazards assumption of the Cox model and that concepts like frailties are helpful in understanding violations of the Cox model, but should not be overrated because their existence is an untestable hypothesis.

Readership

The book is aimed at applied statisticians, particularly biostatisticians, who actively analyze clinical data in collaboration with clinicians. The first sections of each chapter are meant to be practical with a focus on working with real life data and are intended to be of interest to clinical epidemiologists as well.

Each chapter has a closing section, "Additional remarks," that gives additional material, on the interpretation of the models, alternative models or theoretical background. Those sections are intended to stimulate further research on predictive models in survival analysis.

Acknowledgments

Two factors significantly contributed to the conception of the book: i) The yearly fishing emails of Rob Calver to Hans van Houwelingen about writing a book, and ii) the participation of Hans van Houwelingen in the Event History Program led by Odd Aalen and Ørnulf Borgan at the Norwegian Center for Advanced Studies in the academic year 2005-2006.

We are grateful to the reviewers, Per Kragh Andersen, Ewout Steyerberg and Theo Stijnen, of our own department, who provided invaluable feedback.

We acknowledge the European Group for Blood and Marrow Transplantation (EBMT), the European Organisation for Research and Treatment of Cancer (EORTC), the Dutch National Cancer Institute (NKI) and the Benelux CML trial for providing the data.

Hans C. van Houwelingen
Hein Putter

About the Authors

Hans C. van Houwelingen received his Ph.D. in mathematical statistics from the Unversity of Utrecht in 1973. He stayed at the Mathematics Department in Utrecht until 1986. In that time his theoretical research interest was empirical Bayes methodology as developed by Herbert Robbins; see Robbins (1956). His main contribution was the finding that empirical Bayes rules could be improved by monitonization (van Houwelingen 1976).

On the practical side, he was involved in all kinds of collaborations with researchers in psychology, chemistry and medicine. The latter brought him to Leiden in 1986 where he was appointed chair and department head of Medical Statistics at the Leiden Medical School, which was transformed into the Leiden University Medical Center (LUMC) in 1996.

Together, with his Ph.D. students, he developed several research lines in logistic regression, survival analysis, meta-analysis, statistical genetics and statistical bioinformatics. In the meantime, the department grew into the Department of Medical Statistics and Bioinformatics, which also includes the chair and staff in Molecular Epidemiology.

He was editor-in-chief of *Statistica Neerlandica* and served on the editorial boards of Statistical Methods in Medical Research, Lifetime Data Analysis, Biometrics, Biostatistics, Biometrical Journal and Statistics and Probability Letters. He is elected member of ISI, fellow of ASA, honorary member of the Dutch Statistical Society (VVS) and ANed, the Dutch Region of the International Biometric Society (IBS).

Dr. van Houwelingen retired on January 1, 2009. On that occasion he was appointed Knight in the Order of the Dutch Lion.

Hein Putter received his Ph.D. in mathematical statistics from the University of Leiden in 1994, under the supervision of Willem van Zwet, on the topic of resampling methods. After post-doc positions in the Department of Mathematics of the University of Amsterdam and the Free University Amsterdam, and at the Statistical Laboratory of the University of Cambridge, he turned to medical statistics in 1998, working for the HIV Monitoring Fund and the International Antiviral Therapy Evaluation Center (IATEC), based at the Amsterdam Medical Center. In 2000, he was appointed assistant professor in the Department of Medical Statistics and Bioinformatics of the Leiden University Medical Center.

His research interests include: statistical genetics, dynamical models in HIV and survival analysis, in particular, competing risks and multi-state models.

Dr. Putter collaborates closely with the Department of Surgery and the Department of Oncology of the LUMC, and with international organizations like the European Organisation for the Research and Treatment of Cancer (EORTC) and the European Group for Blood and Marrow Transplantation (EBMT). He serves as associate editor of *Statistics and Probability Letters and Statistics in Medicine*, and he was guest editor of special issues of *Biometrical Journal* (Putter et al. 2010) and *Journal of Statistical Software* (Putter 2011). He was one of the initiators of the IBS Channel Network.

In 2010, Dr. Putter was appointed full professor in the Department of Medical Statistics and Bioinformatics of the LUMC.

Part I

Prognostic models for survival data using (clinical) information available at baseline, based on the Cox model

Chapter 1

The special nature of survival data

1.1 Introduction

"Survival analysis" deals with the statistical modeling and analysis of "survival time." The latter term is used to denote the time until a pre-specified event from a given starting point. The event could be anything. In industrial applications (Barlow & Proschan 1975) the event of interest is often the failure of a component or the whole machine or the whole production process. In that setting the term "failure time" is used instead of survival time. Failure time still has a negative connotation for the event of interest, but of course the event can also be positive as will be seen below. Therefore, a more appropriate term that has gained some popularity, especially in the social sciences, would be event history analysis. Since this book will be dealing with medical applications, the term survival analysis will be used.

Examples of starting points and events for (clinical) endpoints are

Starting point	Event	Endpoint
Birth	Death from any cause	Duration of life
Birth	Incidence of disease	Age at onset
Diagnosis of disease	Death from any cause	Overall survival
Diagnosis of disease	Death from the disease	Disease-specific survival
Diagnosis of disease	Progression of disease	Time to progression
Start of treatment	Recurrence of disease or death	Disease-free survival
Admission to college	Graduation	Duration of study
Surgery	Recovery	Time to recover

Generally speaking, this book will only study events that occur once. Events that could happen repeatedly are outside the scope of the book (see Hougaard (2000) and Cook & Lawless (2007) for comprehensive reviews). So, this book will cover models for the incidence of a chronic disease like cancer, but not for repeated incidence of recurrent diseases like influenza or the common cold.

Analysis of survival data takes a unique place in the whole spectrum of statistical modeling and analysis. The most important reason is the following special property

SP 1 *It takes time to observe time*

In clinical trials the implication of (SP1) is that in most cases one cannot wait until the last patient dies. One has to put an end to the clinical trial at some point in time and one has to deal with so-called (right) censored observations of the people still alive at the end of the study, for which only a lower bound for the survival time is known. The censoring of observations asks for special statistical techniques that will be encountered throughout the book. A further consequence is that data are always "old." Data on long term survival of patients is only available from patients that were diagnosed a long time ago. It is a bit dangerous to extrapolate such older data to newer cohorts because all kinds of things might have changed in the meantime. In the same spirit, it is impossible to compute the life expectancy of people born recently. The "official" life expectancies as published by government administrations are always based on such extrapolations and have to be interpreted with care. A similar pitfall is the inclusion in an observational study of all patients that are under follow-up at the start of the study. These patients are selected on still being alive. This phenomenon is called (left) truncation and could be the cause of unexpected fallacies as well.

The need for dynamic prediction that was mentioned already in the preface is also a consequence of (SP1). Due to (SP1) the doctor and the patient will have to wait before they can assess whether the treatment has been beneficial. It will be shown in Chapter 10 that it can take some time before the question "Am I cured?" can be answered positively with sufficient confidence.

A related special property is

SP 2 *The event might never happen*

The clearest example in the list given above is the time till recovery after surgery. Unfortunately, some patients will never recover. Because of special property (SP1) it will always be very hard to tell whether (SP2) applies for any specific example or not. A consequence of (SP1) and (SP2) is that there will always be a natural horizon to the observations. Statements about what might happen beyond the horizon are based upon extrapolation of the available data and cannot be verified empirically.

Another special property of survival data is

SP 3 *You only die once*

The classic example is death from other causes in clinical trials with disease-specific survival as endpoint of interest. Examples of death from other causes are death by a car accident, suicide and death caused by unrelated diseases. The simple solution is to consider them all as a form of censoring. That might be useful in helping to understand the (biological) risk factors for the disease related death. However, it might give a wrong picture of the impact of such risk factors from a public health point of view. Risk factors that influence some disease late in life might be less relevant because most people will already be dead before that risk factor can exert its effect. This leads to the concept of *competing risks*. As will be discussed in Chapter 9, producing dynamic predictions in the case of competing risks is far from straightforward. Matters can be simplified (but relevance lost) by pooling competing risks into an event like "death from any cause."

1.2 Basic statistical concepts

To set up the framework for survival data with right censoring, two random variables need to be defined,

T_{surv} : the survival time,

T_{cens} : the censoring time.

Notice that the censoring time is defined here as a random variable. Different censoring mechanisms can be distinguished. The most common one for the data discussed in this book is so-called *administrative censoring*, where the censoring time is determined by the termination of the study as sketched in Section 1.1. Other mechanisms will not be discussed at this stage, but later in the book, when needed. For most purposes it suffices to consider the censoring to be random. The crucial condition for statistical analysis is that

> Survival time T_{surv} and censoring time T_{cens} are independent.

In the presence of explanatory variables this condition can be weakened to independence of T_{surv} and T_{cens} conditional on the explanatory variables.

For both random variables the cumulative distribution functions

$$\begin{aligned} F_{\text{surv}}(t) &= P(T_{\text{surv}} \leq t) \, , \\ F_{\text{cens}}(t) &= P(T_{\text{cens}} \leq t) \, , \end{aligned}$$

can be defined. The distribution function of the survival time is called the "failure function". In survival analysis it is often more convenient to use the complimentary functions, the survival function $S(t)$ and the censoring function $C(t)$ defined by

$$\begin{aligned} S(t) &= 1 - F_{\text{surv}}(t) = P(T_{\text{surv}} > t) \, , \\ C(t) &= 1 - F_{\text{cens}}(t) = P(T_{\text{cens}} > t) \, . \end{aligned}$$

Throughout the book it will be assumed that T_{surv} has a continuous distribution, implying that the survival function $S(t)$ is continuous and differentiable.

The consequence of (SP1) is that in practice there is always an observation horizon t_{hor} such that $C(t_{\text{hor}}) = 0$. The consequence of (SP2) is that one can never be sure that $S(\infty) = \lim_{t\to\infty} S(t) = 0$.

In summarizing a survival data set the most important information is given by (an estimate of) the survival function, but it is also relevant to show (an estimate of) the censoring function. The censoring function describes the distribution of the follow-up times if no patient would have died. In clinical trials the amount of censoring is often reported by the median follow-up time. It is not always clear how this is defined. The correct definition is the 50%-point of the censoring function. In Appendix A both the survival function and the censoring function are shown for all six data sets used in this book. If there is only administrative censoring, duration of the whole trial, duration of the intake period and minimal follow-up can all easily be derived from the censoring function. A nice example is Data Set 1 containing the results of two clinical trials in advanced ovarian cancer. The right panel of Figure A.1 shows that the minimum follow-up is about 4 years and the maximal follow-up is about 7.5 years. That implies that the last patient was entered about 3.5 years after the first patient. The median follow-up is about 5.5 years. The picture is very similar for all clinical trial data sets. The situation is different for Data Set 6 that contains registry data from the EBMT. As can be seen in the right panel of Figure A.9 the censoring times are more or less uniform between 0 and 14 years, which corresponds to a constant influx of patients over a period of 14 years.

In practice it is mostly impossible to observe both T_{surv} and T_{cens}. The observed "survival time" T is the smallest of the two,

$$T = \min(T_{\text{surv}}, T_{\text{cens}}) \ .$$

Moreover, it is known whether T_{surv} or T_{cens} has been observed. This is indicated by the event indicator D. The usual definition is

$$D = \begin{cases} 0, & \text{if } T = T_{\text{cens}} \ ; \\ 1, & \text{if } T = T_{\text{surv}} \ . \end{cases}$$

So, the information on the survival status is summarized in the pair (T,D).

Kaplan-Meier estimates of the survival function and the censoring function

The starting point for the statistical analysis is a sample of n independent observations

$$(t_1,d_1),(t_2,d_2),...,(t_n,d_n)$$

from (T,D). Of the observed survival times $t_1,\ldots,t_n$, those with $d_i = 1$ are called the *event times*; the set of event times is denoted with $\mathscr{D}$. The observed survival times with $d_i = 0$ are called the *censoring times*, the set of which is denoted by $\mathscr{C}$. Using this information and assuming that T_{surv} and T_{cens} are independent, both $S(t)$

and $C(t)$ can be estimated by versions of the *Kaplan-Meier estimator* (Kaplan & Meier 1958),

$$\hat{S}(t) = \prod_{\substack{t_i \leq t \\ t_i \in \mathscr{D}}} \left(1 - \frac{1}{Y(t_i)}\right), \quad \hat{C}(t) = \prod_{\substack{t_i \leq t \\ t_i \in \mathscr{C}}} \left(1 - \frac{1}{Y(t_i)}\right).$$

Here $Y(t)$ is the size of the *risk set* $R(t)$, defined by those individuals that are still alive and in follow-up at $t-$, that is just prior to t. In formula,

$$R(t) = \{i; t_i \geq t\}.$$

The formulas above are strictly speaking only valid if no two events (or censorings) occur at the same time, that is, in the absence of *ties*. In the presence of ties, similar formulas for $\hat{S}(t)$ and $\hat{C}(t)$ hold. The sets $\mathscr{D}$ and $\mathscr{C}$ should be taken as the sets of *distinct* event times and censoring times, respectively and the 1's in the fractions $1/Y(t_i)$ should be replaced by the number of deaths or censorings at those time-points. In the remainder of this book it will be assumed that the underlying distributions are continuous and hence that ties are not present. When appropriate, modifications will be outlined in the presence of ties.

The Kaplan-Meiers estimates $\hat{S}(t)$ and $\hat{C}(t)$ are not continuous. They have jumps at the observed event times and the observed censoring times, respectively. Per definition, the estimate of $S(t)$ or $C(t)$ is given by the lower value at the jumps. The upper value at jumps is denoted by $\hat{S}(t-)$ and $\hat{C}(t-)$. Strictly speaking, they estimate $S(t-) = P(T_{\text{surv}} \geq t)$ and $C(t-) = P(T_{\text{cens}} \geq t)$, respectively.

The standard errors of these estimates are given by Greenwood's formula (Greenwood 1926),

$$\begin{aligned} \text{se}^2(\hat{S}(t)) &= \hat{S}^2(t) \cdot \sum_{\substack{t_i \leq t \\ t_i \in \mathscr{D}}} \frac{1}{Y(t_i)(Y(t_i)-1)}, \\ \text{se}^2(\hat{C}(t)) &= \hat{C}^2(t) \cdot \sum_{\substack{t_i \leq t \\ t_i \in \mathscr{C}}} \frac{1}{Y(t_i)(Y(t_i)-1)}. \end{aligned}$$

The estimate $\hat{S}(t)$ together with its standard error $\text{se}(\hat{S}(t))$ can be used to construct $(1-\alpha)\cdot 100\%$ pointwise confidence intervals for the survival function $S(t)$. The simplest confidence interval is $\hat{S}(t) \pm z_{1-\alpha/2}\,\text{se}(\hat{S}(t))$, with $z_{1-\alpha/2}$ the $1-\alpha/2$ percentile of a standard normal distribution, which is routinely calculated by most statistical software packages. It might however fall outside the interval $[0,1]$. This can be partly remedied by using the transformation $\ln(S(t))$; $(1-\alpha)\cdot 100\%$ pointwise confidence intervals for $S(t)$ based on the logarithmic transformation are given by $\hat{S}(t)\exp(\pm z_{1-\alpha/2}\,\text{se}(\ln(\hat{S}(t))))$. One possible expression of the standard error of $\ln(\hat{S}(t))$, in line with the formulas above, is given by the expression

$$\text{se}^2(\ln(\hat{S}(t))) = \sum_{\substack{t_i \leq t \\ t_i \in \mathscr{D}}} \frac{1}{Y(t_i)(Y(t_i)-1)}.$$

This is the Greenwood estimate of the variance of the log-survival function. Another widely used estimate is the Aalen estimate, given by

$$\text{se}^2(\ln(\hat{S}(t))) = \sum_{\substack{t_i \le t \\ t_i \in \mathscr{D}}} \frac{1}{Y(t_i)^2} \, . \tag{1.1}$$

This estimate will come back in Section 2.1. The resulting confidence intervals cannot contain negative values, but can still contain values larger than 1. Therefore also confidence intervals based on the complementary log-log function $\ln(-\ln(S(t))$ are in popular use. For details see for instance Section 4.3 of Klein & Moeschberger (2003).

Competing risks

In case of competing risks it is more natural to work with the *cause-specific cumulative incidence function* rather than with its complement the cause-specific survival function. Loosely speaking, it is defined as the probability to have died from the specific cause by time t. To formalize this, the simplest case of two competing risks, denoted by 1 and 2 is considered. Two survival times can be defined $T_{\text{surv},1}$ and $T_{\text{surv},2}$. The failure time is defined as $T_{\text{surv}} = \min(T_{\text{surv},1}, T_{\text{surv},2})$, which can in turn be subject to censoring, so that we observe the survival time

$$T = \min(T_{\text{surv}}, T_{\text{cens}}) \, ,$$

and an event indicator which can take three values,

$$D = \begin{cases} 0, & \text{if } T = T_{\text{cens}}; \\ 1, & \text{if } T = T_{\text{surv},1}; \\ 2, & \text{if } T = T_{\text{surv},2}. \end{cases}$$

The cumulative incidence functions for causes 1 and 2 are defined by

$$\begin{aligned} I_1(t) &= P(T_{\text{surv}} \le t, D = 1) \, , \\ I_2(t) &= P(T_{\text{surv}} \le t, D = 2) \, . \end{aligned}$$

The cumulative incidence functions add up to $I_1(t) + I_2(t) = P(T_{\text{surv}} \le t) = F(t)$. In order to formulate estimators for the cumulative incidence functions, first rephrase the Kaplan-Meier estimate of the overall survival curve (not distinguishing between causes of death) as

$$\hat{S}(t) = \prod_{t_i \le t} \left(1 - \frac{\mathbf{1}\{d_i > 0\}}{Y(t_i)} \right) \, .$$

Here $\mathbf{1}\{A\}$ is the *indicator function* of the event A, which is defined to be one if A is true, and zero otherwise. The probability of a failure by time t is estimated as

$$\hat{F}(t) = \sum_{t_i \le t} \frac{\mathbf{1}\{d_i > 0\}}{Y(t_i)} \cdot \hat{S}(t_i-) \, ,$$

where $\hat{S}(t_i-)$ is the Kaplan-Meier estimate of overall survival $\hat{S}$ evaluated just prior to t_i. The intuition of this formula is as follows: if an event is to occur by time t, then it must occur at some point $t_i \leq t$. For the event to occur at t_i, the individual first has to survive up to just prior to t_i (this has probability $\hat{S}(t_i-)$), and given that this has happened, the individual then has to fail at t_i (probability $\mathbf{1}\{d_i > 0\}/Y(t_i)$). Clearly, because death and being alive are mutually exclusive, the relation $1 - \hat{F}(t) = \hat{S}(t)$ must hold. Although the formulas for $\hat{F}(t)$ and $\hat{S}(t)$ are seemingly different, they can be seen to satisfy $1 - \hat{F}(t) = \hat{S}(t)$ (see Section II.6 of Andersen et al. 1993). For competing risks, it is not difficult to use similar constructions. Now for an event of cause 1 to occur at t_i, the individual first has to survive up to just prior to t_i (probability $\hat{S}(t_i-)$), and conditionally given that this has happened, the individual then has to fail from cause 1 at t_i (probability $\mathbf{1}\{d_i = 1\}/Y(t_i)$). So the formula for $\hat{I}_1(t)$ becomes

$$\hat{I}_1(t) = \sum_{t_i \leq t} \frac{\mathbf{1}\{d_i = 1\}}{Y(t_i)} \cdot \hat{S}(t_i-) \,. \tag{1.2}$$

It differs with respect to the formula for $\hat{F}(t)$ in that only events with $d_i = 1$ contribute. A similar formula for $\hat{I}_2(t)$ only incorporates the competing events with $d_i = 2$. It is easy to see that $\hat{I}_1(t) + \hat{I}_2(t) = \hat{F}(t)$. Equation (1.2) is a special case of the Aalen-Johansen formula (Aalen & Johansen 1978). An example of cumulative incidence functions is given in the left panel of Figure A.9 in Section A.6. The two causes of failure considered there are cause 1 = relapse and cause 2 = death without relapse, also called non-relapse mortality. The top curve of the left panel of Figure A.9 shows one minus the cumulative incidence function of relapse, in other words, the distance between $y = 1$ and the top curve represents $\hat{I}_1(t)$. The distance between the top curve and the lower one represents $\hat{I}_2(t)$, the cumulative incidence function of non-relapse mortality. The lower curve itself represents $1 - \hat{I}_1(t) - \hat{I}_2(t) = \hat{S}(t)$, the *relapse-free survival* curve, the probability of being alive without relapse.

The standard error of $\hat{I}_1(t)$ may be obtained by the Aalen-type (cf. Equation (1.1)) estimator

$$\begin{aligned} \mathrm{se}^2(\hat{I}_1(t)) &= \sum_{t_i \leq t} \hat{S}^2(t_i-) \left(\hat{I}_1(t) - \hat{I}_1(t_i)\right)^2 \frac{\mathbf{1}\{d_i > 0\}}{Y(t_i)^2} \\ &\quad + \sum_{t_i \leq t} \hat{S}^2(t_i-) \left[1 - 2\left(\hat{I}_1(t) - \hat{I}_1(t_i)\right)\right] \frac{\mathbf{1}\{d_i = 1\}}{Y(t_i)^2} \,, \end{aligned}$$

see also Section 10.1 of Martinussen & Scheike (2006) or Section 3.4.5 of Aalen et al. (2008).

1.3 Predictive use of the survival function

The Kaplan-Meier estimate of the survival function $S(t)$ describes the "prognosis" of an individual at $t = 0$, the start of the follow-up. In clinical practice, depending

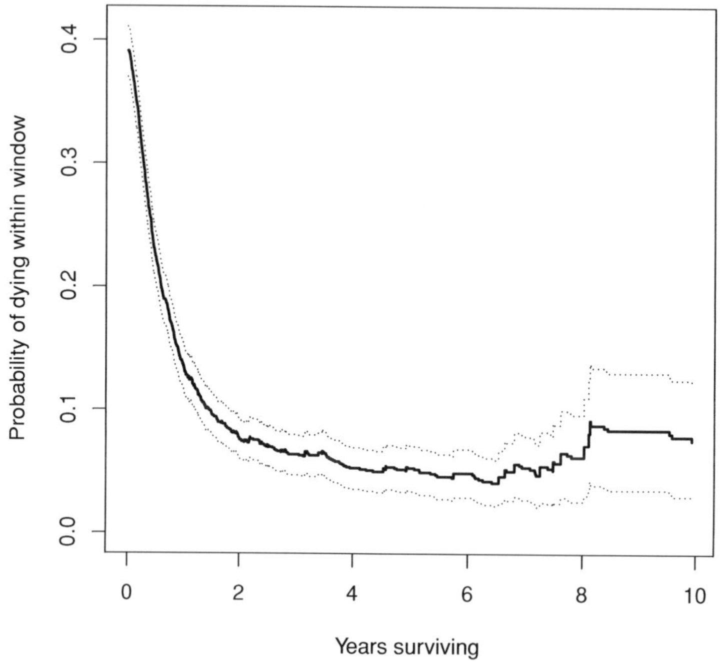

Figure 1.1 *Probability of death or relapse within the next 5 years in Data Set 6; dashed lines represent pointwise 95% confidence intervals*

on the severity of the disease, the information most often given to the patient is the five-year survival probability $S(5)$. However, this does not tell the whole story. Simply giving the 5-year survival probability might give the suggestion that the patient has "conquered" the disease when he will still be alive after 5 years of follow-up. This might not be true. To get a closer look it is very useful to consider the *conditional failure function*

$$S(s|t) = P(T > s \,|\, T \geq t) = \frac{S(s)}{S(t-)} \,,$$

defined for $s \geq t$. This function has two arguments, which makes it hard to display. Useful displays can be obtained by taking a fixed *window of width* w and considering the probability of failure within that window given that you are still alive at time $t-$, i.e., just before t. This *conditional failure function* is defined as

$$F_w(t) = 1 - S(t + w|t) \,,$$

and will be denoted as the *fixed width failure function*. The choice of the window width w depends on the length of the follow-up and the overall prognosis.

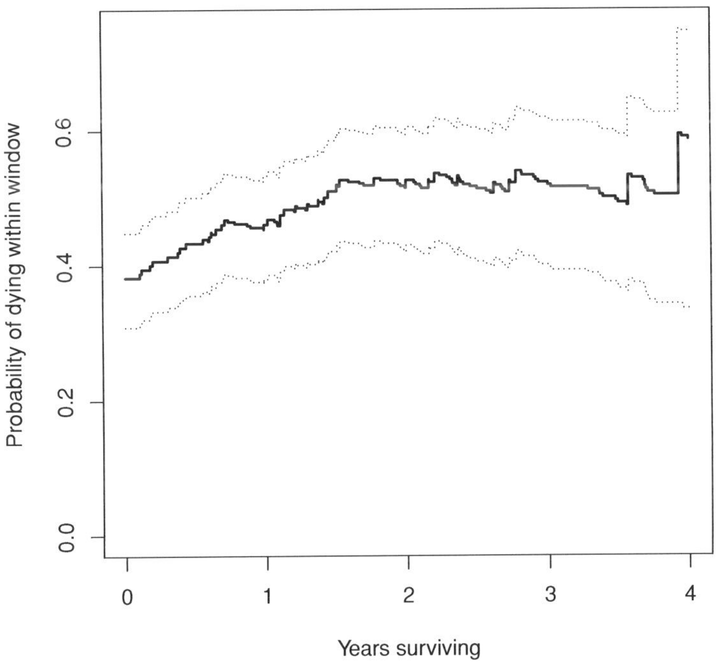

Figure 1.2 *Probability of dying within the next 4 years in Data Set 2; dashed lines represent pointwise 95% confidence intervals*

The insight that can be obtained from such displays can be shown by considering three of the data sets described in the appendix. In Data Set 6, the endpoint is relapse-free survival of patients with Acute Lymphatic Leukemia (ALL). Failure is thus defined as a relapse or a death. The survival function is shown in the left hand panel of Figure A.9. The 5-year survival probability $S(5) = 0.61$. The estimated F_w-curve with a window width $w = 5$ for this data set is shown in Figure 1.1. The good news for the patients is that the probability of failing within the next 5 years very rapidly drops from about 40% to about 10%. If the patient is still alive and relapse-free after two years, the chances are quite good that the disease will not come back. Of course, this is also suggested by the survival function of Figure A.9, but it is hard to estimate such conditional probabilities by eye. It is not hard to compute the standard errors of the estimates in Figure 1.1 and to construct pointwise confidence intervals around it. (The software described in Appendix B allows this.) These confidence intervals are based on the Aalen estimate (cf. Equation (1.1))

$$\mathrm{se}^2(\ln(\hat{F}_w(t))) = \sum_{\substack{t \leq t_i \leq t+w \\ t_i \in \mathscr{D}}} \frac{1}{Y(t_i)^2} \,,$$

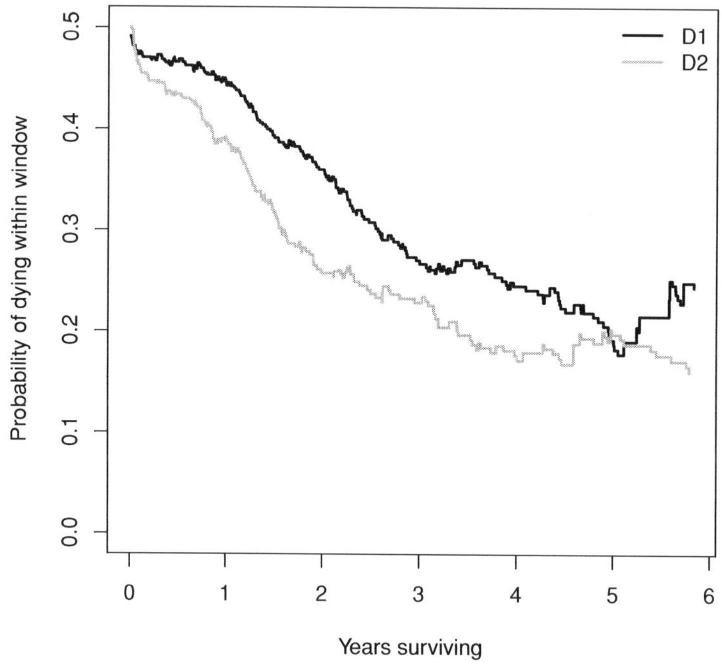

Figure 1.3 *Probability of dying within the next 4 years in Data Set 4*

and shown in thin dashed lines in Figure 1.1; they show that there is a lot of uncertainty about $F_w(t)$ for higher values of t because of the lack of appropriate follow-up data as can be seen from the right hand panel of Figure A.9.

In Data Set 2, where the endpoint is overall survival in patients with Chronic Myeloid Leukemia (CML), the story is completely different. As can be seen in Figure A.2 the survival function falls sharply. The 4 year survival $S(4) = 0.62$, but surviving the first four years does not imply that future survival chances have improved. Figure 1.2 sketches a very grim picture. The probability of dying within the next four years rises slowly. There is very little hope for cure.

Similar displays can also be useful to shed more light on the comparison of an aggressive therapy with a non-aggressive one. In Data Set 4, D1 is the conventional Dutch surgical procedure for gastric cancer surgery, while D2 is the more aggressive Japanese procedure. Figure A.5 shows little difference in survival between the two groups. However, the comparison of the two treatments by the F_w-curves with a window of width $w = 4$ as shown in Figure 1.3 shows a consistent long-term advantage of the D2 procedure for the patients that are willing to accept a higher initial risk. Confidence intervals are not shown because they make the figure too busy.

1.4 Additional remarks

Displaying and summarizing the survival and censoring function

Some software automatically shows the number at risk $Y(t)$ along the time axis of the Kaplan-Meier for the survival function. This suggests that the precision of the Kaplan-Meier estimates at any t is directly related to the value of $Y(t)$. That is not quite true, especially when there is no censoring. Displaying a 95%-confidence band is much more informative. Also, displaying the censoring function gives more information about the potential loss of information due to censoring than the number still at risk.

The Kaplan-Meier estimate is notoriously "jumpy" when the sample size is small. This might look strange, because the true survival function in the total population can be expected to be smooth. However, there is no universally accepted way of smoothing the Kaplan-Meier curve. The Kaplan-Meier is generally accepted despite its jumpiness. The theoretical advantage is that the Kaplan-Meier estimates of survival and censoring function contain all available information.

Useful summary statistics related to Kaplan-Meier curves are the median, quartiles, inter-quartile range and the like. It might happen that the median is not defined because the curve stays above $S(t) = 0.50$. In that case other percentiles like the 90%-percentile might be relevant. Due to the natural follow-up horizon present in most clinical data sets, it does not make sense to compute an estimate of the mean value, called life expectancy in this setting. Attempts of the software to produce such estimates should be ignored.

Life tables

Historically, the study of mortality goes back to the papers of Graunt (1662) and Halley (1693). *Life tables*, sometimes also called *mortality tables* or *actuarial tables* all still used in Official Statistics to compute life expectancies and by insurance companies for life insurances and annuities. In the terminology of this book, the method is based upon dividing the time-axis in intervals and computing the probability of dying in an interval for those alive at the beginning of that interval by the formula

$$p = \frac{d}{r - c/2},$$

where r = number at risk at the beginning of the interval, d = number of deaths within the interval and c = number lost to follow up (censored) within the interval. An example is given in Section 5.4 of Klein & Moeschberger (2003). Using these probabilities the survival up to the end of the ith interval can be computed as $\prod_{j \leq i}(1 - p_j)$. This is very similar to the algorithm for the Kaplan-Meier estimator. For small intervals, the two estimates are very similar and for larger intervals, they are still very similar at the boundaries of the intervals. A rather smooth estimator of the survival function can be obtained by linear interpolation of the survival probabilities at the boundaries of all intervals. Linear interpolation is essential. Drawing step functions, the default in some software, is incorrect, because it suggests that death can only occur at the end of the intervals.

Dynamic cumulative incidence function

Dynamic versions of the cumulative incidence function can be defined by considering all individuals alive at $t-$ and applying the definitions of Section 1.2. That leads to the following dynamic version of a cumulative incidence function for cause k in the case of m competing risks

$$I_k(s|t) = \frac{I_k(s) - I_k(t-)}{S(t-)} = \frac{I_k(s) - I_k(t-)}{1 - \sum_{j=1}^{m} I_j(t-)} \, .$$

Apparently, one needs to know all cumulative incidence functions for the dynamic updating of each individual cumulative incidence function.

Chapter 2

Cox regression model

2.1 The hazard function

In Chapter 1 the "fixed width failure" function $F_w(t) = 1 - S(t + w|t) = P(T \leq t + w \,|\, T \geq t)$ has been defined as a measure that could be used to describe dynamic prediction. For larger values of w it could help to answer the patient's question, "Am I cured?" For small values of w it describes the instantaneous risk of dying for those still alive. This function is known as the *"force of mortality"* or, in more neutral terms, the *hazard function*. Under the assumption of a continuous distribution with differentiable survival function, it can be defined by

$$h(t)dt = P(T < t + dt \,|\, T \geq t) \; .$$

From $P(T > t + dt \,|\, T \geq t) = S(t + dt)/S(t)$ the following alternative definition can be obtained

$$h(t) = -S'(t)/S(t) = -\frac{d\ln(S(t))}{dt} \; .$$

A related concept is the *cumulative hazard function* denoted by $H(t)$ and defined by

$$H(t) = \int_0^t h(s)ds \; ,$$

with inverse relation $h(t) = H'(t)$. Obviously, $H(t)$ and $S(t)$ are closely related. The relation between $H(t)$ and $S(t)$ can be expressed in two ways,

$$H(t) = -\ln(S(t)) \; ,$$

or, inversely, as

$$S(t) = \exp(-H(t)) \; .$$

From the definitions it is clear that the hazard function $h(t)$ is only well defined if the survival function $S(t)$ is differentiable. This fact makes it hard to estimate the hazard function properly, because the Kaplan-Meier estimate of the survival function is a non-differentiable step function and some smoothing is needed before a proper estimate of the hazard function can be obtained.

It is much easier to estimate the cumulative hazard function $H(t)$. One way to do that is to use the link between $H(t)$ and $S(t)$ and to define

$$\hat{H}_{\mathrm{KM}}(t) = -\ln(\hat{S}(t)) = -\sum_{\substack{t_i \leq t \\ t_i \in \mathscr{D}}} \ln\left(1 - \frac{1}{Y(t_i)}\right),$$

where, as defined in Section 1.2, $\hat{S}(t)$ is the Kaplan-Meier estimate, and $Y(t)$ is the size of the risk set $R(t)$, that is the number still alive and in follow-up at $t-$ (just prior to t). The formula above is again assuming the absence of ties, otherwise $\frac{1}{Y(t_i)}$ has to be replaced by $\frac{D_i}{Y(t_i)}$, here and below, with D_i the number of events at t_i. An alternative is the so-called *Nelson-Aalen* estimator (Nelson 1969, Aalen 1975)

$$\hat{H}_{\mathrm{NA}}(t) = \sum_{\substack{t_i \leq t \\ t_i \in \mathscr{D}}} \frac{1}{Y(t_i)}.$$

The standard error of $\hat{H}_{\mathrm{NA}}(t)$ has a particularly simple form, given by

$$\mathrm{se}^2(\hat{H}_{\mathrm{NA}}(t)) = \sum_{\substack{t_i \leq t \\ t_i \in \mathscr{D}}} \frac{1}{Y^2(t_i)}.$$

If the sample size is large, there is very little difference between $\hat{H}_{\mathrm{KM}}(t)$ and $\hat{H}_{\mathrm{NA}}(t)$ or, similarly, between the Kaplan-Meier estimate $\hat{S}(t)$ and $\hat{S}_{\mathrm{NA}}(t) = \exp(-\hat{H}_{\mathrm{NA}}(t))$. The jumps in $\hat{H}_{\mathrm{NA}}(t)$ define a discrete estimate of the hazard concentrated in the event times

$$\hat{h}_{\mathrm{NA}}(t_i) = \frac{1}{Y(t_i)}.$$

Although the hazard function is hard to estimate, it plays an essential conceptual role in thinking about the process of survival. Odd Aalen is one of the most eloquent advocates of the hazard concept. More of his way of thinking can be found in his recent book with Ørnulf Borgan and Håkon Gjessing (Aalen et al. 2008). Revisiting the discussion in Section 1.3, it is noteworthy that the conditional survival function $S(s|t)$ can be expressed as

$$S(s|t) = \exp(-(H(s) - H(t))) = \exp(-\int_t^s h(u)du).$$

The implication is that one only needs to know the hazard on the interval $[t,s]$ to predict survival up to s for those still alive at $t-$. For them, the hazard prior to t is not relevant any more.

The shape of the hazard function determines the long term prospect for a patient. A decreasing hazard function implies that prognosis gets better as you live longer ("old better than new"). This is the case for the ALL patients of Data Set 6

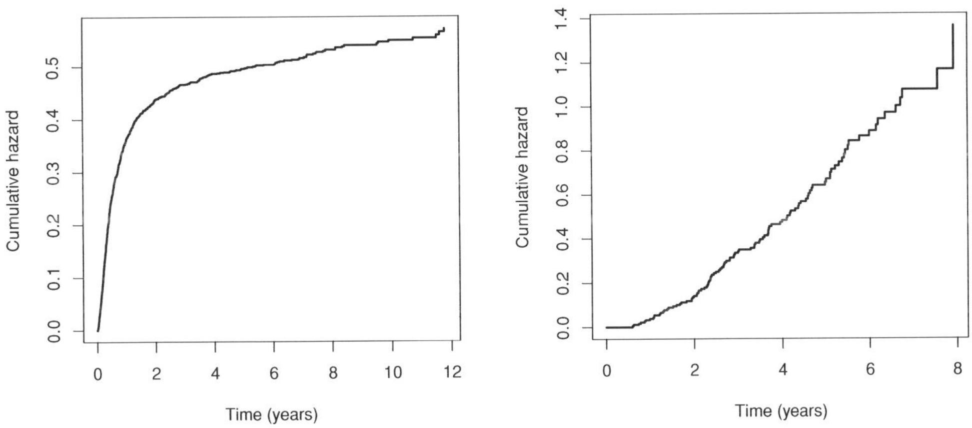

Figure 2.1 *Nelson-Aalen estimates of Data Sets 6 and 2*

as can be seen from Figure 1.1. An increasing hazard function implies that prognosis gets worse as you live longer ("new better than old"). This is the case for the CML patients of Data Set 2 as can be inferred from Figure 1.2. The distinction between decreasing hazards and increasing hazards plays an important role in industrial reliability theory (Barlow & Proschan 1975). Plotting the (estimated) cumulative hazard function for a data set can be a convenient way of detecting an increasing or decreasing hazard function. A convex cumulative hazard function points towards an increasing hazard, while a concave cumulative hazard function goes with a decreasing hazard. Figure 2.1 shows plots of the Nelson-Aalen estimates of death or relapse for Data Set 6 (left), and for death in Data Set 2 (right). The estimated cumulative hazard is clearly concave for Data Set 6, pointing towards an decreasing hazard, consistent with Figure 1.1. For Data Set 2 the estimated cumulative hazard is slightly convex, pointing towards an increasing hazard, again consistent with the "failure within a window" function of Figure 1.2. In case of cure, the cumulative hazard will reach a ceiling.

A popular parametric class of hazard functions (and corresponding distributions) is given by the Weibull distribution that is usually parameterized as

$$H(t|\alpha,\gamma) = \alpha t^{\gamma} .$$

The hazard function $h(t\,|\,\alpha,\gamma) = \alpha\gamma t^{\gamma-1}$ increases if $\gamma > 1$, decreases if $\gamma < 1$, and is constant if $\gamma = 1$. The latter is known as the exponential distribution. An unattractive aspect of the Weibull model is that an increasing hazard always has $h(0) = 0$ and a decreasing hazard always has $h(0) = \infty$. This is often not realistic.

Moreover, the Weibull model always has $H(\infty) = \infty$ implying $S(\infty) = 0$, which is not in line with special property (SP2). This can be remedied by allowing a fraction π for which the event will never happen, leading to

$$S(t) = \pi + (1-\pi)\exp(-\alpha t^{\gamma}) \ .$$

This is an example of a parametric *cure model*. The usefulness of such models will be discussed in Chapters 6 and 10.

The hazard is also a useful concept to describe competing risks and, more generally, multi-state models. The latter are extensively discussed in the Statistics in Medicine Tutorial in Biostatistics of Putter et al. (2007). For competing risks the *cause-specific hazard* for cause k is defined as

$$h_k(t)dt = P(T < t+dt, D=k \,|\, T \geq t) \ .$$

The hazard for death from any cause is given by

$$h(t)dt = P(T < t+dt \,|\, T \geq t) = \sum_k h_k(t)dt \ ,$$

with corresponding cumulative hazard $H(t)$ and survival function $S(t)$. The cumulative incidence function for cause k is given by

$$I_k(t) = \int_0^t h_k(s)S(s-)ds \ .$$

The Aalen-Johansen estimator of Section 1.2 can be recognized as a "discrete" version of this expression.

2.2 The proportional hazards model

To develop methods for survival data in a population of individuals one needs a simple way of describing the variation in survival among individuals. A popular model that fits in well with the interest in dynamic modeling, is to consider the individual specific hazard function $h_i(t)$ and to make the *proportional hazards assumption* that

$$h_i(t) = c_i h_0(t) \ .$$

The constant c_i is called the *hazard ratio* (with respect to the baseline hazard $h_0(t)$) and is denoted by HR_i.

Before discussing how the hazard ratio can be linked to explanatory covariates, it is useful to reflect on how the proportional hazards model translates into differences in survival. So, consider two individuals 1 and 2 with hazard functions $h_1(t)$ and $h_2(t) = \text{HR}_{2|1}h_1(t)$. The symbol $\text{HR}_{2|1}$ denotes the ratio of the hazard of individual 2 with respect to that of individual 1. Consequently, $H_2(t) = \text{HR}_{2|1}H_1(t)$ and

$$S_2(t) = \exp(-H_2(t)) = \exp(-\text{HR}_{2|1}H_1(t)) = S_1(t)^{\text{HR}_{2|1}} \ .$$

To get a better understanding of the effect of the hazard ratio on the absolute risk

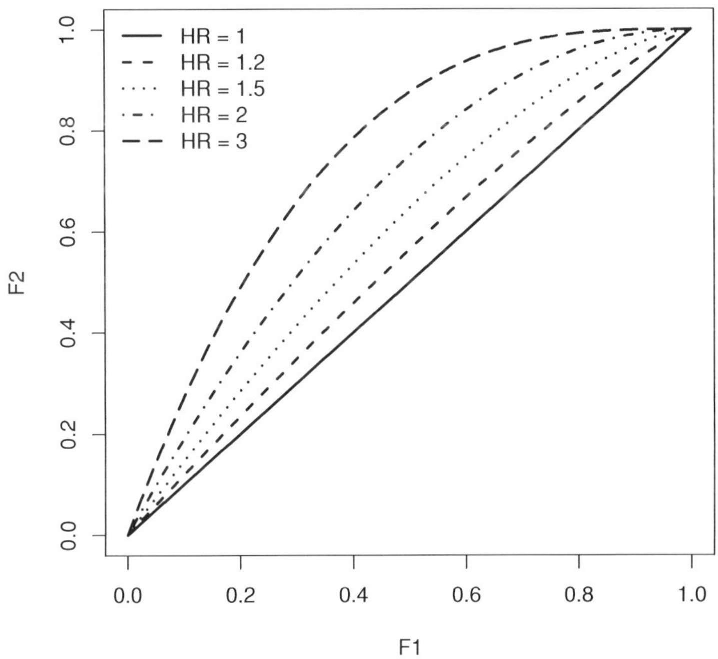

Figure 2.2 *Effect of the hazard ratio on the death risk*

of failure $F(t) = 1 - S(t)$ it is helpful to plot the failure function $P(T_2 \leq t) = F_2(t)$ against $P(T_1 \leq t) = F_1(t)$ as shown in Figure 2.2. The effect of the hazard ratio on the relation between $F_2(t)$ and $F_1(t)$ is not very clear unless the failure function is small as in the lower left corner of the graph. It is not hard to derive that for $F(t)$ close to 0 (implied by small t)

$$F_2(t) \approx \mathrm{HR}_{2|1} F_1(t) \; .$$

In other words, $\mathrm{HR}_{2|1}$ is the slope of the F_2 versus F_1 curve at the bottom-left corner (Figure 2.2). So, the hazard ratio can be interpreted as a relative risk **if** the probability of failure is small. Visually and clinically, the effect of the hazard ratio depends very much on the shape of the hazard function. Figure 2.3 compares the effect of $\mathrm{HR} = 1.5$ for decreasing hazards (left panel) and increasing hazards (right panel). The 5-year survival probabilities for the control group (solid line) are the same for both panels. There is a very clear difference in effect on the life expectancy. This can be seen by comparing the effect on median survival. In the left panel, the median survival is reduced by about 50%, while in the right panel the reduction in median survival is only about 20%. See Section 2.6 for a theoretical explanation.

The effect of covariates on the hazard can conveniently be modeled by taking $\mathrm{HR}_i = \exp(X_i^\top \beta)$ leading to the proportional hazards regression model introduced

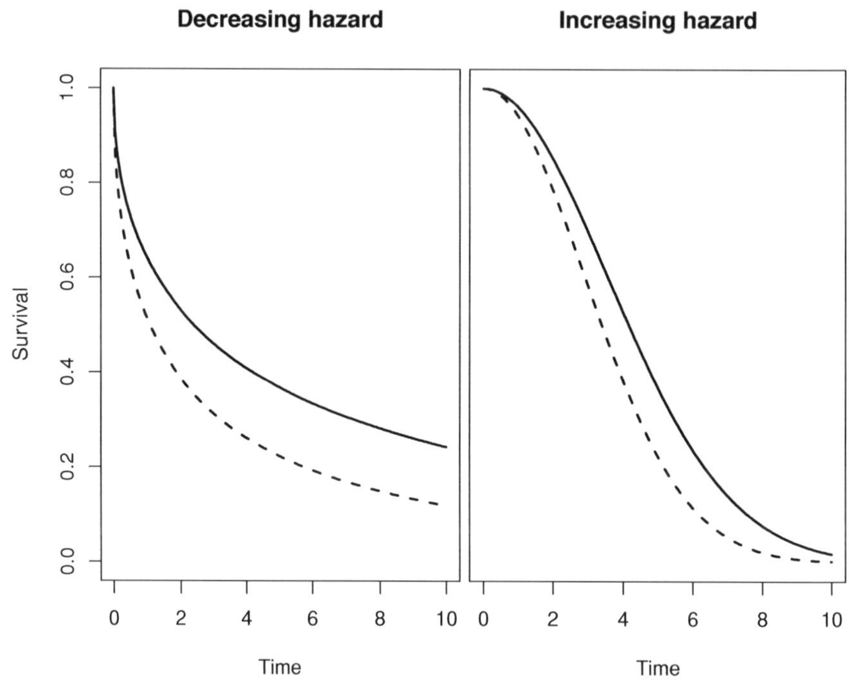

Figure 2.3 *Comparing the effect of HR=1.5 (dotted versus solid) for different types of hazards*

by Cox (1972),

$$h(t|X) = h_0(t)\exp(X^\top\beta)\ .$$

Here, $h_0(t)$ is the so-called *baseline hazard* that determines the shape of the survival function, X is the vector of the covariates of an individual and β is a vector of regression coefficients. It is common practice not to define a parametric model for the baseline hazard. This is in line with the practice to show the Kaplan-Meier estimate of the survival function as a summary of the data. There is no general parametric model for the hazard that fits all (clinical) data well and a model that seems to fit well to the whole data set might show considerable lack of fit when used for dynamic prediction in the way of Chapter 1. As in regression models for other types of data the covariate vector X can contain transformations and interactions of the risk factors. It should be noted that there is no constant term in the regression vector. The constant is absorbed in the baseline hazard. The implication is that centering the covariates, replacing X by $X - \bar{X}$, would change the baseline, but not the regression coefficients. In some software such centering is applied and it is not always easy to figure out what a reported baseline hazard stands for. The survival function implied by the model is given by

$$S(t|X) = \exp(-\exp(X^\top\beta)H_0(t)) = S_0(t)^{\exp(X^\top\beta)}\ .$$

Here, $H_0(t) = \int_0^t h_0(s)ds$ is the *cumulative baseline hazard*, and $S_0(t) = \exp(-H_0(t))$ the baseline survival function. The linear predictor $X^\top\beta$ is known as the *prognostic index* and denoted by PI.

The marginal survival function is obtained by taking the average survival function $\overline{S(t|x)}$ in the population. Since x appears in the exponent, $\overline{S(t|x)}$ is not the same as the estimated survival for the average person $S(t|\bar{x})$ or, more precisely

$$S(t) = E[S(t|X)] \neq S(t|E[X]) \ . \tag{2.1}$$

The difference between $E[S(t|X)]$ and $S(t|E[X])$ can be expected to be small if the variance of the prognostic index $\text{PI} = X^\top\beta$ is small and $S(t|X)$ is not too far from 1. This will be the case in medical applications, where the effect of the covariates are usually limited and the follow-up is quite short.

2.3 Fitting the Cox model

It is most interesting to read the original paper by Cox (1972) and the written discussion following it. The focus of Cox is on the estimation of the regression coefficients using the so-called partial likelihood. The estimation of the baseline hazard has long been neglected. However, as pointed out in the previous section the effect of the hazard ratio can only be fully understood if the baseline hazard is known as well. (The best way of understanding the model is by visualizing the estimated survival curves for representative values of the covariate vector X.) To emphasize the importance of both components (baseline hazard and regression coefficients) the full likelihood of the data will be taken as the starting point for fitting the model.

The available data is a sample of n independent observations from the triple (T,D,X), that is

$$(t_1,d_1,x_1),(t_2,d_2,x_2)...,(t_n,d_n,x_n) \ .$$

It is assumed that there are no ties among the t_i's. The log-likelihood of the data is given by

$$l(h_0,\beta) = \sum_{i=1}^{n}\left(-H_0(t_i)\exp(x_i^\top\beta) + d_i(\ln(h_0(t_i)) + x_i^\top\beta)\right) \ .$$

This expression will be maximized by concentrating all the risk in the event times t_i. This leads to a discrete version of the hazard as discussed in Section 2.1 for which the cumulative hazard is defined as

$$H_0(t) = \sum_{t_i \le t} h_0(t_i) \ .$$

Plugging this into the expression for the log-likelihood and rearranging some terms gives

$$l(h_0,\beta) = \sum_{i=1}^{n}\left(-h_0(t_i)\sum_{j\in R(t_i)}\exp(x_j^\top\beta) + d_i(\ln(h_0(t_i)) + x_i^\top\beta)\right) \ .$$

For fixed value of β this expression is maximal for the so-called Breslow estimator (Breslow 1974),

$$\hat{h}_0(t_i|\beta) = \frac{d_i}{\sum_{R(t_i)} \exp(x_j^\top \beta)} .$$

The resulting maximized (or profile) log-likelihood is

$$l(\hat{h}_0(.|\beta), \beta) = \sum_{i=1}^{n} d_i(-1 + \ln(\hat{h}_0(t_i|\beta)) + x_i^\top \beta) = -\sum_{i=1}^{n} d_i + \mathrm{pl}(\beta) .$$

Here, $\mathrm{pl}(\beta)$ is Cox's partial log-likelihood defined as

$$\mathrm{pl}(\beta) = \sum_{i=1}^{n} d_i \cdot \ln\left(\frac{\exp(x_i^\top \beta)}{\sum_{j\in R(t_i)} \exp(x_j^\top \beta)}\right) . \tag{2.2}$$

Cox did not obtain this expression as a profile likelihood, but used a conditioning argument. The term $\exp(x_i^\top \beta)/\sum_{j\in R(t_i)} \exp(x_j^\top \beta)$ can be interpreted as the probability that individual i is the one that died at event time t_i given the risk set $R(t_i)$ of people still alive and in follow-up just prior to t_i.

So, the computational procedure is to estimate β by maximizing the partial log-likelihood and estimating the baseline-hazard by the Breslow estimator with $\beta = \hat{\beta}$,

$$\hat{h}_0(t_i) = \frac{d_i}{\sum_{j\in R(t_i)} \exp(x_j^\top \hat{\beta})} .$$

The survival function given the covariate x can be estimated either by the analogue of the Nelson-Aalen estimator

$$\hat{S}_{\mathrm{NA}}(t|x,\hat{\beta}) = \exp\left(-\hat{H}_0(t)\exp(x^\top \hat{\beta})\right) ,$$

or the analogue of the Kaplan-Meier estimator, the so-called *product-limit estimator*

$$\hat{S}_{\mathrm{PL}}(t|x,\hat{\beta}) = \prod_{t_i\le t}\left(1 - \exp(x^\top \hat{\beta})\hat{h}_0(t_i)\right) .$$

Most software packages provide $\hat{S}_{\mathrm{NA}}$. In R, both $\hat{S}_{\mathrm{NA}}$ and $\hat{S}_{\mathrm{PL}}$ can be calculated, through the `type` argument of the function `survfit()` in the **survival** package. Note that for some covariate values x, it can happen that $\exp(x^\top \hat{\beta})\hat{h}_0(t_i) > 1$ for some t_i. This then results in negative values of $\hat{S}_{\mathrm{PL}}(t|x,\hat{\beta})$. In practice, there is very little difference between the two methods. It should be pointed out that most of the formulas above still apply in case of ties. It will lead to the Breslow version of the partial likelihood for ties (Breslow 1972, Peto 1972), which is much simpler to handle than the formulas given in Cox's original paper which are meant for the case of truly discrete time in which it is known beforehand that the events can only occur at a limited number of time points. Breslow's approach is perfectly valid if ties are incidental and only due to rounding of the observed times.

It has been shown in the theoretical literature (Tsiatis 1981, Andersen & Gill 1982) that the partial likelihood can be treated as a regular likelihood, in the sense that the estimate $\hat{\beta}$ has an asymptotic normal distribution with mean β and covariance matrix given by the inverse observed Fisher information matrix. Since it will be needed later on, it is useful to give some detail on the first and second order derivatives of the partial likelihood. The first derivative or *score function* is

$$\frac{\partial \mathrm{pl}(\beta)}{\partial \beta} = \sum_{i}^{n} d_i (x_i - \bar{x}_i(\beta)) ,$$

with $\bar{x}_i(\beta)$ the weighted average of the x_j's in the risk set $R(t_i)$, that is

$$\bar{x}_i(\beta) = \frac{\sum_{j \in R(t_i)} x_j \exp(x_j^\top \beta)}{\sum_{j \in R(t_i)} \exp(x_j^\top \beta)} .$$

The *Fisher information* of the partial likelihood is given by

$$I_{\mathrm{pl}}(\beta) = -\frac{\partial^2 \mathrm{pl}(\beta)}{\partial \beta^2} = \sum_i d_i \cdot \mathrm{var}_i(x \,|\, \beta) ,$$

with

$$\mathrm{var}_i(x \,|\, \beta) = \frac{\sum_{j \in R(t_i)} (x_j - \bar{x}_i(\beta))(x_j - \bar{x}_i(\beta))^\top \exp(x_j^\top \beta)}{\sum_{j \in R(t_i)} \exp(x_j^\top \beta)}$$

the weighted covariance matrix in the risk set at event time t_i. Similarly, it can be shown that estimates of individual survival probabilities $\hat{S}(t|x)$ are asymptotically normal with mean $S(t|x)$ and a covariance matrix that can be obtained from the observed Fisher information of the full likelihood $l(h_0, \beta)$. It is immaterial which method (NA or PL) is used to estimate the probabilities, because they are asymptotically equivalent. The asymptotic variance of $\hat{S}(t|x) = \hat{S}(t|x, \hat{\beta})$ is complicated by the fact that it depends on $\hat{\beta}$ both directly, but also indirectly through the dependence of $\hat{H}_0(t)$ on $\hat{\beta}$. The asymptotic variance of $-\ln(\hat{S}(t|x)) = \hat{H}_0(t) \exp(x^\top \hat{\beta})$ may be estimated consistently by

$$\sum_{\substack{t_i \leq t \\ t_i \in \mathscr{D}}} \left(\frac{\exp(x^\top \hat{\beta})}{\sum_{j \in R(t_i)} \exp(x_j^\top \hat{\beta})} \right)^2 + \hat{q}(t|x)^\top I_{\mathrm{pl}}^{-1}(\hat{\beta})\, \hat{q}(t|x) ,$$

with

$$\hat{q}(t|x) = \sum_{\substack{t_i \leq t \\ t_i \in \mathscr{D}}} (x - \bar{x}_i) \frac{\exp(x^\top \hat{\beta})}{\sum_{j \in R(t_i)} \exp(x_j^\top \hat{\beta})} ,$$

with $\bar{x}_i = \bar{x}_i(\hat{\beta})$. These can be used to construct confidence intervals for $\hat{S}(t|x)$ on the ln-scale, or on the probability scale, after applying the delta method, which yields

$$\mathrm{var}(\hat{S}(t|x)) = \hat{S}^2(t|x)\, \mathrm{var}(\ln(\hat{S}(t|x))) .$$

2.4 Example: Breast Cancer II

As an example of using the Cox regression model in dynamic prediction the breast cancer data of Data Set 5 will be analyzed. More information on the data is given in Appendix A.5. The survival and censoring functions are shown in Figure A.7, and the risk factors at baseline and their univariate effect on survival are shown in Table A.5. The regression coefficients in that table stem from a Cox model with only that covariate included in the analysis. For ease of presentation all covariates are categorized. Covariates with 3 categories are analyzed by using two dummy variables.

Since the emphasis in the book is on the predictive use of models and not so much on the assessment of independent relevance of each covariate, a multivariate Cox model for overall survival is fitted with all covariates as coded in Table A.5. For the sake of completeness the regression coefficients and their standard errors are given in Table 2.1. The relevant information is summarized in the centered prognostic index $\mathrm{PI} = (X - \bar{X})^{\top}\hat{\beta}$. It is independent of scaling, centering, and coding of categorical variables (by changing X, also $\hat{\beta}$ will change). The distribution of PI is shown in Figure 2.4. The mean value $\overline{\mathrm{PI}} = 0$ (by construction) and the standard deviation $\mathrm{sd}(\mathrm{PI}) = 0.52$. Since patients with a higher value of PI have a higher risk of dying, the distribution of the prognostic index will shift to the left over time as is shown in Figure 2.4. This shift can also be seen from some simple statistics. Of the 2687 patients, the 2029 patients that were still at risk after eight years of follow-

Table 2.1 *The Cox model for the EORTC breast cancer data (Data Set 5), including all risk factors; chemo = chemotherapy, RT = bradiotherapy*

Covariate	Category	B	SE
Type of surgery	Mastectomy with RT		
	Mastectomy without RT	0.248	0.110
	Breast conserving	-0.074	0.097
Tumor size	< 2 cm		
	2 - 5 cm	0.349	0.096
	> 5 cm	0.891	0.153
Nodal status	Node negative		
	Node positive	0.968	0.110
Age	≤ 50		
	> 50	0.014	0.103
Adjuvant chemo	No		
	Yes	-0.377	0.126
Tamoxifen	No		
	Yes	-0.175	0.115
Perioperative chemo	No		
	Yes	-0.122	0.074

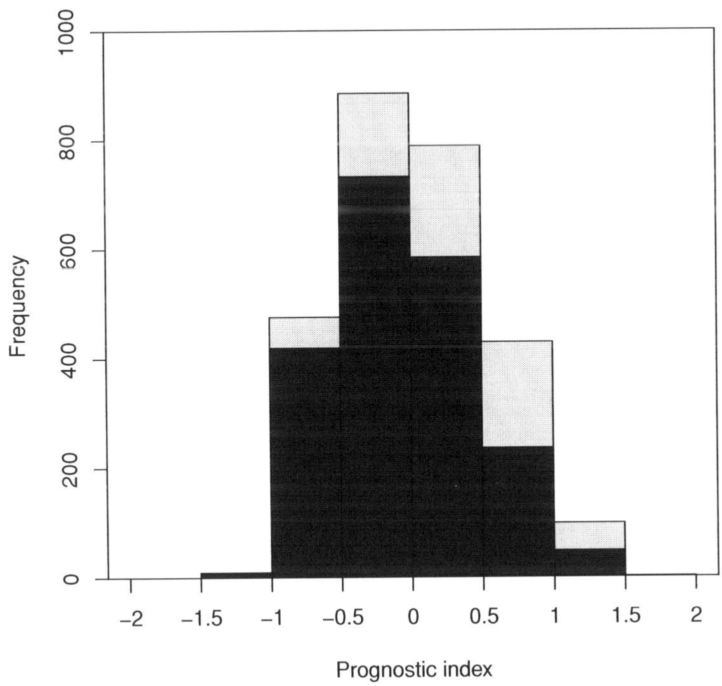

Figure 2.4 *Histogram of the prognostic index PI; in dark-grey patients at risk after eight years, in light-grey patients who died or were lost-to-follow-up within eight years*

up have $\overline{\text{PI}} = -0.08$ and $\text{sd(PI)} = 0.49$, while the 658 patients who died or were lost-to-follow-up within eight years of the start of the follow-up had $\overline{\text{PI}} = 0.24$ and $\text{sd(PI)} = 0.52$.

The estimated survival curves that can be derived from the Cox model are shown in Figure 2.5 for different values that cover the range of the distribution of PI, namely from $\overline{\text{PI}} - 2 \cdot \text{sd(PI)}$ $(= -1.04)$ to $\overline{\text{PI}} + 2 \cdot \text{sd(PI)}$ $(= 1.04)$. The estimated survival curve for a value PI of the prognostic index is given by $S(t|\text{PI}) = S_0(t)^{\exp(\text{PI})}$. The estimated 13-year survival probabilities range from 88% to 36% with an overall (Kaplan-Meier) survival of 68%. Note that the Kaplan-Meier survival curve lies below the predicted curve for $\overline{\text{PI}} = 0$ due to the non-linear effect of the prognostic index PI on the survival probabilities (see also Equation (2.1)).

The dynamic effect of the prognostic index is visualized in Figure 2.6, which shows the probability of dying within the next 5 years for the same range of values as in Figure 2.5. Observe that the curves start to decrease very slowly after 2-3 years of follow-up but are still far from zero, even in the best prognosis group. This means that cure cannot be taken for granted (yet) even after 7 years. Dynamic prediction in this data will be considered in more detail in Chapter 10.

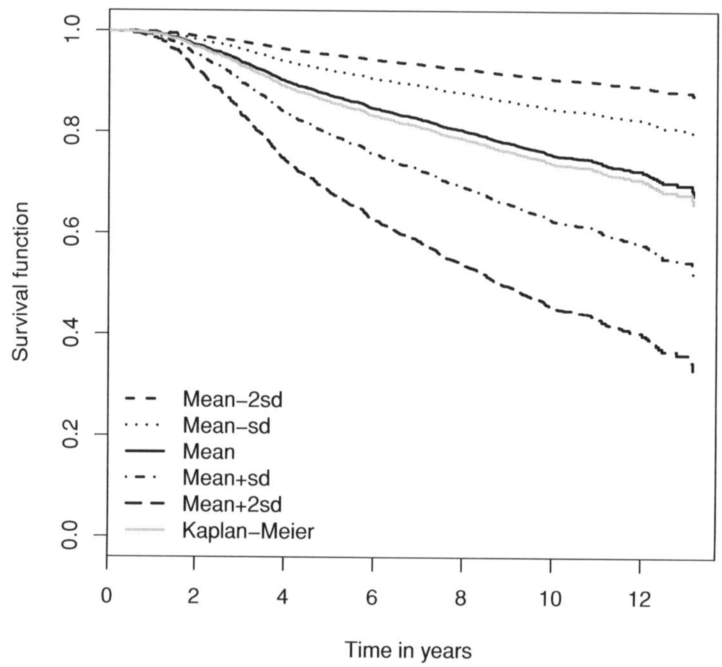

Figure 2.5 *Predicted survival curves for different values of the prognostic index*

2.5 Extensions of the data structure

The data structure of the Cox model considered so far is rather rigid. All patients have to be followed from $t = 0$ onwards, and the covariates are not allowed to change over time. However, the estimation procedure described in Section 2.2 allows relaxation of both conditions.

Delayed entry

Delayed entry (or left truncation) can occur if individuals are included in the data later in the follow-up at some time-point t_{entry}. This implies that such individuals are implicitly selected on being still alive at t_{entry}. As for right censoring the analysis is only valid if the selection is independent of further survival and censoring. Dependence on covariates that are included in the Cox model does not invalidate the analysis. Delayed entry implies that t_{entry} has to be added as an extra piece of information. The data format is $(t_{\text{entry}}, t_{\text{exit}}, d, x)$. The risk set $R(t)$ consists of all individuals with $t_{\text{entry}} < t \leq t_{\text{exit}}$. Consequently, the number at risk $Y(t)$ is no longer a monotonically decreasing function of t. Formally, the estimating procedures are exactly the same as in Section 2.2. This is one of the advantages of modeling the hazard. However, the programming is getting slightly more complicated because

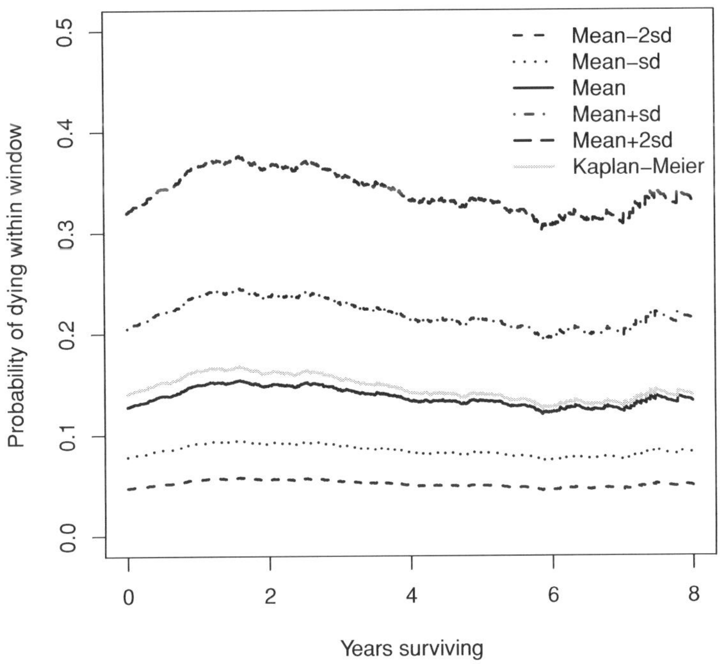

Figure 2.6 *Dynamic prediction curves for different values of the prognostic index*

one needs to take into account when the individuals enter the follow-up. To analyze such data software is needed that allows delayed entry. Examples of such software are SAS, Stata, and R/S-PLUS. Due to the left truncation, there is a danger that the early risk sets are very empty and the early hazards cannot be estimated with enough precision, which has consequences for the precision of the whole survival curve. A minimal number is required of individuals with follow-up from $t = 0$ onwards.

To illustrate the principle of delayed entry, an example from the European Group for Blood and Marrow Transplantation (EBMT) may be instructive. The EBMT is a registry collecting data about blood and marrow transplantations. Interest is in the death hazard for patients after transplantation, where time is measured from the date of diagnosis of leukemia. Since the EBMT is a registry of blood and marrow transplantations, patients for whom transplantation is intended but who die before they are transplanted, will not be recorded in their database. So patients from the EBMT are selected on being alive at time of transplantation. The size of the risk set increases as patients are being transplanted, and decreases, as patients die after transplant or when their follow-up ends. This can be seen from Figure 2.7, which shows the size $Y(t)$ of the risk set, based on data from 246 patients (156 deaths) with myelodysplastic syndrome (MDS), aged 55 – 69, available in the EBMT database.

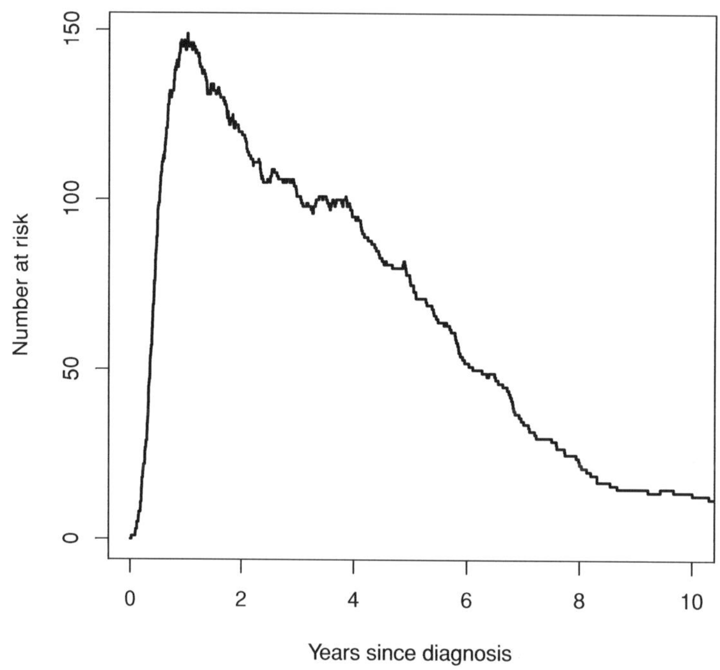

Figure 2.7 *The size of the risk set for survival from diagnosis from 246 MDS patients*

Initially, the size of the risk set increases because patients are transplanted. After reaching a maximum at about 1 year, the size of the risk set decreases, because more patients exit (because they die or are no longer followed up) than enter the risk set. A mistake, commonly found in the medical literature, is to ignore the delayed entry, and to act as if everyone is followed from $t = 0$. At $t = 0$ the size of the risk set is then taken to be the total number in the data set, while it decreases as patients die or are censored. This results in underestimation of the hazard increments $\frac{d_i}{Y(t_i)}$ (and hence overestimation of the survival curve). The resulting bias can be quite severe. The solid line of Figure 2.8 shows the Nelson-Aalen estimate of the cumulative death hazard after transplantation, as obtained from the 246 MDS patients, taking delayed entry into account. Time is measured in years from diagnosis. The Nelson-Aalen estimate ignoring delayed entry (dashed line) clearly underestimates the death hazard. An estimated survival curve can be obtained from the hazard, using $\hat{S}(t) = \exp(-\hat{H}(t))$, but the interpretation of such a survival curve is somewhat awkward; it is an estimate of survival in a hypothetical population of leukemia patients receiving transplantation immediately after diagnosis. Also, note again that the hazard and survival estimates are only unbiased if truncation is independent of survival. This is a quite strong assumption; it means that the decision to transplant early or late should not be based on the condition of the patient.

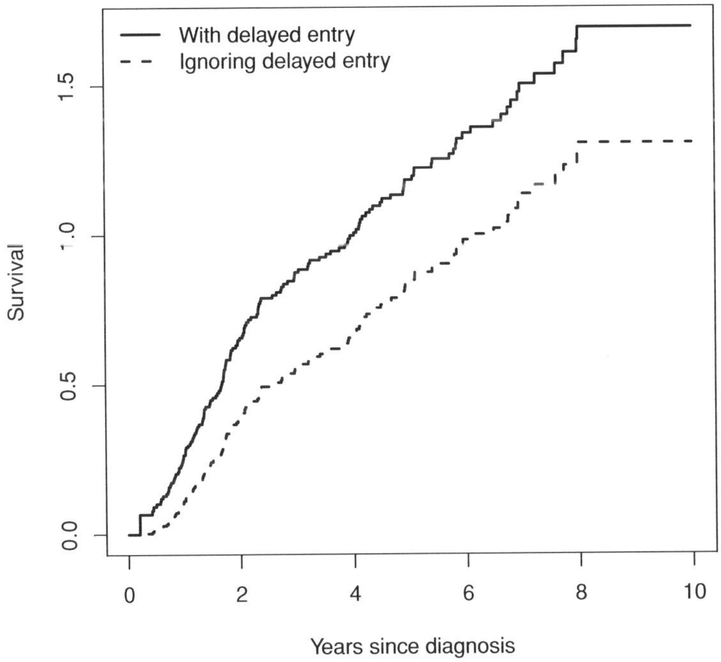

Figure 2.8 *Nelson-Aalen cumulative hazard estimates with and without taking account of delayed entry*

Time-dependent covariates

The second extension is to allow for so-called *time-dependent* covariates, covariates whose value may change over time. These covariates are denoted by $X(t)$. In principal, the formal definitions and estimation procedures remain valid if the covariates are allowed to be time-dependent. The model modifies into

$$h(t|X(t)) = h_0(t)\exp(X(t)^\top\beta) \ .$$

In practice, such data are easy to analyze if $X(t)$ is piecewise constant (and does not change too often), by creating separate records for each period of constant $X(t)$ and using software that allows delayed entry. It is best explained by an example. Suppose that an individual enters at t_0 with covariate vector $X(t_0) = x_0$. That value changes consecutively into x_i at time t_i for $i = 1, 2, \ldots$. At the observed survival time t_{obs} the current value is x_{last} that started at t_{last}. To cover that individual the following records for $(t_{entry}, t_{exit}, d, x)$ have to be entered:

Entry time	Exit time	Status	Covariate
t_0	t_1	0	x_0
t_1	t_2	0	x_1
$\vdots$	$\vdots$	$\vdots$	$\vdots$
t_{last}	t_{obs}	d	x_{last}

It is not hard to check that this will give the correct contribution to the log-likelihood for this individual. If $X(t)$ changes very often, or indeed if it is a continuous function of t, one should realize that in an actual data set only the values of $X(t)$ at the *event times* are needed. This means that one record for each event time $\leq t$ suffices for an individual with survival time t. It is doable to create such enlarged data sets if the number of events is not too big (the **survival** package in R contains the function `survSplit()` to do just that). Otherwise, information on survival time, status and fixed covariates should be combined with information on time-dependent covariates "internally" in the software to obtain the correct likelihoods.

Although models with time-dependent covariates can be fitted rather easily, it should be stressed that in first instance a Cox model with time-dependent covariates is of no predictive use, unless the distribution of future values of $X(t)$ is known. The main purpose of this book is to show how time-dependent information still can be used without creating very complicated models and procedures. Time-dependent covariates are encountered in Data Sets 2, 5 and 6. In Data Set 2 it concerns the repeatedly measured White Blood Cell count of CML patients, in Data Set 5 it concerns the occurrence during the follow-up of local recurrence and/or distant metastasis in breast cancer patients, and in Data Set 6 time-dependent covariates are implied by the occurrence of platelet recovery and/or acute graft-versus-host disease in ALL patients.

2.6 Alternative models

Accelerated failure time (AFT) model

An alternative way of modeling survival is by using the regression paradigm. Since survival time is positive, a convenient model is obtained by a traditional regression model for $\ln(T)$ that can be written as

$$\ln(T) = X^\top \beta + \sigma\varepsilon \ ,$$

with ε an *"error"* term with cumulative distribution $F(e) = P(\varepsilon \leq e)$. Popular choices are the normal model $F(e) = \Phi(e)$ leading to the *log-normal model*

$$S(t|X) = 1 - \Phi((\ln(t) - X^\top \beta)/\sigma) \ ,$$

and the extreme value distribution $F(e) = 1 - \exp(-\exp(e))$ leading to the *Weibull model*

$$S(t|X) = \exp(-\exp((\ln(t) - X^\top \beta)/\sigma)) \ .$$

These two models (and similar models) have a long tradition in industrial applications. They are quite easy to fit using some variant of the EM algorithm. Software has been provided by the PROC(edure) LIFEREG of SAS even before the introduction of Cox's proportional hazards model.

More insight is gained by writing the basic model as a multiplicative model

$$T = \exp(X^\top \beta) \cdot \exp(\varepsilon)^\sigma \, .$$

Loosely speaking, the effect of the covariates (compared to the value $X = 0$) is that life is prolonged by a factor $\exp(X^\top \beta)$. Notice that in this formulation $X^\top \beta$ has a positive effect on survival in contrast to the proportional hazards model in which $X^\top \beta$ has a negative effect. The multiplicative model is known as the *accelerated failure time (AFT) model* although "decelerated" would be more appropriate, because the model describes the lengthening of survival that comes with slowing down the failure process. The σ parameter affects the shape of the distribution. It is proportional to $\mathrm{sd}(\ln(T))$, the standard deviation of $\ln(T)$, if the latter exists. The survival function of the model can be written as

$$S_\sigma(t|X) = S_{0,\sigma}(t/\exp(X^\top \beta)) = S_{0,\sigma}(t\exp(-X^\top \beta)) \, ,$$

where $S_{0,\sigma}(t) = 1 - F(\ln(t)/\sigma)$. It is interesting to compare this model with the proportional hazards model. This can be done by looking at the hazard of the AFT model

$$h_\sigma(t|X) = \exp(-X^\top \beta) h_{0,\sigma}(t\exp(-X^\top \beta)) \, .$$

If and only if $h_{0,\sigma}(t\exp(-X^\top \beta))$ is constant, which is the case for the exponential distribution, the term $\exp(-X^\top \beta)$ plays the role of the hazard ratio. Or, the other way around, the hazard ratio can be only be interpreted as the factor by which life is shortened if the hazard is constant. This fact is related to the observations made in Figure 2.3 that the interpretation of the hazard ratio depends on the shape of the hazard function. For the Weibull model as parameterized above, the baseline survival function is given by $S_{0,\sigma}(t) = t^{1/\sigma}$, leading to

$$h_{\text{Weibull},\sigma}(t|X) = \exp(-X^\top \beta/\sigma) \frac{1}{\sigma} t^{\frac{1}{\sigma}-1} \, .$$

So, the Weibull model leads again to a proportional hazards model, but the hazard ratio is not equal to the shortening factor $\exp(-X^\top \beta)$. Inversely, the relation could be written as

$$\text{Life Shortening Factor} \;=\; [\text{Hazard Ratio}]^\sigma \, .$$

This is precisely what can be observed in Figure 2.3. The left hand panel comes from a Weibull model with $\sigma = 2 > 1$, which has a decreasing hazard and the right hand panel comes from a Weibull model with $\sigma = 0.5 < 1$, which has an increasing hazard.

The appealing aspect of the AFT model is the easy interpretation of the regression model in terms of the lengthening of life. However, it needs a parametric model for the survival, and there is no general model that fits well in all applications. This is partly because most parametric models implicitly assume that $S(\infty) = 0$ violating special property (SP2). It is tempting to define a general semi-parametric AFT model as

$$h(t|X) = \exp(-X^\top\beta)h_0(t\exp(-X^\top\beta)) .$$

Unfortunately, such a model is much harder to fit than the Cox model. A non-technical explanation is that each individual has his own time scale and convenient concepts like the risk set and the partial likelihood are not useful anymore. A more technical explanation is that the counting process approach (Andersen & Borgan 1985, Andersen et al. 1988, 1993) that helped deriving the statistical properties of the Cox model cannot be applied to this model. Another (related) drawback of the AFT model is that dynamic prediction is more complicated than for the Cox model.

Proportional odds model

Another interesting approach is the *proportional odds model* (Pettitt 1982, Bennett 1983*a*,*b*, Murphy et al. 1997), defined by

$$S(t|X) = \frac{1}{1+\exp(X^\top\beta + A(t))} ,$$

for some non-decreasing function $A(t)$ with $A(0) = -\infty$. This is a logistic regression model that is very convenient for cross-sectional (current status) data for which it is only observed once whether an individual has died before the moment of observation. It is also suited for predictions from $t = 0$ up to some fixed horizon t_{hor}. The model has interesting symmetry features. It is invariant under reversal of time and the interchanging of “life” and “death.” However, from a predictive point of view it is less convenient than the Cox model.

Additive hazards model

Finally, there is the *additive hazards model* proposed by Odd Aalen (Aalen 1980, 1989). It is defined as

$$h(t|X) = X^\top\beta(t) + h_0(t) .$$

The attractive feature is the additivity of the model that makes it easier to estimate and to interpret (in some, not all, situations). Semi-parametric versions of the additive hazards model, where some or all of the elements of $\beta(t)$ are taken to be constant over time, have been proposed by Lin & Ying (1994) and McKeague & Sasieni (1994). These have the complicating feature that $h(t|X)$ can become negative. The impossibility of giving a simple quantification of the effect of a covariate has made the model not very suited for clinical modeling and prediction. Nevertheless, the model can be quite useful in exploring time-varying effects of covariates, which will be discussed in part II of this book. The cumulative regression functions $B_j(t) = \int_0^t \beta_j(u)du$ can be estimated non-parametrically, along with the cumulative baseline hazard $H_0(t)$, using formulas very similar to ordinary least squares regression. Include a row vector of 1's in the design matrix X to estimate $H_0(t)$. For each event time t, regress the event indicators $N_i(t)$ at that time on X, for the individuals in the risk set $R(t)$. (The $N_i(t) = \mathbf{1}\{T_i \leq t, D_i = 1\}$ are the so-called *counting processes*.) The estimators $d\hat{B}_j(t)$ of $\beta_j(t)$ obtained in this way can behave quite wildly, but the cumulative sums $\hat{B}_j(t) = \int_0^t d\hat{B}_j(u) = \sum_{t_i \leq t} d\hat{B}_j(t_i)$ behave much more regularly. Pointwise confidence intervals for the $\hat{B}_j(t)$'s may be derived using martingale theory. For a more technical and detailed discussion of the additive hazards model see also Section 4.2 of Aalen et al. (2008) and Chapters 5 and 7 of Martinussen & Scheike (2006).

2.7 Additional remarks

Schoenfeld residuals

As pointed out by Schoenfeld (1982), the contributions of each risk set to the score function of Section 2.3 can be used to check the validity of the proportional hazards assumption. Let

$$s_i = x_i - \bar{x}_i(\hat{\beta})$$

be the contribution of the ith risk set to the score. It is known as the *Schoenfeld residual*. If the proportional hazards model holds true, the expected value of s_i equals zero. The validity of the model for the jth covariate can be checked by plotting the jth component of s_i versus t_i. The mean value of the residual equals zero by construction. Time trends in the residuals are an indication of violation of the proportional hazards assumption. Visual inspection of the plots can be followed by a formal test of the proportional hazards assumption.

This approach has been refined by Grambsch & Therneau (1994, 1995) and described in Section 6.2 of their book (Therneau & Grambsch 2000). Their proposal is to plot the components of the so-called *scaled Schoenfeld residual*

$$s_i^* = \hat{\beta} + V_i^{-1} s_i \ ,$$

where

$$V_i = \mathrm{var}_i(x \,|\, \hat{\beta})$$

is the contribution of the ith risk set to the Fisher information. They point out that s_i^* is an estimate of the local value of the regression coefficient at t_i in the case of time-varying coefficients. Such models will be discussed in Section 6.1.

Stratified models

The Cox model can be extended into a *stratified Cox model* by considering a categorical stratification variable, G say, with values $1, \ldots, K$. The stratified model allows the baseline hazard to depend on the stratum, that is

$$h(t|X,g) = h_{g0}(t)\exp(X^\top \beta)\ .$$

This model assumes that the effect of the covariates in X is the same in each stratum, but allows the baseline hazard to depend on the stratum. The model can be fitted in the same way as the Cox model. The technical difference is that now the risk sets are stratum specific. In the setting of prediction models, the main application of the stratified model is to include categorical covariates for which it can be expected a priori that the PH assumption will not hold. In that situation the stratum specific baseline hazard is an essential part of the prediction model and the model can only be reliable if each stratum has enough events to obtain a decent estimate of the baseline hazard. A simple check is to draw stratum specific Kaplan-Meiers. There is little hope to obtain good predictive models for the strata in which the Kaplan-Meiers are too "jumpy."

An application in epidemiological setting is correction for confounding. If, for example, age is a confounding factor, adjustment for age can be obtained by using age as a stratification variable after proper categorization. This is very similar to conditioning (matching) on age categories in logistic regression for binary data. The advantage is an unbiased view on the effect of the covariates other than age. The disadvantage is that it does not produce a reliable prediction model.

Understanding the standard error of $\hat{S}(t|x)$

The formula for the variance of $-\ln(\hat{S}(t|x)) = \hat{H}_0(t)\exp(x^\top\hat{\beta})$ in Section 2.3 looks quite mysterious. Actually it is simply built on $\text{var}(\hat{\beta}) = I_{\text{pl}}^{-1}$ and $\text{var}(\hat{h}_0(t_i)) = 1/[\sum_{j\in R(t_i)}\exp(x_j^\top\hat{\beta})]^2$. The trick is that $\hat{\beta}$ and $\hat{h}_0(t_i)$ are asymptotically independent **if** X is centered at $\bar{x}_i$ for each risk set separately. This can be seen from computing the mixed derivatives $\frac{\partial^2 l(h_0,\beta)}{\partial\beta_i\partial h_0(t_j)}$ in the second formula for $l(h_0,\beta)$ in Section 2.3. It is left as an exercise to derive the variance formula by the standard delta-method using the variances of $\hat{\beta}$ and $\hat{h}_0(t_i)$ and the independence after proper centering.

Chapter 3

Measuring the predictive value of a Cox model

3.1 Introduction

After the development of a prognostic index and a corresponding prognostic model, different questions may arise in the assessment of the predictive value of the model. For a general discussion see the paper of van Houwelingen & le Cessie (1990), and the books by Harrell (2001) and Steyerberg (2009).

In this chapter it will first be discussed how to assess the performance of a pre-specified model. Issues to be considered are the visualization of the model, its discriminative ability and the prediction error of the model. Assessing the predictive value of a model in the data set from which it was derived, might give an assessment that is too optimistic about the predictive value of the model. Some form of cross-validation is needed to correct the over-fitting. This will be discussed in the second part of this chapter.

The example used throughout this chapter is the ovarian cancer data of Data Set 1. See Table A.1 for description of the covariates and Figure A.1 for the Kaplan-Meier estimates of the survival and censoring function. A Cox model is fitted with a linear effect for Karnofsky as a continuous covariate and all other predictors as categorical covariates. The fitted model is given in Table 3.1. No attempt is made at this stage to refine the model by selecting covariates, joining categories, allowing interactions and the like. The centered prognostic index PI obtained this way has mean $\overline{\text{PI}} = 0$, by definition, and standard deviation $\text{sd}(\text{PI}) = 0.60$.

3.2 Visualizing the relation between predictor and survival

The standard way of presenting the relation between a predictor and survival is a plot like Figure 2.5 where model based survival functions are shown for different values of the predictor covering the whole range of the distribution of the predictor. An alternative that stays closer to the data is to group the patients on the basis of the predictor, for example in four groups of equal size, and to show the Kaplan-Meier survival curve in each subgroup. The usefulness of a predictor is then assessed by the apparent differences between the survival curves. Examples of this type

Table 3.1 *The fitted Cox model for Data Set 1. Karnofsky is recoded as 6 ($\leq$ 60) to 10 (100)*

Covariate	Category	B	SE
FIGO	III		
	IV	0.504	0.136
Diameter	Microscopic		
	$<$ 1 cm	0.247	0.323
	1-2 cm	0.665	0.326
	2-5 cm	0.785	0.315
	$>$ 5 cm	0.836	0.304
Broders	1		
	2	0.573	0.243
	3	0.520	0.234
	4	0.323	0.274
	Unknown	0.650	0.265
Ascites	Absent		
	Present	0.272	0.155
	Unknown	0.205	0.213
Karnofsky	Continuous	-0.173	0.054

of visualization are abundant in the medical literature, especially in the field of oncology. Some examples are van Houwelingen et al. (1989), Thorogood et al. (1991) and van Nes et al. (2010). The data in van Houwelingen et al. (1989) are a subset of Data Set 1. The conclusion in the paper that the survival of patients with advanced ovarian cancer is "predictable" is far too optimistic as will be shown later in this section.

Two interesting alternative ways of visualizing the relation between predictor and survival outcome can be found in Henderson et al. (2001) and Royston (2001). The former paper proposes to show 80% tolerance intervals for the survival time, given the predictor, while the latter suggests to impute the censored survival times and simply show the scatter plot of (imputed) survival time against predictor.

To formalize both constructions let $S(t|x) = P(T > t|x)$ be the predicted survival curve for an individual with predictor $X = x$ and let $q_\alpha(x)$ be the α-quantile, that is the solution of $1 - S(t|x) = \alpha$. Generally, the $(1-\alpha)$-tolerance interval is given by $(q_{\alpha/2}(x), q_{1-\alpha/2}(x))$. In survival analysis there are two complications alluded to in Chapter 1 that hamper the computation of tolerance intervals. Firstly, there is the incompleteness of the follow-up implying that $S(t|x)$ is only available up to some horizon t_{hor}. Secondly, the survival curves may reach a plateau because the event might never happen, for example because the patient is cured. In technical terms this is described as a defective survival distribution for which $S(\infty|x) = \lim_{t\to\infty} S(t|x) > 0$. The effect of either complication is that the right hand side of the tolerance interval cannot always be determined. Sometimes it is only known that it falls beyond the horizon, which leads to right-censored tolerance intervals. A visualization of the predictive value of the predictor is obtained by plotting the tolerance intervals against the predictor. Henderson et al. (2001) plot lower bound and (possibly censored) upper bound against the predictor. A modification used here is to show the tolerance interval for each individual in a high-low plot. The advantage is that such a plot also shows the distribution of the predictor. In that sense, it resembles a traditional scatterplot.

The visualization proposed by Royston (2001) is an adaptation of the scatterplot for survival data. Such an adaptation is needed because the ordinary scatter plot is impossible due to the presence of censored data. Royston fits the parametric log-normal model to the data and imputes censored observation by randomly drawing a value from the conditional distribution $S(t|x)/S(t_{\text{cens}}|x)$. This approach can lead to very unrealistic imputations if the survival distribution is defective. A modification applied here is to use the Cox model to obtain $S(t|x)$ and to impute censored values using this model with administrative censoring at the horizon t_{hor}.

Both visualizations can be used for individual covariates and for prognostic indices that summarize the information.

The survival curves based on the Cox model of Table 3.1 are shown in Figure 3.1. The curves are very similar to those in van Houwelingen et al. (1989).

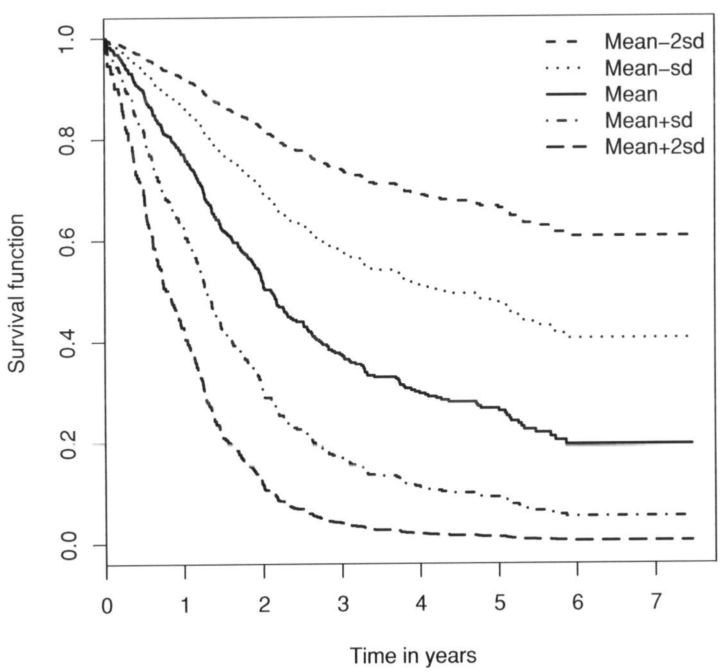

Figure 3.1 *Predicted survival curves for different values of the prognostic index*

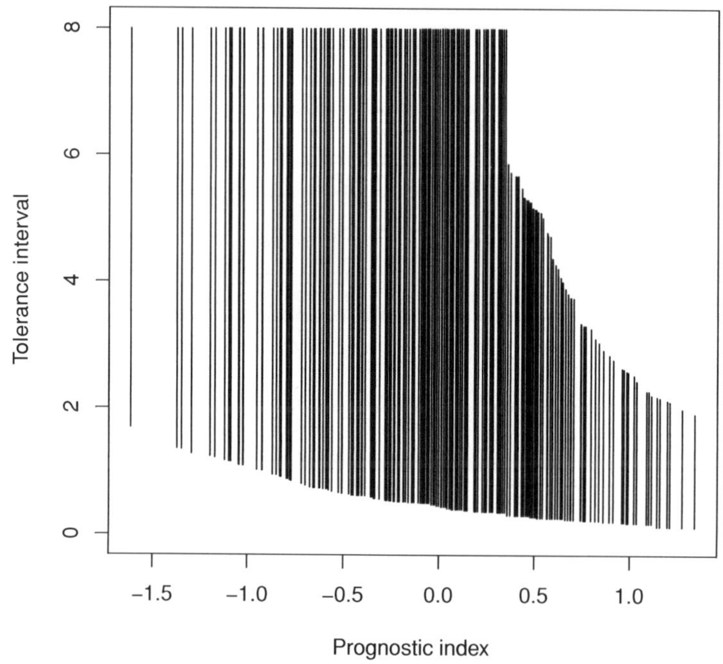

Figure 3.2 *(Censored) 80% tolerance intervals for different values of the prognostic index*

The survival curves are quite far apart and survival looks quite predictable. The alternative visualizations are shown in Figure 3.2 and Figure 3.3, respectively. The tolerance intervals in Figure 3.2 and the imputations in the imputed scatterplot Figure 3.3 are censored at $t_{\text{hor}} = 8$ years. Both plots convey the same message that the association between the predictor and survival is far from impressive. In Figure 3.3 a simple linear regression is plotted. The explained variation is $R^2 = 0.207$. This is not the most appropriate quantification, but it is an indication that survival for these patients is not as predictable as suggested in van Houwelingen et al. (1989).

3.3 Measuring the discriminative ability

After visualizing the relation between a predictor and the survival outcome as discussed in the previous section, the next step is to obtain some measure of the strength of the relation, like the correlation coefficient R or its square R^2 in linear regression. In models for a binary outcome B, like diagnostic testing and case-control studies, a well established methodology is comparing the distribution of the (ordinal) predictor X in the subgroups of "cases" ($B = 1$) and "controls" ($B = 0$) by means of the *Receiver Operation Characteristic (ROC) Curve* that plots the *True Positive Rate* $\text{TPR}(x) = P(X > x|B = 1)$ (*sensitivity*) against the *False Positive Rate* $\text{FPR}(x) = P(X > x|B = 0)$ (1-*specificity*). A useful summary statistic is the *Area un-*

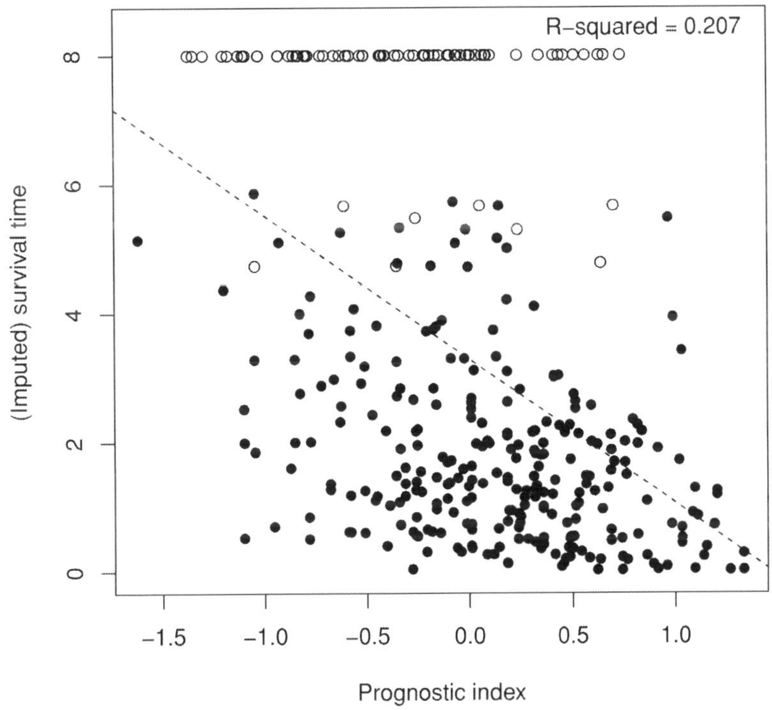

Figure 3.3 *Partly imputed survival time versus prognostic index. Imputed values are represented by open circles, observed survival times by solid circles*

der the Curve (AUC) which is equal to $P(X_1 > X_0) + 0.5 \cdot P(X_1 = X_0)$, where X_1 is drawn from the distribution of $X|B = 1$, and X_0 is independently drawn from the distribution of $X|B = 0$. Based on a sample of n_0 controls with values x_{0i}, and n_1 cases with values x_{1j}, the AUC may be estimated by

$$(n_0 n_1)^{-1} \sum_{i=1}^{n_0} \sum_{j=1}^{n_1} \left(\mathbf{1}\{x_{1j} > x_{0i}\} + 0.5 \cdot \mathbf{1}\{x_{1j} = x_{0i}\} \right) . \tag{3.1}$$

An extensive discussion of the properties and the usefulness of the ROC curve can be found in the book by Krzanowski & Hand (2009).

Notice that the ROC curve makes the implicit assumption that X is positively associated with B. Moreover, the ROC curve is not affected by monotonic transformations of X. This makes it clear that the AUC-measure is not based upon a model for the relation between B and X, but is just a measure of the strength of the association. It is a very useful tool to compare different X's. In diagnostic testing the comparison might correspond with different tests. In case-control studies AUC might help compare the effects of different continuous covariates (risk factors) or different prognostic indices based on the whole set of risk factors. AUC runs from 0 to 1. If B and X are independent, the value AUC $= 0.5$ is obtained, which can be used as a reference value for "no discriminative ability". An absolute interpretation is hard to give because it depends heavily on the context for which the model is developed.

Interesting ideas about the use of ROC curves in assessing discriminative ability in survival models can be found in the paper by Heagerty & Zheng (2005). An obvious application of the ROC concept described in the paper, called *cumulative/dynamic*, is to consider a fixed time-point t_0 and to use ROC and AUC to measure the discrimination between those who die before t_0 and those who die after t_0. Technically, this can be described by considering the dichotomous outcome $B = \mathbf{1}\{T > t_0\}$. A practical problem is censoring before t_0.

From the perspective of dynamic prediction the dichotomy $T > t_0$ versus $T < t_0$ is not very interesting. A much more interesting proposal in the paper, called *incident/dynamic*, is to compare at each time point those who die and those who do not. A practical problem here is that, in the absence of ties, at each time point only one person will die. Heagerty and Zheng solve that problem by using a survival model like the Cox model to "impute" more deaths at the same time point. This approach complicates matters and makes the approach more model dependent. It will be shown below how a simplification of their approach can be very useful to describe the discriminative ability over time and to obtain a summary measure that is very closely linked to the *Concordance Index (C-index)* advocated by Harrell et al. (1996). This simplification closely resembles the construction of Stare et al. (2011).

Let $R(t)$ be the risk set of those still alive at some event time t (as defined in Section 1.2) and $Y(t)$ the size of the risk set. Let individual i be the one that dies at that particular time-point and let x_i be the value of the covariate or the prognostic index x. A "diagnostic test" can be imagined that diagnoses each individual in the risk set as "doomed" (about to die) when the value of x is greater than or equal to a cut-off value c. Again assuming no ties in the event time points, the sensitivity of this diagnostic test is completely determined by the only individual that dies, and hence equals 0 for $c < x_i$ and 1 for $c \geq x_i$. The AUC at time t is given by Equation (3.1) (with $n_1 = 1$ and $n_0 = Y(t) - 1$)

$$\mathrm{AUC}(t) = \frac{\#\{j \in R(t); x_j < x_i\} + 0.5 \cdot \#\{j \in R(t), j \neq i; x_j = x_i\}}{Y(t) - 1} \,. \tag{3.2}$$

As observed by Stare et al. (2011), the definitions above extend immediately to the case where X changes over time. A plot of $\mathrm{AUC}(t)$ against t (at the event times) for the prognostic index of the model of Table 3.1 is shown in Figure 3.4. The plot is very similar to the plot of the Schoenfeld residuals $x_i - \bar{x}_i$ (see Section 2.3). To the plot a lowess smoothing curve has been added, thus providing a model-free alternative for the model-based imputations of Heagerty and Zheng.

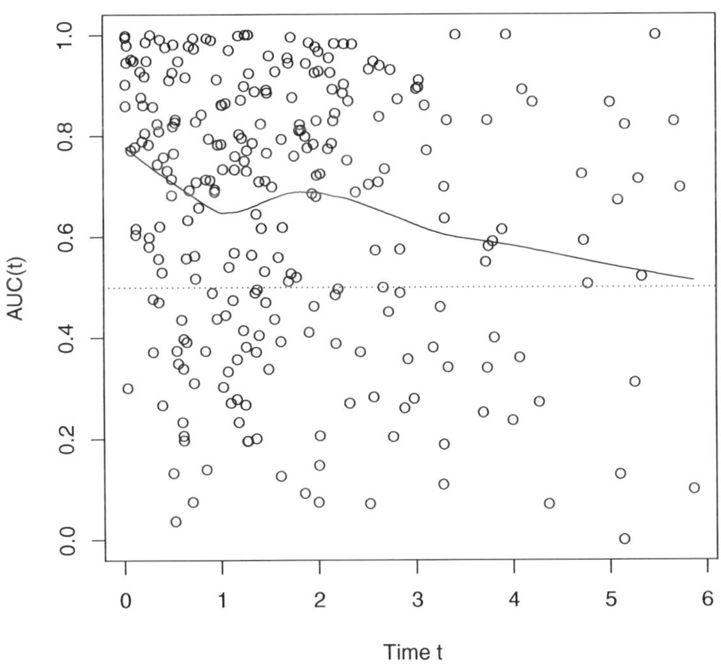

Figure 3.4 *AUC(t) for the Cox model of the ovarian cancer data*

An AUC-type summary measure can be obtained by computing some weighted average. A simple measure with weights proportional to $Y(t) - 1$ would lead to

$$\begin{aligned} C &= \sum_{i \in \mathscr{D}} \left[(Y(t_i) - 1) \cdot \mathrm{AUC}(t_i) \right] / \sum_{i \in \mathscr{D}} (Y(t_i) - 1) \\ &= \frac{\sum_{i \in \mathscr{D}} (\#\{j \in R(t_i); x_j < x_i\} + 0.5 \cdot \#\{j \in R(t_i), j \neq i; x_j = x_i\})}{\sum_{i \in \mathscr{D}} (Y(t_i) - 1)} . \end{aligned} \tag{3.3}$$

This is precisely Harrell's concordance measure (C-index) (Harrell et al. 1996), originally defined as the proportion of pairs of observations for which the order of survival times and model predictions are concordant. The pairs for which the order of the true survival times can be determined are the pairs (i, j) for which the smallest observed time is an event time. The pair is concordant if the one that dies first has the largest x-value. Its value for the prognostic index of the ovarian cancer data set is $C = 0.666$. Dynamic versions of C can be obtained by averaging over all event times within the window $[t, t + w]$,

$$C_w(t) = \frac{\sum_{\substack{i \in \mathscr{D} \\ t \le t_i \le t+w}} (\#\{j \in R(t_i); x_j < x_i\} + 0.5 \cdot \#\{j \in R(t_i), j \neq i; x_j = x_i\})}{\sum_{\substack{i \in \mathscr{D} \\ t \le t_i \le t+w}} (Y(t_i) - 1)} . \tag{3.4}$$

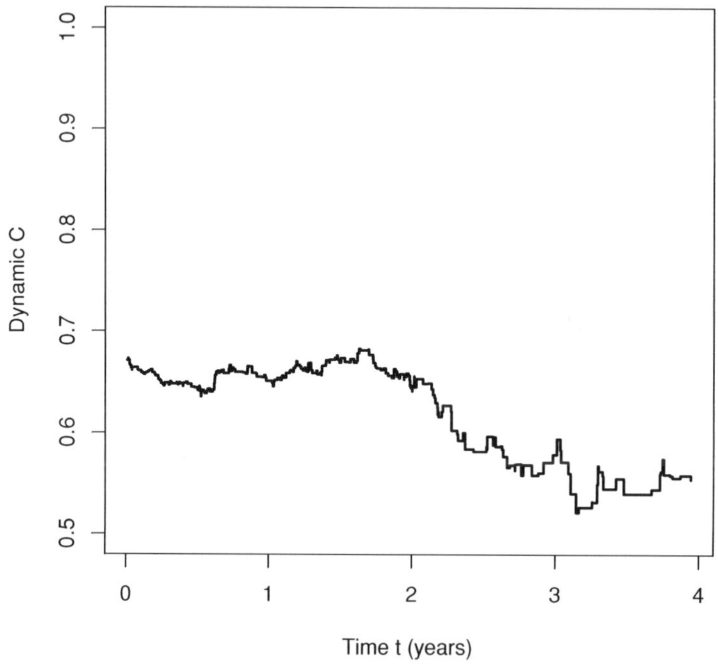

Figure 3.5 *$C_w(t)$ with $w = 2$ for the Cox model of Table 3.1. The y-axis runs from 0.5 to 1*

This is shown in Figure 3.5 for the window $w = 2$. Both Figures 3.4 and 3.5 show that the discriminative ability slowly decreases over time.

3.4 Measuring the prediction error

The AUC-type measures of the previous section only measure the strength of the association between the true survival time T and the predictor X. The Cox model, or any similar model, is only used to obtain a prognostic index $\text{PI} = X^\top \hat{\beta}$ for the risk of dying, that summarizes the information on all covariates. A more ambitious goal is to measure predictive performance by measuring "prediction error". There is a huge literature and little agreement on how to measure prediction error (Korn & Simon 1990, O'Quigley & Flandre 1994, Schemper & Stare 1996, Graf et al. 1999, Schemper & Henderson 2000, Schemper 2003, Royston & Sauerbrei 2004, O'Quigley et al. 2005, Gerds & Schumacher 2006). The most popular version appears to be the Brier score as developed by Graf et al. (1999). The idea behind their measure, following the same line of thought as in the previous section, is to focus first on the prediction of survival beyond some fixed time point t_0, define a prediction error for the prediction of the dichotomy $Y = \mathbf{1}\{T > t_0\}$ and generalize that to a prediction error for the survival time T itself. As in all this literature, prediction is not to be understood as an absolute yes/no prediction whether a patient will survive beyond t_0 or not, but as a *probabilistic prediction* quantifying the probability of survival beyond t_0.

Let $\hat{S}(t_0|x)$ be the model-based probabilistic prediction for the survival of an individual beyond t_0 given the predictor x and let $y = \mathbf{1}\{t > t_0\}$ be the actual observation (ignoring censoring for the time being). Following van Houwelingen & le Cessie (1990), there are three ways to define prediction error

$$\begin{aligned}
\text{AbsErr}(y, \hat{S}(t_0|x)) &= |y - \hat{S}(t_0|x)| \,, \\
\text{Brier}(y, \hat{S}(t_0|x)) &= (y - \hat{S}(t_0|x))^2 \,, \\
\text{KL}(y, \hat{S}(t_0|x)) &= -[y \ln(\hat{S}(t_0|x)) + (1-y)\ln(1-\hat{S}(t_0|x))] \,.
\end{aligned}$$

The first one, absolute error, was proposed in this setting by Schemper & Henderson (2000). However, it has some unpleasant properties that makes it less suitable. The problem is that it is not "proper", which means that its expected value is not minimized by the true value of $S(t|x_0)$, but by the dichotomization $S_{\text{dich}}(t|x_0) = \mathbf{1}\{S(t|x_0) > 0.5\}$. Applying a similar dichotomization to $\hat{S}(t_0|x)$ could improve the prediction error. Such a dichotomization makes sense in classification problems like giving a diagnosis, but is not very suitable for probabilistic prediction. Another way of phrasing the problem is by saying that absolute error is not able to recognize biases in the prediction model. For that reason, absolute error will not be considered further and attention is restricted to the *Brier score* and the *Kullback-Leibler (KL) score*. The latter is referred to in Graf et al. (1999) as the logarithmic score.

The Brier score, the second measure, is much more well-behaved. This can be seen from computing its expected value with respect to a new observation Y_{new} under the true model $S(t_0|x)$. It is not hard to see that

$$E[\text{Brier}(Y_{\text{new}}, \hat{S}(t_0|x))] = S(t_0|x)(1 - S(t_0|x)) + (S(t_0|x) - \hat{S}(t_0|x))^2 \,.$$

So, the Brier score can be seen to consist of two components: the "true variation" $S(t_0|x)(1 - S(t_0|x))$ and the "model error" $(S(t_0|x) - \hat{S}(t_0|x))^2$ due to misspecification of the model. A perfect prediction is only possible if $S(t_0|x) = 0$ or $S(t_0|x) = 1$. When comparing two different models, the best model is the one with the smallest model error or, equivalently, the best fit. Unfortunately, in practice the two components cannot be separated since the true $S(t_0|x)$ is unknown. A partial solution to this problem is to scale the Brier score by the score of a prediction based on a model without covariates. The natural candidate in the survival setting is the Kaplan-Meier estimator $\hat{S}_{\text{KM}}(t_0)$ defined in Section 1.2.

The Kullback-Leibler score KL is based on the general principle that predictive performance of any predictive model can be measured by the log-likelihood of the prediction model evaluated at the observations. The advantage is that it is closely connected to maximum likelihood estimation and likelihood ratio tests, which have been very extensively studied. It is also closely connected with Akaike's Information Criterion AIC, to be discussed in Section 3.6, The main disadvantage is that it does not look very natural and, therefore, is harder to "sell" to the people who will use the model in practice. Another disadvantage is that it might explode, $\mathrm{KL} = \infty$, if the observed value has predicted probability equal to 0, which occurs when $\hat{S}(t_0|x) = 0$ and $y = 1$ or $\hat{S}(t_0|x) = 1$ and $y = 0$. Its expected value under the true model is given by

$$E[\mathrm{KL}(Y_{\mathrm{new}}, \hat{S}(t_0|x))] = -[S(t_0|x)\ln(\hat{S}(t_0|x)) + (1 - S(t_0|x))\ln(1 - \hat{S}(t_0|x))] \, .$$

It is not hard to show that this expression is minimal for $\hat{S}(t_0|x) = S(t_0|x)$. Computing the second derivative at the minimum gives the following Taylor-expansion

$$E[\mathrm{KL}(Y_{\mathrm{new}}, \hat{S}(t_0|x))] = \mathrm{KL}(S(t_0|x), S(t_0|x)) + 0.5 \cdot \frac{(S(t_0|x) - \hat{S}(t_0|x))^2}{S(t_0|x)(1 - S(t_0|x))} + \dots \, .$$

It shows that, like the Brier score, the Kullback-Leibler score can be split into a measure of true variation $\mathrm{KL}(S(t_0|x), S(t_0|x)) = -[S(t_0|x)\ln(S(t_0|x)) + (1 - S(t_0|x))\ln(1 - S(t_0|x))]$ and a nonnegative term due to "model error" which is approximately equal to $0.5 \cdot (S(t_0|x) - \hat{S}(t_0|x))^2 / (S(t_0|x)(1 - S(t_0|x)))$ if the prediction model is close to the true model. A nice property of KL is that the contribution of the model error is already scaled. Its distribution is not very sensitive to the true value of $S(t_0|x)$. This can be seen in the simplest situation where $\hat{S}(t_0|x)$ is based on an estimate in a training sample of size n_x from the population with $X = x$. In that case $E[\mathrm{KL}(S(t_0|x), \hat{S}(t_0|x))] \approx \mathrm{KL}(S(t_0|x), S(t_0|x)) + 1/n_x$, independent of $S(t_0|x)$.

The choice between Brier score and KL score is mainly a matter of taste. In this book both scores will be used, leaving the choice to the reader.

In order to assess the predictive performance of a prediction rule in actual data by means of either the Brier score or the KL score, we need to deal with those observations that are censored before t_0. Following Graf et al. (1999) this can be done by skipping such observations and multiplying the usable ones by $1/\hat{C}(t^*|x)$. Here $t^* = t$ for those with event time $t < t_0$ and $t^* = t_0$ for those still at risk at t_0 and $\hat{C}(t|x)$ is an estimate of $P(T_{\mathrm{cens}} > t|x)$. This weighting scheme is known as *inverse probability of censoring weighting (IPCW)*. It can be shown that it leads to an unbiased estimate of the average prediction error that would have been obtained if all true survival times before t_0 could be observed. The simplest estimate of the probability of not being censored is the Kaplan-Meier estimate of the censoring distribution as defined in Section 1.2, which does not depend on the predictor x. Using IPCW leads to the following formulae for an estimate of the *average prediction error* of the prediction model $\hat{S}(t|x)$ at $t = t_0$:

$$\mathrm{Err}_{\mathrm{Score}}(\hat{S}, t_0) = \frac{1}{n} \sum_i \mathbf{1}\{d_i = 1 \vee t_i > t_0\} \frac{\mathrm{Score}(\mathbf{1}\{t_i > t_0\}, \hat{S}(t_0|x_i))}{\hat{C}(\min(t_i-, t_0)|x_i)} \, . \tag{3.5}$$

Here, Score is either the Brier or the Kullback-Leibler (KL) score for the prediction error.

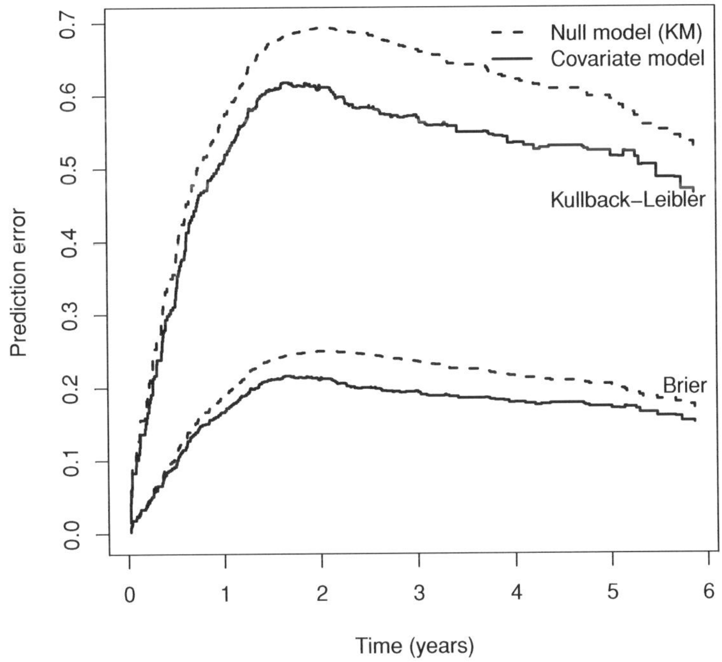

Figure 3.6 *Prediction error curves*

Figure 3.6 shows the average prediction error (both Brier and Kullback-Leibler) over time of the prognostic model of Table 3.1 for the ovarian cancer data, together with the average prediction error of the null model, given by the Kaplan-Meier estimate. Notice that the shape of the curves is mainly determined by the overall survival probability as estimated by the KM survival curve of Figure A.1. Figure 3.7 shows the relative error reduction

$$\frac{\text{Prediction error}_{\text{Null model}} - \text{Prediction error}_{\text{Prognostic model}}}{\text{Prediction error}_{\text{Null model}}}$$

achieved by the prognostic model. Both measures behave similarly, but KL shows a more stable reduction over time, presumably because it is less sensitive to the actual survival probability.

Dynamic prediction error

Instead of considering the "cross-sectional" errors as shown in Figures 3.6 and 3.7 it might be more interesting to look into the dynamic errors, that is, the errors in predicting survival up to $t+w$ for those still alive at time t. The obvious general-

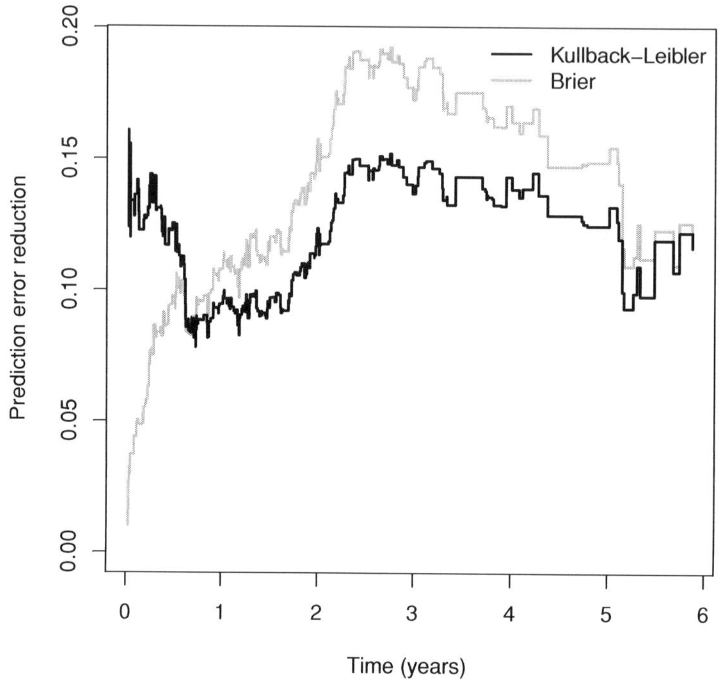

Figure 3.7 *Relative error reduction curves*

ization of the definitions above are

$$\mathrm{Err}_{\mathrm{Score},w}(\hat{S},t) = \frac{1}{Y(t)} \sum_{i \in R(t)} \mathbf{1}\{d_i = 1 \vee t_i > t+w\} \cdot \\ \cdot \frac{\mathrm{Score}(\mathbf{1}\{t_i > t+w\}, \hat{S}(t+w|t,x_i))}{\hat{C}(\min(t_i-, t+w)|t,x_i)}. \quad (3.6)$$

Here, $\hat{C}(t|s,x) = \hat{C}(t|x)/\hat{C}(s-|x)$ is defined, similarly as $\hat{S}(t|s,x)$, as the conditional probability of no censoring before t for those still at risk at $s-$.

The dynamic analogues of Figures 3.6 and 3.7 are shown in Figure 3.8 for the window $w = 2$ years. Again these graphs show that the model loses its predictive value later in the follow-up.

Global prediction error

A final issue to be discussed in this section is the definition of a single overall measure of prediction error. Common practice is to take the integrated (Brier) score over the whole range up the horizon $\int_0^{t_{\mathrm{hor}}} \mathrm{Err}_{\mathrm{Score}}(\hat{S},t_0)dt_0$. However, little motivation is given for this measure. An appropriate measure of global average prediction error

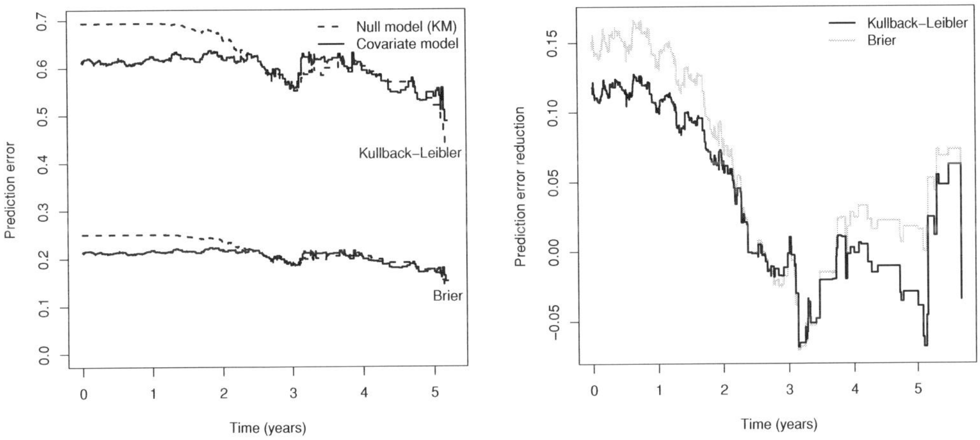

Figure 3.8 *Dynamic prediction error curves ($w = 2$) for Kullback-Leibler and Brier scores*

would assess how well the actual survival can be predicted by the survival model. As observed above, a necessary condition for a good prediction model is a good fit with the actual survival distribution. For example, for a good prediction model the tolerance intervals from the previous section should have the correct coverage.

For a model that completely specifies the full density of the outcome of interest, a global measure can be based on the predicted log-likelihood evaluated at the observations. However, the Cox model does not specify a density because the estimated baseline hazard is degenerate, and the Cox model is not able to look beyond the horizon of the follow-up. The actual situation is better described by dividing the time axis in a set of $L+1$ intervals $I_1, I_2, ..., I_{L+1}$, defined as $[t_0 = 0, t_1), \ldots, [t_{L-1}, t_L = t_{\mathrm{hor}}), [t_{\mathrm{hor}}, \infty]$. The intervals could be the first L years of the follow-up plus *"anything beyond the first L years"*. A good prediction model should be able to specify estimates $\hat{p}_l(x)$ of $p_l(x) = P(T \in I_l|x)$ for all l and a global measure of prediction error should capture the prediction errors related to those probabilities.

Ignoring censoring for the time being, given survival time t and covariate x, a global measure in the spirit of the Brier score could be based on

$$\mathrm{Brier}_{\mathrm{global}}(t, \hat{p}(x)) = \sum_{l=1}^{L+1} (\mathbf{1}\{t \in I_l\} - \hat{p}_l(x))^2 .$$

Notice that for $L = 1$, this score is twice the Brier score for dichotomies as defined above. The integrated Brier score as mentioned above would be very close to taking

$$\mathrm{Brier}_{\mathrm{integrated}}(t, \hat{p}(x)) = \sum_{l=1}^{L} (\mathbf{1}\{t > t_l\} - \sum_{m=l+1}^{L+1} \hat{p}_m(x))^2 .$$

This expression is not exchangeable in the intervals. It is hard to figure out what the relation is with the global Brier score above.

A global version of the Kullback-Leibler score is given by the negative multinomial log-likelihood

$$\mathrm{KL}_{\text{global}}(t,\hat{p}(x)) = -\sum_{l=1}^{L+1} \mathbf{1}\{t \in I_l\} \ln(\hat{p}_l(x)) .$$

This global KL measure has a nice dynamic interpretation. Let

$$h_l(x) = P(T \in I_l | T \geq t_{l-1}, x) = p_l(x) / \sum_{m=l}^{L+1} p_m(x) \text{ for } l = 1,..,L .$$

This is an "interval version" of the hazard. Let $\hat{h}_l(x)$ be its estimate derived from $\hat{p}(x)$. The global KL score can be rewritten as

$$\begin{aligned} \mathrm{KL}_{\text{global}}(t,\hat{h}(x)) &= -\sum_{l=1}^{L} \left[\mathbf{1}\{t \in I_l\} \ln(\hat{h}_l(x)) + \mathbf{1}\{t \geq t_l\} \ln(1-\hat{h}_l(x))\right] \\ &= \sum_{l=1}^{L} \left[\mathbf{1}\{t \geq t_{l-1}\} \cdot \mathrm{KL}(\mathbf{1}\{t \in I_l\}, \hat{h}_l(x))\right] . \end{aligned}$$

This can be interpreted as:

"The global KL prediction error is the sum of the KL prediction errors of the interval hazards."

It can be generalized beyond KL into a global "dynamic" score based on the interval hazards

$$\mathrm{Score}_{\text{dynamic}}(t,\hat{h}(x)) = \sum_{l=1}^{L} \left[\mathbf{1}\{t \geq t_{l-1}\} \cdot \mathrm{Score}(\mathbf{1}\{t \in I_l\}, \hat{h}_l(x))\right] .$$

Here, Score can be KL or Brier or any other score for a dichotomous prediction error. This dynamic score can be used to define an average dynamic prediction error. Censoring can be handled by deleting censored observations whenever their interval score cannot be determined and weighting the observations by the probability of not being censored. If the censoring does not depend on the covariates, this leads to the following definition in terms of the dynamic error of Equation (3.6).

$$\mathrm{Err}_{\text{Score,dynamic}}(\hat{h}(x)) = \sum_{l=1}^{L} \hat{S}_{\text{KM}}(t_{l-1}) \mathrm{Err}_{\text{Score},t_l-t_{l-1}}(t_{l-1}) . \tag{3.7}$$

If the intervals up to t_{hor} all have the same width w, the average dynamic prediction error is closely related to $\int_0^{t_{\text{hor}}} \hat{S}_{\text{KM}}(t) \mathrm{Err}_{\text{Score},w}(t) dt$. Generally speaking, some

Table 3.2 *Interval-specific and total prediction errors; KM=Kaplan-Meier; Total is calculated as in Equation (3.7)*

			Brier			Kullback-Leibler		
Interval	KM (start)	Hazard	KM	Model	Relative Reduction	KM	Model	Relative Reduction
0-1	1.000	0.249	0.188	0.168	0.109	0.564	0.510	0.096
1-2	0.751	0.327	0.220	0.200	0.089	0.632	0.584	0.076
2-3	0.508	0.249	0.189	0.178	0.058	0.565	0.537	0.051
3-4	0.380	0.169	0.144	0.141	0.023	0.464	0.453	0.024
4-5	0.316	0.092	0.090	0.090	0.004	0.325	0.324	0.005
5-6	0.287	0.225	0.175	0.173	0.009	0.534	0.530	0.007
Total			0.582	0.540	0.073	1.757	1.647	0.063

form of integrating the dynamic error plot shown in Figure 3.8 seems to be more appropriate in defining a global average prediction error than integrating the pointwise error plot of Figure 3.6.

Application of this concept to the example data leads to the results of Table 3.2. Both error measures show the same picture, namely that the model does a poor job in predicting the dynamic survival in the later intervals. Further analysis showed that refining the intervals leads to increased errors and limited reduction by the model. The explanation is that the non-parametric aspect of the baseline hazard in the Cox regression model implies that the hazard is estimated poorly. Estimation of the hazard can be improved when the hazards are smoothed somehow. Considering a set of intervals that contain "enough" events per interval introduces an implicit smoothing that reduces random error and is able to show the usefulness of the predictive information.

Summary

The main findings of this long section are:

- Brier score and KL behave quite similar.
- Prediction error is mainly driven by the marginal distribution. Error reduction gives more insight.
- Dynamic prediction error is more informative than cross-sectional error.
- Global prediction error could best be measured by integration of the dynamic prediction errors over time, just like the C-index in Section 3.4 that integrates the dynamic AUC(t).

3.5 Dealing with overfitting

The problem with the approach in the previous sections is that the same data are used both to obtain the model and to assess its predictive performance. This could lead to so-called overfitting. The issue of overfitting is best understood from a hypothetical experiment. Suppose that as many completely random and independent predictors are generated as there are individuals in the study and that all those predictors are used in a model like the Cox model. The result would be a model that fits the data perfectly. Each individual is predicted to live exactly as long as his observed survival time. The *apparent error rate* would be zero. However, using that model for new individuals would lead to predictions that are not correlated at all with the actual survival with a very large *actual error rate*. (See van Houwelingen & le Cessie (1990) for more formal definitions of apparent error rate, actual error rate and the *optimism*, being the difference between actual and apparent error rate). So, ideally, predictive performance should be assessed on new independent data. A distinction could even be made between data from the same population (*internal* assessment) and data from other populations (*external* assessment).

A popular approach to internal assessment is the so-called *split sample* approach where the data are split into two parts, one part for model development (*the training set*) and one part for assessment of the performance of the model (*the test set*). Usually two-thirds are used for the training set and one third for the test set. An example can be found in van Houwelingen & Thorogood (1995). This approach is fine in very large data sets, but for smaller data sets it is somewhat wasteful of resources to "throw away" one-third of the data. Moreover, the results might depend on the split and different splits could lead to different conclusions. A more systematic way is *k-fold cross-validation* where the data set is split in k data sets of roughly the same size. The results in each subset are predicted using the other $(k-1)$ subsets. This allows an assessment of the predictive performance in each subset (and for each individual) and yields an estimate of the overall predictive value of the model. The simplest approach (but potentially computationally demanding) is to take $k = n$, also known as *leave-one-out* cross-validation. (If the term cross-validation is used without further specification, it is always meant to be leave-one-out cross-validation.) The advantage is that it stays as close as possible to the original data and that there is no ambiguity on how to split the data, which is still there if $k < n$.

Applying cross-validation in the setting of the Cox model yields estimates $\hat{\beta}_{(-i)}$ and $\hat{h}_{0,(-i)}(t)$ of the regression coefficients and the baseline hazard leaving individual i out, allowing the computation of predictive survival probabilities

$$\hat{S}_{(-i)}(t|x) = \exp\left(-\hat{H}_{0,(-i)}(t) \cdot \exp(x^\top \hat{\beta}_{(-i)})\right) .$$

It also yields prognostic indices $\mathrm{PI}_{j,(-i)} = x_j^\top \hat{\beta}_{(-i)}$ that could be useful, if exploited with care. The technical problem is that $\hat{S}_{(-i)}(t|x)$ does not depend on scaling of the x's, neither do differences between $\mathrm{PI}_{j,(-i)}$'s for fixed i, but the differences in prognostic index $\mathrm{PI}_{j,(-i)}$ for different i's do depend on the scaling of the x's. This problem virtually disappears if the covariates (including dummy variables for categorical covariates) are all centered beforehand in the full data set.

In order to obtain a cross-validated assessment of the discriminative ability as expressed in the C-index in the running example, the simplest approach is to replace the prognostic index $\mathrm{PI}_i = x_i^\top \hat{\beta}$ as used in Section 3.3 by the cross-validated prognostic index $\mathrm{PI}_{i,(-i)}$ based on centered covariates. This yields an estimate $C = 0.643$, slightly smaller than the "apparent" $C = 0.666$. The access over the reference value $C = 0.5$ is reduced by a factor of 0.86. A more complicated approach is to perform two assessments of concordance for each pair (i, j), namely one using $\mathrm{PI}_{i,(-i)}$ and $\mathrm{PI}_{j,(-i)}$ and one using $\mathrm{PI}_{i,(-j)}$ and $\mathrm{PI}_{j,(-j)}$. Both comparisons are based on the results using the same leave-one-out set and, therefore, are not sensitive to the scaling of the x's. Application to the ovarian cancer example leads to a slightly higher $C = 0.657$. A more intuitive approach, but much more computationally intensive, is to do pair-wise cross-validation. To do so, we leave each of the $\binom{n}{2}$ pairs out and base their comparison on $\mathrm{PI}_{i,(-i,-j)}$ and $\mathrm{PI}_{j,(-i,-j)}$. The result is $C = 0.645$. The general conclusion is that the concordance measure is not very sensitive to optimism bias. Intuitively, that could be expected because it only measures the association between predictor and outcome and does not need a fully quantified predictive model.

Similarly, an estimate of the error rate corrected for bias due to overfitting is obtained by using cross-validated survival probabilities in the computation of either the Brier score or the Kullback-Leibler score. Figure 3.9 repeats the Brier and Kullback-Leibler prediction error curves for the null model and the covariate model of Figure 3.6. To these are added in grey the Brier and Kullback-Leibler prediction error curves using cross-validated survival probabilities. Both for Brier and Kullback-Leibler, the cross-validated prediction error curves fall about two thirds between the null model and the covariate model curves, indicating that roughly two thirds of the error reduction of the covariate model is retained. On closer inspection it turns out that the fraction of prediction error retained gradually decreases over time.

3.6 Cross-validated partial likelihood

A more generic way of measuring the quality of any model is by looking at the likelihood of the fitted models. The increase in the log-likelihood when going from a simple model to a complex model can be used to test the significance of the complex model. The likelihood ratio test statistic, defined as twice the difference in log-likelihoods, usually has a χ^2-distribution with degrees of freedom (df) equal to the increase in the number of parameters. The performance of a prediction model can be assessed in a similar way by looking at the *predictive* likelihood of the new data under the prediction model. Interestingly, it can even be "predicted" what the predictive log-likelihood will be without having access to new data. It has been shown by Akaike (1974) that the optimism, that is the expected difference between the fitted (apparent) log-likelihood and the (actual) log-likelihood in (hypothetical) new observations from the same individuals, is simply equal to the dimension of

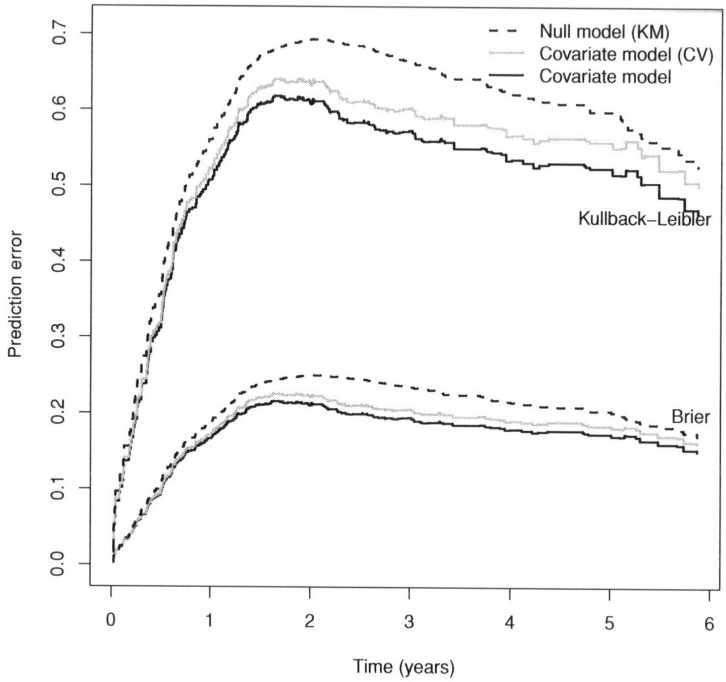

Figure 3.9 *Prediction errors with cross-validation*

the model, that is the numbers of free parameters in the model. This observation has led to the introduction of *Akaike's Information Criterion (AIC)* defined as

$$\text{AIC} = -2 \cdot \log(\text{likelihood}) + 2 \cdot \text{dimension}$$

that can be used as a very general error measure. In general, there is no straightforward absolute interpretation of AIC, which makes it hard to explain in (clinical) applications. Nevertheless it is well suited for the comparisons of different models and the selection of the best predictive model. The theoretical result derived by Akaike holds if the model is correctly specified and the number of observations is large. If there exists any doubt about the correctness of the model or if the sample size is small, an estimate of the predictive log-likelihood can be obtained by cross-validation.

Unfortunately, the application of AIC is limited to parametric models with well-defined densities, which renders it unsuited for the Cox model with its model-free baseline hazard. However, it can still be applied to the partial likelihood that does not contain the baseline hazard. AIC-corrected partial log-likelihood can be used to compare models. The partial likelihood somehow resolved the problem of the baseline hazard, but in doing so it lost the simplicity of the independent observations. In the partial likelihood the observations are dependent, which makes it hard to de-

Table 3.3 *Partial log-likelihoods with and without cross-validation; KM = Kaplan-Meier stands for the partial likelihood with $\beta = 0$; PL = partial likelihood; CVPL = crossvalidated partial likelihood; CV-Cox model = cross-validated Cox model*

Model	PL	Increase PL	CVPL	Increase CVPL
KM	−1414.314	0	−1679.563	0
Cox model	−1374.716	39.598	−1639.801	39.761
CV-Cox model			−1653.775	25.787

fine proper cross-validation. The solution proposed by Verweij & van Houwelingen (1993) is to use the so-called cross-validated partial log-likelihood CVPL defined as

$$\text{CVPL} = \sum_{i=1}^{n} \left[\text{pl}(\hat{\beta}_{(-i)}) - \text{pl}_{(-i)}(\hat{\beta}_{(-i)})\right] .$$

Here $\text{pl}(\beta)$ is the partial log-likelihood of Equation (2.2), and $\text{pl}_{(-i)}(\beta)$ is defined similarly, but excluding individual i. Thus $\text{pl}(\beta) - \text{pl}_{(-i)}(\beta)$ represents the "independent" contribution of individual i to the partial log-likelihood. Application of this concept to the ovarian cancer prognostic index leads to the results of Table 3.3. The left-hand side of this table shows that fitting the Cox model increased the partial log-likelihood by 39.598. The number of regression parameters in this model equals dim $= 12$. So, with a chi-square test statistic $\chi^2_{[12]} = 79.2$ ($P < 0.0001$), the improvement of the Cox model is very significant. The right-hand side shows that using the CVPL construction with the overall estimate $\hat{\beta}$ instead of the cross-validated estimate, that is $\text{CVPL} = \sum_{i=1}^{n}[\text{pl}(\hat{\beta}) - \text{pl}_{(-i)}(\hat{\beta})]$, yields a very similar increase, namely 39.761. However, proper cross-validation reduces the increase to 25.787, that is by a factor of 0.65. It is no coincidence that this is very similar to the two thirds reduction shown in Figure 3.9. Another observation is that the reduction of the increase $39.761 - 25.787 = 13.974$ is close to Akaike's theoretical reduction dim $= 12$. What is observed here, is the general *"rule of thumb"*

$$\text{Actual Error Reduction} = \left(1 - \frac{2 \cdot \text{dim}}{\chi^2_{\text{model}}}\right) \cdot (\text{Apparent Error Reduction}) .$$

Apparently, using AIC on the partial log-likelihood gives a good impression of the amount of overfitting and the optimism of the Cox model. It is noteworthy that that the reduction factor for the prediction error (0.65) is much smaller than for the C-index (0.86). This confirms that the C-index is not able to detect overfitting.

In the example an "overfitted" model has been used where all covariates except Karnofsky are taken categorical. This has been done on purpose to show the effect of overfitting. Unfortunately, model building techniques like stepwise regression do not fully help to combat overfitting. The model building process generates a certain amount of overfitting by itself. See van Houwelingen & Thorogood

(1995) and Varma & Simon (2006) for a further discussion. The effect is that cross-validation after model selection cannot be used to obtain an unbiased estimate of the prediction error. To obtain such an unbiased estimate, double cross-validation would be needed. That does not invalidate model selection. Throwing out covariates that do not show any prognostic relevance can be very useful.

Finally, citing van Houwelingen & le Cessie (1990), a warning should be issued that cross-validation based estimates of prediction error can only produce an estimate of the prediction error averaged over all possible training sets. This could be far off from the prediction error of the model based on one particular data set. That prediction error could only be assessed by collecting new data.

3.7 Additional remarks

Censoring correction for the C-index

The C-index is sensitive to censoring because it influences the selection of pairs for which the ordering of their survival can be determined. A version of C corrected for censoring can be obtained by Inverse Probability Weighting as applied to the Brier and KL scores. In case the censoring does not depend on the covariates, the summands in Equations (3.3) and (3.4) have to be divided by the censoring function, that is the probability of no censoring before t. This might become relevant if $\text{AUC}(t)$ decreases with time as seen in Figure 3.2. See also Koziol & Jia (2009).

Limitation of correction for censoring

Correction for censoring gets very instable if the observed censoring distribution $\hat{C}(t)$ gets close to zero. A sensible time horizon t_{hor} should be selected beforehand based on the graph of $\hat{C}(t)$. It should be underlined that correction for censoring does not imply that censoring has no influence on the analysis.

Concordance in the Cox model

It can be shown that for two individuals with covariate vectors x_1 and x_2, and corresponding prognostic indices PI_1 and PI_2 respectively, the probability $P(T_1 < T_2)$ in the Cox model is obtained from the simple formula

$$P(T_1 < T_2) = \frac{\exp(\text{PI}_1)}{\exp(\text{PI}_1) + \exp(\text{PI}_2)} ,$$

provided that they both can be observed and both $P(T_1 < \infty) = 1$ and $P(T_2 < \infty) = 1$. The corresponding interpretation of the C-index in relation to the Cox model is given by its expected value

$$E[C] = E\left[\frac{\exp(|\text{PI}_1 - \text{PI}_2|)}{\exp(|\text{PI}_1 - \text{PI}_2|) + 1}\right] .$$

Here, PI_1 and PI_2 are independently drawn from the distribution of PI. Notice that this formula implies that the concordance will always be smaller than 1 if $P(X_1 = X_2) > 0$, which will be the case if X is categorical.

However, in practice there is always an observation limit t_{hor} such that it is impossible to determine the ordering of T_1 and T_2 if both of them are beyond t_{hor}. It can be shown that the formula still holds conditional on either $T_1 < t_{\text{hor}}$ or $T_2 < t_{\text{hor}}$. However, since this selection depends on the prognostic index, the expected value of C will in general depend on t_{hor} and, more generally, on the censoring distribution.

Interpretation of the CVPL

As observed in Section 3.4, the full log-likelihood cannot be used for cross-validation because the Cox model does not have a proper density. Actually, when the underlying distribution is continuous, all new observations are impossible under the Cox model with its hazard concentrated in the event times of the observed data. It can be shown that CVPL circumvents this problem by an approach equivalent to the following two-stage procedure:

1. Estimate the leave-one-out parameter $\hat{\beta}_{(-i)}$ as usual,
2. Use that to estimate the baseline hazard for the *whole data set* by Breslow's estimator, yielding $\hat{h}_0(t_j|\hat{\beta}_{(-i)}) = d_j / \sum_{k \in R(t_j)} \exp(x_k^\top \hat{\beta}_{(-i)})$,
3. Use the full log-likelihood (see Section 2.3) in that model to compute the contribution of individual i.

Extension of the CVPL

The CVPL construction easily extends to the case of k-fold cross-validation, which may be preferable for practical reasons (large data sets) or for theoretical reasons. See Bøvelstad et al. (2007) for an application.

Using a piecewise exponential model

An alternative to the Cox model is the piecewise exponential model that takes the baseline hazard to be constant in intervals. The piecewise exponential model could be used as an alternative for the global KL-based error of Section 3.4. Using the same interval construction the hazard of individual i in interval l could be taken to be $h(t|x_i) = \ln(1 - h_l(x_i))/(t_l - t_{l-1})$ for $t \in I_l$. Here $h_l(x)$ is the "interval hazard". This would lead to a global error measure that is very similar to the KL-measure of Section 3.4. The advantage would be that the actual time of the event can be used and censored observations are used up to the moment of censoring. However, censoring would still have an effect on the expected value of the error measure and some correction for censoring might be appropriate.

Prediction error of AFT-models

In the AFT models of Section 2.6 where $\ln(T) = X^\top\beta + \sigma\varepsilon$, a measure of explained variation could be simply defined as

$$R^2_{\text{AFT}} = \frac{\text{var}(X^\top\beta)}{\text{var}(X^\top\beta) + \sigma^2\text{var}(\varepsilon)} .$$

This would be a useful measure if there is neither a horizon effect nor death from competing risks. It is not clear how to adjust R^2_{AFT} for horizon effects. Consequently, such a measure has to be interpreted with care. There is an interesting link between the Cox model and the AFT-models that could be used to obtain a measure in the same spirit. Let $H_0(t)$ be the cumulative baseline hazard of the Cox model and assume that the model is not "defective," that is $H_0(\infty) = \infty$. Then the Cox model can be rewritten as

$$\ln(H_0(T)) = -X^\top\beta + \varepsilon$$

where ε has the extreme value distribution with

$$P(\varepsilon > e) = \exp(-\exp(e)) .$$

The variance of ε is given by $\text{var}(\varepsilon) = \pi^2/6 = 1.645$. This leads to the R^2-measure of Kent & O'Quigley (1988) for the Cox model

$$R^2_{\text{KOQ}} = \frac{\text{var}(X^\top\beta)}{\text{var}(X^\top\beta) + 1.645} ,$$

see also Royston (2006). The appealing property of this measure is that it solely depends on the variance of the prognostic index var(PI) $= \text{var}(X^\top\beta)$, which does make sense when comparing different Cox models for the same data. However, it has no absolute interpretation and ignores the problem of the horizon completely.

Chapter 4

Calibration and revision of Cox models

4.1 Validation by calibration

After developing a prognostic model like the Cox model the question is often asked how the *validity* of the model can be assessed. Unfortunately, it is not easy to define validity in a concise and precise way. There is even ambiguity about the meaning of the term "model." For an interesting discussion, see Altman & Royston (2000). Following van Houwelingen (2000), a statistical model is defined as a rule to compute (survival) probabilities given the observation of the covariates included in the model. With this definition of a model a statistical definition of validity can be based on the concept of *calibration* ("validation by calibration").

To get a better understanding of the concept, suppose that the prognostic information X is summarized by a grouping variable G defining K subgroups $g_1,...,g_K$. This is in line with the common practice in clinical applications to make subgroups running from *"very good prognosis"* to *"very bad prognosis"*. The model completely specifies the survival probability by $S_{\text{model}}(t|g_k)$ for each subgroup g_k as the average $S_{\text{model}}(t,x)$ over all x's in g_k. The model is well calibrated if the true survival probabilities do not differ from the modeled ones, that is $S_{\text{true}}(t|g_k) = S_{\text{model}}(t|g_k)$. It is important to note that, generally speaking, the subgroups will not be homogeneous with respect to prognostic information X. The definition of calibration as used here does not imply that $S_{\text{true}}(t|x) = S_{\text{model}}(t|x)$ for all x. To be precise the definition requires that $E[S_{\text{true}}(t|X)|G = g_k] = S_{\text{model}}(t|g_k)$. For readers familiar with logistic regression it is noteworthy that this is precisely the claim tested by the popular Hosmer-Lemeshow test (Hosmer & Lemeshow 1980). See also Hosmer et al. (1997). This test is commonly known as a goodness-of-fit test. Actually, that is a misnomer. A correct term would be a goodness-of-calibration test. A goodness-of-fit test, like those developed in the PhD-research of le Cessie, attempt to test the hypothesis $S_{\text{true}}(t|x) = S_{\text{model}}(t|x)$ for all x; see le Cessie & van Houwelingen (1991, 1995).

To generalize beyond forming prognostic subgroups, let $Z(X)$ be a one-dimensional prognostic index based on X, such that the prognostic model specifies the survival probabilities as functions of $Z(X)$, that is $S_{\text{model}}(t|X) = S_{\text{model}}(t|Z(X))$. The model is correctly calibrated if

$$E[S_{\text{true}}(t|X)|Z(X) = z] = S_{\text{model}}(t|z) \ .$$

To check calibration or even to test the null hypothesis of proper calibration, it is useful to define a so-called *calibration model*

$$S_{\text{cal}}(t|z, \theta) \ ,$$

for which $S_{\text{model}}(t|z) = S_{\text{cal}}(t|z, \theta_{\text{model}})$. In this setting a test for calibration boils down to testing $\theta = \theta_{\text{model}}$. For logistic regression such a regression calibration goes back to Cox (1958). In that paper new observations are regressed on $\text{logit}(p_{\text{model}}(z)) = \ln(p_{\text{model}}(z)/(1 - p_{\text{model}}(z)))$. So the calibration model is a simple logistic model with $\text{logit}(p_{\text{model}}(z))$ as the only covariate. For a well calibrated prediction model, the intercept of the calibration model should be zero and the slope should be one. See also van Houwelingen & le Cessie (1990).

In this chapter, calibration of survival models will be discussed focusing mainly on the Cox model.

4.2 Internal calibration

As discussed in Chapter 3, models with a large number of predictors tend to suffer from overfitting. The apparent prediction error in the data set in which the model was obtained is smaller than the actual prediction error as can be estimated through cross-validation. As observed by Copas (1983, 1987) and iterated by van Houwelingen & le Cessie (1990), the performance of a linear prognostic index can be improved by *shrinkage towards the mean*, that is by using

$$Z_{\text{shrink}}(c) = \overline{Z} + c(Z - \overline{Z}) \ ,$$

with appropriate $0 \le c \le 1$. As pointed out in the same papers, the best value of c can be obtained theoretically or by cross-validation. Theoretical arguments lead to the so-called heuristic shrinkage factor of van Houwelingen & le Cessie (1990)

$$\hat{c}_{\text{heur}} = 1 - \frac{\text{dim}}{\chi^2_{\text{model}}} \ .$$

Here, dim is the dimension of the prediction model, and $\chi^2_{\text{model}} = 2(\text{ll}_{\text{model}} - \text{ll}_0)$ (ll denoting log-likelihood) is the likelihood ratio test statistic. A cross-validation based shrinkage factor $\hat{c}_{\text{cal}}$ is obtained by regressing the observations on their centered cross-validated predictions. The usefulness of both variants of shrinkage for prediction models has been extensively studied by Steyerberg (Steyerberg et al. 2000, 2001, 2004, Steyerberg 2009).

Generally speaking, a slightly more subtle way of obtaining an estimate of the optimal shrinkage factor is by using the factor that minimizes the cross-validated log-likelihood. For Cox regression this can be achieved by using the cross-validated partial log-likelihood, that is by taking as $\hat{c}_{\text{cvpl}}$ the value of c that maximizes

$$\text{CVPL}(c) = \sum_{i=1}^{n} \left[\text{pl}(c\hat{\beta}_{(-i)}) - \text{pl}_{(-i)}(c\hat{\beta}_{(-i)}) \right] .$$

The advantage of this approach is that no centering of the covariates is needed and that the result is independent of the scaling and the coding of the covariates. Moreover it allows a quantification of the improvement of the predictive performance yielded by this internal calibration. After obtaining the shrinkage factor the prediction model is completed by adjusting the baseline hazard as well by using the obvious adaptation of the Breslow estimator

$$\hat{h}_{0,\text{shrink}}(t_i) = \frac{d_i}{\sum_{j \in R(t_i)} \exp(\hat{c} \cdot x_j^\top \hat{\beta})} .$$

The internal calibration by shrinkage can nicely be demonstrated on the ovarian cancer data analyzed in Chapter 3. From Table 3.3 it can be derived that the heuristic shrinkage factor is given by $\hat{c}_{\text{heur}} = 1 - 12/(2 \cdot 39.598) = 0.848$. Applying Cox regression with the cross-validated prognostic index yields $\hat{c}_{\text{cal}} = 0.802$ with standard error 0.110. Applying calibration within the cross-validation partial likelihood leads to virtually the same estimate $\hat{c}_{\text{cvpl}} = 0.801$, with a maximized CVPL of -1652.148. The subtle difference between the two approaches is shown in Figure 4.1. It should be noted that the improvement of 1.628 in CVPL induced by the calibration is very modest and hardly significant (one-sided P=0.036).
There is a link between the heuristic shrinkage $\hat{c}_{\text{heur}}$ and the AIC-based reduction correction factor $(1 - 2 \cdot \text{dim}/\chi^2_{\text{model}})$ mentioned in the "rule of thumb" in Section 3.6, namely $1 - 2 \cdot \text{dim}/\chi^2_{\text{model}} \approx \hat{c}^2_{\text{heur}}$. The intuitive explanation is that application of a shrinkage factor c reduces the variance of the prognostic index predictor by a factor c^2. The rule of thumb "predicts" how much explained variation will be left after correction for overfitting. This correction can be achieved by shrinkage. This yields a modest reduction of the predictor error. Moreover it removes the inherent biases in the predictions of an over-fitted model. Since the shrinkage factor affects all covariates in the same way, shrinkage is no alternative for model selection. See the observation in Section 3.6 that *"Throwing out covariates that do not show any prognostic relevance can be very useful."*

4.3 External calibration

Once a model has been obtained (and properly calibrated) in a particular population the question arises how it generalizes to other populations. This can be seen from the point of a view of a clinician (or a group of clinicians or an institution) that wonders whether a prognostic model published in the literature applies to their own patients. This requires that similar data has been collected. The naive approach is to develop a new model using only the new data and compare the new model with the published model. That might be a disappointing experience. If some "model selection" is applied, that is selection of covariates to be included in the model, different covariates might show up and even if the same covariates are used, the coefficients might look quite different. Looking at the "prediction formula" is not the way to go. If a new model is developed it should be checked whether it produces the same prediction as the existing model when applied to new individuals. The predictions might be much closer than the formulae would suggest. The calibration approach allows to check directly whether the literature model applies to the new data set without having to build a new model. Moreover, the model obtained after calibration might be more reliable than using a model based on the new data alone. Notice that calibration has a bit of a Bayesian flavor in the sense that external (or prior) information is used to model the new data, but it does not use the Bayesian formalism.

Starting point for the calibration methodology of this section is a data set $(t_1, d_1, x_1), \ldots, (t_n, d_n, x_n)$ of size n and an external Cox model $S^*(t|x) =$

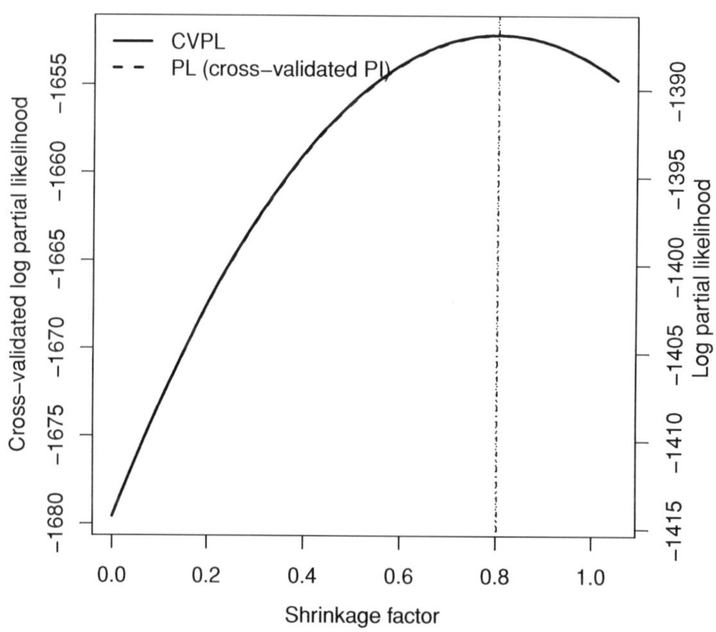

Figure 4.1 *Log-likelihood as function of the shrinkage factor. The two curves virtually overlap*

$\exp(-\exp(Z^*)H_0^*(t))$ with $Z^*(x) = x^\top \beta^*$. The corresponding cumulative hazard is given by $H^*(t|x) = \exp(Z^*(x))H_0^*(t)$. Unfortunately, the baseline hazard is often not reported in publications on prediction models. One might try to reconstruct the model as done in van Houwelingen (2000) or switch to the methods discussed in Section 4.5. For the time being it is assumed that the external cumulative baseline hazard is known and differentiable, allowing a baseline hazard $h^*(t|x) = \exp(Z^*(x))h_0^*(t)$. Calibration can be done at different levels of complexity. The most general calibration model can be formulated as

$$\ln(H_{\text{cal}}(t|x)) = \theta_0 + \theta_1 Z^*(x) + \theta_2 \ln(H_0^*(t)) \ .$$

The simplest calibration model takes $\theta_1 = \theta_2 = 1$. That model presumes that the external model is correct but for a general shift that multiplies the hazard for everybody by the factor $\exp(\theta_0)$. Using the general likelihood of Section 2.3, it is not hard to show that the maximum likelihood estimator of θ_0 is given by

$$\hat{\theta}_0 = \ln\left(\frac{\sum_i d_i}{\sum_i H^*(t_i|x_i)}\right) ,$$

with standard error

$$\text{se}^2(\hat{\theta}_0) = \frac{1}{\sum_i H^*(t_i|x_i)\exp(\hat{\theta}_0)} = \frac{1}{\sum_i d_i} \ .$$

This is no surprise for those familiar with the epidemiological concept of the *Standardized Mortality Rate* (SMR). They will recognize $\hat{\theta}_0 = \ln(\text{SMR})$ and the well-known formula $\text{se}^2(\ln(\text{SMR})) = 1/\sum_i d_i$. The estimate of the parameter and its standard error can be obtained by fitting the *Poisson calibration model*

$$d_i \sim \text{Poisson}(\exp(\theta_0)H^*(t_i|x_i)) \ .$$

Some care is needed here. The claim is not that d_i has a Poisson distribution. The precise claim is that the likelihood of the survival model is proportional to the likelihood of the Poisson model and, hence, application of the Poisson model gives the correct estimate, standard error and likelihood ratio test.

At the next level of complexity both θ_0 and θ_1 are free parameters and only θ_2 is fixed at $\theta_2 = 1$. This models checks both the general level and the correct specification of the effect of the prognostic index Z^*. It can be checked that the parameters of this model can be estimated by fitting the Poisson model

$$d_i \sim \text{Poisson}(\exp(\theta_0 + \theta_1 Z^*)H_0^*(t_i)) \ .$$

To test overall calibration ($\theta_0 = 0$, $\theta_1 = 1$) it is convenient to rewrite the model as

$$d_i \sim \text{Poisson}(\exp(\theta_0 + (\theta_1 - 1)Z^*)H^*(t_i|x_i)) \ .$$

Finally, the most general model with all three parameters also tests whether the shape of the baseline hazard is correct. As explained in van Houwelingen (2000), the parameters of this model can be estimated by transforming the time-scale by taking

$$\tilde{t} = H_0^*(t)$$

and fitting the *Weibull calibration model* with log cumulative hazard

$$\ln(H(\tilde{t})|x) = \theta_0 + \theta_1 Z^* + \theta_2 \ln(\tilde{t})$$

to the transformed data. This would mean that $\tilde{T} = H_0^*(T)$ follows a Weibull distribution with scale parameter $\exp(\theta_0 + \theta_1 Z^*)$ and shape parameter θ_2. If a Weibull model is fitted using software for Accelerated Failure Time (AFT) models, then this is usually parametrized as

$$\ln(\tilde{T}) = \beta_0 + \beta_1 Z^* + \sigma W ,$$

with W following the extreme value distribution with survival function $P(W > w) = \exp(-e^w)$. The relation between $(\theta_0, \theta_1, \theta_2)$ and $(\beta_0, \beta_1, \sigma)$ is given by $\theta_2 = \sigma^{-1}$ and $(\theta_0, \theta_1) = -(\beta_0, \beta_1)\sigma^{-1}$. See also the remarks in Section 2.6 on the Weibull as AFT model.

It is interesting to observe that in order to carry out the analysis above one only needs to know the cumulative baseline hazard $H_0^*(t)$ and not its derivative $h_0^*(t)$. So there is no direct need for smoothing. However, some smoothing is recommended before fitting the Weibull model to prevent the occurrence of many ties in the observations on the $\tilde{t}$ scale and very early observations that are tied at zero. The simplest smoother is to use a linear interpolation of $H_0^*(t)$. More detail is given in the example below.

Example

The calibration methods will be illustrated using the ALL patients of Data Set 6, see Appendix A.6. Figure 4.2 shows relapse-free survival (RFS) curves for the three cohorts. It can be seen that relapse-free survival is higher for the two last cohorts 1990-1994 and 1995-1998 compared to the first cohort 1985-1989. The univariate log hazard ratios of the 1990-1994 and 1995-1998 cohorts compared to the first are shown in Table A.6. For the remainder of this chapter the two last cohorts 1990-1994 and 1995-1998 will be combined to a single cohort 1990-1998, referred to as cohort 2, and the first cohort 1985-1989 will be referred to as cohort 1. The log hazard ratio for relapse-free survival of cohort 2 with respect to cohort 1 is -0.307 with a standard error of 0.071 ($P < 0.0001$). A prognostic index, based on a multivariate Cox model using cohort 1 is shown in Table 4.1. Figure 4.3 shows Breslow's estimate of the baseline cumulative hazard. The dynamic prediction errors of this model in the first cohort for a window of $w = 0.5$ years, with and without cross-validation are shown in Figure 4.4. The small window was chosen because the

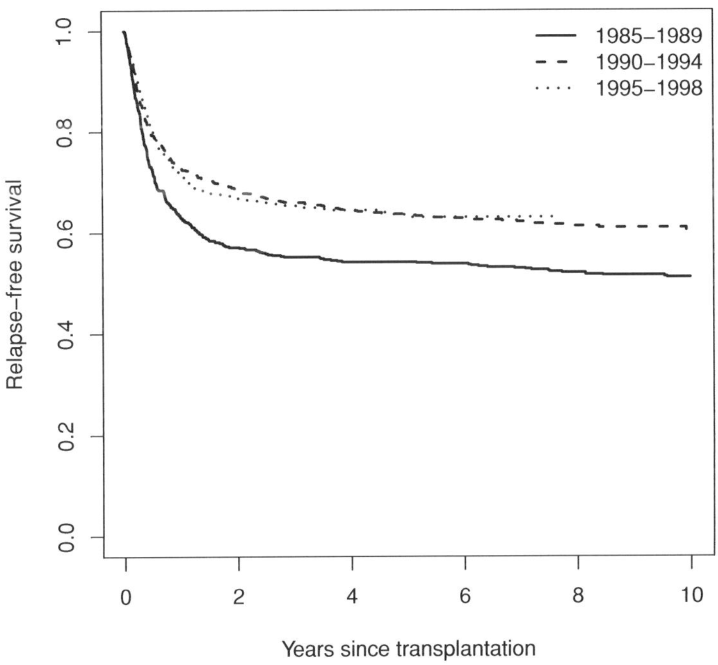

Figure 4.2 *Relapse-free survival curves for each of the three ALL cohorts*

Table 4.1 *Prognostic index for relapse-free survival based on cohort 1 of the ALL data*

Covariate	Category	B	SE
Donor recipient	No gender mismatch		
	Gender mismatch	0.119	0.128
GvHD prevention	No TCD		
	TCD	0.122	0.115
Age at transplant	≤ 20		
	20-40	0.401	0.139
	> 40	0.433	0.197

majority of the RFS-events happen within the first two years, see also Figure 4.3. It can be seen from Figure 4.4 that the predictive ability of the prognostic index is very modest indeed. The value of Harrell's C-index was a modest 0.546.

The prediction model, defined by the prognostic index $x^\top \beta^*$ of Table 4.1 and $H_0^*(t)$ of Figure 4.3, will now be validated on cohort 2. The number of RFS events in cohort 2, $\sum d_i$, equals 588, and $\sum_i H^*(t_i|x_i) = 803.82$, which yields an SMR of 0.732, and a log SMR of -0.313 with a standard error of 0.041. The log SMR is quite close to the log hazard ratio of the Cox model, but the standard error is

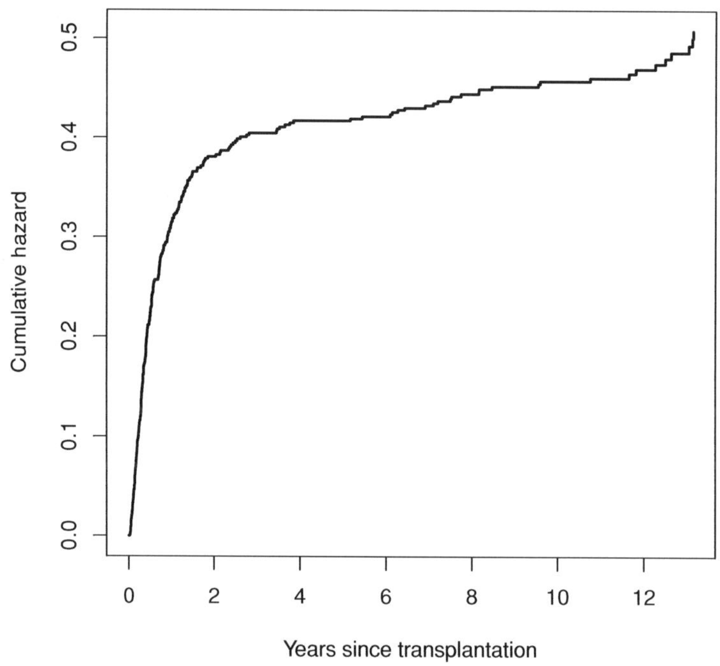

Figure 4.3 *Estimated cumulative baseline hazard of the prediction model*

smaller because the components of the prognostic index are considered fixed rather than estimated and because the baseline hazard of cohort 2 is not estimated either. (As always there is the danger that this reduction in variance is balanced by an increased bias, particularly if the "shape" of the survival model in cohort 2 differs from the one in cohort 1.) Poisson regression with the PI added reveals that θ_1 is estimated as 1.022 with a standard error of 0.238, not significantly different from 1 ($P = 0.93$). So, apart from the difference in baseline risk of the two cohorts, the prognostic index seems to be valid in the second cohort.

For the Weibull AFT regression a smoothed version $\tilde{H}_0^*(t)$ of $H_0^*(t)$ was first obtained by linear interpolation. More precisely, $H_0^*(t)$ was linearly interpolated through the midpoints of the jumps. Figure 4.5 shows this graphically for the first six event times. At the last event time point and beyond, the largest value of the original step function was taken. The results of the Weibull AFT regression with $\ln(\tilde{H}_0^*(T))$ as outcome and the PI as covariate are shown in the left column, termed "Basic", of Table 4.2. Correct calibration here should show intercept and log scale of zero, and a coefficient for PI equal to -1. The scale parameter σ was estimated as $\exp(-0.008) = 0.992$, values of β_0 and β_1 were estimated as 0.307 and -1.004, implying values of -0.309 (SE=0.108) and 1.012 (SE=0.236) for θ_0 and θ_1, respectively. The standard errors were derived using the delta method.

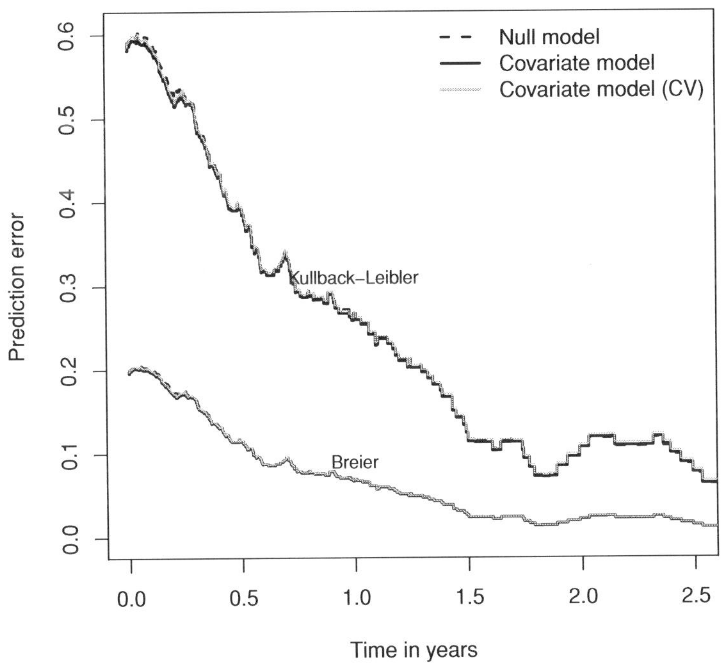

Figure 4.4 *Dynamic prediction errors (window width $w = 0.5$) with cross-validation for the prognostic index of Table 4.1 in the first cohort of the ALL data*

Table 4.2 *Calibration and revision of the prognostic index*

Covariate	Basic	Expanded		
Intercept (β_0)	0.307 (0.113)	0.306 (0.113)	0.302 (0.113)	0.275 (0.111)
PI (β_1)	-1.004 (0.238)	-0.983 (0.247)	-0.975 (0.254)	-1.240 (0.580)
Log(σ)	-0.008 (0.039)	-0.008 (0.039)	-0.008 (0.039)	-0.010 (0.039)
Gender mismatch		-0.030 (0.097)		
TCD			-0.036 (0.110)	
Age 20-40				0.243 (0.263)
Age > 40				-0.013 (0.284)

Figure 4.6 shows the results of the calibration effort. Three patients were selected, with the smallest (left) and largest (right) values of the prognostic index defined by Table 4.1 and in the middle for the mean value of the PI in the data. The smallest (baseline), average, and largest PI values were 0, 0.369 and 0.674, respectively. The relapse-free survival curves for the original model are shown in solid lines, and those for the calibrated model in dashed lines. Since in our example the estimates of θ_1 and θ_2 were very close to zero, the only effect clearly seen is the

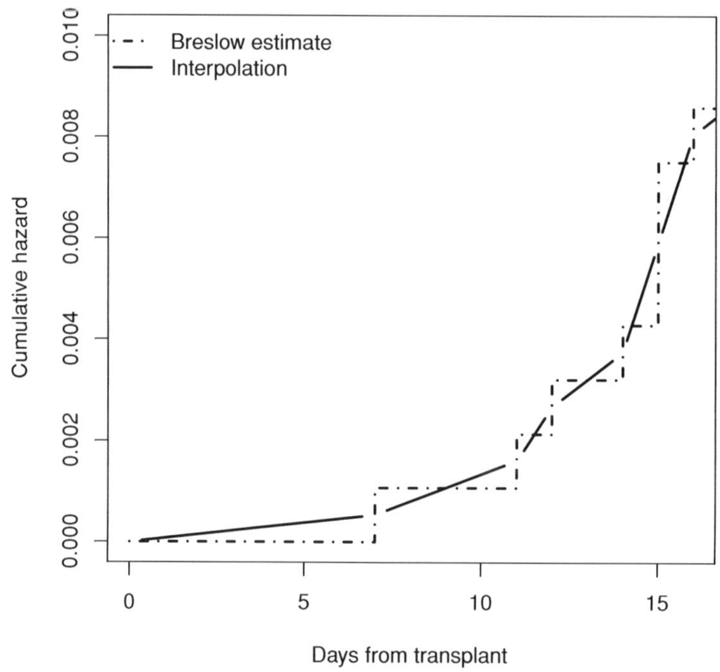

Figure 4.5 *Linear interpolation of the cumulative hazard illustrated graphically for the first six event times*

correction for the overall level of the hazard; the calibrated curves are clearly much higher than the original ones throughout the range of PI values. This result is not unexpected; it was already seen in the beginning of the example (see Figure 4.2) that relapse-free survival is higher for cohort 2 compared to cohort 1.

These ideas have also been applied in a prognostic model for predicting waiting list mortality for heart transplant candidates (Smits et al. 2003). In that data the parameter θ_2 turned out to be essential in the calibration. Generally, this might be expected if the shape of the survival function in the "training" data set differs from the one in the "calibration" data set, like the difference between the left and right panel in Figure 2.3.

4.4 Model revision

As pointed out it might happen that a well functioning prognostic model needs *model revision* because either the effect of some covariate has changed or new important covariates have been introduced. An interesting example of the former possibility is a change in the tumor grading due to new surgical procedures for a tumor as discussed in Hermans et al. (1999). Such changes in definitions might occur for Karnofsky and Broders in Data Set 1 (Table A.1), histological grade in

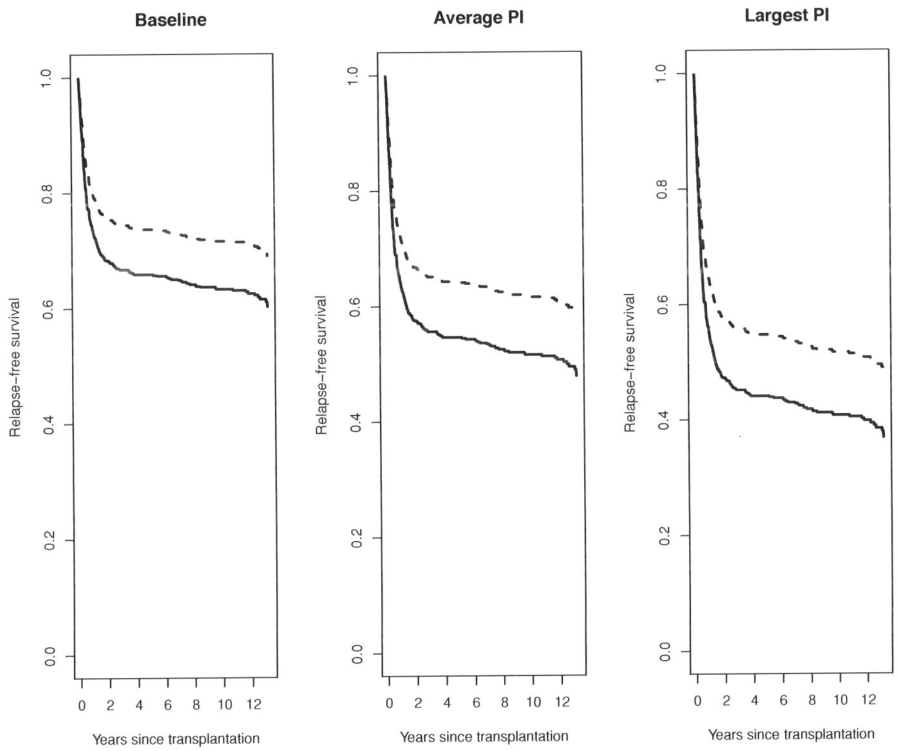

Figure 4.6 *Calibrated relapse-free survival curves for three selected patients; solid lines are the original, dashed lines the calibrated relapse-free survival curves*

Data Set 3 (Table A.3), T stage in Data Set 4 (Table A.4) and nodal stage in Data Set 5 (Table A.5). An example of new important covariates is given by the introduction of tumor gene-expressions as in Data Set 6. This all leads to the need to replace the existing index Z^* by a new one, $\tilde{Z}$, that can be written as

$$\tilde{Z} = \eta_1 Z^* + \eta_2 \tilde{x} \ ,$$

where $\tilde{x}$ is either one of the covariates included in Z^* or a new covariate. This kind of model revision has been extensively studied by Steyerberg et al. (2004) for binary outcomes. For survival data it could easily be achieved by allowing the new covariate in the calibration model of Section 4.3, leading to

$$\ln(H_{\text{cal}}(t|x)) = \theta_0 + \theta_1 Z^*(x) + \theta_2 \ln(H_0^*(t)) + \theta_3 \tilde{x} \ .$$

Such an introduction of an extra covariate definitely affects θ_1. It could also affect θ_2. It might be convenient to check beforehand the potential added value of $\tilde{x}$ by fitting a Cox model with Z^* and $\tilde{x}$ as covariates. If this leads to an improvement, the full calibration model can be fitted as described above.

Continued example

For illustration, the three covariates making up the prognostic index of Table 4.1 are added to the Weibull regression model, one by one. The results are shown in the right columns of Table 4.2. There is no evidence that the PI needs to be revised.

4.5 Additional remarks

Calibration of alternative models

The calibration model of Section 4.3 could easily be modified to models with a similar generalized linear model structure, that somehow separates covariates and time. Examples discussed in Section 2.6 are the proportional odds and the accelerated failure time models. The latter behaves very much like an ordinary linear regression model. Calibration for a very general model could be attempted by a calibration model like

$$\ln(H_{\text{cal}}(t|x)) = \theta_0 + \theta_1 \ln(H^*(t|x)) \; .$$

However, time and covariates cannot be separated in such a calibration model, which makes it much harder to interpret. Technically, it can be fitted by a Weibull model after application of a covariate specific transformation. After such a transformation, the censoring depends heavily on the covariate, which makes it a bit more tricky to fit such models.

Calibration of a Cox model if no baseline hazard has been reported

If a reliable Cox model is available, but no baseline hazard, the simplest solution is to validate/calibrate the prognostic index and to re-estimate the baseline hazard. This could easily be done by fitting a Cox model in the new data with the prognostic index as single covariate. However, this could lead to an unprecise estimate of the baseline hazard and the model survival probabilities if the new data set is small. If an estimate of the overall survival function, usually as a Kaplan-Meier curve, in the "training" set is available, that information could be used to get some feeling for the shape of the baseline hazard. It would be even close to the baseline survival function if the prognostic index had been censored. Unfortunately, the distribution of the prognostic index might not have been reported either. A way of using the Kaplan-Meier $\hat{S}_{\text{KM}}(t)$ is by taking the corresponding cumulative hazard $\hat{H}_{\text{KM}}(t) = -\ln(\hat{S}_{\text{KM}}(t))$ as an uncalibrated estimate of the cumulative baseline hazard and applying the Weibull calibration model on $\tilde{t} = \hat{H}_{\text{KM}}(t)$.

Often, the overall Kaplan-Meier is not given but only Kaplan Meiers in k prognostic subgroups, even without specifying the Cox model in full detail. As in van Houwelingen (2000) one might try to fit a proportional hazards model to those curves and use that as external model. That might be not so easy in practice. Alternatively, the estimated survival curves $\hat{S}(t|g)$ in each group might be used directly for calibration/validation. One way is using the calibration model $\ln(H_{\mathrm{cal}}(t|x)) = \theta_0 + \theta_1 \ln(H^*(t|x))$ mentioned above, but that would require group-specific time transforms. A safer way could be a two-stage approach by obtaining estimates of θ_0 and θ_1 and their standard errors and correlations for each prognostic subgroup separately and analyzing those. This approach is similar to meta-analysis and the methods of van Houwelingen et al. (2002) could be used for such an analysis. Simply plotting the estimates plus confidence intervals against the index of the subgroup might help to identify a need for calibration and suggest a calibration model.

Combination of models

If different external sources give prediction models for the outcome of interest, one might be interested in integrating such models. In Generalized Linear Models, this is known as model averaging. In recent work van der Laan et al. (2007) showed that such a *super learner* might be used to obtain better predictions. For linear models one could easily obtain a combined model by regressing the outcome on the predictions obtained from the different models. For survival data it is not straightforward how to combine Cox models with different prognostic indices and different baseline hazards. The problem has been addressed in van Houwelingen (2000), which suggested three different ways of combining two models, $S_1(t|x)$ and $S_2(t|x)$ say, namely:

1. Mixture model $S_{\mathrm{com}}(t|x) = \pi S_1(t|x) + (1-\pi)S_2(t|x)$,
2. Additive cumulative hazard model $H_{\mathrm{com}}(t|x) = \alpha_1 H_1(t|x) + \alpha_2 H_2(t|x)$,
3. Multiplicative cumulative hazard model $\ln(H_{\mathrm{com}}(t|x)) = \alpha_1 \ln(H_1(t|x)) + \alpha_2 \ln(H_2(t|x))$.

The first approach is inspired by Bayesian model averaging; see Hoeting et al. (1999) for an overview. The second approach is directly related to Aalen's additive hazard model and the third approach fits in the calibration models discussed in this chapter. However, these models have found little application in the area of prognostic model building. A slightly different application to be discussed in Chapter 11 concerns combining clinical and genomic information in a prognostic model for survival of breast cancer patients in Data Set 3.

Part II

Prognostic models for survival data using (clinical) information available at baseline, when the proportional hazards assumption of the Cox model is violated

Chapter 5

Mechanisms explaining violation of the Cox model

5.1 The Cox model is just a model

In physics there is a clear distinction between theoretical laws like the laws of motion, that are based on theoretical arguments, and empirical laws that are purely descriptive. A very good example of the last one is Hooke's Law, described as follows in Wikipedia:

> *In mechanics, and physics, Hooke's law of elasticity is an approximation that states that the extension of a spring is in direct proportion with the load added to it as long as this load does not exceed the elastic limit.*

In essence this law only states that it can be assumed that the extension is a differential function of the load. As all differentiable functions, the local behavior is linear. That is all there is to Hooke's Law.

The Cox model, as most regression models in statistics, is of the same nature as Hooke's Law. It has no mathematical underpinning, and it only holds approximately. To understand the Cox model from this perspective one only needs to make the following assumptions:

1. The prognostic information of the centered covariates X can be summarized through a linear predictor $Z = X^\top \beta$;
2. The hazard $h(t|z)$ is well-defined and well-behaved for $t > 0$. More precisely: $0 < h(t|0) < \infty$, $\ln(h(t|z))$ is differentiable with respect to z and $\partial \ln(h(t|z))/\partial z|_{z=0}$ is continuous in t with $\lim_{t\downarrow 0} \partial \ln(h(t|z))/\partial z|_{z=0} = \gamma$.

These assumptions lead to the local approximation ($t \approx 0$, $z \approx 0$)

$$\ln(h(t|z)) = \ln(h(t|0)) + z\gamma + \ldots . \quad (5.1)$$

Without loss of generality γ can be taken to be equal to 1. The Cox model makes the assumption that this approximation is precise and holds for large t and z far away from zero. It is known as the *Proportional Hazards (PH) assumption.* However, there is no theorem in statistics or probability that prescribes that this assumption should hold. It is only very convenient if it does.

Since assumptions 1 and 2 above hold for many generalized linear models, like the proportional odds model of Section 2.6, all such models are locally ($t \approx 0, z \approx 0$) equivalent and differences between such models can only be shown statistically if the predictor has a large effect on survival, or if there is long follow-up or there are many events. In clinical studies the effect of the predictor is usually not very big. Therefore, violation of the Cox model can only be expected to show up in large studies with long follow-up.

It should be stressed that the emphasis in this part of the book is on violation of the proportional hazards assumption and not on the linear structure of the model formulated in assumption 1. That does not take away that a simple linear model in the covariates could be a gross simplification. Optimal scaling of covariates through regression splines (Eilers & Marx 1996) or fractional polynomials (Royston & Sauerbrei 2008) and prudent introduction of interactions can substantially improve the performance of a model. There is much to be learned from the excellent book on statistical learning by Hastie et al. (2009) for those interested in clinical prediction models.

Since the PH-based Cox model has played such an important role in modelin survival for nearly 40 years now, it is of interest to explore how violations of the model could be understood from relatively simple extensions of the model itself. This will be explored in the next sections, focusing on the following mechanisms:

- Heterogeneity between individuals: frailties and the like
- Measurement error in covariates
- Dynamic behavior of covariates
- Informative dropout; competing risks

5.2 Heterogeneity

Heterogeneity in general

Heterogeneity among individuals in a population is a bit of a controversial issue in statistics. The standard statistical formulation of random variation among a population of n individuals is that the observations $X_1, ..., X_n$ are a random sample from a distribution F. Heterogeneity means to say that each individual $i = 1, .., n$ has his own distribution F_i. The most debated example is where X is a $0/1$ variable. The standard model states that $P(X = 1) = \pi$, while heterogeneity would imply that $P(X_i = 1) = \pi_i$. For the dichotomous case there is no way to distinguish the two approaches if only X can be observed. For the more general case such a distinction is only possible if assumptions are made about the individual distributions F_i. For example, if the outcome X concerns count data and the individual X_i are supposed to follow a Poisson distribution, $F_i = \text{Poisson}(\mu_i)$, heterogeneity leads to *overdispersion* meaning that the variance $\text{var}(X)$ is larger than the expectation $E[X]$, which is impossible under the homogeneous Poisson model. Similarly, for continuous data, violation of the popular normal (Gaussian) model could be explained by a model where $F_i = N(\mu_i, \sigma^2)$, and the distribution of the $\mu_i's$ is not normal. However, there is no law that prescribes that outcomes should follow a normal distribution. The central limit theorem is often cited in defense of the normal distribution. However, the central limit speaks about averages and not about single outcomes. (There is some theoretical foundation for the Poisson distribution for count data, see Feller (1950, Section VI.6).)

Heterogeneity in survival

For survival data the story is very similar. Heterogeneity can be expressed in terms of the survival function $S_i(t)$, the cumulative hazard function $H_i(t)$ or the hazard function itself $h_i(t)$. If there is no further information, this cannot be distinguished from a "standard" model with $S(t) = \frac{1}{n}\sum_{i=1}^{n} S_i(t)$. (There is a subtlety here: the overall (cumulative) hazard is not equal to the average (cumulative) hazard, that is $H(t) = -\ln(S(t)) \neq \frac{1}{n}\sum_{i=1}^{n} H_i(t)$ and similarly for $h(t)$.) If one dares to assume that $h_i(t)$ is constant, $h_i(t) \equiv h_i$ say, the overall hazard $h(t)$ is decreasing (unless all h_i are the same) and the distribution of the h_i, up to a multiplicative factor, can be inferred from $h(t)$. However, for clinical data there is no law that prescribes that the hazards should be constant.

A special form of heterogeneity is the so-called *frailty* model that specifies a particular form of heterogeneity at the level of the hazard function, namely

$$h_i(t) = Z_i\tilde{h}(t) .$$

Notice the similarity with the proportional hazards model as introduced in Section 2.2. The frailty model can be seen as a "latent" proportional hazards model in which the covariates cannot be observed and $Z_i = \exp(X_i^\top \beta)$. It is convenient to interpret Z as a random variable and to rewrite the frailty model as

$$h(t|Z) = Z\tilde{h}(t) .$$

The random variable Z is called the frailty of the individual. There is huge literature on the concept. Noteworthy are the review papers by Aalen (1994), Keiding et al. (1997), and the books by Hougaard (2000), Duchateau & Janssen (2008) and Aalen et al. (2008).

To understand the effect of the unobservable frailty Z, it is of interest to look into the relation between the overall or marginal frailty $h(t)$ and the "baseline" or conditional hazard $\tilde{h}(t) = h(t\,|\,Z = 1)$. Starting point is that

$$S(t) = E[\exp(-Z\tilde{H}(t))] = \int_0^\infty \exp(-z'\tilde{H}(t))f(z')dz' .$$

Here $\tilde{H}(t) = \int_0^t \tilde{h}(s)ds$ as usual. For notational convenience it is assumed that Z has density f(z). The conditional distribution of Z given $T \geq t$ is given by

$$f(z|T \geq t) = \frac{P(T \geq t|z)f(z)}{P(T \geq t)} = \frac{\exp(-z\tilde{H}(t)f(z)}{\int_0^\infty \exp(-z'\tilde{H}(t)f(z')dz'} .$$

Under the assumption that Z has finite expectation, it follows that

$$h(t) = \frac{-S'(t)}{S(t)} = \frac{\int_0^\infty z\tilde{h}(t)\exp(-z\tilde{H}(t)f(z)dz}{\int_0^\infty \exp(-z\tilde{H}(t)f(z)dz} = E[Z|T \geq t]\tilde{h}(t) .$$

This could also be understood intuitively, because $h(t)$ is the average of the conditional hazard $Z\tilde{h}(t)$ taken over all individuals still alive at $T = t-$. Similarly, the following result holds under the assumption that $\mathrm{var}(Z)$ is finite as well

$$\frac{dE[Z|T \geq t]}{dt} = -h(t)\mathrm{var}(Z|T \geq t) .$$

The implication is that the ratio $h(t)/\tilde{h}(t)$ is a monotonically decreasing function of t. For small values of t (or more precisely, for small values of $\tilde{H}(t)$) the following approximation holds

$$h(t) \approx \tilde{h}(t)(E[Z] - \mathrm{var}(Z)\tilde{H}(t)) .$$

Two frailty distributions are of special interest: the gamma distribution and the two-point mixture.

Gamma frailty distribution

The gamma distribution is a popular model for frailties, mainly because of mathematical convenience. The density of the gamma(α,β) distribution is given by

$$f(z|\alpha,\beta) = \frac{\beta^\alpha}{\Gamma(\alpha)} z^{\alpha-1}\exp(-\beta z) .$$

Its first two moments are $E[Z] = \alpha/\beta$ and $\mathrm{var}(Z) = \alpha/\beta^2$. Since the conditional distribution of Z given $T \geq t$ has density proportional to $f(z|\alpha,\beta)\exp(-z\tilde{H}(t))$, it is immediate that this conditional distribution is again a gamma distribution with parameters α and $\tilde{\beta} = \beta + \tilde{H}(t)$. This leads to

$$h(t) = \frac{\alpha\tilde{h}(t)}{\beta + \tilde{H}(t)} , \quad S(t) = (1 + \tilde{H}(t)/\beta)^{-\alpha} .$$

Without loss of generality Z can be centered at $E[Z] = 1$. A convenient reparametrization is to take $\alpha = \beta = 1/\xi$, leading to $E[Z] = 1$, $\mathrm{var}[Z] = \xi$, $h(t) = \tilde{h}(t)/(1 + \xi\tilde{H}(t))$ and $S(t) = (1 + \xi\tilde{H}(t))^{-1/\xi}$.

Two point mixture

The simplest model for the frailty distribution one can think of is the two-point mixture that allows two values for Z, namely ζ_0 and ζ_1 with $P(Z = \zeta_0) = 1 - \pi$ and $P(Z = \zeta_1) = \pi$. Its expectation is given by $E[Z] = (1-\pi)\zeta_0 + \pi\zeta_1$ and its variance by $\text{var}(Z) = (\zeta_1 - \zeta_0)^2 \pi(1-\pi)$. The conditional distribution of Z given $T \geq t$ is determined by

$$P(Z = \zeta_1 | T \geq t) = \frac{\pi P(T \geq t | Z = \zeta_1)}{S(t)} = \frac{\pi \exp(-\zeta_1 \tilde{H}(t))}{S(t)} ,$$

where

$$S(t) = (1-\pi)\exp(-\zeta_0 \tilde{H}(t)) + \pi \exp(-\zeta_1 \tilde{H}(t)) .$$

This formula can be used to obtain the marginal hazard $h(t)$. The formulas are not as transparent as for the gamma distribution, but easy to implement. The special case $\zeta_1 = 0$ leads to a form of the cure model as mentioned in Section 2.1 and discussed in Section 6.2 and Chapter 10.

Frailties in the Cox model

Replacing the general baseline hazard in the formulas above by a hazard coming from a Cox model with covariate vector X leads to a model that can be interpreted as a Cox model in which one or more covariates are omitted (not observed) that are independent of the covariates included in the model. This leads to a frailty term $Z = \exp(X_{\text{omit}} \beta_{\text{omit}})$. (If the omitted covariate is not independent of the other covariates denoted by X_{other}, the only part that shows up in the frailty term is the residual $X_{\text{omit}} - E[X_{\text{omit}} | X_{\text{other}}]$.) The approximation for small values of t (or more precisely, for small values of $\tilde{H}(t)$) now reads

$$h(t|x) \approx \tilde{h}(t)\exp(x^\top \beta)\left(E[Z] - \text{var}(Z)\tilde{H}(t)\exp(x^\top \beta)\right) . \tag{5.2}$$

It is clear that the regression coefficients β of the "complete" model still apply for $t = 0$ or very close to zero. The regression coefficients shrink towards zero with increasing t for small t, but it is not clear what happens later on.

As an example, the very simple case is considered of a single dichotomous covariate ($X = 0$ or $X = 1$) with conditional hazard ratio under the complete model $\text{HR} = \exp(\beta) = 2$, baseline hazard $\tilde{H}(t) = t$ and a frailty distribution with $E[Z = 1] = 1$ and $\text{var}(Z) = 0.25$. Figure 5.1 shows the hazard ratio $h(t|x=1)/h(t|x=0)$ for the two groups under the gamma frailty model and under the two point mixture model with $\pi = 0.5$, $\zeta_0 = 0.5$ and $\zeta_1 = 1.5$. The first observation is the two models agree for small values of t, as expected. The second observation is that for the gamma model the hazard $h(t|x=1)/h(t|x=0)$ monotonically decreases from $\text{HR} = 2$ to $\text{HR} = 1$, while for the mixture model HR returns to $\text{HR} = 2$ for large t. The explanation is that in the mixture model the conditional distribution of Z given $T \geq t$ for large t will be concentrated in ζ_0 independent of x. The difference between the two models looks more dramatic than it actually is, because in the high-risk group the probability of survival beyond $t = 2$ is quite small as shown in Figure 5.2 and the models do agree well before $t = 2$.

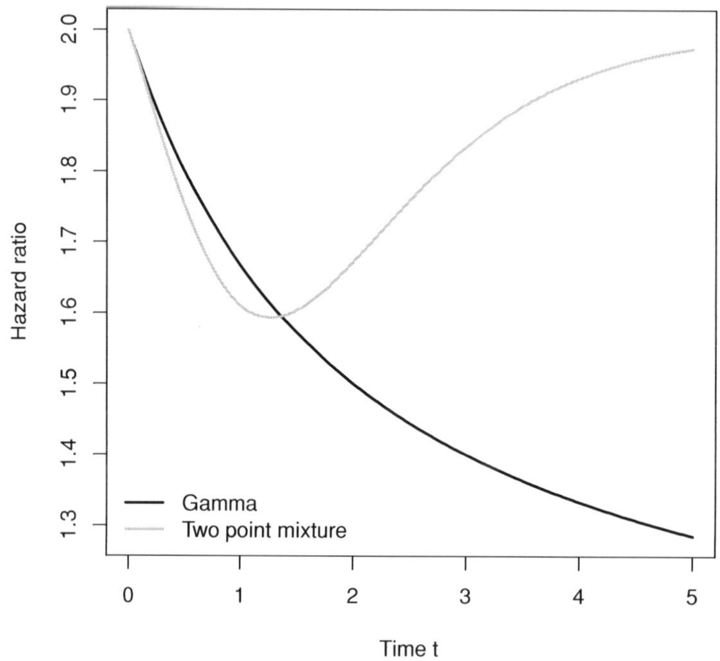

Figure 5.1 *Hazard ratio for simple model for gamma and mixture frailty distribution*

Dynamic interpretation

In the setting of the theme of this book it is interesting to observe that the frailty models allow a nice dynamic interpretation. In order to compute the predictive distribution given that an individual is still alive at $T = t-$, one only needs to update the distribution of the frailty by

$$f(z|T \geq t, x) = f(z)\exp(-zH(t|x)) / \int_0^\infty f(z')\exp(-z'H(t|x))dz' \,.$$

Shared frailties

Another very important application of the concept of frailty is its use to model dependence between survival times within groups. A (shared) frailty term, specific for the group, automatically generates such a dependence. In that situation the frailty distribution can be identified even when covariates are absent. These different applications of the frailty concept should be well distinguished to avoid confusion.

Cure model and similar extensions

Extensions of the frailty model can be obtained by letting the distribution of Z depend on the covariates X as well. Such an extension of the gamma model is

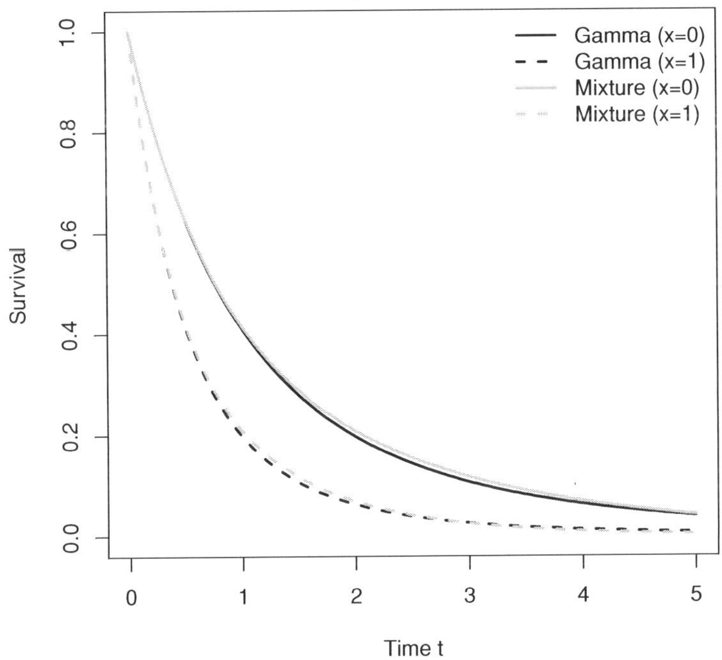

Figure 5.2 *Marginal survival in the two groups under the two models*

discussed in Barker & Henderson (2004). An extension of the two point mixture is to let $\pi = P(Z = \zeta_1)$ depend on X. Taking a logistic regression model for $\pi(X)$ and $\zeta_0 = 0$, $\zeta_1 = 1$ leads to the traditional cure model of Kuk & Chen (1992). However, it is beyond the scope of this section to go into much detail. For this section it suffices to note that all such models lead to violations of the Cox model.

5.3 Measurement error in covariates

Simple measurement error

A mechanism that is often overlooked, but might happen quite frequently, is measurement errors in one or more covariates. In this section a brief review is given and the consequences for survival analysis are discussed. A comprehensive treatment can be found in the book by Carroll et al. (2006). To get an idea of the effect of measurement error it suffices to consider the case of two covariates X_1 and X_2 and a perfect Cox model

$$h(t|x_1,x_2) = h_0(t)\exp(\beta_1 x_1 + \beta_2 x_2) \; .$$

The covariates are taken to be independent; the first one X_1 could be any type of covariate, while the second one X_2 has a normal distribution with mean zero and variance $\sigma^2_{X_2}$. Unfortunately, X_2 can only be observed with some error. Instead of X_2 the variable $V = X_2 + \varepsilon$ is observed, where ε is normally distributed with mean zero and variance σ^2_ε. Consequently $\text{var}(V) = \sigma^2_{X_2} + \sigma^2_\varepsilon$. The question of interest is how the hazard depends on the observable covariates X_1 and V. To answer the question it is useful to invert the relation between V and X_2. Standard theory gives that

$$X_2 = \lambda V + \varepsilon^* ,$$

with $\lambda = \sigma^2_{X_2}/\sigma^2_V$, ε^* independent of V, normally distributed with mean zero and $\text{var}(\varepsilon^*) = \lambda\sigma^2_\varepsilon$. So, the true hazard can be rewritten as

$$h(t|x_1, v, \varepsilon) = h_0(t)\exp(\beta_1 x_1 + \beta_2 \lambda v + \beta_2 \varepsilon^*) .$$

There are two observations to be made. First, the regression coefficient of V is equal to $\lambda\beta_2$ and not equal to β_2. This is called attenuation in measurement error theory. Secondly, there is the unobservable extra term $\beta_2\varepsilon^*$. The consequence of that extra term is that the effect of observable covariates X_1 and V cannot be described by a Cox model, because of the presence of the frailty term $Z = \exp(\beta_2\varepsilon^*)$, which has a log-normal distribution. For t close to zero Cox model still holds, with regression coefficients β_1 and $\lambda\beta_2$, respectively, but for larger values of t the model is distorted. There are no explicit formulas for the effects of log-normal frailties on the survival and the hazard function, but they can be approximated by the effects of gamma frailties with the same means and variances.

Ageing covariates

A mechanism that is related to measurement error is that a fixed covariate measured at (or just before) $t = 0$ is a "snap-shot" of the state of the person in the study. Such a covariate is actually time-dependent, but only observed at $t = 0$. An example of such a covariate is the Karnofsky performance index in the ovarian cancer data set (Data Set 1).

To understand what is going on, it suffices to consider a situation with two covariates: a time-fixed covariate X_1 and a time-dependent covariate $X_2(t)$. The covariates are independent; the first could be any type of covariate, while the second is a Gaussian stochastic process with mean zero and covariance function $C(s,t) = \text{cov}(X(s), X(t))$. The "perfect" model is a simple model in which the effect of the covariate process $X_2(.)$ on the hazard is through its current value $X_2(t)$, that is,

$$h(t|x_1, x_2(t)) = h_0(t)\exp(\beta_1 x_1 + \beta_2 x_2(t)) .$$

The question is how the hazard depends on X_1 and $X_2(0)$. Regressing $X_2(t)$ on $X_2(0)$ leads to the representation

$$X_2(t) = \lambda(t)X_2(0) + \varepsilon(t) .$$

Here, $\lambda(t) = C(0,t)/C(0,0)$ and $\varepsilon(t)$ is a Gaussian stochastic process with mean zero and covariance function $C^*(s,t) = C(s,t) - C(0,s)C(0,t)/C(0,0)$. The hazard can be rewritten as

$$h(t|x_1, x_2(0), \varepsilon(t)) = h_0(t)\exp(\beta_1 x_1 + \beta_2\lambda(t)x_2(0) + \beta_2\varepsilon(t)) \ .$$

From this it can be concluded that one can expect that the effect of $X_2(0)$ can strongly vary over time, depending on the covariance structure of the $X_2(.)$ process. A second conclusion is that also the effect of X_1 (and all other covariates) will be influenced by the presence of $\varepsilon(t)$. The effect will be negligible for small t because the variance of $\varepsilon(t)$ will be very small for small t.

To get a better understanding of the phenomenon more insight is needed in the effect of "time-dependent frailties". That is getting very technical and beyond the scope of the book. An approximation in the spirit of the effect of a simple gamma frailty as discussed in Section 5.2, is given in Perperoglou, van Houwelingen & Henderson (2006).

5.4 Cause specific hazards and competing risks

In Appendix A.4 the results are presented from a clinical trial in gastric cancer comparing two different surgical procedures denoted by D1 and D2. Figure A.5 shows crossing survival curves for the two treatments and Figure A.6 shows that the hazard ratio of D2 with respect to D1 is definitely larger than 1 for small t and smaller than 1 for large t. This outspoken violation of the PH assumption can be understood through the concepts of competing risks and cause-specific hazards. A model for this data would be to distinguish between causes

1. death caused by the surgery;
2. death caused by the cancer itself.

With the treatment $x = \mathbf{1}\{\text{treatment} = \text{D2}\}$ as covariate, separate Cox models for the two cause-specific hazards may be specified,

$$h_1(t|x) = h_{10}(t)\exp(x\beta_1) \ , \ h_2(t|x) = h_{20}(t)\exp(x\beta_2) \ .$$

The total hazard is simply given by $h(t|x) = h_1(t|x) + h_2(t|x)$. It is not hard to construct a hazard ratio graph like that of Figure A.6 by taking $\beta_1 > 0$, $\beta_2 < 0$, $h_{10}(t)$ concentrated in the first year and $h_{20}(t)$ concentrated in the later years. Figure 5.3 shows the result for $\beta_1 = 1$, $\beta_2 = -0.5$, $h_{10}(t) = 0.2\exp(-t)$ and $h_{20}(t) = 0.1$. It is not difficult to show (see (5.1)) that the log hazard ratio for overall survival is approximated by

$$\frac{\partial}{\partial x}\ln(h(t\,|\,x)) = \pi(t)\beta_1 + (1 - \pi(t))\beta_2 \ ,$$

with $\pi(t) = h_1(t)/(h_1(t) + h_2(t))$ the relative hazard of cause 1. It can be seen that the overall log hazard ratio is a time-weighted average of the two cause-specific log

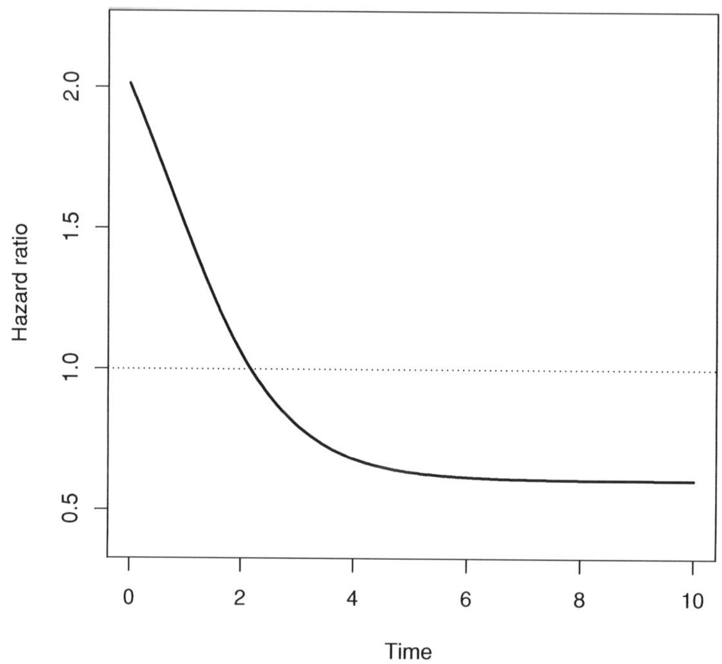

Figure 5.3 *Hazard ratio for "imitation" of Data Set 5*

hazard ratios that will only be time-constant when the two cause-specific hazards are proportional to each other. The concepts of relative and total hazard can also be used directly to model competing risks data, see Nicolaie et al. (2010).

Another interesting violation of the Cox model arises from the combination of frailties and competing risks. In theory it is thinkable that there exists separate frailties, Z_1 and Z_2 say, for the two cause-specific hazards. If these frailties are correlated interesting phenomena can be observed. The approximation given in Section 5.2 for a single Cox model generalizes to

$$h_1(t|x) \approx h_{10}(t)\exp(x^\top \beta_1)\cdot \\ \cdot (EZ_1 - \text{var}(Z_1)H_{10}(t)\exp(x^\top \beta_1) - \text{cov}(Z_1, Z_2)H_{20}(t)\exp(x^\top \beta_2))$$

for the first hazard and similarly for the second one. This shows that, if the frailties are positively correlated, a covariate that has no effect on the first cause hazard in the "full" model and a negative effect on the second cause hazard might pop up with a positive effect on the first cause hazard later in the follow-up. This is coined *False Positivity* by Di Serio (1997) and also discussed in Aalen et al. (2008). The intuition behind is that the risk set at time t is formed by those who are still event free. Individuals with high values of $x^\top \beta_2$ can only survive up to t if they have

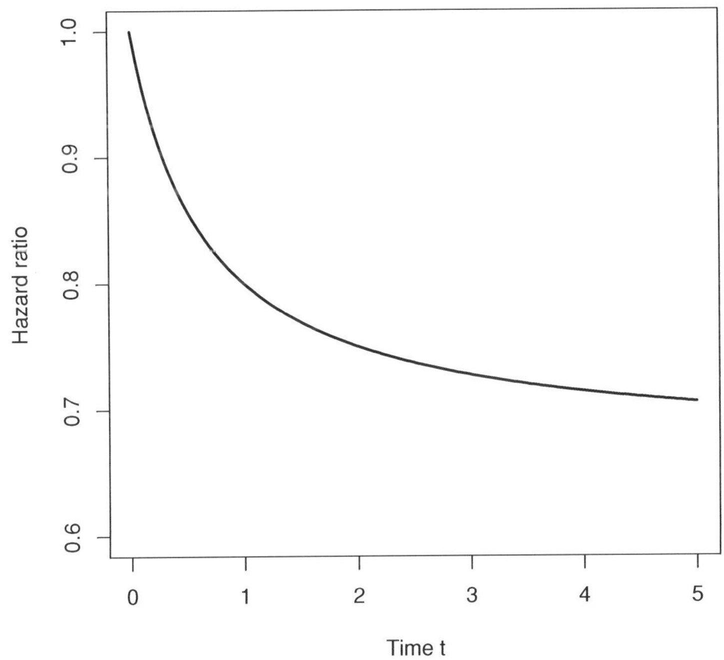

Figure 5.4 *Hazard ratio for the marginal cause-specific hazard of cause 1, for the shared gamma frailty model with variance 0.5; conditional hazard ratios are 1 for cause 1 and 2 for cause 2*

a small value of Z_2. If so, the value of Z_1 must be small as well, leading to the observed positive effect.

As an example consider the case of $Z_1 = Z_2 = Z$ coming from a gamma distribution with $E[Z] = 1$ and $\mathrm{var}(Z) = \xi$. In that case the following exact results holds:

$$h_1(t|x) = \frac{h_{10}(t)\exp(x^\top\beta_1)}{1+\xi(\exp(x^\top\beta_1)H_{10}(t)+\exp(x^\top\beta_2)H_{20}(t))}.$$

In the simple case that $x = 0, 1$, $\beta_1 = 0$, $\exp(\beta_2) = \mathrm{HR}$, $H_{10}(t) = H_{20}(t) \to \infty$, $h_1(t|1)/h_1(t|0)$ decreases monotonically from 1 to $2/(1+\mathrm{HR})$. Figure 5.4 shows the cause-1 specific hazard ratio $h_1(t|1)/h_1(t|0)$ for $\mathrm{HR} = 2$, a frailty variance of 0.5, and $H_{10}(t) = H_{20}(t) = t$. As expected, X seems to have a protective effect on death from cause 1 for larger t.

If censoring is not purely administrative, it could be considered as a competing risk for the event of interest. The results of this section give some idea what might go wrong if there is *informative censoring*, meaning that censoring is not independent of survival, conditional on the covariates. Strange results of covariates that have only late effects could be explained by unaccounted informative censoring.

5.5 Additional remarks

If there are no covariates, there is no way to infer the frailty distribution without making very strong assumptions on the shape of the conditional hazard given the frailty. If there are covariates, the story is more subtle. In econometrics there is a whole literature on the identifiability of the frailty model for simple survival data and for competing risks data, see Elbers & Ridder (1982), Kortram et al. (1995), Abbring & van den Berg (2003) for details. The claim is that the distribution of a single frailty in a Cox model and the bivariate distribution of the pair of frailties in a competing risks model can be identified from the data, if the moments of the distribution are finite. The argument is related to the approximation (5.2) in Section 5.2. In principle, the moments of Z can be estimated from the derivatives of $\ln(h(t|x))$ with respect to x in $x = 0$, $t = 0$ and from those moments the distribution can be derived. As discussed in Hougaard (2000), this is not true in general. A frailty with a positive stable distribution will not lead to violation of the Cox model, and, therefore cannot be identified.

But, even if the moments of the frailty distribution exist, the distribution can only be retrieved if the model given the frailty is fully correct. As discussed above, there are many mechanisms leading to violation of the PH model. The frailty model is a way of explaining violation of the PH model. However, the inverse claim that violation of the PH model implies heterogeneity in the population is false, because it is only one of many explanations.

Frailties can only be identified if they are shared among individuals and even then it is too ambitious to retrieve the whole frailty distribution from the data.

Chapter 6

Non-proportional hazards models

6.1 Cox model with time-varying coefficients

The main effect of the mechanisms distorting the Cox model discussed in Chapter 5 is the violation of the proportional hazards assumption. The standard extension of the Cox model that allows non-proportional hazards but keeps the linear effects of the predictors, is the following model

$$h(t|x) = h_0(t)\exp(x^\top \beta(t)) \ .$$

Actually, this only becomes a model if it is specified how the regression coefficients in $\beta(t)$ can depend on t. Unlike Aalen's additive hazards model briefly discussed in Section 2.6, it is impossible to let $\beta(t)$ completely free, because it would lead to exploding parameter estimates. For categorical covariates the problem could be solved by stratification, briefly mentioned in Section 2.7. Strata are defined by (combinations) of categorical covariates and each stratum g has its own baseline hazard leading to

$$h_g(t|\tilde{x}) = h_{g0}(t)\exp(\tilde{x}^\top \tilde{\beta}(t)) \ .$$

The covariate vector $\tilde{x}$ cannot contain the categorical covariate(s) defining the strata, because they are constant within the strata, but they can still contain interaction between those categorical covariates and other risk factors. However estimating baseline hazards in all strata separately affects the stability of the estimates and the reliability of the prediction model. Therefore, it is wiser to use a parametric model for the time-varying regression coefficients. This can be done by considering a set of m basis functions $f_1(t),...,f_m(t)$ and taking

$$\beta(t) = \sum_{j=1}^{m} \gamma_j f_j(t) \ .$$

Here, each γ_j is a vector of the same length as β, namely the number of covariates. It is helpful for the interpretation of the parameters if the basis functions are defined in such a way that

$$f_1(t) \equiv 1 \ , \quad f_j(0) = 0, \text{ for } j = 2,\ldots,m \ .$$

The interpretation is then that the effect of the covariates at $t = 0$ is given by γ_1 ($\beta(0) = \gamma_1$), while the other γ's describe the violation of the PH-model for each of the covariates. The effect at $t = 0$ is of special importance because it is not influenced by the presence of frailties or unobserved covariates, see the discussion in Section 5.2. The global null hypothesis that the PH assumption is satisfied is equivalent to testing $H_0: \gamma_2 = \ldots = \gamma_m = 0$. A popular choice for the second basis function is $f_2(t) = \ln(1+t)$, which starts with $f_2(0) = 0$, has derivative $f_2'(0) = 1$ and slows down later in the follow-up. If the interest is only in testing the PH-assumption for each covariate, it suffices to take $m = 2$ and $f_2(t) = \ln(1+t)$. This is very close to the suggestion in Cox's original paper (Cox 1972), who takes $f_2(t) = \ln(t)$ itself. If very early events can occur, Cox's proposal could put too much emphasis on early events, because $\ln(t)$ "explodes" for t close to zero.

Models with time-varying coefficients can be handled in the same way as time-dependent covariates. They are in fact equivalent, because the set of time-dependent covariates $Xf_1(t), \ldots, Xf_m(t)$ yields exactly the same model. Some packages allow to do this internally, e.g., SPSS and SAS, while others, such as R/S-PLUS, require restructuring the database as discussed in Section 2.5.

An example, the ovarian cancer data set

To get more feeling for the (im)possibilities of the time-varying effect model such a model is fitted to the ovarian cancer data of Data Set 1, taking $m = 2$ and $f_2(t) =$

Table 6.1 *Extension of the Cox model for Data Set 1 with time-varying effects; Karnofsky is recoded as 6 (≤ 60) to 10 (100) and used as a continuous covariate*

Covariate	Category	$\hat{\beta}_{\text{fixed}}$ (SE)	$\hat{\gamma}_1$ (SE)	$\hat{\gamma}_2$ (SE)
FIGO	III			
	IV	0.504 (0.136)	0.588 (0.289)	-0.147 (0.305)
Diameter	Microscopic			
	< 1 cm	0.247 (0.323)	0.881 (0.967)	-0.461 (0.766)
	1-2 cm	0.665 (0.326)	1.654 (0.957)	-0.761 (0.770)
	2-5 cm	0.785 (0.315)	2.212 (0.931)	-1.293 (0.756)
	> 5 cm	0.836 (0.304)	1.712 (0.915)	-0.648 (0.730)
Broders	1			
	2	0.573 (0.243)	0.044 (0.525)	0.558 (0.499)
	3	0.520 (0.234)	0.020 (0.505)	0.527 (0.478)
	4	0.323 (0.274)	-0.304 (0.611)	0.696 (0.573)
	Unknown	0.650 (0.265)	0.545 (0.557)	0.037 (0.557)
Ascites	Absent			
	Present	0.272 (0.155)	0.318 (0.373)	-0.024 (0.343)
	Unknown	0.205 (0.213)	0.774 (0.467)	-0.676 (0.475)
Karnofsky	Continuous	-0.173 (0.054)	-0.459 (0.115)	0.352 (0.125)

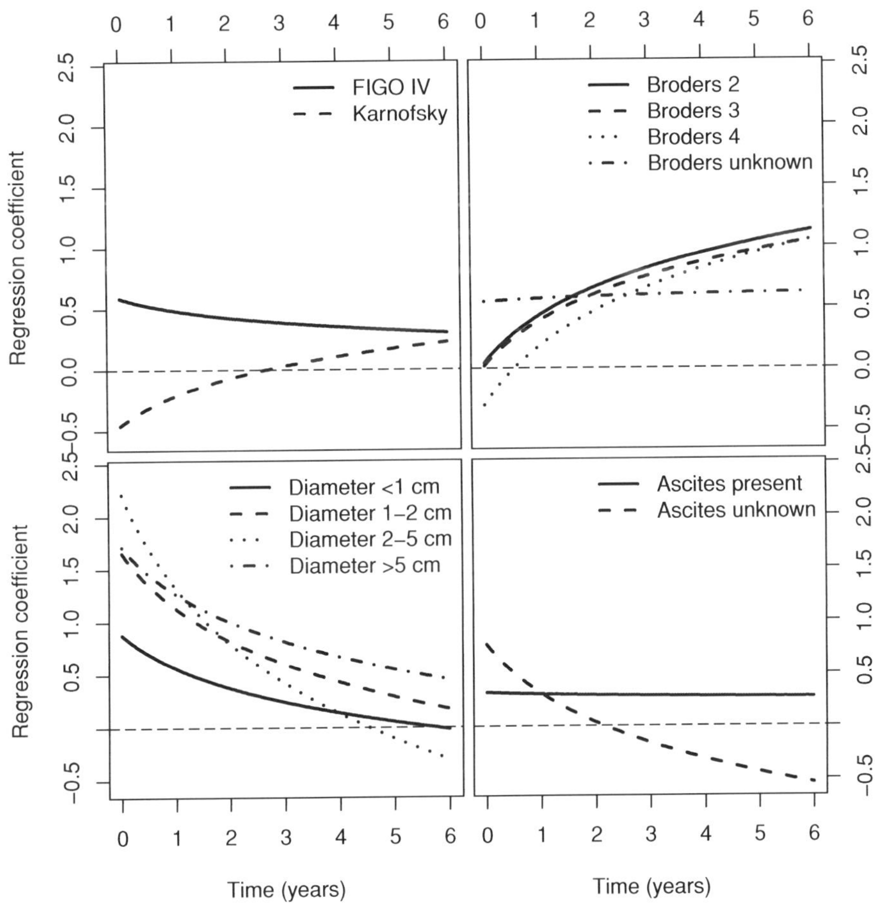

Figure 6.1 *The time-varying coefficients of the model of Table 6.1*

$\ln(1+t)$. The results together with the results for the time-constant model are given in Table 6.1 and a plot of all the time-varying coefficients is shown in Figure 6.1.

Observe that nearly all the regression coefficients of Karnofsky, FIGO IV and Diameter shrink towards zero over time. Some of them even change sign, but that could be due to spurious effects of the $\ln(1+t)$-model. The picture is less clear for the categorical covariates Broders and Ascites. It strongly depends on the choice of the base line category.

In a model like this the prognostic index Z varies over time as well, that is $Z = Z(t)$. The index at $t = 0$ correlates well with the prognostic index (Z_{fixed}) in the simple Cox model with time-fixed effects: $\text{cor}(Z(0), Z_{\text{fixed}}) = 0.91$, but the standard deviation of $Z(0)$ is about twice as large ($\text{sd}(Z(0)) = 1.12$ versus $\text{sd}(Z_{\text{fixed}}) = 0.59$), an indication of a larger effect in the early stage. Over time, $\text{sd}(Z(t))$ rapidly decreases to $\text{sd}(Z(3.2)) = 0.463$ and increases slowly after that. It is remarkable that later values of $Z(t)$ have a slight negative correlation with $Z(0)$, which is implied by the choice of the $\ln(1+t)$-model and the observation

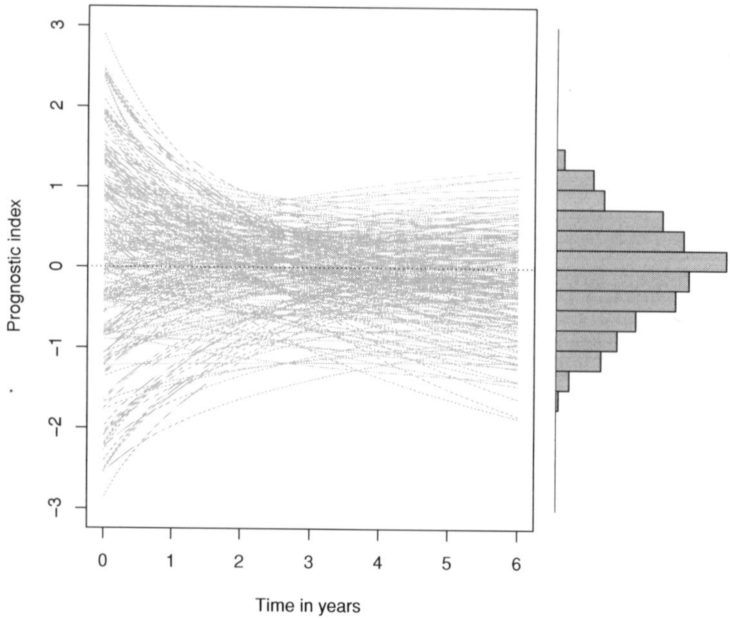

Figure 6.2 *The time-varying prognostic indices $Z(t)$ for each of the patients in the ovarian cancer data; the right panel shows a histogram of the time-fixed prognostic indices of these patients*

that $\mathrm{cor}(Z(0), Z'(0)) = \mathrm{cor}(x^\top \hat{\gamma}_1, x^\top \hat{\gamma}_2) = -0.91$. Figure 6.2 shows the individual trajectories in the data of the $Z(t)$'s, along with a histogram of Z_{fixed}.

This model is a clear case of overfitting. The number of parameters has doubled while the partial likelihood has not improved much. The model could be pruned by significance testing. The extension of the model with all time-varying effects is significant ($\chi^2_{[12]} = 24.148$, $P = 0.020$), but a stepwise procedure would only retain the time-varying effect of Karnofsky ($\chi^2_{[1]} = 12.525$, $P = 0.0004$). The danger of starting from a time-constant model and checking the significance of the extension to a time-varying model, is that covariates showing a time-varying effect that changes from positive to negative (or the other way around) might be missed. Inspired by the gastric cancer example of Data Set 4 a strategy for model building that would detect such switching effects is presented in Putter et al. (2005). The strategy proposed there consisted of a forward selection procedure in which each of the covariates together with their interaction with time was considered. The covariate was included in the model together with the covariate by time interaction if the likelihood-ratio test for the model with both covariate and covariate by time indicated a significantly better fit compared to the model without. In a subsequent pruning step, each of the covariate by time interactions were considered and removed from the model in case the interaction was not significant. Similar strategies are developed by Sauerbrei et al. (2007).

The model as presented here might even be too simple as a starting model because it has a very simple model for the time-varying effects. The "constant + $\ln(1+t)$" model might not be adequate for covariates that have a very outspoken time-varying effect. The basis of time functions needs to be extended to detect such effects. Some variation of *natural cubic splines* as advocated in Harrell's book (Harrell 2001) could do the job. However, if there is no prior knowledge about the covariates for which such a strong time-variation could be expected, this would lead to even bigger starting models and an enlarged danger of overfitting. As mentioned before in Section 3.6 and documented in van Houwelingen & Thorogood (1995), stepwise procedures do not completely remedy overfitting. Generally speaking, it might very hard to control the prediction error in these time-varying coefficient models.

Dynamic predictions are obtained through the formula

$$\hat{S}(t|x,s) = \exp(-\sum_{s \leq u \leq t} \hat{h}_0(u)\exp(x^\top \hat{\beta}(u))) \,. \tag{6.1}$$

Figure 6.3 shows how the time-varying effect of Karnofsky affects (dynamic) survival curves. Based on the models of Table 6.1, two individuals are considered, with Karnofsky scores 7 and 10; for the other covariates the mean is taken. The left panel shows ordinary failure functions, the right panel shows the fixed width failure functions using a $w = 2$ year window. In grey survival probabilities are shown based on the time-fixed Cox model, in black based on the time-varying Cox model. The curves for Karnofsky $=$ 7 and Karnofsky $=$ 10 grow closer for the time-varying model, because the effect of Karnofsky decreases over time. The dynamic curves for Karnofsky $=$ 7 and Karnofsky $=$ 10 even cross after two years. Figure 6.4 shows dynamic prediction error curves of the time-fixed prognostic index of Chapter 3, as well as the time-varying prognostic index as defined in Table 6.1. The cross-validated curves are based on the cross-validated dynamic predictions

$$\hat{S}_{(-i)}(t|x,s) = \exp(-\sum_{s \leq u \leq t} \hat{h}_{0,(-i)}(u)\exp(x^\top \hat{\beta}_{(-i)}(u)))$$

instead of (6.1), with $\hat{\beta}_{(-i)}(t)$ and $\hat{h}_{0,(-i)}$ the time-varying regression coefficients and baseline hazard estimates based on the data with individual i removed. The cross-validated dynamic prediction errors of the time-varying model are much higher than the dynamic prediction errors themselves, even higher than the dynamic prediction errors of the null model after two years, indicating the overfitting of this model, especially later in the follow-up.

An "unpleasant" property of this type of models, hampering the clinical interpretation, is that the one-dimensional prognostic index is lost. In the example above, there are two linear combinations of the covariates $X^\top \hat{\gamma}_1$ and $X^\top \hat{\gamma}_2$ that drive the prediction. Increasing the dimension of the basis of time-functions automatically increases the dimension of the prognostic index and the model loses all transparency. A technical problem with these models is that it is getting harder to

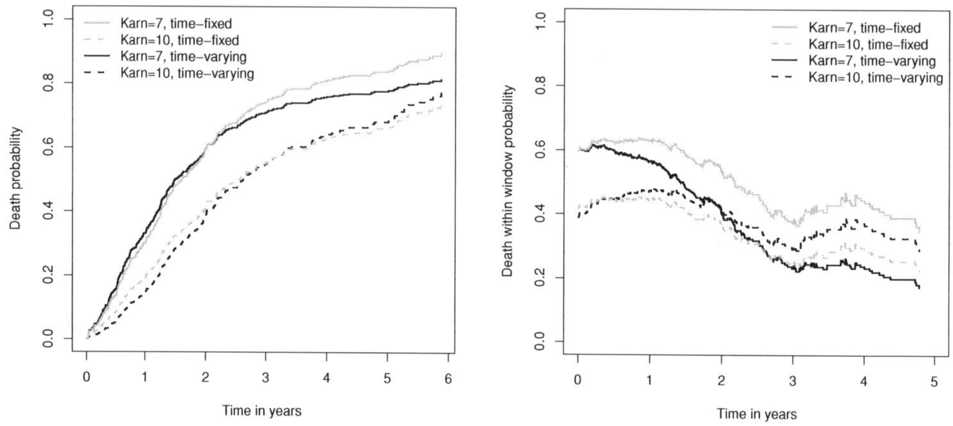

Figure 6.3 *Model-based failure functions (left) and dynamic fixed width failure functions with $w = 2$ (right) for two patients with Karnofsky scores 7 and 10, other covariates at mean values*

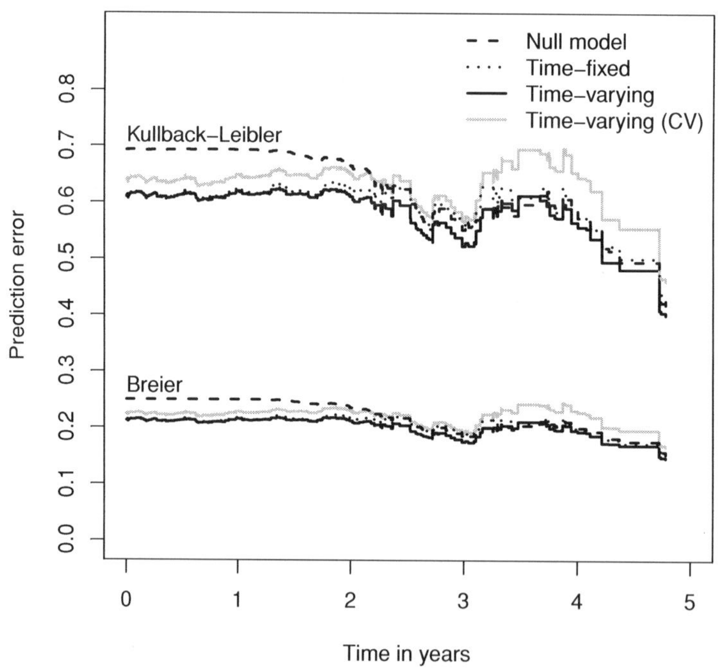

Figure 6.4 *Dynamic prediction error curves, with and without cross-validation, of the time-fixed and the time-varying models*

obtain standard errors of the predictive models. This issue is discussed further in Section 6.4.

6.2 Models inspired by the frailty concept

The Burr model and relaxations

In Section 5.2 frailties were discussed as mechanisms that lead to violation of the proportional hazards model. That also implies that frailty models could be used as an extension of PH models. Given some parametric family of frailty distributions, this would lead to a very parsimonious extension yielding more robust prediction models than the time-varying coefficients models of Section 6.1. In this subsection attention is focused on the gamma frailty model, which is the most popular model mainly because the resulting marginal distribution has an elegant formula. It is known as the *Burr model* (Burr 1942) and given by

$$h(t|x) = \frac{h_0(t)\exp(x^\top\beta)}{1+\xi H_0(t)\exp(x^\top\beta)}\,,\quad S(t|x) = (1+\xi H_0(t)\exp(x^\top\beta))^{(-1/\xi)}\,.$$

Here, ξ is the variance of the gamma frailty distribution, centered at $E[Z] = 1$. In this model, when $\xi > 0$, the hazard ratio for any pair of individuals converges to one as $H_0(t) \to \infty$. (This need not be the case. It is possible that $H_0(\infty) < \infty$.) As observed in Section 5.2 and shown in Figure 5.1 this convergence does not necessarily hold true for general frailty distributions.

The model looks simple, but it is a bit hard to fit directly because the hazard term $h_0(t)$ in the numerator is linked to the cumulative hazard term $H_0(t)$ in the denominator. Exploiting properties of the gamma frailty distribution, the model can be fitted using an EM-algorithm (Nielsen et al. 1992, Klein 1992) or penalized log-likelihood (Ripatti & Palmgren 2000, Therneau et al. 2003) as implemented in `coxph()` of the **survival** package in R. Fitting the model for the ovarian cancer data using `coxph()` gives the results in the column "Burr" of Table 6.2. The standard error of the frailty variance is not returned by the software; the methods of Andersen et al. (1997) could be used to obtain it. The full log-likelihood as defined in Section 2.3 is not returned either.

The very nice feature of this model and all similar frailty models is that the dynamic prediction is very simple to compute and driven by a single prognostic index.

In Perperoglou, van Houwelingen & Henderson (2006) a modification of the Burr model is proposed that loosens the link between $h_0(t)$ in the numerator and $H_0(t)$ in the denominator. This so-called *relaxed Burr model* is defined as

$$h(t|x) = \frac{h_0(t)\exp(x^\top\beta)}{1+F(t|\theta)\cdot\exp(x^\top\beta)}\,.$$

Here, $F(t|\theta)$ can be any nonnegative function. The only condition needed to assure

Table 6.2 *Burr and relaxed Burr models applied to the ovarian cancer data*

Covariate	Category	Burr B (SE)	Relaxed Burr B (SE)
FIGO	III		
	IV	0.744 (0.230)	1.077 (0.417)
Diameter	Microscopic		
	<1 cm	0.402 (0.436)	0.346 (0.476)
	1-2 cm	1.163 (0.461)	1.341 (0.606)
	2-5 cm	1.425 (0.440)	1.537 (0.542)
	>5 cm	1.387 (0.416)	1.756 (0.589)
Broders	1		
	2	0.644 0.3759	0.777 (0.471)
	3	0.674 0.3630	0.826 (0.459)
	4	0.253 0.4216	0.110 (0.527)
	Unknown	0.937 0.4228	1.114 (0.561)
Ascites	Absent		
	Present	0.408 (0.241)	0.467 (0.308)
	Unknown	0.549 (0.351)	0.576 (0.483)
Karnofsky	Continuous	-0.321 (0.088)	-0.492 (0.184)
Frailty variance		1.171	
θ			1.170 (0.691)

identifiability, is that $F(0|\theta)=0$. In Perperoglou, van Houwelingen & Henderson (2006) the function is linked to a dynamic frailty model with a general autocorrelation structure. A more down to earth interpretation is that it gives a way to define a low-dimensional interaction between the prognostic index and time. A very simple model is something like $F(t|\theta)=\theta t$. The interesting feature is that this model could be fitted using the Cox partial likelihood. It is not very complicated to extend any Newton-Raphson algorithm for the simple Cox model. A disadvantage of the model is that there is no explicit formula for the survival function. The baseline hazard can be estimated using a Breslow-type estimator. Predicted survival functions have to be computed by summation over the observed survival times, similar to the formula at the end of Section 6.1. The result of applying this very simple model to the ovarian cancer data is shown in Table 6.2, in column "Relaxed Burr". Note that, in contrast to the Burr model, the variance of θ in the relaxed Burr model with $F(t|\theta)=\theta t$ is straightforward to obtain.

Cure models

The cure model that was briefly discussed in Section 5.2 has gained some popularity, partly because of the magical term "cure." Such an interpretation should be handled with care, but the model is an interesting extension of the Cox model. The

Table 6.3 *Results for the cure models for the ovarian cancer data compared with the Cox model of Table 3.1*

Covariate	Category	Pure cure	Pure Cox	Cox + cure	
				Cure	Cox
		B (SE)	B (SE)	B (SE)	B (SE)
Intercept		0.177 (2.715)		-1.048 (1.558)	
FIGO	III				
	IV	-3.098 (1.399)	0.504 (0.136)	-0.958 (0.450)	0.381 (0.140)
Diameter	Microscopic				
	< 1 cm	-0.952 (0.604)	0.247 (0.323)	0.773 (0.590)	0.571 (0.322)
	1-2 cm	-5.456 (2.480)	0.665 (0.326)	0.127 (0.662)	0.867 (0.326)
	2-5 cm	-2.677 (0.781)	0.785 (0.315)	0.509 (0.609)	1.252 (0.316)
	> 5 cm	-4.840 (1.100)	0.836 (0.304)	-0.627 (0.612)	0.890 (0.305)
Broders	1				
	2	-1.899 (0.679)	0.573 (0.243)	-1.267 (0.456)	0.030 (0.243)
	3	-2.191 (0.686)	0.520 (0.234)	-1.650 (0.453)	-0.098 (0.235)
	4	0.086 (0.738)	0.323 (0.274)	-1.162 (0.546)	-0.253 (0.275)
	Unknown	-19.725 (NA)	0.650 (0.265)	-1.352 (0.552)	0.132 (0.266)
Ascites	Absent				
	Present	-0.439 (0.528)	0.272 (0.155)	0.056 (0.360)	0.411 (0.156)
	Unknown	1.972 (0.830)	0.205 (0.213)	0.595 (0.491)	0.486 (0.217)
Karnofsky	Continuous	0.122 (0.268)	-0.173 (0.054)	0.059 (0.146)	-0.212 (0.054)

model can be written as

$$S(t|x) = \pi(x) + (1-\pi(x))S_f(t|x) \ .$$

Here $\pi(x)$ is the probability of “cure” (frailty $Z=0$), and $S_f(t|x) = \exp(-H_f(t|x))$ the survival function of those who are not cured (frailty $Z=1$). The subscript “f” stands for “fatal” or “failure”. The resulting hazard can be written as

$$h(t|x) = \frac{(1-\pi(x))S_f(t|x)}{\pi(x) + (1-\pi(x))S_f(t|x)} h_f(t|x) \ .$$

Popular choices for the two components of the model are the logistic model and the PH model, respectively, leading to

$$\pi(x) = \frac{\exp(\beta_{c0} + x^\top \beta_c)}{1+\exp(\beta_{c0} + x^\top \beta_c)} \ , \qquad h_f(t|x) = h_{f0}(t)\exp(x^\top \beta_f) \ .$$

The problem with this model is that there is a lot of redundancy in the full model where the covariates are allowed to influence both $\pi(x)$ and $h_f(t|x)$. This can be demonstrated on the ovarian cancer data. The results, using the **semicure** package for R of Peng (2003), are presented in Table 6.3 together with the Cox model of Table 3.1.

The table shows that the pure cure model degenerates. This is no surprise because the mortality among these patients is high, and it can be expected that the probability of "cure" is estimated to be zero in subgroups like "Broders unknown" leading to degenerated regression coefficients. Notice that despite its degeneracy, the model can still yield sensible estimates of the survival probabilities.

Notice also that the patterns of regression coefficients in the pure cure model and the pure Cox model are very similar but with opposite signs. High probability of cure in the former model corresponds with low hazard ratios in the latter model. Unfortunately, the models cannot be compared on their log-likelihood because that is not returned by the software.

The last column shows the full model with cure and proportional hazard components. Observe that the signs of the regression coefficients are not very consistent anymore, which hampers the interpretation. It can be conjectured that the likelihood of this model is very flat due to the inherent redundancies in the model, but such information is not available from the software. Correlations between the estimated coefficients in the two components are not produced. Altogether, it must be concluded that the "Cox+cure" model can only be used for exploratory analysis and not for predictive modeling, although it has the convenience of an easy way of computing a dynamic update of predictive models.

6.3 Enforcing parsimony through reduced rank models

Rank=1 model

The models of the previous section have some two interesting features: i) the predictive models are easy to update, ii) they require only a few extra parameters to capture the violation of the PH assumption. The model of Section 6.1 has the pleasant property that it gives a transparent model for the effect of the covariates on the hazard at each time point as shown in Figure 6.1. However, the model needs very many parameters to do so. This section will show how the number of parameters in the time-varying coefficients model can be greatly reduced by exploiting the fact that most covariates show very similar patterns over time.

A starting point that gives an impression how such a model is obtained by the following very simple two-stage procedure

1. Fit a simple Cox yielding $Z_{\text{fixed}} = x^\top \hat{\beta}_{\text{fixed}}$;
2. Fit a time-varying model with Z_{fixed} as single covariate.

Application of this approach to the example of Section 6.1 with the time-functions $f_1(t) \equiv 1$ and $f_2(t) = \ln(1+t)$ and $\hat{\beta}_{\text{fixed}}$ as specified in the first column of Table 6.1 gives $\hat{\gamma}_1 = 1.647$ and $\hat{\gamma}_2 = -0.709$. This extension of the model is highly significant ($\chi^2_{[1]} = 8.148$, $P = 0.004$). The ratio $\hat{\gamma}_2/\hat{\gamma}_1 = -0.430$ can be seen as a kind of average value of the covariate specific ratios in Table 6.1 The covariate specific time-varying coefficients are obtained as $\hat{\beta}_{\text{fixed}} \cdot \hat{\gamma}_1 + \hat{\beta}_{\text{fixed}} \cdot \hat{\gamma}_2 \cdot \ln(1+t)$. For FIGO IV, Karnofsky and Diameter they are very much in line with the results in Table 6.1 as shown in Figure 6.1. The difference with the model of Table 6.1 is that all time-varying coefficient curves have the same shape, being proportional to $1-0.43\ln(1+t)$ which changes sign at $t=9.23$, well beyond the range of the follow-up.

The slightly heuristic two-stage approach can be formalized by the model

$$h(t|x) = h_0(t)\exp\left(x^\top\beta \cdot \left(\sum_{i=1}^{m} \gamma_i f_i(t)\right)\right) .$$

Here, the same restriction applies to the time-functions as in Section 6.1, namely $f_1(t) \equiv 1$ and $f_j(0) = 0$, for $j = 2,\dots,m$. This model is known as a *reduced rank model* of rank $= 1$. Reduced rank is an established methodology for parsimonious interaction models in analysis of variance and linear regression. It has been introduced in the present context by Perperoglou, le Cessie & van Houwelingen (2006*a*,*b*), who also developed the software to fit such models. The simplest way to fit such models is a technique known as *Alternating Least Squares (ALS)*: for fixed β, the parameters in γ can be estimated by using software for time-varying effects and for fixed γ, the parameters in β can be estimated by using software for time-dependent covariates $X \cdot \sum_i \gamma_i f_i(t)$. However, such alternating schemes do not give the correct standard errors of the time-varying effects as produced by Perperoglou's software. For details on both reduced rank models and on ALS, see Anderson (1951) and Gifi (1990).

The model above is a rank $= 1$ model because all covariate specific time-varying effect functions $\beta(t)$ lie in the one-dimensional subspace spanned by $\sum_{i=1}^{m} \gamma_i f_i(t)$. The rank $= 1$ model is a very convenient tool for data-analysis because it allows testing for the overall presence of time-varying effects using only a few extra parameters and can give more insight in the overall shape of the time-varying effect. Therefore, it will be explored in more detail before switching to more general models.

The first observation is that there is scale redundancy in the model above: dividing β by c and multiplying γ by the same c does not alter the model. This redundancy is repaired by the restriction $\gamma_1 = 1$. Fitting this model with the same time functions as in the two-stage approach yields $\hat{\gamma}_2 = -0.488$, which is very close to the ratio -0.430 seen above. Also the estimated $\hat{\beta}$ is very similar. This is related to a very modest gain in likelihood ($\Delta(\text{ll}) = 1.036$) with respect to the two-stage approach. The interesting feature is that the variation over time can be explored with a larger set of basis functions at the price of adding just a few parameters. Extending the set with $f_3(t) = (\ln(1+t))^2$ and $f_4(t) = (\ln(1+t))^3$ gives a substantial further increase of the likelihood ($\Delta(\text{ll}) = 3.517$) and a most interesting graph of $\sum_{i=1}^{m} \gamma_i f_i(t)$ (Figure 6.5) showing that the effect of the prognostic index decreases much more rapidly than anticipated by the choice of $\ln(1+t)$ as the second basis function. The corresponding estimated effects at $t=0$ in this model are all larger than in the $m=2$ model.

The rank $= 1$ model can also be useful in testing time-varying effects for ordinal or categorical covariates and continuous covariates that show a non-linear effect. In

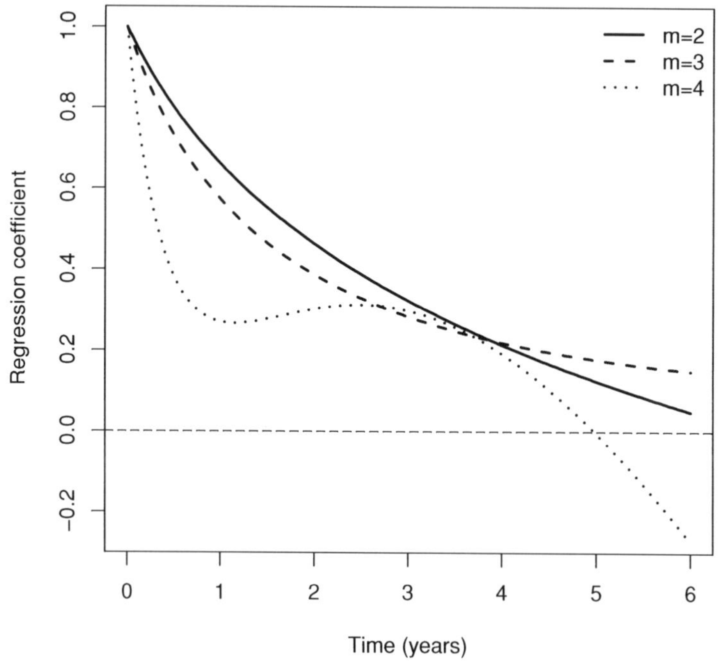

Figure 6.5 *Shape of the time variation of the rank=1 model for extended bases*

both situations a single risk factor gives rise to a set of covariates $X_1,...,X_p$. For ordinal or categorical covariates these covariates will be a set of dummy variables describing contrasts, while for a continuous covariate this might be a set of (fractional) polynomials. A (univariate) test for main effect and the time-varying effect of such a risk factor that is independent of the particular choice of the covariates representing the risk factor, is provided by the rank $= 1$ model with only that set of covariates. An example is provided by the ordinal risk factor Diameter in Table 6.1. The "constant $+\ln(1+t)$" model yields $\hat{\gamma}_2 = -0.504$ and an improvement of log-likelihood from -1395.480 to -1392.418 ($\Delta(\text{ll}) = 3.064$).

Rank=2 model

The general *rank* $=$ *r model* allows the time-varying coefficients to lie in an r-dimensional subspace of the basis functions. The general model requires complex mathematical notation. Therefore, attention is restricted to the rank $= 2$ model. It is given by

$$h(t|x) = h_0(t)\exp\left[x^\top\beta_1\cdot(\sum_{i=1}^{m}\gamma_{1i}f_i(t)) + x^\top\beta_2\cdot(\sum_{i=1}^{m}\gamma_{2i}f_i(t))\right].$$

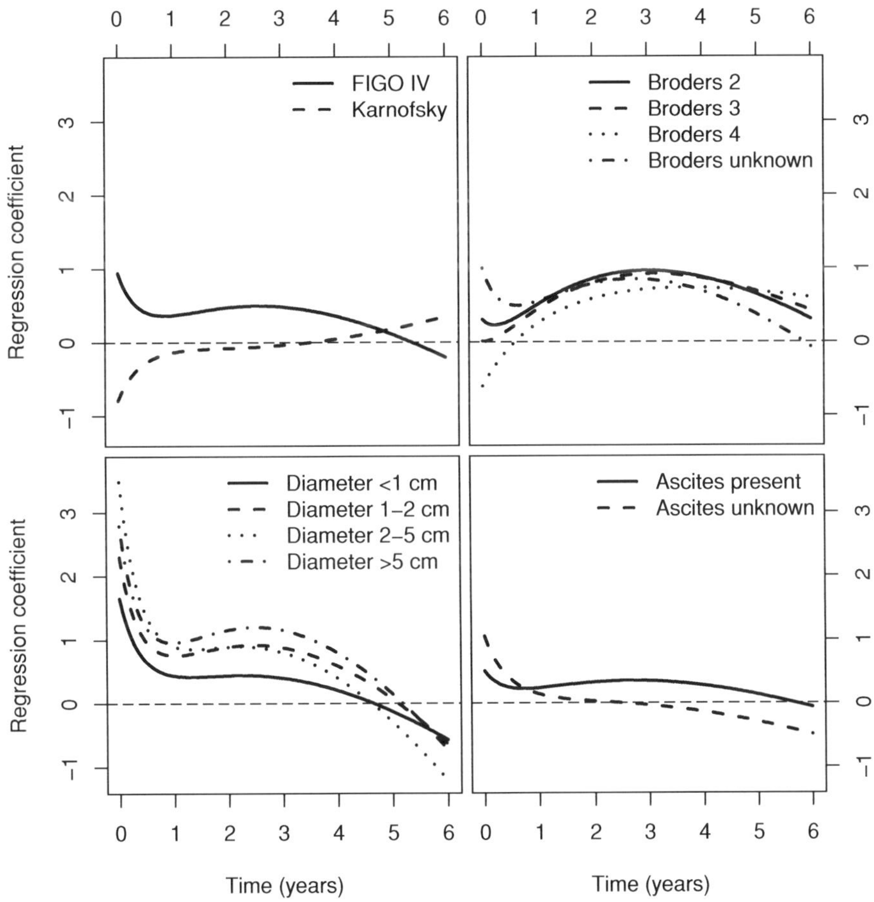

Figure 6.6 *Time-dependent regression effects for the rank* $= 2$, $m = 4$ *model*

The redundancy in this model can be remedied by the restrictions

$$\gamma_{11} = 1,\ \gamma_{21} = 0,\ \gamma_{22} = 1\ .$$

Application of the rank $= 2$ model with $m = 4$ to the example improves the log-likelihood by 6.997 with respect to the rank $= 1$ model with $m = 4$ at the cost of 13 parameters and yields the time-varying effects shown in Figure 6.6.

A disadvantage of the model is that it does not separate "main effects" of the covariates and "interaction effects" due to time-varying effects. Modifications of the reduced rank models in this direction will be discussed in Section 6.4. Table 6.4 shows the log-likelihoods of different time-fixed and time-varying models fitted on the ovarian cancer data. From the AIC point of view the reduced rank model with *rank* $= 1$ and $m = 4$ performs best.

Table 6.4 *Partial log-likelihoods of different time-fixed and time-varying models fitted on the ovarian cancer data*

Model	Rank	Number of time functions m	Log-likelihood	Number of parameters	AIC
Time-fixed			−1374.716	12	−1386.716
Relaxed Burr			−1369.018	13	−1382.018
Two-stage		2	−1370.642	13	−1383.642
Reduced rank	1	2	−1369.606	13	−1382.606
	1	3	−1369.431	14	−1383.431
	1	4	−1366.089	15	−1381.089
	2	2	−1362.642	24	−1386.642
	2	3	−1361.798	26	−1387.798
	2	4	−1359.092	28	−1387.092

6.4 Additional remarks

Cox-type models versus frailty-type models

The time-varying coefficient models of Sections 6.1 and 6.3 are direct extensions of the Cox model. Together with the relaxed Burr model of Section 6.2, they fit in the class of *Cox-type models* defined by

$$h(t|x) = h_0(t)\exp(v(x|t,\eta))\ .$$

Here, $v(x|t,\eta)$ is a parametric function describing the ln(hazard ratio), depending on the fixed-dimensional parameter η. In Section 6.1, η contains the γ-parameters, in Section 6.3 the β- and the γ-parameters and in the relaxed Burr model the β-parameters and the single θ-parameter. To avoid identifiability problems, the function $v(x|t,\eta)$ should satisfy $v(0|0,\eta)=0$.

The parameters of the model can be fitted by the same procedure as the Cox model. Maximizing the partial likelihood will yield the maximum likelihood estimator of η and the baseline hazard is estimated by the Breslow estimator of Section 2.3. As mentioned in Section 6.2 for the relaxed Burr model, maximizing the partial likelihood is not much harder than for the Cox model. Partial log-likelihood and full log-likelihood differ by the number of events and model comparison on partial log-likelihood or full log-likelihood are equivalent.

The cure model and the Burr model of Section 6.2 could be coined *frailty-type* models. They can be fitted by EM-type algorithms. Generally speaking, baseline and time-varying hazard ratios cannot be disentangled. If the software does not produce an estimate of the baseline hazard, there is no easy way of obtaining it by something like the Breslow estimator. Once the model has been fitted and all components have been obtained, the partial likelihood can still be computed, but there is no simple relation with the full likelihood. Hence, it is not quite fair to compare frailty-type models with Cox-type models on the basis of the partial likelihood. For example, for the models of Table 6.2 the relaxed Burr model beats the Burr model on the partial log-likelihood (data not shown), but it might be the other way around on the full log-likelihood. For that reason the Burr model and the cure models have not been included in Table 6.4 which gives the partial log-likelihoods for the Cox-type models.

In summary, frailty-type models have some intuitive appeal, because they are directly linked to the frailty concept, but they are quite inconvenient as statistical models. Moreover, the intuition can be misleading as discussed in Section 5.5. The Cox-type models are easier to fit and easier to analyze. For example, the standard errors of prediction are not too hard to obtain, as discussed below.

Standard errors of predictions for the Cox-type models

The standard error of the predicted curve is given by a direct generalization of the formula in Section 2.3. The asymptotic variance of $-\ln(\hat{S}(t|x)) = \hat{H}_0(t)\exp(v(x|t,\hat{\eta})$ may be estimated consistently by

$$\sum_{\substack{t_i \leq t \\ t_i \in \mathscr{D}}} \left(\frac{\exp(v(x|t_i,\hat{\eta}))}{\sum_{j\in R(t_i)} \exp(v(x_j|t_i,\hat{\eta}))} \right)^2 + \hat{q}(t|x)^\top \Sigma(\hat{\eta})\, \hat{q}(t|x)\,,$$

with

$$\hat{q}(t|x) = \sum_{\substack{t_i \leq t \\ t_i \in \mathscr{D}}} (s(x|t_i,\hat{\eta}) - \bar{s}_i) \frac{\exp(v(x|t_i,\hat{\eta}))}{\sum_{j\in R(t_i)} \exp(v(x_j|t_i,\hat{\eta}))}\,,$$

$$s(x|t,\eta) = \frac{\partial v(x|t,\eta)}{\partial \eta}\,,$$

$$\bar{s}_i = \frac{\sum_{j\in R(t_i)} s(x_j|t_i,\hat{\eta}) \exp(v(x_j|t_i,\hat{\eta}))}{\sum_{j\in R(t_i)} \exp(v(x_j|t_i,\hat{\eta}))}\,,$$

and $\Sigma(\hat{\eta}) = I_{\mathrm{pl}}^{-1}(\hat{\eta})$, the inverse of the observed information matrix for η in the partial likelihood, which estimates the covariance matrix of $\hat{\eta}$. After all, this is not hard to program in the context of a partial likelihood maximization routine. It would be great if software developers provided this as a standard option.

Suitable basis functions for time-varying effects

It is hard to give recommendations for the choice of the basis function in the time-varying model of Section 6.1. As remarked there, it is helpful for the interpretation of the parameters if $f_1(t) \equiv 1$ and $f_j(0) = 0$ for $j = 2, \ldots, m$. The fractional polynomials suggested by Sauerbrei et al. (2007) are not well controlled for t close to zero. Therefore, they might not be the best choice for modeling time-varying effects. It is not hard impose the restrictions on *regression splines (B-splines)* if the construction of Harrell (2001) is followed. The problem with splines is that they can get quite wobbly if there are too many knots. This can be observed in Perperoglou, le Cessie & van Houwelingen (2006*a*,*b*) and Perperoglou et al. (2007). The polynomials in $\ln(1+t)$ as used in this chapter have a stable behavior, but may lead to some bias in later follow-up. However, it should be noted that $\ln(1+t)$ is not independent of the time-scale. A more general version is given by $\ln(1+c\cdot t)/c$.

Finally, it should be noticed that it is possible to choose different time-functions for different covariates. This could be useful in an extensive exploratory analysis, which make sense in large data sets, but could lead to spurious results in smaller data sets.

Extensions of the reduced rank procedure

The drawback of the simple rank $=1$ model of Section 6.3 is that none or all covariates show a time-varying effect. There is no way to distinguish between covariates with constant effects and time-varying effects. A way out could be a hybrid model

$$h(t|x) = h_0(t)\exp\left(x_1^\top\beta_1 + x_2^\top\beta_2\cdot\left(\sum_{i=1}^{m}\gamma_i f_i(t)\right)\right) .$$

Here, x_1 stands for the group of covariates with constant effects and x_2 for the covariates with a time-varying effect. It is not very hard to extend the rank $=1$ model in this direction. However it is much harder to decide to which group a covariate belongs. This could be done by a preliminary test on the presence of time-varying effects for each covariate separately. However, this approach might miss rank $=1$ time-varying effects that are not so strong on individual covariates, but show a notable time-varying effect on the prognostic index. An alternative model is given by

$$h(t|x) = h_0(t)\exp\left(x^\top\beta_1 + x^\top\beta_2\cdot\left(\sum_{i=1}^{m}\gamma_i f_i(t)\right)\right) .$$

In this model no a priori distinction is made between the covariates. It allows to test $\beta_2 = 0$ for each covariate. The disadvantage of the model is that the number of regression parameters is doubled and that the proportionality of the time-varying effects is lost. It can be seen as a special case of the rank $=2$ model with restrictions on γ_1. These models have not been tried in practice yet. They make only sense in large data sets with long follow-up. The numbers of parameters might get quite large and the analysis quite slow.

Chapter 7

Dealing with non-proportional hazards

7.1 Robustness of the Cox model

Some theory about the time-varying Cox model

In Chapter 6 different models have been presented that can all be considered as extensions of the simple Cox model. The models agree on how to combine covariates through linear predictors, but they differ in the way they model the violation of the proportional hazards model. All these models are considered and compared in Perperoglou et al. (2007), using long term survival data of a large cohort of breast cancer patients. The interesting phenomenon observed in that paper is that there appears to be a large discrepancy between the models when looking at the estimated hazard functions for different covariate patterns, while the differences between the estimated survival functions are much smaller. To get a better understanding of this phenomenon it is helpful to have a closer look at the time-varying coefficient model of Section 6.1,

$$h(t|x) = h_0(t)\exp(x^\top \beta(t)) .$$

The survival function $S(t|x)$ is directly related to the cumulative hazard $H(t|x) = -\ln(S(t|x))$. The following approximation is crucial for our understanding of the model.

$$\frac{H(t|x)}{H_0(t)} = \frac{\int_0^t h_0(s)\exp(x^\top \beta(s))ds}{\int_0^t h_0(s)} \approx \exp(x^\top \bar{\beta}(t)) ,$$

where $\bar{\beta}(t)$ is defined by

$$\bar{\beta}(t) = \frac{\int_0^t h_0(s)\beta(s)ds}{\int_0^t h_0(s)ds} ,$$

This approximation is of the type $E[\exp(Y)] \approx \exp(E[Y])$, which holds true if $\mathrm{var}(Y)$ is small. Translated to the application here, it can be concluded that

$$H(t|x) \approx H_0(t)\exp(x^\top \bar{\beta}(t)) ,$$

if

$$\frac{\int_0^t h_0(s)(x^\top(\beta(s)-\bar{\beta}(t)))^2 ds}{\int_0^t h_0(s)ds}$$

is small. This requires that $\beta(s)$ does not vary too much over time. The condition is easier to satisfy if the covariates are centered at $\bar{x}=0$.

The lesson to be learned from this approximation is that the effect of a covariate on the survival function $S(t|x)$ goes through the average of the time-varying effect $\beta(s)$, averaged over the interval $[0,t]$ with weights proportional to the baseline hazard. This observation can explain why models that appear to have very different $\beta(s)$-functions can produce very similar survival probabilities. It is the cumulative hazard that matters and not the hazard itself. It is tempting to develop models with time-varying coefficients for the cumulative hazard itself like in the approximation above. This is done by Royston & Parmar (2002), but the problem with such models is that it is hard to retain the monotonicity of $H(t|x)$ as function of t.

The finding that apparent differences in hazards do not necessarily translate into differences in survival could also be explained from a more theoretical point of view. Like density functions, hazard functions are very hard to estimate non-parametrically. The familiar $1/\sqrt{n}$ convergence rate cannot be achieved for densities. In contrast, cumulative hazard functions (like cumulative density functions) can be estimated easily by the Nelson-Aalen estimator or the Kaplan-Meier estimator (see Section 2.1) and those estimators converge quickly (at $1/\sqrt{n}$ rate) to the unknown true value. So in larger data sets, the overall cumulative hazard function and the cumulative hazard function in (large) subgroups are close to the true value and all reasonable (well fitting) models should yield very similar cumulative hazard functions, while the hazard functions themselves, that are so much harder to estimate, might differ considerably between models. This is also the explanation why Aalen's additive hazard models (see Section 2.6) are always presented by showing plots of the integrated coefficients $B(t)=\int_0^t \beta(s)ds$ instead of the coefficients $\beta(t)$ themselves. The integrated estimated coefficients converge at $1/\sqrt{n}$ rate while the estimated coefficients themselves are very erratic.

Since there is no particular model that can serve as the generally accepted extension of the Cox model, it might be interesting to investigate the question what happens if a Cox model is fitted even when it is known that $\beta(t)$ varies over time. An approximation is given in van Houwelingen (2007), using methodology of Struthers & Kalbfleisch (1986), Hjort (1992), and Xu & O'Quigley (2000). The results are given here without further discussion of the technical details. If the data are used with administrative censoring at t_{hor} and random censoring before t_{hor}, the regression coefficient obtained from a Cox model converges to a limiting value that is approximately given by

$$\tilde{\beta}_{\text{Cox}} \approx \left(\int_0^{t_{\text{hor}}} S(t)C(t)h(t)\text{var}(X|T=t)dt\right)^{-1} \cdot \cdot \int_0^{t_{\text{hor}}} S(t)C(t)h(t)\text{var}(X|T=t)^\top \beta(t)dt.$$

Here, $S(t)$ is the marginal survival function, $C(t)$ the marginal censoring function, $h(t)$ the marginal hazard and $\text{var}(X|T=t)$ the weighted covariance matrix of X in the risk set at timc t that also shows up in the Fisher information matrix of the partial log-likelihood in Section 2.3. The approximation is valid under the condition that at each t, $\tilde{\beta}_{\text{Cox}}$ does not differ to much from the true $\beta(t)$, which is equivalent to requiring that $\beta(t)$ does not vary too much over the interval $[0,t_{\text{hor}}]$. A simplification of the approximation is obtained if it can be assumed that $\text{var}(X|T=t)$ is constant over the interval. This will be true if the effects of the covariates are small and/or t_{hor} is not too far away. Under those conditions

$$\tilde{\beta}_{\text{Cox}} \approx \frac{\int_0^{t_{\text{hor}}} S(t)C(t)h(t)\beta(t)dt}{\int_0^{t_{\text{hor}}} S(t)C(t)h(t)dt} .$$

If the t_{hor} is small indeed, $C(t) \approx 1$, $S(t) \approx 1$ and $h(t) \propto h_0(t)$. The implication is that

$$\tilde{\beta}_{\text{Cox}} \approx \bar{\beta}(t_{\text{hor}}) ,$$

if t_{hor} is small and/or $\beta(t)$ does not vary too much and/or $\beta(t)$ is small.

Finally, it can be shown that under the same conditions the Breslow estimator of the baseline hazard in the Cox model converges to

$$h_{\text{Cox},0}(t) \approx h_0(t)\exp(E[X|T=t]^\top(\beta(t)-\bar{\beta}(t_{\text{hor}}))) .$$

From the same arguments as in the beginning of the section, it follows immediately that

$$H_{\text{Cox},0}(t_{\text{hor}}) = \int_0^{t_{\text{hor}}} h_{\text{Cox},0}(t)dt \approx H_0(t_{\text{hor}}) ,$$

and hence

$$H_{\text{Cox}}(t_{\text{hor}}|x) \approx H(t_{\text{hor}}|x) .$$

So, the Cox model gives (approximately) correct predictions of surviving up to t_{hor} even if the true effect of the covariates is time-varying, provided that $S(t)$ and $C(t)$ stay close to 1 and $\beta(t)$ does not vary too much. Or, the other way around, a decent estimate of $S(t_0|x)$ can be obtained by a *Cox model stopped at* t_0, that is by a model fitted to the data with additional administrative censoring at t_0. This does not imply that the predictions will be correct over the whole interval $[0,t_0]$.

An example

The phenomenon can be demonstrated by an example that is inspired by the Gastric Cancer Data of Data Set 4. A situation is considered with a single dichotomous covariate X with $P(X = 1) = P(X = 2) = 0.5$. Survival in the two groups is given by

$$S_1(t) = S(t|x = 1) = \exp(-(t/5)^{4/3}) \; , \; S_2(t) = S(t|x = 2) = \exp(-(t/5)^{3/4}) \; .$$

The two curves intersect at $t = 5$. The hazard in the first group increases, the hazard in the second group decreases and the two hazard curves intersect at $t = 1.86$ well before $t = 5$. A huge data set with $n = 50,000$ is generated from this model to study the fit of different Cox models. All observations are censored at $t = 10$.

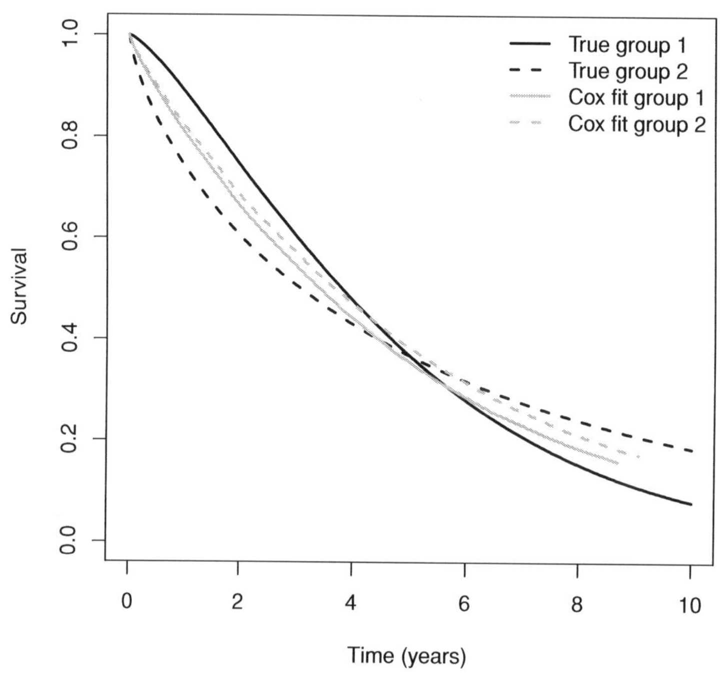

Figure 7.1 *The survival curves of the example and simple Cox model fit*

The survival curves for the two groups and the fitted simple Cox model are shown in Figure 7.1. Observe that the predictions are correct at about $t = 6$. For larger values of t, especially for $t = 10$ the predictions are not perfect, but at least in the right order. For smaller values of t, the predictions are off the mark and in the wrong order.

The predictions based on the stopped Cox model are shown in Figure 7.2. These predictions are obtained by fitting simple Cox models with additional administrative censoring at t for a dense grid of t-values on the interval [0,10]. The graphs

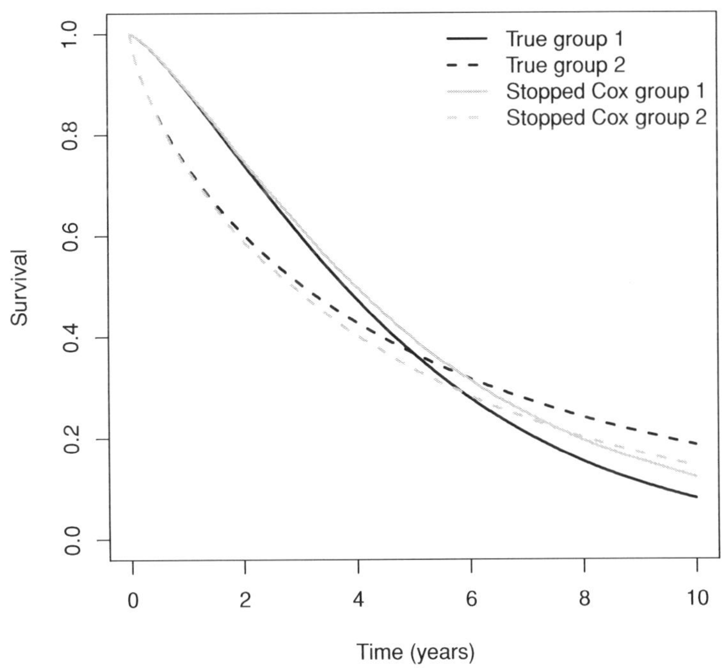

Figure 7.2 *Survival predictions based on "stopped Cox models" for each time-point*

show that the predictions based on the stopped Cox model are very accurate for $t < 5$, but loose accuracy if the follow-up is too long, or, more precisely, if the overall survival $S(t)$ gets too small. This is completely in line with the approximations given before. To get a good prediction, $\beta(s)$ should be averaged over $s \in [0,t]$ with weights proportional to $h(s)$, while the stopped Cox model gives an average weighted by the density $h(s)S(s)$, provided there is little censoring before t.

7.2 Obtaining dynamic predictions by landmarking

Sliding landmarking

As discussed in Section 7.1, fitting a simple Cox model might give a reasonable prediction of survival up to some horizon t_{hor}, even if the proportional hazards assumption is violated. However, using such a model for dynamic prediction might be disastrously wrong. This can be nicely demonstrated on the example of Section 7.1. Figure 7.3 shows the fixed width failure function, the dynamic probability of death within the next three years, $F_3(t|x)$, as defined in Section 1.3, for the true model and the simple Cox model. It is clear that the Cox model is not able to pick up the dynamic differences at all.

The technical explanation in terms of the approximations discussed in Section 7.1 is that the weighted average of $\beta(s)$ over the interval $[t, t+w]$ is needed to obtain a reasonable PH prediction model instead of the average over the whole follow-up range $[0, t_{\text{hor}}]$. This can be achieved by combining stopping with *landmarking*. The latter concept can be described as follows:

> *In the landmarking approach dynamic predictions for the conditional survival after $t = t_{LM}$ should be based on current information of all patients still alive just prior to t_{LM}.*

The landmark concept was introduced in Anderson et al. (1983) as a way for properly handling the time-dependent covariate "tumor response" in survival models. The practice was to take "tumor response" as a fixed covariate in the Cox model. Since it takes time before "tumor response" can be assessed, this creates a substantial *immortal time bias* in favor of "tumor response." The remedy proposed in the paper is to take a fixed time-point (t_{LM}), define "tumor response" as "response before t_{LM}," and use that in a Cox model for survival after t_{LM}. That approach circumvents the computational complications of fitting a time-dependent covariate. In this book landmarking plays an essential role for two reasons: i) it keeps the models as transparent as possible; ii) it leads to robust predictions that are not sensitive to unchecked assumptions.

In van Houwelingen (2007), landmarking is introduced as a tool to obtain predictions of survival up to a fixed horizon t_{hor}. As discussed in Chapter 1 it makes more sense taking a sliding window of width w and focusing on prediction from t_{LM} up to $t_{\text{hor}} = t_{\text{LM}} + w$. In order to obtain such a prediction the *sliding landmark model* is defined as the simple Cox model

$$h(t|x, t_{\text{LM}}, w) = h_0(t|t_{\text{LM}}, w)\exp(x^\top \beta_{LM}) \,, \quad t_{\text{LM}} \leq t \leq t_{\text{LM}} + w \,. \tag{7.1}$$

This model applies for all individuals at risk at $t = t_{\text{LM}}$ and ignores any event after $t = t_{\text{LM}} + w$. Estimates of both β_{LM} and $h_0(t|t_{\text{LM}}, w)$ can be obtained by fitting a Cox model to the data in the *sliding landmark data set* obtained by truncation at t_{LM} and administrative censoring at $t_{\text{LM}} + w$. Such a model can be used to obtain a reliable estimate of $F_w(t_{\text{LM}}|x) = 1 - S(t_{\text{LM}} + w|x, t_{\text{LM}})$ through

$$H(t_{\text{LM}} + w|t_{\text{LM}}, x) = \exp(x^\top \beta_{LM}) H_0(t_{\text{LM}} + w|t_{\text{LM}}, x) \,.$$

It should be stressed that *it is not claimed* that the PH model is correct for all $t_{\text{LM}} \leq t \leq t_{\text{LM}} + w$. The only claim is that it is a very convenient and useful way to obtain a dynamic prediction without having to fit a model with complicated time-varying effects.

Fitting this model to the simulated data set of our example yielded a nearly perfect fit for $F_3(t|x)$ (graphs not shown) in contrast to the misfit of the simple Cox model of Figure 7.3.

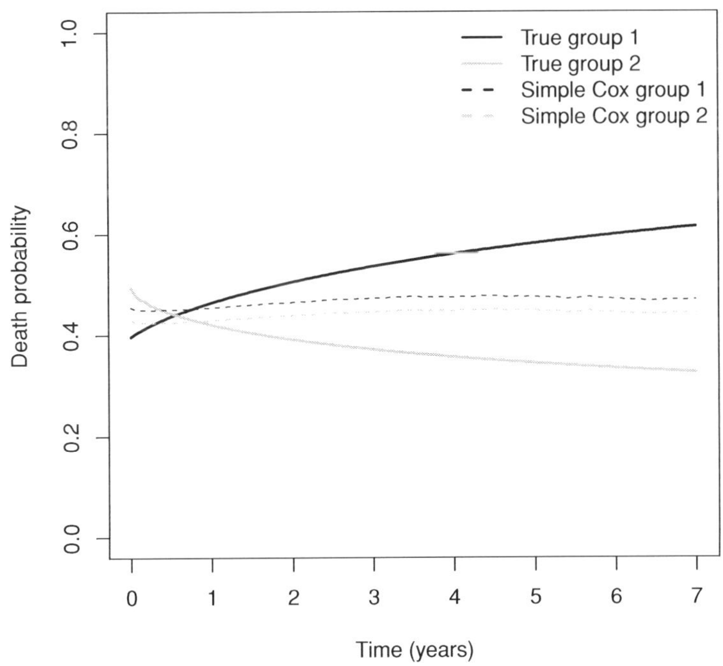

Figure 7.3 *Dynamic predictions with a window of 3 years comparing a simple Cox model with the true model for the simulation example of Section 7.1*

Applications, first round

Figure 7.4 shows the dynamic prediction based on a simple Cox model and on the sliding landmark model with $w = 4$ years for the Dutch gastric cancer data. The sliding landmark predictions are very close indeed to the non-parametric estimates based on the Kaplan-Meier curves; the difference is only visible after 5 years.

The landmarking supermodel

The approach sketched so far requires a separate Cox model to be fitted at each time-point t_{LM} for which a prediction is required. This is not very practical and hard to communicate to clinical users. Some form of smoothing and simplification is needed. This can be achieved by computing all the separate prediction models and smoothing them by local smoothers like "loess" or by fitting some regression model to the predictions with the landmark t_{LM} as explanatory variable. This would require ad hoc software and ad hoc choices. In van Houwelingen (2007) a different approach is presented that allows the use of existing survival software to obtain a prediction model that can be applied over a range of prediction times. It is based on the following construction of a "super prediction data set:"

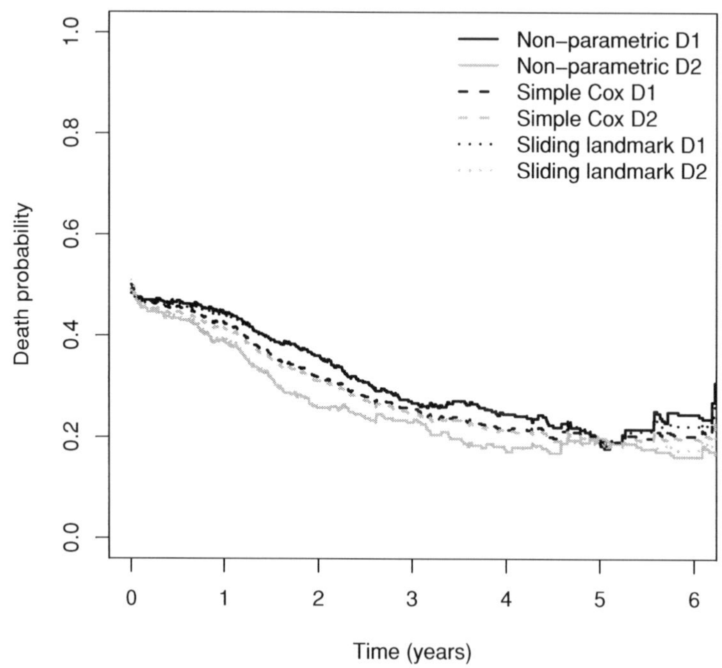

Figure 7.4 *Dynamic predictions based on the Kaplan-Meier curves, a simple Cox model and on the sliding landmark model with a 4 year window. The sliding landmark D1/D2 coincide with the non-parametric D1/D2 up to 5 years*

1. Fix the prediction window w;
2. Select a set of prediction time points $\{s_1, ..., s_L\}$;
3. Create a prediction data set for each $t_{\mathrm{LM}} = s_l$ by truncation and administrative censoring;
4. Stack all those data sets into a single "super prediction data set."

The selection of the set of prediction time-points implicitly defines a weighting of the prediction time points in the model to be developed. The simplest approach is selecting an interval $[s_1, s_L]$ and taking an equidistant grid of points on the interval. The grid can be quite coarse. A value of L between 20 and 100 will be sufficient. The choice should not depend on the actual event times. In this large data set the subsets corresponding to a given prediction time $t_{\mathrm{LM}} = s_l$ are labeled as "strata." The risk set $R(t_i)$ for some event time t_i is present in all strata with $s_l \leq t_i \leq s_l + w$. Passing from one stratum to the next one corresponds to sliding the window over the time range. Risk sets are "lost" if s_l passes an event time t_i and "gained" if $s_l + w$ passes an event time t_i. This explains why the "crude" regression coefficients $\hat{\beta}_{LM}$ as shown in Figure 7.5 show only small variations when s moves continuously from s_1 to s_L. They will even be constant as long as s and $s+w$ do not pass an event time.

A first step to a "super" prediction model is to let the regression coefficients β_{LM} depend on $t_{\mathrm{LM}} = s$ in a smooth way and to model that in a linear way. This means that

$$h(t|x, t_{\mathrm{LM}} = s, w) = h_0(t|s,w)\exp(x^\top \beta_{LM}(s)) \ , \quad s \leq t \leq s+w \ , \tag{7.2}$$

where

$$\beta_{LM}(s) = \sum_{j=1}^{m_b} \gamma_j f_j(s) \ .$$

Purposely, the same notation is used as for the time-varying effect model in Section 6.1. The subscript b in m_b for $\beta_{LM}(s)$ is needed to distinguish m_b from m_h defined below. The subtle, but very important, difference is that the regression parameters depend on the *prediction* time $t_{LM} = s$ and not on the *event* time $t_i = t$. A convenient standardization is to take $f_1(s) \equiv 1$, and $f_j(s_1) = 0$ for $j = 2, .., m_b$. As in the time-varying effects model, the number m_b of time functions is allowed to vary across covariates. The model can be fitted by applying a Cox model with stratification on s and inclusion of interaction terms $X * f_j(s)$. This can be seen as the maximization of the *integrated partial log-likelihood ipl* (integrated over s) introduced in van Houwelingen (2007),

$$\mathrm{ipl}(\gamma) = \sum_{s \in s_1, \ldots s_L} \mathrm{pl}(\beta_{LM}(s|\gamma)) = \sum_{i=1}^{n} d_i \sum_{s|s \leq t_i \leq s+w} \ln\left(\frac{\exp(x_i^\top \beta_{LM}(s|\gamma))}{\sum_{j \in R(t_i)} \exp(x_j^\top \beta_{LM}(s|\gamma))}\right) .$$

This is a *pseudo partial log-likelihood* implying that it yields consistent estimates, but that the standard errors cannot be obtained through the second derivative.

This approach based on a stratified analysis produces nice smooth landmark dependent effects. However, it gives separate estimated baseline hazards for each stratum

$$\hat{h}_0(t_i|s,w) = \frac{1}{\sum_{j \in R(t_i)} \exp(x_j^\top \hat{\beta}_{LM}(s))} \ , \qquad \text{for } s \leq t_i \leq s+w \ .$$

The baseline hazard function depends on s via the smooth functions $\hat{\beta}_{LM}(s)$. This smoothness could be modeled directly by letting

$$h_0(t|s,w) = h_0(t)\exp(\theta(s)) \ , \tag{7.3}$$

where

$$\theta(s) = \sum_{j=1}^{m_h} \eta_j g_j(s) \ ,$$

for proper basis functions $g_j(s)$ standardized by

$$g_j(s_1) = 0 \ ,$$

because of the indeterminacy of the baseline hazard. Notice that this excludes the constant function. A convenient choice is $g_j(s) = f_{j+1}(s)$.

This model can be fitted by applying a Cox model without stratification with main effects for the stratum variable s modeled by $\theta(s)$ and interaction of s with the covariates modeled by $\beta(s)$. It is advisable to center the covariates before fitting this model The software should allow for left truncation or delayed entry as described in Section 2.5. This leads to a different pseudo partial log-likelihood, denoted by ipl^*, and given by

$$\mathrm{ipl}^*(\gamma,\eta) = \sum_{i=1}^{n} d_i \ln\left(\frac{\sum_{\{s|s\leq t_i\leq s+w\}} \exp(x_i^\top \beta_{LM}(s|\gamma)+\theta(s|\eta))}{\sum_{\{s|s\leq t_i\leq s+w\}} \sum_{j\in R(t_i)} \exp(x_j^\top \beta_{LM}(s|\gamma)+\theta(s|\eta))}\right).$$

The corresponding estimate of the baseline hazard is given by

$$\hat{h}_0^*(t_i) = \frac{\#\{s|s\leq t_i\leq s+w\}}{\sum_{\{s|s\leq t_i\leq s+w\}} \sum_{j\in R(t_i)} \exp(x_j^\top \beta_{LM}(s|\gamma)+\theta(s|\eta))},$$

for $s_1 \leq t_i \leq s_L + w$. Predictions in the ipl-model are obtained from the stratum specific models. Predictions in the ipl*-model are obtained for *all* $s \in [s_1, s_L]$ by

$$\hat{H}(s+w|x,t_{\mathrm{LM}}=s) = \exp(x^\top \hat{\beta}_{LM}(s)+\hat{\theta}(s))\,(\hat{H}_0^*(s+w)-\hat{H}_0^*(s-)) .$$

Although both models (ipl and ipl*) can be fitted by standard software, some special effort is needed to obtain the correct baseline hazard(s) and the correct robust standard errors of the coefficients and the predictions computed from this model. Significance testing and model building can be based on the robust covariance matrix of the estimated coefficients, using the sandwich estimators of Lin & Wei (1989) implemented in for instance `coxph()` in the **survival** package for R. Appendix B contains more details. Obtaining standard errors of the predictions is not implemented in software. The issue is further discussed in Section 7.3. The whole modeling process is still a bit cumbersome. Data exploration can better be done by standard software. The strategy chosen in the examples in this chapter is like the two-stage approach of Section 6.3. The covariates are condensed into risk scores as much as possible before fitting the landmark models. This also has also the advantage of more clarity in tables and graphs.

Applications second round

Dutch gastric cancer trial

As a first example, the dynamic treatment effects of the Dutch gastric cancer trial will be illustrated. As an initial "univariate" landmark analysis, the sliding landmark effects for overall survival of D2 with respect to D1 surgery for $w = 4$ years in the Dutch gastric cancer trial are shown in Figure 7.5. "Crude" refers to estimates obtained in the separate landmark data sets defined by (7.1) at each of the time points where $\hat{F}_4(t \mid x)$ changes value. "Supermodel" refers to the stratified supermodel (7.2) with landmarks chosen on an equidistant grid from $s_1 = 0$ to $s_L = 6$ years with distance 0.1. The time functions chosen were $f_1(s) \equiv 1$, $f_2(s) = s/6$ and $f_3(s) = (s/6)^2$. The resulting regression coefficients are shown in Table 7.1, under "Treatment only, stratified." In Putter et al. (2005), a multivariate model was con-

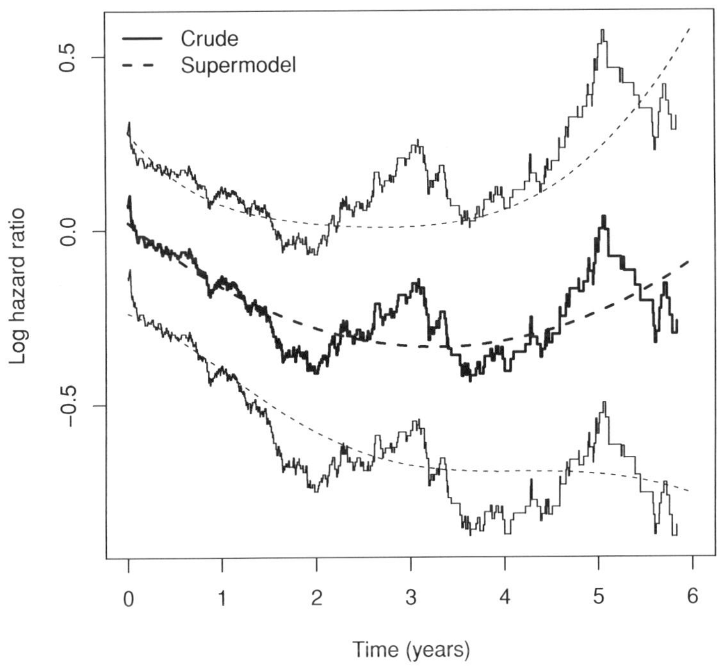

Figure 7.5 *Sliding landmark effects and pointwise 95%-confidence intervals for "treatment" in the gastric cancer data using a window of width* $w = 4$

sidered based on (possibly time-varying) effects of the clinical risk factors shown in Table A.4. It was found there that also residual tumor showed a time-varying effect while for the remaining risk factors no time-varying effect was found. "Risk score" is defined here as the combined effects of all the significant factors in that multivariate model without time-varying effects, i.e., as 0.785*(nodal status = "positive") + 0.545*(age = "¿65") + 0.448*(T-stage = "T2") + 0.853*(T-stage = "T3") - 0.347*(type of resection = "partial"), see the right column of Table IV of Putter et al. (2005). Based on treatment, risk score and residual tumor, backward selection based on Wald tests was used to first exclude landmark-dependent effects of these three covariates. The effect of residual tumor was not found to be time-varying ($p = 0.89$) and the interactions with f_2 and f_3 were subsequently removed. The time-varying effect of treatment was then found to be only trend-significant $p = 0.096$, while the time-varying effect of the risk score was significant, $p = 0.011$.

At first sight this seems to contradict the results of Putter et al. (2005). Notice, however, that the major changes in the effect of treatment take place within the very first prediction window [0,4]. This can be seen from Figure A.6. Hence, the variation of the treatment effect with the landmark time-point shown in Figure 7.5 is much smaller than the variation with time shown in Putter et al. (2005). This explains why the treatment by time effect is less significant. A similar explanation applies to the disappearance of the significant R1 by time interaction. The popping up of the risk score by time interaction in the landmark analysis, while the components of the score did not show such interaction in Putter et al. (2005), can be explained by the congruence of the interactions with time for all components similar to the rank=1 model of Section 6.3. In such a situation, there is much more power to detect the interaction with time for the risk score than for the individual components.

Because interest is in the time-varying effect of treatment in the first place, the time-varying effects of both treatment and risk score were retained. The resulting model is shown in Table 7.1, under "Multivariate, stratified." The left panel of Fig-

Table 7.1 *Sliding landmark model for the gastric cancer data set*

Part of model	Covariate	Time function	Coefficient	SE
Treatment only, stratified				
$\beta_{LM}(s)$	Treatment	1	0.021	0.132
		$s/6$	−1.309	0.779
		$(s/6)^2$	1.200	0.854
Multivariate, stratified				
$\beta_{LM}(s)$	Treatment	1	0.049	0.140
		$s/6$	−1.648	0.779
		$(s/6)^2$	1.439	0.856
	Risk score	1	1.155	0.107
		$s/6$	0.297	0.539
		$(s/6)^2$	−1.137	0.600
	R1	1	1.030	0.184
Multivariate, proportional hazards				
$\beta_{LM}(s)$	Treatment	1	0.042	0.140
		$s/6$	−1.649	0.777
		$(s/6)^2$	1.437	0.847
	Risk score	1	1.159	0.107
		$s/6$	0.256	0.513
		$(s/6)^2$	−1.088	0.559
	R1	1	1.029	0.184
$\theta(s)$		$s/6$	0.069	0.152
		$(s/6)^2$	−0.114	0.170

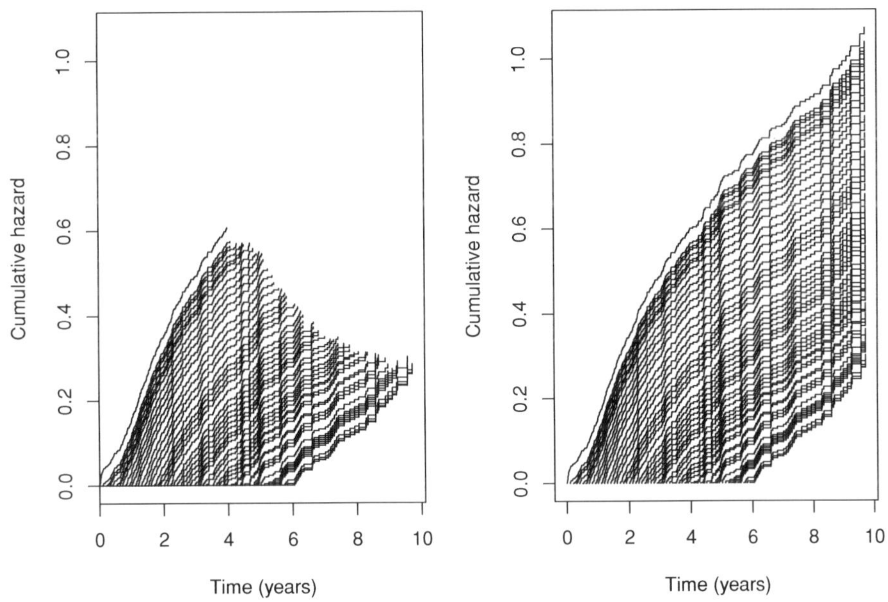

Figure 7.6 *Baseline hazards in stratified landmark supermodel (left) and in the proportional baselines landmark supermodel (right)*

ure 7.6 shows the baseline hazards in the resulting stratified landmark model. The stratified landmark model gives information on the regression effects, but can only be used for prediction at the landmark time points $s_1, \ldots, s_L$. For other time points s, the baseline hazard $h_0(t \mid s, w)$ in (7.2) is not defined. Dynamic prediction from s outside these landmark time points is possible for the ipl$*$ model, which replaces the stratified baseline hazards by a single baseline hazard with additional effects of time functions $g_j(s)$. Fitting the ipl$*$ model with $g_1(s) = s/6$ and $g_2(s) = (s/6)^2$, gives the results in Table 7.1, under "Multivariate, proportional hazards." Figure 7.7 shows the estimated baseline hazard for $s = 0$, as well as the hazard ratio $\exp(\theta(s))$. Together these define the baseline hazards for each value of $s \in [s_1, s_L]$, the result of which is shown in the right panel of Figure 7.6 for the landmark time points $s_1, ..., s_L$. The baseline hazards in the proportional baselines landmark supermodel ipl* in the right panel are defined on the whole range from t_{LM} to the end of follow-up, because of the proportional hazards assumption that connects them, while the baseline hazards of the ipl model on the left are defined only on $[t_{\mathrm{LM}}, t_{\mathrm{LM}} + w]$. It should be noted, however, that the parts of the baseline hazards that are defined for the ipl* and not for the ipl models are not relevant for the predictions. The resulting dynamic prediction is shown in Figure 7.8. The 10% and 90% quantiles of the risk score were chosen, 0 and 1.836, referred to as low risk and high risk, respectively. Figure 7.8 shows the 4-year dynamic probability of dying, based on the

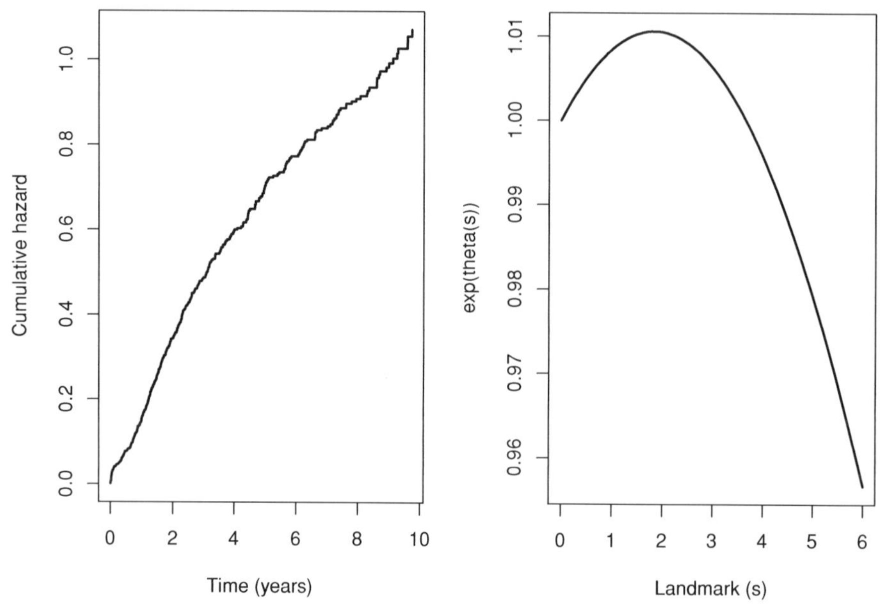

Figure 7.7 *Baseline hazard and landmark effects in proportional baselines landmark supermodel*

proportional baselines landmark supermodel of Table 7.1, for each combination of low/high risk, R0/R1, D1/D2.

Ovarian cancer

The second example concerns the ovarian cancer data with the prognostic score of Table 3.1 in Section 3.2. In Section 6.3 it was seen that the effect of this prognostic score was time-varying. To illustrate the use of landmark analysis, we choose a window of $w = 3$ years. Figure 7.9 shows the sliding landmark effects along with pointwise 95%-confidence intervals for the prognostic index. "Crude" refers to the estimates obtained in the separate landmark data sets defined by (7.1) at each of the time points where $F_3(t\,|\,x)$ changes value. "Supermodel" refers to the stratified supermodel (7.2) with landmarks chosen on an equidistant grid from $s_1 = 0$ to $s_L = 5$ years with distance 0.1. The time functions chosen were $f_1(s) \equiv 1$, $f_2(s) = s/5$ and $f_3(s) = (s/5)^2$. The resulting regression coefficients are shown in Table 7.2, under "Stratified." Figure 7.10 shows the 3-year dynamic probability of dying $F_3(t\,|\,x)$, based on the proportional baselines landmark supermodel of Table 7.2, for the mean value of the prognostic score, for x equal to the mean value $\pm$ 1 standard deviation, and for the mean value $\pm$ 2 standard deviations. It can be seen from Figure 7.10 that at $t = 0$, the estimates of $F_3(t\,|\,x)$ differ appreciably across values of x, while for t close to 5 years, $F_3(t\,|\,x)$ does not seem be influenced by x. This is in

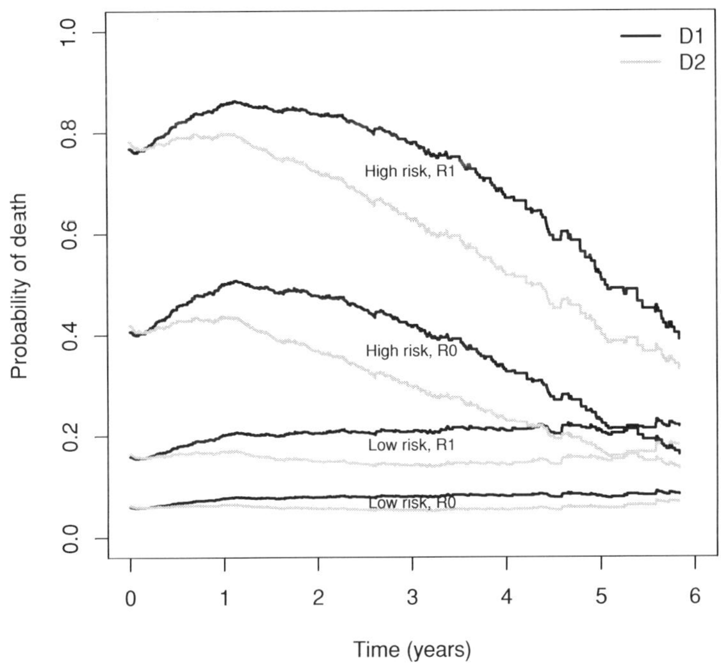

Figure 7.8 *Dynamic predictions (fixed width failure functions with $w = 4$ years) in the proportional baselines landmark supermodel of Table 7.1; predictions are shown for each of the two treatments, D1 and D2, for four groups of patients, defined by each combination of high/low risk and R0/R1*

Table 7.2 *Sliding landmark model for the ovarian cancer data*

Part of model	Covariate	Time function	B	SE
Stratified				
$\beta_{LM}(s)$	Risk score	1	1.081	0.151
		$s/5$	−0.722	0.881
		$(s/5)^2$	−0.383	1.250
Proportional hazards				
$\beta_{LM}(s)$	Risk score	1	1.080	0.149
		$s/5$	−0.711	0.834
		$(s/5)^2$	−0.395	1.191
$\theta(s)$		$s/5$	0.147	0.144
		$(s/5)^2$	−0.324	0.335

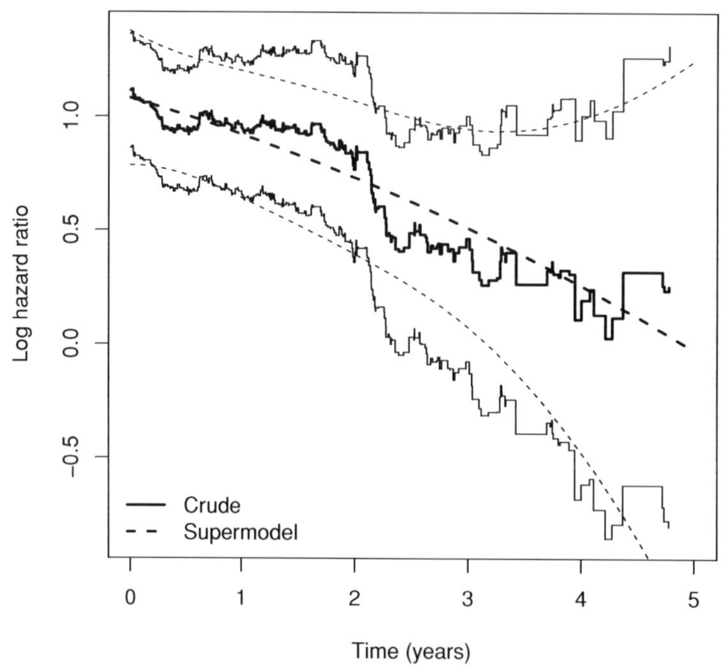

Figure 7.9 *Sliding landmark effects and pointwise 95%-confidence intervals for the prognostic index of Table 3.1 in the ovarian cancer data using a window of width $w = 3$*

accordance with the results of Table 7.2, where the proportional baselines landmark supermodel (ipl*) has $\hat{\beta}_{LM}(0) = 1.080$ and $\hat{\beta}_{LM}(5) = -0.026$.

7.3 Additional remarks

Standard errors of the ipl-model*

The ipl*-model can be seen as a special case of the Cox-type model of Section 6.4 with

$$v(x|t,\gamma,\eta) = \ln\left(\frac{\sum_{\{s|s\leq t\leq s+w\}}\exp(x^\top\beta_{LM}(s|\gamma)+\theta(s|\eta))}{\#\{s|s\leq t\leq s+w\}}\right).$$

It is not hard to check that the maximum likelihood estimates in this model are identical to the estimates for the regression parameters and the baseline hazards given in Section 7.2. The covariance matrix of the regression parameters (γ,η) can be obtained from the information matrix of the partial likelihood or from robust estimators in the spirit of Lin & Wei (1989). The standard errors of the predictions can be obtained from the formulas in Section 6.4. In those formulas a robust estimator of the covariance matrix can replace the theoretical based on the partial likelihood. The standard errors and the testing procedures reported in van Houwelingen (2007)

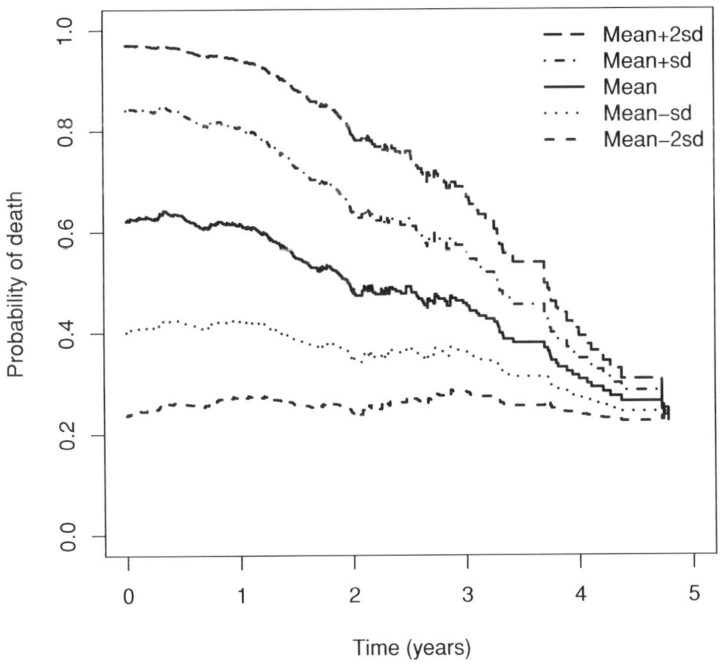

Figure 7.10 *Three year dynamic probabilities of dying based on the proportional baselines landmark supermodel for the ovarian cancer data*

are based upon ad hoc software for this Cox-type model. They agree well with the standard errors obtained from the standard software mentioned in Section 7.2. Unfortunately, no implementation of the formulas for the prediction error is available at the time of writing.

Some more theory

It should be pointed out that the super model of Section 7.2 does not satisfy the *coherence* condition formulated in Jewell & Nielsen (1993). The condition is defined for models with time-dependent covariates to be discussed in Chapter 8, but is also relevant here. Loosely speaking, their requirement is that a prediction model at any time-point t_{LM} should be consistent with earlier models. In the situation of this chapter this boils down to the requirement that all future models can be obtained from the model at $t = 0$ by conditioning on being alive at t_{LM}. This is true for the models of Sections 7.1 and 7.2 but not for the "sliding landmark super model" of this section. Since all models are only approximations of the truth, the best approximation for those alive at $t = 0$ need not be coherent with the best approximation at a later t_{LM}. The "super model" is not a comprehensive probability model, but a sequence of best models indexed by t_{LM}. That is, the model is fitted by maximizing pseudo likelihoods and not by using regular likelihoods.

In earlier versions of the landmark approach as described in the PhD thesis of de Bruijne (2001), landmark models were defined be resetting the clock at $t_{\mathrm{LM}} = s$ leading to models like

$$h(s+u|t_{\mathrm{LM}}=s,x) = h_0(u)\exp(x^\top \beta_{LM}(s) + \theta(s)) .$$

The implication is that event times and the risk sets change with s because the events are occurring at $u_{\mathrm{event}} = t_{\mathrm{event}} - s$. This complicates matters considerably. The counting process approach does not apply any more and it is not clear how statistical properties of the estimators in such "super models" could be obtained.

Relation with Aalen's additive model

As mentioned in Section 2.6, Aalen's additive model allows estimation of time-varying coefficients $\beta(t)$, but for several reasons the model is presented by plots of the integrated coefficients $B(t) = \int_0^t \beta(s)ds$. Insight in the predictive relevance and the influence of the covariates $F_w(t) = P(T \leq t+w|T \geq t)$ can be obtained by plotting $B(t+w) - B(t)$ as function of t for fixed width w. That would lead to plots similar to Figure 7.5.

Direct models for the survival function

The "stopped Cox model" of Section 7.1 is introduced as a way to obtain an estimate of $S(t_0|x)$ that is not sensitive to the proportional hazards assumption of the Cox model. In the literature this is known as *direct modeling*. Well-known examples are the pseudo-value approach of Klein and Andersen (see Andersen et al. 2003, Klein & Andersen 2005, Andersen & Klein 2007) and the direct modeling of Scheike and colleagues (Scheike & Zhang 2007, Scheike et al. 2008). Both methods use binomial regression models for the binary variable $\mathbf{1}\{T > t_0\}$ and both are confronted by the problem of censoring before t_0. Klein and Andersen solve that problem by an ingenious application of the *jackknife* leading to so-called pseudo variables. Scheike and colleagues solve the problem by inverse probability of censoring weighting (see Section 3.3). Both methods consider a whole range of t_0-values. To obtain parameter estimates and their standard errors Klein and Andersen use *Generalized Estimating Equations* (Liang & Zeger 1986), while Scheike and colleagues use pseudo likelihoods as in Section 7.2.

The typical graphical presentation of the results of both approaches is a graph showing how the covariate x influences $S(t|x)$ over time. The difference with the landmark approach introduced in this chapter is that only predictions at baseline are obtained and no dynamic predictions later on in the follow-up. Nevertheless, direct modeling of $P(T > s+w|T \geq s)$ for sliding s could be an alternative to the sliding Cox models of Section 7.2.

Part III

Dynamic prognostic models for survival data using time-dependent information

Chapter 8

Dynamic predictions using biomarkers

8.1 Prediction in a dynamic setting

Introduction

In Parts I and II, attention has been focused on prediction models based on covariate information available at the start of follow-up. In real life the situation is often more complicated. A schematic presentation of the type of information that becomes available during the follow-up of patients is given in Figure 8.1.

All kinds of dynamic information can be collected that is directly related to the disease of the patient and the changes in therapy that might be a response to progression (or cure) of the disease. This is the kind of information that will be stored in the *Electronic Patient Record*, the panacea of modern health care. The hope is that such information can be used for better individualized treatment depending on a dynamic assessment of the prognosis of the patient. A poor prognosis might lead to more aggressive treatment while a good prognosis might be interpreted as the patient being cured and not needing any further treatment. For such assessments dynamic prediction models are needed. Part III of this book is dedicated to the development of such models. Unfortunately, choice of therapy based on prognostic models will directly affect the validity of such models, but nevertheless prognostic models are needed even if they are self-destroying. In some sense, not only the information on the patient is dynamic, but so is any prognostic model, because it should be based on current and not on past treatment protocols.

In very general terms, the information can be described as a multi-dimensional *internal dynamic covariate* $X(t)$. The term internal means that the information is somehow generated by the patient himself, in contrast to external dynamic covariates, like the daily changing weather; for a more extensive discussion see Kalbfleisch & Prentice (2002), Chapter 6. A dynamic prediction model at time s should give an assessment of the future survival probabilities given a patient is alive at $T = s$ based on all covariate history information available at $T = s$, denoted by $X^*(s)$. Mathematically speaking, the definitions should be more precise and "at $T = s$" should be replaced by "just prior to s" or at $T = s-$, but "at s" will be used throughout for the sake of simplicity.

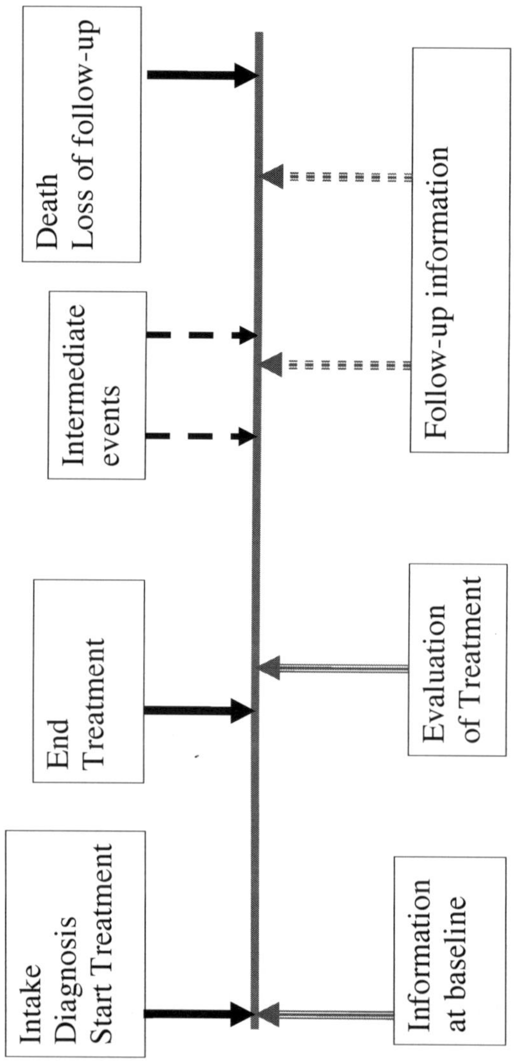

Figure 8.1 *Follow-up scheme for clinical data*

The traditional statistical approach is to develop a joint model for $(T, X(t))$ and to use that to obtain the desired prognostic probabilities. There are several approaches to handle the dependence between T and $X(t)$, that will be briefly sketched below.

Forward modeling: $X(t) \rightarrow T$

Here the approach is to model the conditional distribution of $T|X^*(t)$ by means of the hazard $h(t|X^*(t))$ and to obtain the marginal distribution of $X(t)$ under a *Missing at Random* assumption, with missingness caused by death or censoring. Strictly speaking, the Missing at Random assumption would require continuous observation of $X(t)$ and noninformative censoring. Obtaining predictions under this approach is rather awkward, because it involves integration over all possible futures of $X(t)$ given $X^*(s)$. Nevertheless, modeling the two components could be very helpful in getting more insight in the underlying process. Examples of this approach can be found in the papers of Pawitan & Self (1993), De Gruttola & Tu (1994), Tsiatis et al. (1995), Wulfsohn & Tsiatis (1997), Hogan & Laird (1997*b*).

Backward modeling: $T \rightarrow X(t)$

This approach, related to pattern mixture modeling as discussed in Brant et al. (2003) and Fieuws et al. (2008), models the conditional distribution of $X^*(t)|T = t$ and the marginal distribution of $T = t$. With these components available the prognostic probabilities can be derived by application of Bayes' theorem. Computationally it is less involved than the forward approach, provided that there is no censoring. If there is censoring, unobserved survival times have to be imputed to fit the model, which is hard to handle if the survival distribution is defective. Nevertheless, the idea is appealing that one looks at the process $X(t)$ backwards from $T = t$ and "predicts" future survival by horizontal alignment of the observed $X(t)$. A further discussion and an example can be found in Chapters 5 and 6 of De Bruijne's PhD thesis (de Bruijne 2001) and will be briefly touched upon in Section 8.3.

Latent variable models: $X(t) \perp T$ *given* Z

This approach makes the assumption that the association between $X(t)$ and T could be explained by some latent (unobserved) patient characteristic denoted by Z. To be more precise, $X(t)$ and T are conditionally independent given Z. For the survival part of the data, Z plays a similar role as the unobserved frailty in Section 5.2. For the covariate part, Z could be something like a random effect in a mixed model for repeated measurements. The easiest models to fit are the latent class models in which Z is a categorical outcome with a limited number of categories. Such models can be easily fitted by the EM-algorithm. The ultimate case is the two-point mixture model that was also discussed in Section 5.2. The latent class models are quite popular, see Hogan & Laird (1997*a*), Henderson et al. (2000), Proust-Lima

& Taylor (2009). The drawback, however, that it is hard to tell from such a latent class model how the covariate process actually influences the survival.

Prediction by landmarking

If the main interest is dynamic prediction of future survival given all covariate information at $T = s$, there is no need to build complex joint models. As argued in van Houwelingen (2007), landmarking can do the job in a much easier way. See also Zheng & Heagerty (2005), where a similar approach is described as partly conditional survival modeling. Part III of this book describes applications of the landmark approach in different situations; the present chapter deals with dynamic prediction using repeatedly measured bio-markers using Data Set 2, Chapter 9 considers dynamic prediction in multi-state models using Data Set 6 and Chapter 10 focuses on the dynamic prediction in chronic disease using the breast cancer data of Data Set 5.

8.2 Landmark prediction model

The starting point is the Cox model that allows time-dependent covariates and time-varying effects

$$h(t|x(t)) = h_0(t)\exp(x(t)^\top \beta(t)) \ .$$

In this model the effect of the covariate process can have different components. Some of them may be time-constant and some of them may vary over time. The truly time-dependent covariates, *biomarkers* are supposed to have continuous trajectories over time. (Discrete processes will be discussed in the next chapter.) The effect of time-dependent components is through their current value. That is no loss of generality, because relevant components of the covariate history could be included by expending the covariate processes with components like earlier values or recent changes. Generally speaking the regression parameters will change over time rather slowly.

Although such a model cannot be used for prognostic purposes unless the distribution of $X(t)$ is known, fitting such models can give very useful insight in the biological mechanisms, i.e., which aspects of the covariate process drive the hazard, and whether the effects on the hazard vary with time.

The situation changes when we switch to a landmark (conditional) model that uses only the information available at time $t_{LM} = s$. For a single landmark s we can postulate the model

$$h_s(t|x(s)) = h_{s,0}(t)\exp(x(s)^\top \beta_s(t)) \ , \quad \text{for } t \geq s \ .$$

It is of interest to reflect on the question how the regression parameter $\beta_s(t)$ in the landmark model is related to the regression parameter in the original Cox model. This resembles very much the situation of "ageing" covariates discussed in Section 5.3. The landmark value $x(s)$ is a proxy for the true $x(t)$ and, therefore, might rapidly lose its relevance. As a first approximation we will have that

$$x(s)^\top \beta_s(t) \approx E[X(t)^\top \beta(t)|X(s) = x(s)] \; .$$

For a one-dimensional process this implies

$$\beta_s(t) \approx \frac{\mathrm{cov}(X(s), X(t))}{\mathrm{var}(X(s))} \beta(t) \; .$$

If the process $X(t)$ shows rapid variation, this attenuation effect might be very dramatic. It should be explored at suitable landmarks before prognostic super-models are defined and fitted.

Super model 1: ipl

As in Chapter 7, this model defines a model for a sequence of landmark models in the range $[s_1, s_L]$ postulating a model for $\beta_s(t)$, but allowing landmark specific baseline hazards. In the light of the approximation above based on the covariance function of the $X(t)$ process, a plausible model is given by

$$\beta_s(t) = \beta_{LM}(s) * \lambda(t-s) \; ,$$

where $\beta_{LM}(s)$ is modeled by a linear model as in Section 7.2, $\lambda(t-s)$ is monotonically decreasing from $\lambda(0) = 1$ to $\lambda(\infty) = 0$ and $*$ stands for coordinate-wise multiplication. A suitable choice would be $\lambda(t-s) = \exp(-\nu(t-s))$ but this is hard to fit because of its non-linear dependence on ν. A more practical approach it to approximate $\beta_s(t)$ by a polynomial in s and $t-s$ or a similar linear model, though the monotonicity is lost then. If the focus is purely on prediction of survival from s to $s+w$, such an approximation hardly affects the predictions and makes them much easier to compute.

This supermodel with linear models for $\beta_s(t)$ could be fitted by creating large stacked data sets similar to those discussed in Chapter 7. In such a data set there are separate records for each individual at each combination of landmark point s and event time t at which the individual is at risk. The covariate value at such a record is the value at the landmark point $x(s)$. The current time t is used to define time-varying effects in the supermodel. Once the stacked data set has been created, the analysis (including the computation of predictions) is not different from the one discussed in Chapter 7.

*Super model 2: ipl**

The second supermodel also postulates a model for the landmark specific baseline

$$h_{s,0}(t) = h_0(t)\exp(\theta(s)) \; ,$$

similar to the one in Section 7.2. However, one should realize that landmark specific baseline hazards for the model without a λ-term, estimated by

$$\hat{h}_{s0}(t_i) = \frac{1}{\sum_{j \in R(t_i)} \exp(x_j(s)^\top \beta_{LM}(s))}, \quad \text{for } s \leq t_i \leq s+w$$

depend on s through the regression coefficients $\beta_{LM}(s)$ *and* the current value $x(s)$. While in Chapter 7 centering of the covariates x might help to control the $\theta(s)$-term, it can be useful for this kind of data to remove trends in the covariates before fitting the models. Centering and detrending need not be precise (mean $=$ 0, no trend) but can be led by convenience in communicating the results.

8.3 Application

Description of the data

As an example the data of Data Set 2 will be (re)analyzed. These are extensively discussed in de Bruijne et al. (2001) and are also used in van Houwelingen (2007). The data come from a clinical trial in Chronic Myeloid Leukemia (*CML*) that has been reported in The Benelux CML Study Group (1998). The data set contains data on 190 patients followed up to 8 years. In the first stage of the study white blood cell counts (*WBC*) are measured every one to two months. Details are given in Table 8.1.

Table 8.1 *Number of WBC measurements by year*

Year	Number of patients with observations	Mean number of observations per patient
1	190	12.52
2	155	7.59
3	117	7.31
4	74	7.30
5	52	6.42
6	30	6.13
7	16	6.00
8	5	7.00

The problem of this data set is that at some stage WBC is not measured any more. The reason is that patients go "off study," mainly because of treatment failure. They are still followed for survival, but WBC is not recorded anymore. The exact date at which a patient goes off study is not available in the data set. For the purpose of this analysis an "approximate" off study date is defined as "date of last WBC measurement + 3 months." This leads to the definition of a first event that can

Table 8.2 *Cross-tabulation of first event status and survival status*

	Survival status		
First event	Censored	Death	Total
Censored	28	0	28
Off study	53	82	135
Death	0	27	27
Total	81	109	190

either be "censored" or "off study" or "death." The relation with overall survival is given in Table 8.2. Time until the first event can be considered to be the "failure-free survival time." The Kaplan-Meier estimates of the survival function and the censoring function for failure-free survival are shown in Figure 8.2. They differ from the graphs in Figure A.2 because those are based on actual survival and not on failure-free survival.

The WBC measurements are very skewed. For the analysis they are transformed and centered into

$$\text{LWBC} = {}^{10}\log(\text{WBC}) - 0.95\,.$$

The trajectories of the transformed data for all patients are shown in Figure 8.3. The overall mean of all observations is 0 with a standard deviation of 0.35. The data do

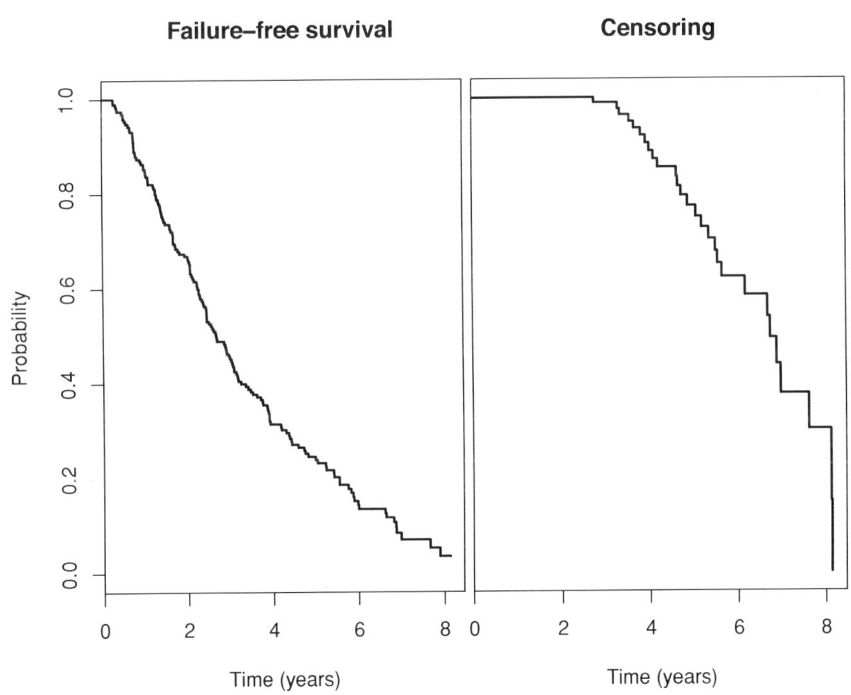

Figure 8.2 *Survival and censoring for failure-free survival in the Benelux CML data*

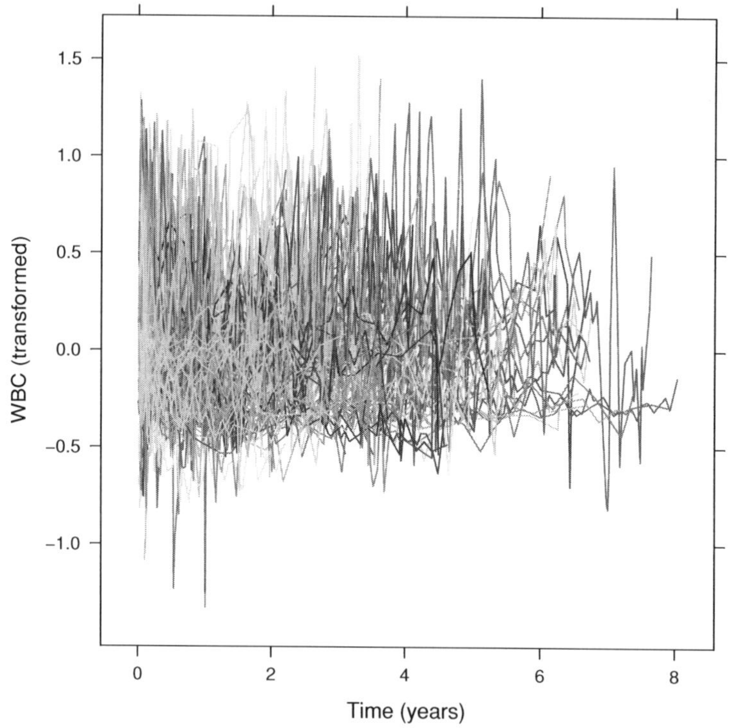

Figure 8.3 *LWBC trajectories of all patients in the Benelux CML data*

not show much of a trend. There is considerable variation within the observations of the same patient. The overall intra class correlation ICC is given by ICC=0.30. When considered per year the ICC rises to about ICC=0.58 in the third year and decreases thereafter. An interesting graph is obtained by plotting LWBC in reverse time, that is against time until failure for those patients who are not censored before failure. This is shown in Figure 8.4. The appealing aspect of this plot is that it could be used as a simple tool to predict failure-free survival given the current LWBC value, by reading the time until failure from (a smoothed version) of the mean curve in the graph. This was called the pattern mixture approach in Section 8.1. However, it is far from easy to quantify the uncertainty of such a prediction.

Exploratory analysis by Cox models with time-dependent covariates

The traditional analysis of such bio-marker data is by means of Cox models with time-dependent covariates as in de Bruijne et al. (2001). As pointed out by Andersen & Liestøl (2003), there are many pitfalls in such an analysis. Most of those pitfalls can be circumvented by the landmark analysis to be discussed later. Nevertheless, a time-dependent Cox-model can give relevant insight in the data.

In order to perform such an analysis the current value of the time-dependent covariate is needed for each patient at each event time the patient is at risk. Since the

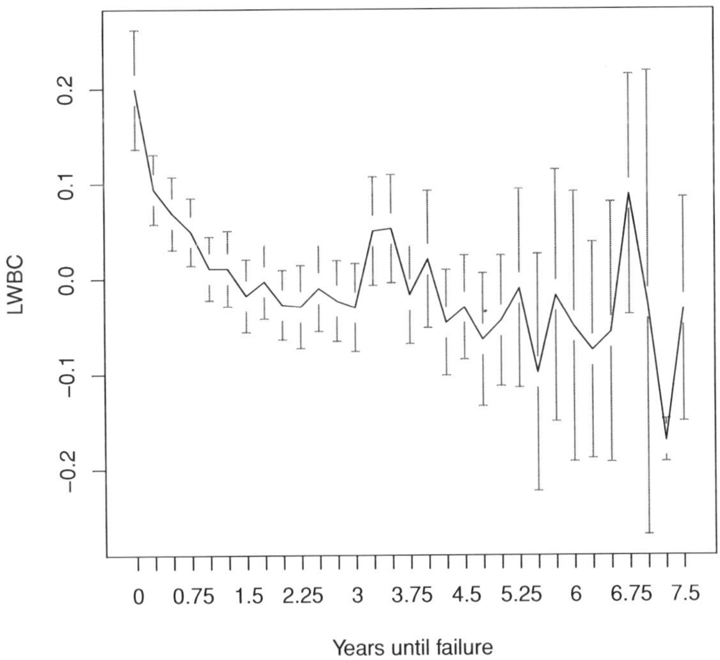

Figure 8.4 *Mean trajectories of LWBC in reverse time*

current value is not available, the last observation is taken as a proxy. Moreover, the information about the "age of the observation" defined as *Time Elapsed since Last measurement* and denoted by $\text{TEL}(t)$ is available at each event time. Analyses can be carried out for failure-free survival and for overall survival. Besides information on the biomarker WBC, also the fixed covariates age (at start of the trial) and Sokal (a clinical risk score for CML patients, which also includes age as one on the risk factors) will be included in the analysis.

In Table 8.3 the results of three different analyses are given: 1) with death as endpoint, where "off study" is censored, 2) with "off study" as endpoint, where death is censored, and 3) with both endpoints combined, whichever comes first. There are 27 deaths while on study and 135 patients went off study alive. Although the number of deaths is small, it looks like LWBC has more effect on "death" than on "off study." The intrinsic process $\text{TEL}(t)$ is not included, because it does not make sense to consider $\text{TEL}(t)$ in relation to "off study" since it is used in the definition of "off study," and it has no significant effect on "death."

The analysis of overall survival is reported in Table 8.4. This analysis comprises patients "on study" and "off study." The effect of $\text{TEL}(t)$ might be a proxy for the fact that "off study" patients have a bad prognosis, while the interaction term $\text{TEL}(t) * \text{LWBC}(t)$ indicates that "old" values of $\text{LWBC}(t)$ might be less relevant.

Table 8.3 *Time-dependent Cox regression for different endpoints*

	Death		Off study		Any event	
Covariate	B	SE	B	SE	B	SE
Sokal	−0.026	0.462	0.347	0.154	0.302	0.144
Age/10	0.473	0.200	−0.158	0.068	−0.079	0.063
LWBC(t)	3.009	0.479	1.573	0.226	1.815	0.201

Table 8.4 *Time-dependent Cox regression for overall survival*

Covariate	B	SE	B	SE
Sokal	0.459	0.181	0.398	0.186
Age/10	0.164	0.081	0.197	0.082
LWBC(t)	1.534	0.216	1.987	0.280
TEL(t)			0.387	0.078
TEL(t) $*$ LWBC(t)			−0.426	0.162

However, this model cannot be used for predictions unless the joint distribution of LWBC(t) and TEL(t) is known. In principle, this could be worked out if the true "off study" status is available, but in practice it is more convenient and more robust to switch to the landmark approach.

Prediction by landmark models

The question that could be sensibly answered by the data is the calculation of future survival probabilities for patients that are still on study. In the data such patients can be characterized as having TEL(t) ≤ 0.25 (or any suitable cut-off). Landmark data sets will only contain such patients and will be used to predict their survival given their recent LWBC(t) together with Sokal and age. If a patient has a gap of more than 0.25 years between WBC measurements he/she will temporarily be excluded from the landmark analysis. Carrying out such a landmark analysis is not much different from the analysis for time-varying effects. The data were analyzed with $t_{LM} = s$ running from $s_1 = 0.5$ to $s_L = 3.5$ with steps of 0.1. The prediction window was set at $w = 4$.

Table 8.5 shows the results for a simple model and for a model extended with time-varying effects for LWBC. The latter might be more realistic, but the former is more convenient for predictions. Observe that the effects of Sokal and age/10 are very similar to those in the time-dependent Cox analysis. However, the effect of LWBC appears to be smaller, presumably due to the restriction to patients still under study. The time-varying effect of LWBC is shown in Figure 8.5. Note that, as expected, the effect decreases more or less monotonically over the prediction window. The effect rising again after 2.5 years is implausible and presumably an artifact of the time dependent effect taken to be quadratic. Figure 8.6 shows the

Table 8.5 *Results of landmark analyses for the WBC counts; s runs from 0.5 to 3.5 with steps of 0.1; window width w = 4*

		Simple		Extended			
		ipl		ipl		ipl*	
Covariate		B	SE	B	SE	B	SE
$\beta_s(t)$	Sokal	0.582	0.194	0.550	0.191	0.555	0.191
	Age/10	0.111	0.099	0.119	0.099	0.118	0.099
	LWBC	0.823	0.256	2.309	0.479	1.884	0.391
	LWBC*$(t-s)$			−1.558	0.531	−1.107	0.465
	LWBC*$(t-s)^2$			0.291	0.139	0.194	0.125
$\theta(s)$	$s/4$					−0.042	0.630
	$(s/4)^2$					−1.028	0.705

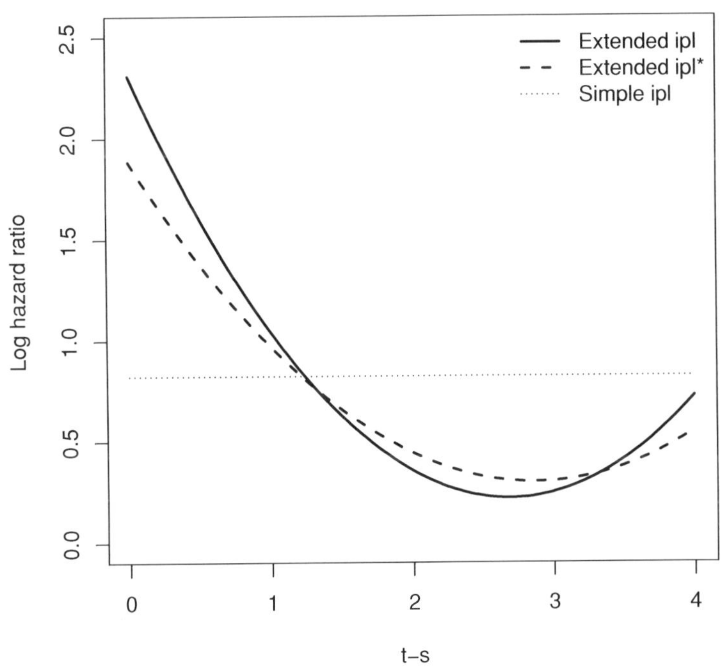

Figure 8.5 *Time-varying effect of LWBC for the ipl and the ipl* models of Table 8.5*

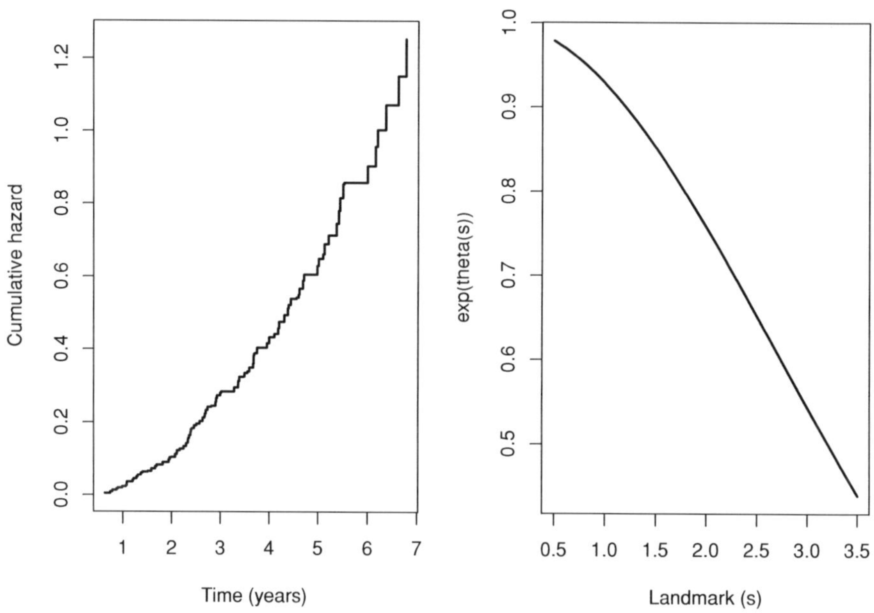

Figure 8.6 *Baseline hazard and landmark effects in proportional baselines landmark supermodel*

baseline hazard and the landmark effects in the proportional baselines landmark supermodel (ipl* in Table 8.5). To illustrate the effects of LWBC on survival, we consider three patients, with mean values for Sokal and age, which differ in terms of the behavior of their LWBC $x(t)$; the first patient has $x(t) \equiv 0$, the second has $x(t) \equiv 0.5$, while for the third patient $x(t)$ increases linearly from 0 at $t = 0$ to 1 at $t = 3.5$ years. Figure 8.7 shows the dynamic predictions of death within a window of $w = 4$ years for these three patients. The general picture seems to be that increasing LWBC increases the probability of dying within 4 years. The curve in gray shows the dynamic predictions of death within a window of $w = 4$ years for a patient with mean values for Sokal and age in a Cox model not taking into account LWBC at all.

8.4 Additional remarks

As pointed out by Andersen & Liestøl (2003) an important problem in the interpretation of classic time-dependent Cox models for biomarkers is that often unscheduled measurements are taken when patients experience some "crisis" that might lead to the death of a patient. That would inverse the possible causal relation between the marker and death. For that reason, they use a time-lagged version of the biomarker $x(t-\Delta)$ with $\Delta = 7$ days in their data analysis. In the landmark analysis as discussed in this chapter the danger of a bias due to measurements taken

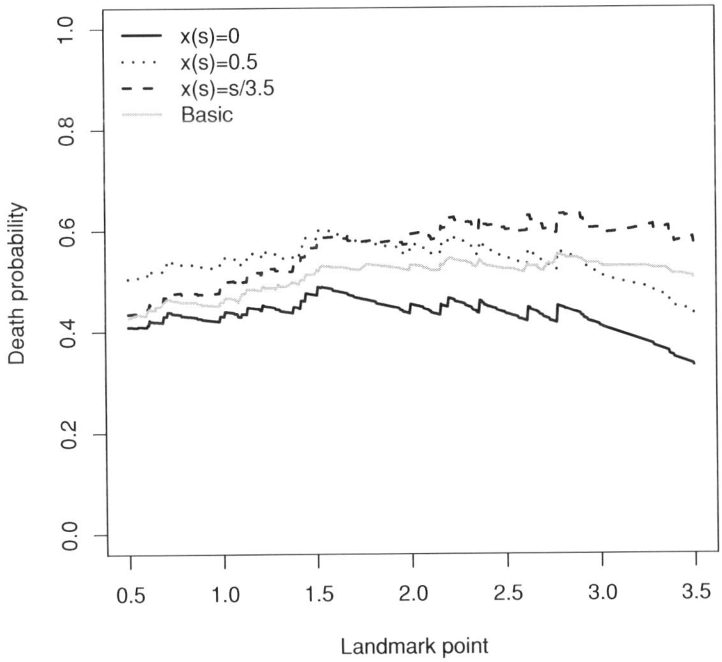

Figure 8.7 *Dynamic predictions of death within w = 4 years in the proportional baselines landmark supermodel for different trajectories of LWBC; basic refers to dynamic predictions in a Cox model not taking into account LWBC*

because of a "crisis" is much smaller, since the landmark time-point s determines which value will be used. However, an extra safeguard can be built in by introducing some "lagging" as well, for example by taking the last value of $x(t)$ before $s-\Delta$. The main concern of Andersen & Liestøl (2003) is the "ageing" of the biomarker if measurements are too far apart. This definitely plays a role in the landmarking modeling, but as observed in Table 8.5, it could easily be remedied by allowing the effect to vary with $t-s$, that is time reset at the moment of prediction.

The landmark approach not only gives direct prediction without the need for complex modeling, it also comes much closer to a causal interpretation of the biomarker. The approach of Gran et al. (2010) toward causal modeling for time-dependent covariates is very much in the same spirit. Their sequential Cox models are close to the ipl-models of this chapter.

Chapter 9

Dynamic prediction in multi-state models

9.1 Multi-state models in clinical applications

Multi-state models have become increasingly popular as a convenient tool to describe the follow-up of patients or, more generally, individuals at risk for certain events. Per Kragh Andersen has been one of the main promotors of multi-state models. It is quite interesting to read the section on multi-state models in his contribution to the Tenth Anniversary Issue of Statistics In Medicine in 1991 (Andersen 1991), where he describes multi-state models as one of the important developments in the 1980's. Comprehensive reviews are given by Commenges (1999), Hougaard (1999), Andersen & Keiding (2002), Putter et al. (2007). Interesting applications in the field of bone marrow transplantation are given in the papers by the group around Klein, Keiding and Andersen (Klein et al. 1994, Keiding et al. 2001, Klein & Shu 2002). The first paper in this area from the Leiden department of Medical Statistics is Hansen et al. (1994) on multi-state models for liver transplant patients.

In the familiar multi-state models individuals can only be in one state at a time and transitions between states can only be in one direction (if they are possible at all) and states can only be visited once. Technically, this is known as a directed a-cyclic graph. Usually there is one *initial state*, and there can be more *final states*, also known as *absorbing states*. All states between the initial state and the absorbing states are *transient states*. Multi-state configurations can be built from three elementary forms: consecutive events, splits due to competing events and merges due to transitions into the same state. Examples of possible configurations are given in Figure 9.1. The simplest multi-state model, the so-called illness-death model is shown in Figure 9.2.

An example of an illness-death model is given by the liver transplant multi-state model of Hansen et al. (1994). The population consists of all patients that undergo a liver transplant. The three states stand for

- State 0: Alive with functioning graft (transplanted liver)
- State 1: Alive without functioning graft
- State 2: Dead

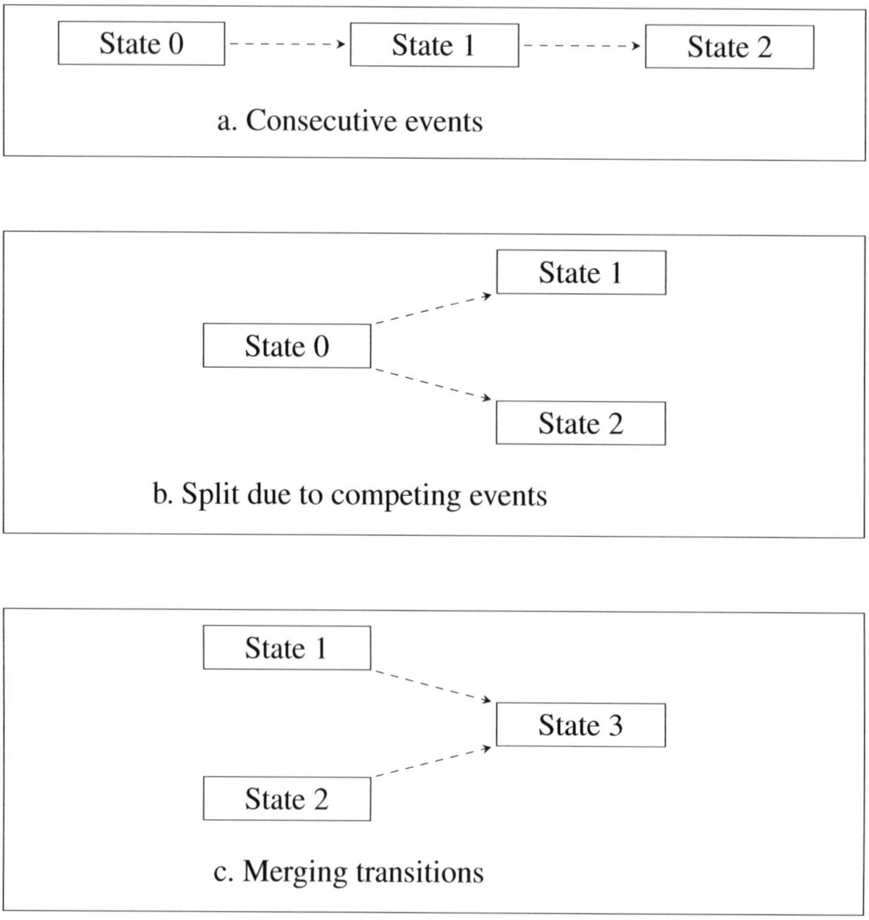

Figure 9.1 *The building blocks for multi-state models*

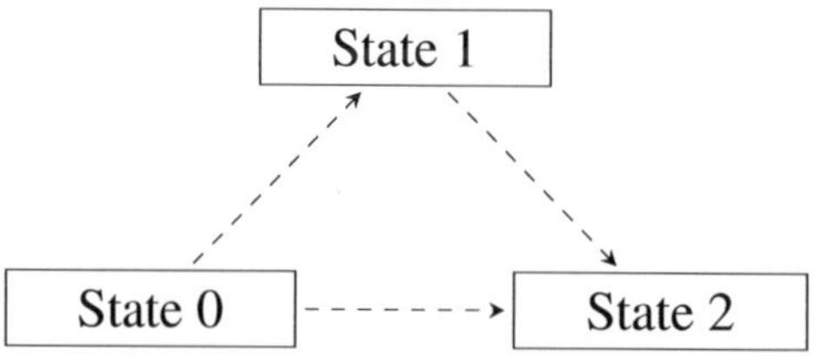

Figure 9.2 *The illness-death model*

The three transitions stand for the following events:

- Transition $0 \to 1$: Failure of the graft
- Transition $0 \to 2$: Death with functioning graft
- Transition $1 \to 2$: Death after graft failure

Actually, this model is a simplification of the true situation. First of all, the patient might die during or within one day after the operation. This is called technical failure. Secondly, after failure of the graft, it is possible to give the patients a second transplant. This leads to the much more complex multi-state model of Figure 9.3. It is even more complicated because a third transplant after failure of the second one is not excluded.

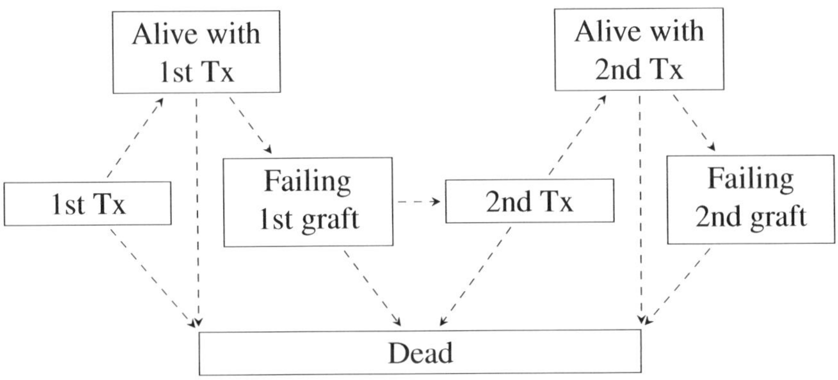

Figure 9.3 *Multi-state model for liver transplant; Tx=transplant*

The liver transplant example is complicated because of the repeated occurrence of graft failure. The examples to be considered in the sequel are of a simpler nature. There are two data sets in the appendix that have a multi-state nature: Data Set 5 and Data Set 6. Data Set 5 involves breast cancer patients with long term follow-up and information on three different events, local recurrence (LR), distant metastasis (DM) and death (D). Evidently, death is an absorbing state and follow-up stops when death occurs. The slight complication in the data set is that the exact moment of LR and DM is not known. It is known when they are diagnosed, but they might have actually occurred before. Strictly speaking this is a case of interval censored data. That aspect will be ignored in the analysis. As a consequence, LR and DM can occur simultaneously. A multi-state model for this data is given in Figure 9.4.

The transitions in this multi-state model correspond to the following events:

Event	Transition(s)
Diagnosis of LR	$0 \to 1, 2 \to 3$
Diagnosis of DM	$0 \to 2, 1 \to 3$
Simultaneous diagnosis of LR and DM	$0 \to 3$
Death	$0 \to 4, 1 \to 4, 2 \to 4, 3 \to 4$

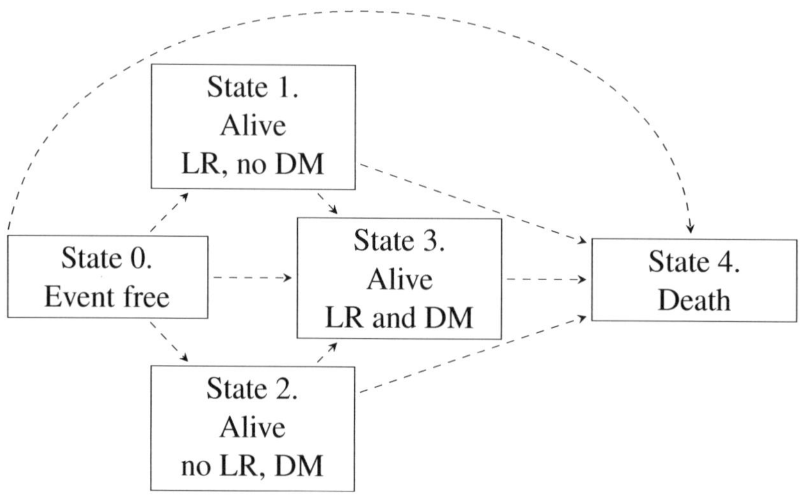

Figure 9.4 *Multi-state model for Data Set 5*

An alternative representation that underlines the sequential nature of the model is given in Figure 9.5.

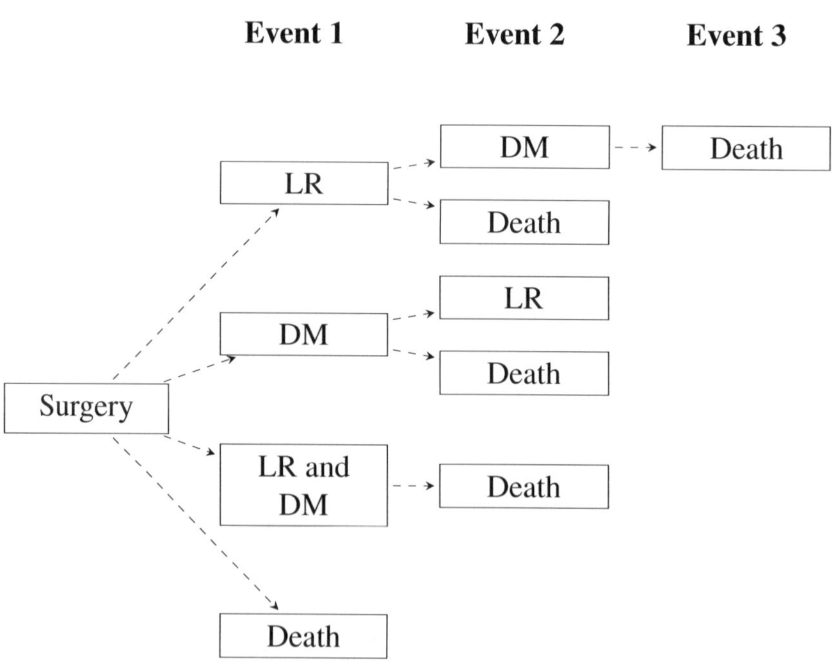

Figure 9.5 *Sequential representation of Figure 9.4*

The second data set to be used in the sequel is Data Set 6 coming from the EBMT registry. It deals with acute lymphoid leukemia (ALL) patients who had an allogeneic bone marrow transplantation from an HLA-identical sibling donor.

Events recorded during the follow-up of these patients are: acute graft-versus-host disease (AGvHD), platelet recovery (PR, the recovery of platelet counts to normal level), relapse and death. Although there is some information in the data on survival after relapse, that information is not systematically available because relapse is considered as an endpoint in the EBMT (after a relapse a new treatment protocol starts and the patient is, in some sense, considered to be a new patient). More information on the data can be found in Section A.6 and in Fiocco et al. (2008) and van Houwelingen & Putter (2008). The multi-state model for this data set is given in Figure 9.6. It looks quite complicated. This is because of the competing risks "Relapse" and "Death" and the presence of "AGvHD" and "PR." It could be argued whether the latter two should be considered as states. They might rather be considered as time-dependent dichotomous covariates AGvHD(t) and PR(t) that have to be taken into account when formulating (dynamic) predictions for the true (competing) endpoints "Relapse" and "Death." Figure 9.7 gives a slightly more abstract representation of the model, that gives a better insight in the structure of the data.

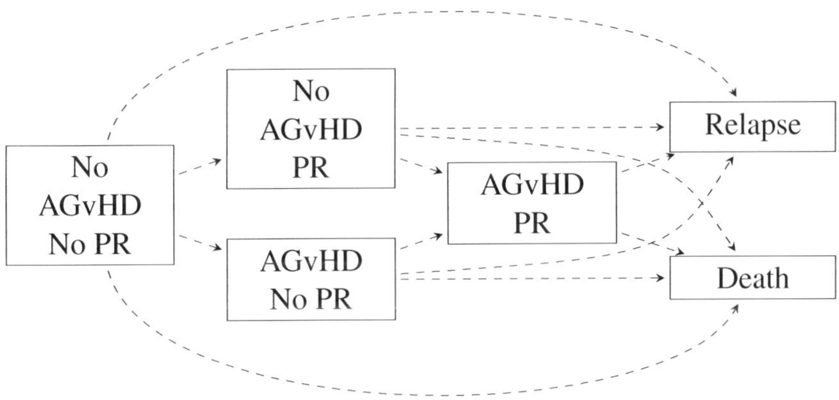

Figure 9.6 *Multi-state model for Data Set 6*

9.2 Dynamic prediction in multi-state models

Given a multi-state model as described in Section 9.1, the simplest prediction questions that can be asked concern prediction at time $t_{\text{pred}} = 0$, the moment an individual enters the initial state 0. The probabilities of interest are defined as

$$P_i(t) = \text{Probability of being in state } i \text{ at time } t\ .$$

Knowing these probabilities for all possible states gives a complete specification of the probability distribution of the state of the individual at time t. Plotting $P_i(t)$ against t for all i, including $i = 0$ gives useful insight of the possible developments

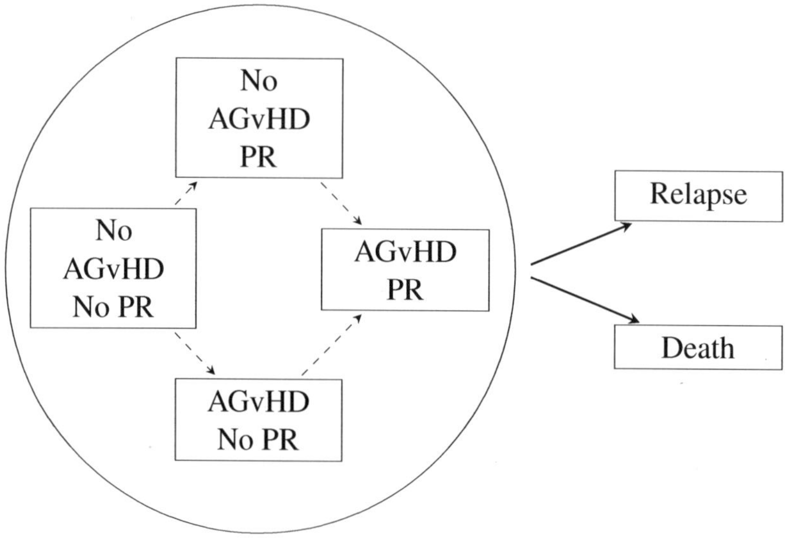

Figure 9.7 *Alternative representation of the multi-state model for Data Set 6*

over time. Notice that $P_i(t)$ will be decreasing for the initial state $i = 0$ and increasing for absorbing states. Actually, for absorbing states these curves are just the cumulative incidence functions defined in Section 2.1. For transient states the curves will show one or more peaks. Of course, these probabilities will depend on covariate information available at $t_{\text{pred}} = 0$.

Dynamic prediction deals with prediction later on in the follow-up, at $t_{\text{pred}} = s$, say. Generally, the probabilities of interest not only depend on the current state i, say, but also on the history up to s, denoted by $H(s)$. The probabilities of being in state j at time t, conditional on being in state i at time s, with history $H(s)$, can be denoted as

$$P_{ij}(t|s,H(s)) \ .$$

Obviously, these probabilities will again depend on the covariate information at $t = 0$. However, they might also depend on other (biomarker) information collected before s that is not included in the multi-state model. For the time being the existence of such time-dependent covariates will be ignored. This issue will be discussed in Section 9.4.

Prediction based on the full multi-state model

The classic approach is based on modeling the hazards of the transitions from state i to state j defined as

$$\lambda_{ij}(t)dt = P(\text{in state } j \text{ at } t+dt \,|\, \text{in state } i \text{ at } t) \ .$$

The simplest model is the so-called *Markov model* where $\lambda_{ij}(t)$ does not depend on the history prior to t. This is a very convenient assumption, but not very plausible in practice. The Markov model allows formal computation of probabilities of interest given all transition hazards through the general formula of Aalen & Johansen (1978), but even under this model the computations get quite tedious if the multi-state model is complex. Examples of such computations can be found in Andersen et al. (1993), Klein et al. (1994), Putter et al. (2007), and they are implemented in the **mstate** package for R (de Wreede et al. 2010). When the Markov assumption is not satisfied, it is often much easier to compute the probabilities by simulating a large number of individual trajectories as done in Dabrowska et al. (1994), Fiocco et al. (2008) and in Section 9.3.

The "minimal" extension of the Markov model is to let $\lambda_{ij}(t)$ depend on the moment the individual entered state i, denoted by u. A very simple model is to take

$$\lambda_{ij}(t|u) = \lambda_{ij}(t-u) \ .$$

This amounts to resetting the clock at the moment a new state is entered. In some situations like repeated measurements, this might be a natural thing to do, but in most cases having different time scales in one model complicates further modeling as discussed below and makes the mathematics of the large sample distribution theory very hard because the counting process approach of Andersen & Borgan (1985), Andersen et al. (1993) does not apply anymore. Therefore, the approach of choice is to stick to a single timescale t and to model $\lambda_{ij}(t|u)$ by approaches allowing delayed entry at u and inclusion of u as a history-dependent covariate. Within the setting of the Cox model this would lead to a model like

$$\lambda_{ij}(t|u,x) = \lambda_{ij,0}(t)\exp(x^\top \beta_{ij} + f_{ij}(u|\theta_{ij})) \ .$$

Here, $f_{ij}(u|\theta_{ij})$ is a parsimonious (linear) parametric model for the effect of u. This approach leads to many non-parametric baseline hazards, namely one for each transition. These baseline hazards might be hard to estimate if the number of observed transitions of a certain type is small and, consequently, predictions could become quite unreliable. Substantial gain in parsimony and robustness can be made by assuming proportional hazards for transitions related to the same event, like death. In such a model the β_{ij}'s corresponding to the same event can be taken identical or be modeled by parsimonious interaction models as in the reduced rank models described by Fiocco et al. (2005, 2008). Given such models, dynamic probabilities can be computed (by simulation) provided the dynamic conditioning specifics the moment of entry into the current state. The R package **mstate** yields such probabilities together with their standard errors.

Prediction by landmarking

The essential element of landmarking in the context of multi-state models is that it focuses on direct dynamic prediction for the absorbing states given all information about (fixed) covariates and intermediate events at the moment of prediction (t_{LM}).

"Direct" means ignoring possible intermediate events in the future. In the simplest situation there is only a single absorbing state "Death" as in Data Set 5, but competing risks as in Data Set 6 can be handled as well. The landmarking approach is not restricted to Markov models. All information about the individual's history can be included. Landmark predictions are easy to obtain, even in the case of competing risks, because there is no need to consider future intermediate events.

9.3 Application

As an application the ALL data of Data Set 6 will be analyzed in detail (the breast cancer data of Data Set 5 will be analyzed extensively in Chapter 10). Data Set 6 was also analyzed in van Houwelingen & Putter (2008). However in that analysis the two competing risks were merged into a single relapse-free survival endpoint, as in Section 4.3. In this section the competing risks of relapse and death before relapse will be considered together with the merged event "Relapse or Death."

Cox models for the transition probabilities

Table 9.1 *Cox regression for death and/or relapse*

Covariate	Category	Relapse		Death before relapse		Relapse or death	
		B	SE	B	SE	B	SE
Donor recipient	No mismatch						
gender mismatch	Mismatch	0.054	0.121	0.180	0.098	0.126	0.076
GvHD prevention	No TCD						
	TCD	0.243	0.121	0.093	0.101	0.154	0.077
Year of	1985-1989						
transplantation	1990-1994	0.028	0.132	-0.506	0.107	-0.290	0.082
	1995-1998	-0.070	0.146	-0.414	0.117	-0.281	0.091
Age at transplant	≤ 20						
	20-40	0.002	0.125	0.552	0.128	0.289	0.089
	> 40	-0.001	0.155	0.954	0.142	0.538	0.102
AGvHD		-0.183	0.107	0.927	0.104	0.410	0.073
recent – AGvHD		-0.309	0.555	-0.579	0.199	-0.229	0.179
PR		0.138	0.111	-0.507	0.103	-0.216	0.074
recent – PR		0.162	0.380	-1.509	0.376	-1.093	0.257

To get more insight in the data, traditional Cox models are fitted with the fixed covariates of Table A.6 and time-dependent covariates $\text{AGvHD}(t) = \mathbf{1}\{\text{AGvHD before } t\}$, $\text{recent} - \text{AGvHD}(t) = \mathbf{1}\{\text{AGvHD in last month before } t\}$, $\text{PR}(t) = \mathbf{1}\{\text{platelet recovery before } t\}$ and finally $\text{recent}-\text{PR}(t) = \mathbf{1}\{\text{platelet recovery in last month before } t\}$. The covariates $\text{recent} - \text{AGvHD}(t)$ and $\text{recent} - \text{PR}(t)$

are introduced to capture time-varying effects of these time-dependent covariates. Note carefully that the coding of for instance AGvHD(t) and recent – AGvHD(t) is such that at the time of AGvHD, both AGvHD(t) and recent – AGvHD(t) change value from 0 to 1, and one month later recent – AGvHD(t) changes value back from 1 to 0. Thus, recent – AGvHD(t) serves as a contrast between the effect of AGvHD in the first month and later on.

Table 9.1 shows the results of these Cox models. It shows that the covariates have a bigger effect on the hazards of death than on the hazards for relapse. Moreover, platelet recovery has a big immediate protective effect, that dies out later on, while the deleterious effect of AGvHD is more stable.

For further understanding of the data it is also relevant to see what the effect of the covariates is on the incidence of AGvHD and platelet recovery. As mentioned in Section A.6 both events happen very early in the follow-up. The effects in a Cox model for both transitions are shown in Table 9.2. Noteworthy is the big effect of year of transplantation on platelet recovery. It would be an interesting exercise to disentangle the causal effects of fixed and time-dependent covariates, but that is beyond the scope of this book. Finally, to complete the picture the baseline hazards

Table 9.2 *Cox regression for AGvHD and platelet recovery*

Covariate	Category	AGvHD		Platelet recovery	
		B	SE	B	SE
Donor recipient	No mismatch				
gender mismatch	Mismatch	-0.052	0.070	-0.065	0.068
GvHD prevention	No TCD				
	TCD	-0.274	0.076	-0.216	0.074
Year of	1985-1989				
transplantation	1990-1994	0.008	0.077	0.446	0.080
	1995-1998	-0.116	0.082	0.652	0.083
Age at transplant	≤ 20				
	20-40	0.155	0.076	-0.093	0.070
	> 40	0.156	0.090	0.035	0.083
AGvHD				0.563	0.129
recent – AGvHD				-0.607	0.140
PR		0.243	0.234		
recent – PR		0.053	0.242		

for all the Cox models (with all time-fixed covariates at their reference values and all time-dependent covariates set to 0) are shown in Figure 9.8.

The Cox models shown above (together with the baseline hazards) completely specify all transition hazards in the multi-state model of Figure 9.6. Note that these Cox models define a somewhat simplified multi-state model, where the effects of

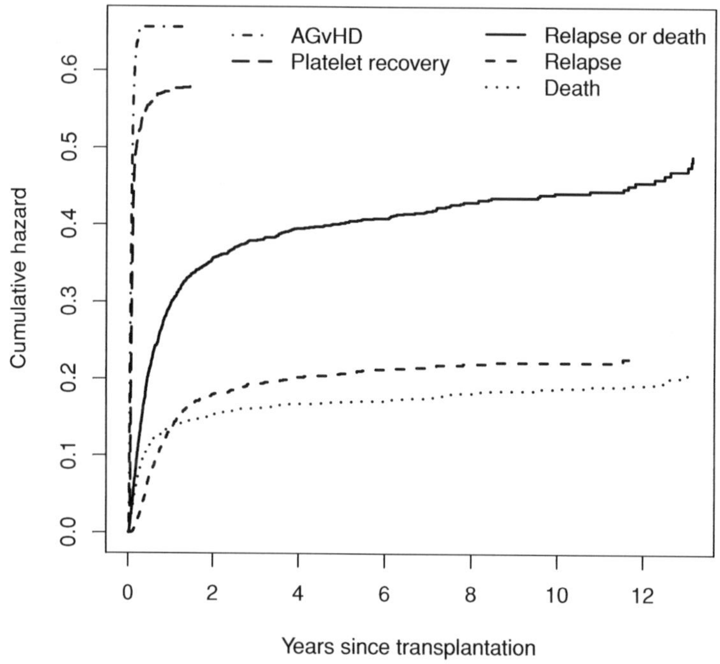

Figure 9.8 *Cumulative baseline hazards for AGVHD, platelet recovery, and for relapse and/or death*

covariates are assumed to be unchanged after the occurrence of the intermediate events acute Graft versus Host Disease and platelet recovery. Different covariate effects before and after AGvHD for instance can be accommodated by including interactions of the covariates with the time-dependent covariate AGvHD, but this is not pursued here. The Cox models allow dynamic predictions given any history at any moment. It should be observed that for such predictions explicit information is needed about the time an individual acquired AGvHD and/or platelet recovery. Such predictions are easy for those who experienced both AGvHD and platelet recovery, but are a bit tedious for those who have had no AGvHD or platelet recovery yet. The fact that a distinction is made between for instance recent-AGvHD and AGvHD more than one month ago implies that the resulting multi-state model is no longer Markov (given AGvHD for instance, the future depends on the past through whether this AGvHD was recent or not). No general software exists that can deal with this type of non-Markovian multi-state models. However, it is very well possible to approximate dynamic prediction probabilities by simulating many paths through the multi-state model.

Prediction by landmarking

In order to apply the landmark approach it is required to create landmark data sets for a grid of landmark prediction points. Since the intermediate events occur within one year, the interesting period for dynamic prediction is the first year. In the analysis below an equally spaced grid is used of 101 points ranging from $t_{\mathrm{LM}} = 0$ to $t_{\mathrm{LM}} = 1$. The endpoint for the analysis is relapse-free survival or cumulative incidence of relapse or death before relapse in a window of width $w = 5$ from the moment of prediction. A landmark supermodel with proportional baseline hazards (ipl* model) of Equation (7.3) was fitted. No interactions of the covariates with the landmark time-points were included in the model. The result of the analysis is shown in Table 9.3.

The effects of the fixed covariates are very similar to those in the Cox regression of Table 9.1. However, the covariates related to the intermediate events AGvHD and PR have a different interpretation. Now, "AGvHD" stands for having experienced acute graft-versus-host disease *before the landmark moment of prediction* t_{LM} and "recent-AGvHD" stands for having experienced AGvHD within one month before t_{LM}, with similar definitions for "PR" and "recent-PR." The effects of "recent-AGvHD" and "recent-PR" are now supposed to hold for the whole win-

Table 9.3 *Landmark supermodel with proportional baseline hazards for death and/or relapse, based on an equally spaced set of landmark time-points from 0 to 1 with distance 0.01 and window width $w = 5$ years*

Covariate	Category	Relapse		Death before relapse		Relapse or death	
		B	SE	B	SE	B	SE
Donor recipient	No mismatch						
gender mismatch	Mismatch	0.001	0.136	0.272	0.129	0.132	0.095
GvHD prevention	No TCD						
	TCD	0.262	0.138	0.233	0.130	0.248	0.096
Year of	1985-1989						
transplantation	1990-1994	-0.006	0.148	-0.561	0.140	-0.266	0.103
	1995-1998	-0.099	0.018	-0.374	0.148	-0.235	0.111
Age at transplant	≤ 20						
	20-40	-0.033	0.139	0.592	0.173	0.209	0.108
	> 40	-0.165	0.175	1.092	0.188	0.408	0.125
AGvHD		-0.104	0.092	0.736	0.076	0.348	0.059
recent – AGvHD		0.021	0.053	0.094	0.069	-0.036	0.045
PR		0.164	0.121	-0.336	0.127	-0.056	0.087
recent – PR		-0.023	0.054	-0.176	0.071	-0.130	0.043
$\theta(s)$	s	-1.023	0.306	-4.036	0.414	-2.320	0.244
	s^2	-0.300	0.269	2.097	0.352	0.726	0.211

dow $[t_{\mathrm{LM}}, t_{\mathrm{LM}} + w]$. This explains why the effects of both "recent" terms are much smaller.

Actually, the same effects could be described as well by removing the "recent" terms and introducing interactions of AGvHD and PR with t_{LM}. The reason is that AGvHD and PR are both recent by definition for $t_{\mathrm{LM}} \approx 0$ and not recent anymore for $t_{\mathrm{LM}} \approx 1$ (because both mainly happen within the first half year). For example, the model for relapse-free survival (last column in Table 9.3) implies that the effect for AGvHD increases from $0.348 - 0.036 = 0.312$ at $t_{\mathrm{LM}} = 0$ to 0.348 at $t_{\mathrm{LM}} = 1$ and the effect for PR changes from $-0.056 - 0.130 = -0.186$ to -0.056. Similar effects are indeed observed if a model with linear interactions is used. Since the model with "recent" terms is closer to the disease process, that model is preferred in the sequel.

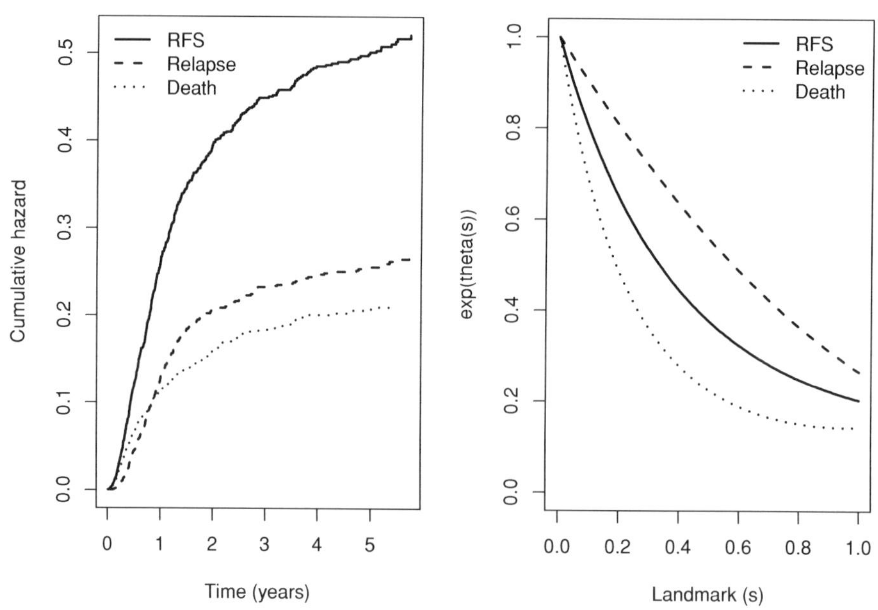

Figure 9.9 *Baseline hazard and landmark effects in the proportional baselines landmark supermodel of Table 9.3*

Together with the baseline hazards, shown in Figure 9.9, the models of Table 9.3 allow prediction of relapse-free survival and cumulative incidences of relapse and death before relapse at $t = t_{\mathrm{LM}} + w$ for patients still alive in remission at $t = t_{\mathrm{LM}}$. Such predictions are shown in Figure 9.10 for relapse-free survival, for patients in the three age categories with "no mismatch," "no TCD," transplanted in 1995 or later, and for all combinations of AGvHD and PR information. Here, "recent AGvHD" means AGvHD within one month prior to the time of prediction, and "past AGvHD" means AGvHD more than one month prior to the time of predic-

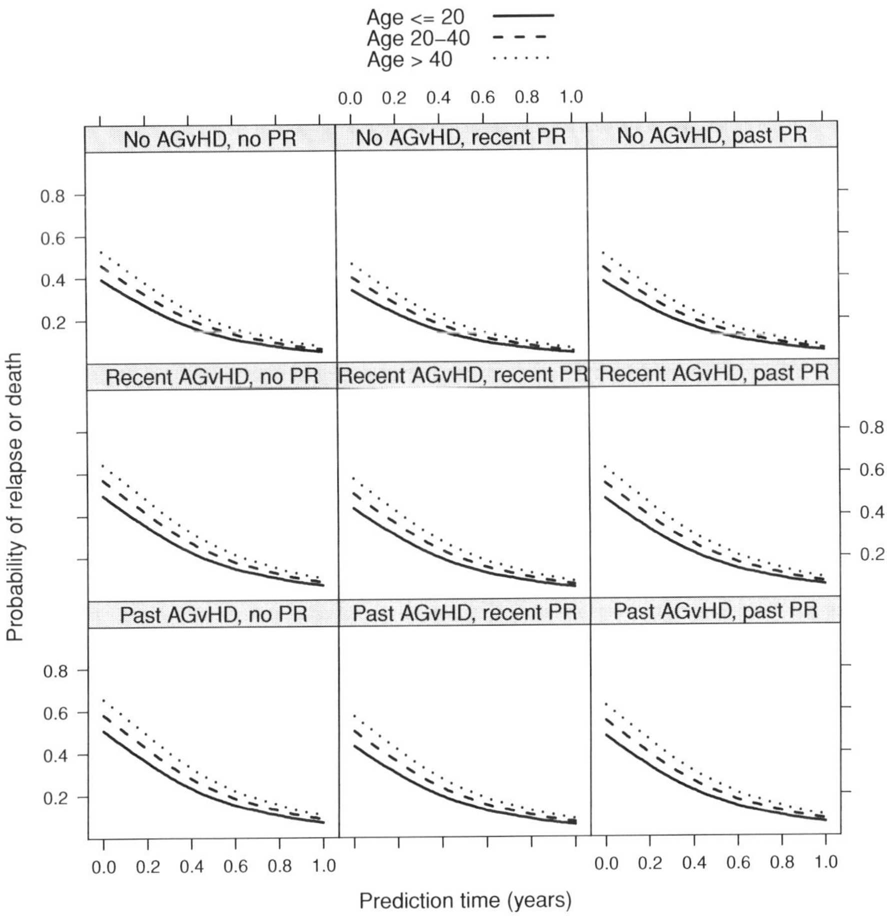

Figure 9.10 *Landmark dynamic fixed width ($w = 5$) prediction probabilities of relapse or death for all combinations of AGvHD and PR information*

tion, and similarly for "recent PR" and "past PR." Note that the landmark approach allows predictions also for combinations of AGvHD and PR that can in fact not occur. The landmark supermodel allows for instance dynamic predictions at $s = 0$ for "past AGvHD," while this is impossible in reality since "past AGvHD" by definition cannot be in effect until one month after transplantation. Similarly, "recent AGvHD" cannot be a present state for $s > 100$ days, since by definition AGvHD has to appear before 100 days post-transplant. In Figure 9.10 these predictions are shown nonetheless.

Figure 9.11 shows an example of dynamic fixed width prediction, based on the landmark model, for a patient in the oldest age group, over 40 years of age (and with reference values for the other covariates), who experienced platelet recovery at 30 days post-transplant, and AGvHD at 80 days post-transplant. This patient initially follows the predicted probabilities of the "No AGvHD, no PR" cell of Figure 9.10, until 30 days (0.08 years), when he/she switches to the predictions from the "No

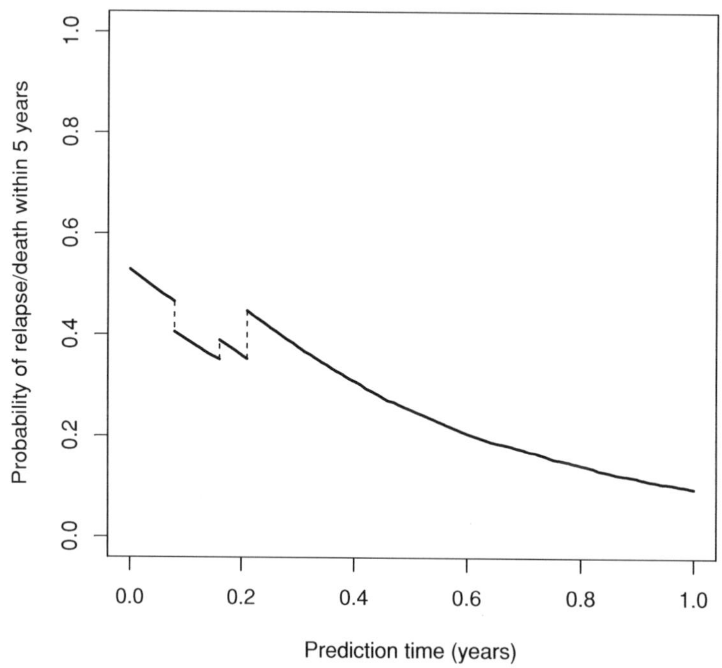

Figure 9.11 *Landmark dynamic fixed width ($w = 5$) prediction probabilities of relapse or death for a patient as intermediate clinical events occur*

AGvHD, recent PR" cell. One month later (at 0.17 years), the platelet recovery is no longer recent, so he/she switches again, now to the predictions from the "No AGvHD, past PR" cell, until at 80 days (0.22 years), the AGvHD occurs. From that time on, until one month later (at 0.30 years), the predictions from the "Recent AGvHD, past PR" cell are followed, until, finally, the patient switches to the "Past AGvHD, past PR" one month later (this last switch is not visible in Figure 9.11 since the dynamic predictions from "Recent AGvHD, past PR" and "Past AGvHD, past PR" are very similar).

Prediction by simulation from the multi-state model

Dynamic prediction probabilities of relapse or death, similar to those of Figure 9.10, now based on simulations from the Cox models of Tables 9.1 and 9.2, are shown in Figure 9.12. For this purpose 100,000 paths were generated through the multi-state model, starting at the transplant state at $t = 0$. For any prediction time point s and combination of AGvHD and PR status, to be used as starting point for the prediction, those paths were selected that satisfied the constraints defined by that combination of AGvHD and PR status at s. Subsequently, from this subset, the relative frequencies of paths within the subset that corresponded to a relapse or

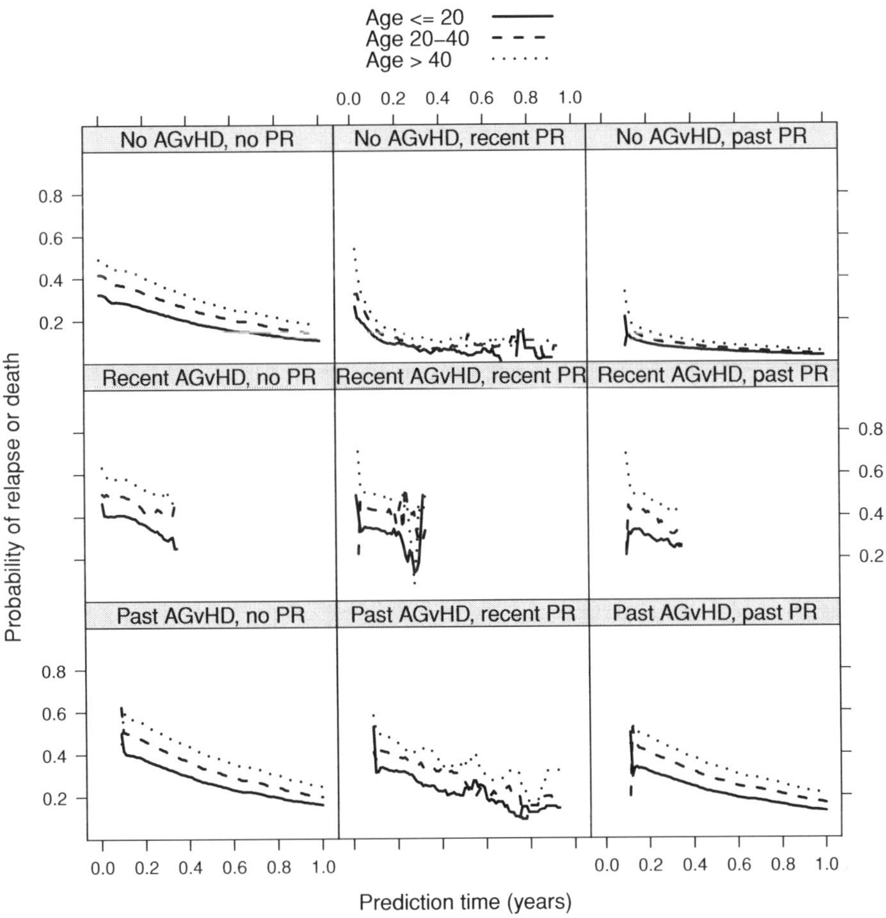

Figure 9.12 *Dynamic fixed width ($w = 5$) prediction probabilities of relapse or death for all combinations of AGvHD and PR information, simulated from the time-dependent Cox models of Tables 9.1–9.2*

death within 5 years after the prediction time point defined the dynamic fixed width predictions.

Comparing the two approaches

Comparing the predictions of the landmark approach of Figure 9.11 with those based on the multi-state model approach of Figure 9.12, a number of issues become apparent.

First, the simulation approach shown in Figure 9.12 does not allow prediction in situations that cannot occur in reality. Such situations are simply not present in the "database" containing the 100,000 simulated paths from which the predictions are obtained. For example, "Recent AGvHD, PR" cannot occur before one month, since "PR" means PR more than one month before the prediction time point, and it cannot occur later than 100 days plus one month, since AGvHD by definition occurs before 100 days, and hence one month later AGvHD can no longer be recent.

A second, related point is that at the boundaries of where combinations of (recent) AGvHD and/or (recent) PR can occur, these combinations will be quite rare, and as a result predictions based on simulations from the multi-state model will be erratic. Figure 9.13 shows the proportion of the 100,000 sampled paths that could be assigned for each of the combinations of AGvHD and PR over time; multiplying these with 100,000 gives the effective sample size on which the predictions of Figure 9.12 are based. Clearly, the predictions for "No AGvHD, no PR" are quite

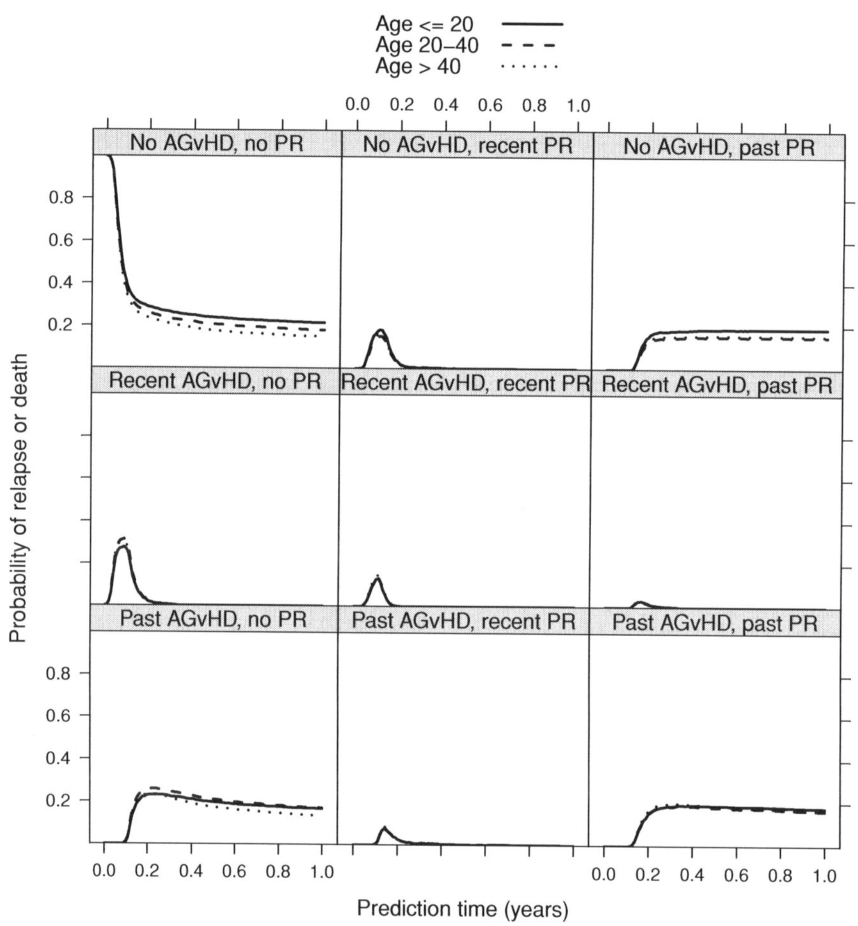

Figure 9.13 *Frequencies for all combinations of AGvHD and PR information, simulated from the time-dependent Cox models of Tables 9.1–9.2*

accurate (more precisely have small simulation error), but for instance the erratic behavior of the predictions from "Recent AGvHD, recent PR" (Figure 9.12) after 0.2 years is the result of considerable simulation error.

Other than that, the predictions obtained by landmarking and the multi-state model are qualitatively the same.

9.4 Additional remarks

Multi-state versus time-dependent covariates

The landmark approach treats prediction in multi-state models very similar to prediction based on biomarkers. Acute graft-versus-host disease and platelet recovery are treated as time-dependent covariates AGvHD(t) and PR(t), respectively. The landmark uses the history of these "biomarkers" up to the moment of prediction. The whole set-up is not different from the modeling in Chapter 8. This flexibility is an important advantage of the landmark approach. It has no problem with combining biomarkers like LWBC(t) in Chapter 8 with multi-state information like AGvHD(t) and PR(t) in this chapter. Developing joint models for data with various types of time-dependent information gets quite complicated and getting predictions out of such models even more so.

Landmarking and competing risks

Table 9.3 specifies landmark models for the cause-specific hazards of relapse and death before relapse, respectively. These could be used to obtain (sliding) dynamic predictive models for the cumulative incidences of relapse and death before relapse. In theory this is easy, but in practice it could be a bit cumbersome. Direct models for cumulative incidence functions are given by the methodology of Fine & Gray (1999). Their model could be fitted for the landmark data set at each prediction time point as discussed in Cortese & Andersen (2010). Alternatively, the pseudo-value approach of Andersen and Klein could be applied. However, it is not quite clear how to combine different landmark data sets into a "super data set" to be analyzed by "super models."

Chapter 10

Dynamic prediction in chronic disease

10.1 General description

> *"Chronic diseases are diseases of long duration and generally slow progression. Chronic diseases, such as heart disease, stroke, cancer, chronic respiratory diseases and diabetes, are by far the leading cause of mortality in the world, representing 60% of all deaths."*

This quote comes from the website of the WHO, the World Health Organization. It gives a definition of chronic disease and an impression of the importance of chronic diseases in health care.

This chapter is dedicated to dynamic prediction models in chronic disease. Such models are important for treatment choice and patient counseling and can also be helpful in understanding the time course of the disease. In a simplified model of chronic disease different stages of the disease can be defined $S_0, S_1, .., S_k$ where $S_0 =$ "No disease," $S_{k-1} =$ "End-stage disease" and $S_k =$ "Death." This could be seen as a special case of a multi-state model, where the states can be ordered according to the severity of the disease coming with worsening prognosis. In the simplest (and most pessimistic) situation a patient progresses through the consecutive stages to die eventually. Actually, this might not be true for two reasons. The first, positive, reason is that the patient might respond to treatment which will bring her/him back to a less severe stage or even to $S_0 =$ "No disease." The second, negative, reason is that the patient progresses so quickly that intermediate stages are not observed.

In theoretical statistical terms the individual disease process is a discrete stochastic process that "stops" when stage $S_k =$ "Death" is reached. The history of a patient can be characterized by the moments of transitions from one stage to another stage. Unfortunately, the process cannot be observed continuously. Transitions might go unnoticed since there is only information available about the current stage at control visits to the clinic.

Dynamic prediction in general terms attempts to predict the future course of the disease conditional on the current history. A full scale prediction would require a complete probabilistic model of the disease. Developing such a model would require a data set with accurate long term disease histories for a large representative group of patients. The availability of such data sets is utopian in most countries and even if they are available, they will be historic data sets by definition and their relevance for dynamic prediction of new patients might be questionable. That does not mean that dynamic prediction is hopeless. However, the prediction goals should be modest and realistic. A condition is that the question could be answered directly from the data and does not rely on extrapolation beyond the horizon of the available data. Landmarking as described in Chapters 8 and 9 can be a very useful tool to answer questions like, *"What is the probability of still being alive 10 years from now?"* or *"What is the probability of no further progression in the next 5 years?"* The question, *"What is the probability that the patient is cured?"* does not classify as "realistic" because it extrapolates beyond the horizon of available data. Such a question can only be answered on the basis of so called cure models that implicitly makes such extrapolations. See Sections 5.2 and 6.2 for definition and discussion of cure models.

In this chapter it will be explored to what extent landmarking can be used for dynamic prediction in chronic disease. No attempt will be made to do this in general abstract terms. The approach will be to demonstrate the possibilities by an in-depth analysis of Data Set 5, the EORTC breast cancer data. The important feature of that data set is that it contains follow-up information on local recurrence and distant metastasis, which can be seen as the next two stages in breast cancer and even in cancer in general.

10.2 Exploration of the EORTC breast cancer data set

Information on the data set can be found in Section A.5, which gives more detail on the survival and censoring function, distribution of the fixed covariates and their univariate effects and some information on the prevalence of local recurrence (LR) and distant metastasis (DM).

A first analysis using only the covariates at baseline is given in Section 2.4 including a Cox model in Table 2.1 and dynamic prediction in Figure 2.6. Multi-state model graphs for this data are shown in Figures 9.4 and 9.5 without further elaboration.

Role of LR and DM

The first step in the further exploration of the data is to get more insight in the effect of LR and DM on survival. For that purpose counts of the observed number of transitions in Figure 9.5 are shown in Table 10.1. One striking observation is that of the 729 patients in the data set that died 622 (85%) had DM. The outlook after diagnosis of DM is very grim but not completely hopeless as demonstrated in Figure 10.1. The median survival after DM is about 2.5 years. In this graph time is measured from the moment of DM (clock reset). That is a sensible thing to do, but for the quantification of the effect of DM and for further modeling it is convenient to stick to the original time scale of time since surgery. To quantify the effect of DM, a Cox model can be used in which the effect of the time-dependent covariate

Table 10.1 *Counts of the events in Figure 9.5*

First event	n	Second event	n	Third event	n
Censored	1542				
LR	221	Censored	116		
		DM	82	Censored	24
				Death	58
		Death	23		
DM	739	Censored	243		
		LR	40	Censored	12
				Death	28
		Death	456		
LR+DM	101	Censored	21		
		Death	80		
Death	84				

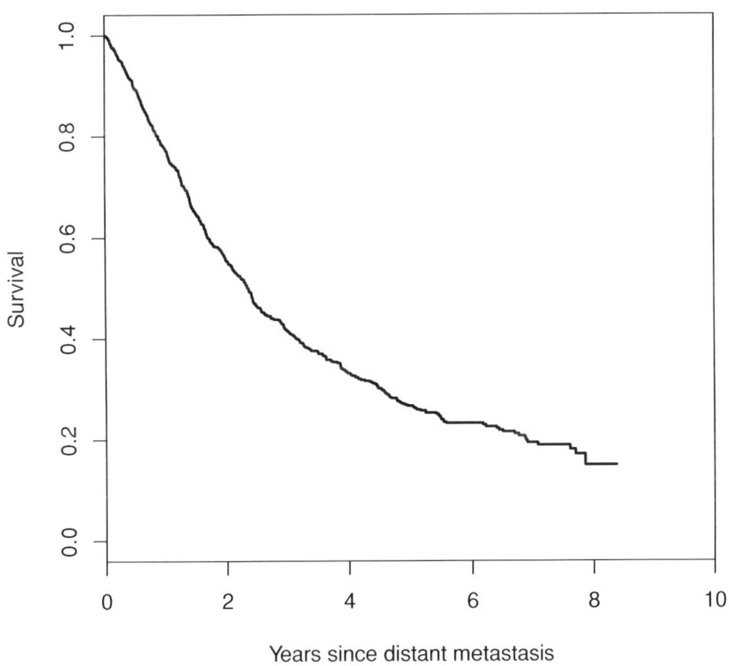

Figure 10.1 *Survival after distant metastasis*

$\text{DM}(t) = \mathbf{1}\{t_{\text{DM}} < t\}$, with t_{DM} thc moment of diagnosis of DM, is allowed to depend both on t_{DM} and the current time t. Some care is needed because there might be confounding between t_{DM} and t due to the high mortality after t_{DM}. An analysis with simple linear terms shows a slow decrease of the $\text{DM}(t)$ effect ($\hat{\beta} = -0.091$, SE $= 0.035$) over time t and no significant effect of t_{DM}.

Next, the role of LR is explored defining $\text{LR}(t) = \mathbf{1}\{t_{\text{LR}} < t\}$ with t_{LR} the moment of diagnosis of LR. When fitting a Cox model for the time-dependent covariates $\text{LR}(t)$ and $\text{DM}(t)$ and their interaction, the finding is first of all that LR has a strong effect as well, although not as big as the effect of DM. Moreover, there appears to be a very strong mutual interaction of $\text{LR}(t)$ and $\text{DM}(t)$. The effect of LR only ($\text{LR}(t) = 1, \text{DM}(t) = 0$) is much bigger than the added effect of LR when DM is present. Moreover, there is significant dependence of all effects on current time t. This is quantified in Table 10.2. There appeared to be no significant interaction with t_{LR} or t_{DM} (data not shown).

Table 10.2 *Effects of time-dependent covariates in a Cox model*

Covariate	B	SE	B	SE
LR only	2.159	0.237	3.566	0.470
DM only	4.044	0.122	4.587	0.237
LR and DM	4.553	0.137	5.115	0.275
(LR only) * t			-0.281	0.088
(DM only) * t			-0.113	0.039
(LR and DM) * t			-0.117	0.045

To complete the picture of the role of DM and LR it is of importance to explore how they affect each other. There is an outspoken effect of $\text{LR}(t)$ on the hazard for DM. The estimated coefficient in the Cox model is $\hat{\beta} = 1.260$ with SE $= 0.119$. In the other direction there is a smaller, but significant effect of $\text{DM}(t)$ on the hazard for LR. The statistics are $\hat{\beta} = 0.577$ with SE $= 0.175$. In both analyses the simultaneous occurrence of DM and LR is considered as a different type of event and handled as censored data as far as the effect of LR on DM and vice versa is concerned.

In summary, this exploratory analysis warrants a representation of breast cancer as a chronic disease with stages LR and DM in that order. To be more precise, the following stages can be discerned are

S_1: no LR and no DM

S_2: LR, no DM

S_3: DM (irrespective of LR status)

S_4: Death

The effect of being at stage 2 or 3 is well described in a Cox model with time-dependent covariates showing slowly varying effects as in Table 10.2. It should be stressed that the LR and DM stand for ever having experienced LR or DM respectively. Therefore the stages are irreversible by definition. If it were possible to assess cure of LR or DM, the situation would be completely different. Similarly, there is no S_0 stage because there is no objective way of assessing cure from breast cancer.

Role of the fixed covariates

A first insight in the effects of the fixed covariates was already given in the Cox model for survival of Table 2.1. In that model, age is categorized in two groups, $<$ 50 years and $>=$ 50 years. Such a categorization can loose important information. Luckily the age at diagnosis is available in the data set as well. The histogram is shown in Figure 10.2. Notice the age limit for the clinical trial where the data are coming from. Some exploration of the data showed a U-shaped effect of age that is better captured by a quadratic relation. A similar U-shaped effect of age has been observed in Sauerbrei (1999).

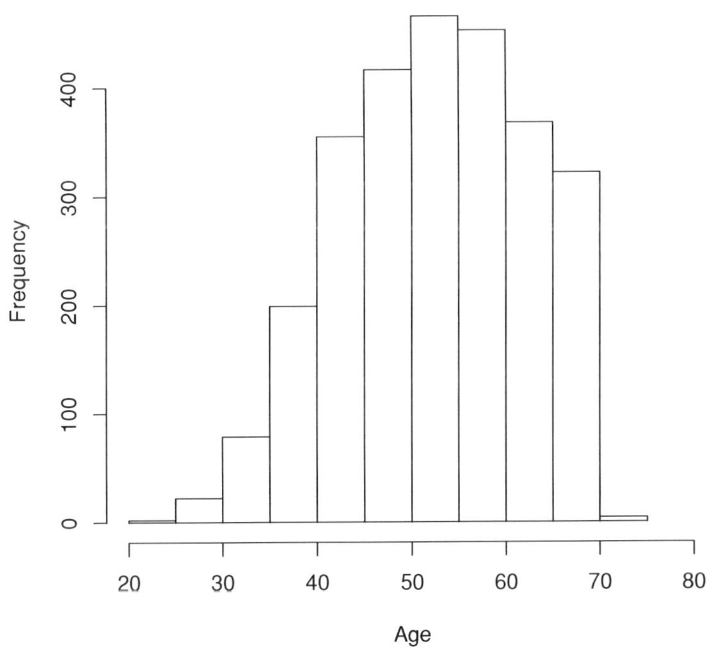

Figure 10.2 *Histogram of age*

Table 10.3 shows the effects in a Cox model for survival allowing for time-varying effects. A stepwise procedure showed significant time-varying effects of tumor size and nodal status. Unfortunately in this data set information on tumor size and nodal status has already been categorized and the raw data are not readily available; relevant information might be lost. The analysis shows that the effects of tumor size and nodal status decay at the same rate, in the sense that the ratio $B_1/B_0 \approx -0.3$ for all of them. Some type of reduced rank model as discussed in Section 6.3 could be useful, but will not be pursued here.

Table 10.3 *The Cox model for overall survival in the EORTC breast cancer data (Data set 5); CT = chemotherapy, RT = radiotherapy, Mast = mastectomy, BCS = breast conserving surgery, AgeC = $(Age-50)/10$, AgeC2 = AgeC^2, B: fixed effects, $B_0 + B_1 * \ln(1+t)$: time-varying effects; significant effects ($P < 0.05$) are in* ***bold***

		Time-fixed		Time-varying			
				Constant		$\ln(1+t)$	
Covariate	Category	B	SE	B_0	SE	B_1	SE
Type of	Mast, RT+						
surgery	Mast, RT-	**0.255**	0.110	**0.250**	0.110		
	BCS	-0.083	0.097	-0.082	0.097		
Tumor size	< 2 cm						
	2 - 5 cm	**0.353**	0.096	**0.857**	0.377	-0.278	0.200
	> 5 cm	**0.900**	0.153	**2.236**	0.518	**-0.797**	0.298
Nodal status	Negative						
	Positive	**0.949**	0.109	**1.925**	0.320	**-0.556**	0.168
Adjuvant	No						
CT	Yes	**-0.401**	0.121	**-0.385**	0.121		
Tamoxifen	No						
	Yes	-0.120	0.113	-0.107	0.112		
Perioperative	No						
CT	Yes	-0.119	0.074	-0.118	0.074		
AgeC		-0.062	0.047	-0.058	0.047		
AgeC2		**0.068**	0.034	**0.067**	0.034		

The question can be raised whether the effects of the fixed covariates arise through their effects on LR and DM. Especially effects of the covariates on the occurrence of DM can lead to a big indirect effect. To study this question in more detail, the effects of the covariates on the hazards for LR and DM are obtained through appropriate Cox models for the hazards of DM, LR and the simultaneous occurrence of LM and DR, respectively. The results are shown in Table 10.4.

Next, Table 10.5 shows the effects on the hazard for death in the three disease stages: S_1 = no LR, no DM, S_2 = no LR, no DM and S_3 = DM. In this model there is a common baseline hazard for death with effects for the fixed covariates, stage and the interaction of the fixed covariates with stage, if present. In a forward stepwise procedure, only tumor size and age showed a significant interaction with stage.

Table 10.4 *The effects on the hazards for LR, DM and LR+DM, respectively; significant effects ($P < 0.05$) are in* **bold**

		LR		DM		LR+DM	
Covariate	Category	B	SE	B	SE	B	SE
Type of	Mast, RT+						
surgery	Mast, RT-	**0.511**	0.225	-0.082	0.110	0.107	0.299
	BCS	**0.556**	0.196	-0.111	0.092	-0.077	0.254
Tumor size	< 2 cm						
	2 - 5 cm	-0.022	0.138	**0.329**	0.087	**0.939**	0.296
	> 5 cm	0.509	0.285	**0.826**	0.148	**0.987**	0.485
Nodal status	Negative						
	Positive	0.020	0.187	**0.526**	0.107	**0.795**	0.299
Adjuvant	No						
CT	Yes	-0.249	0.212	**-0.264**	0.121	-0.215	0.311
Tamoxifen	No						
	Yes	0.262	0.201	0.001	0.112	0.024	0.307
Perioperative	No						
CT	Yes	**-0.324**	0.125	-0.013	0.070	**-0.448**	0.204
AgeC		**-0.407**	0.072	-0.059	0.044	-0.245	0.134
AgeC2		0.055	0.056	0.050	0.032	-0.080	0.100
DM		**0.568**	0.176				
LR				**1.307**	0.121		

It is not easy to summarize the whole picture. Nearly all fixed covariates show a mixed bag of direct and indirect effects. Tamoxifen is the only covariate that does not show any effect at all, while positive nodal status is the only factor that consistently implies bad prognosis. Also noteworthy is the observation that the effects of almost all covariates are no longer significant for overall survival from Stage 3 (the significant effect of "Stage" for $S3$ in Table 10.5 merely indicates that the baseline death rate increases considerably after distant metastasis). This finding, that the prognosis for a breast cancer patient is very grim after distant metastasis, irrespective of the value of the covariates, was also found in Putter et al. (2006), where the same data were analyzed through a multi-state model. For a more extensive interpretation more clinical expertise is required. However, the age effects need some statistical explanation because the coefficients in the tables have to be translated into the effect of age with respect to the chosen baseline if Age = 50. This is shown in Figure 10.3. The left plot of Figure 10.3 shows the log hazard ratios of age with respect to Age = 50 for LR, DM, and LR+DM of Table 10.4, while the right plot of Figure 10.3 shows the log hazard ratios of age for death in the three disease stages (Table 10.5). In a nutshell, the disease is more aggressive for younger patients, while older patients show a higher frailty.

Table 10.5 *The effects of the covariates in a Cox model for survival when taking LM and DR into account. Significant effects ($P < 0.05$) are in* **bold**

Covariate	Category	B	SE				
Type of surgery	Mast, RT-						
	Mast, RT+	**0.348**	0.108				
	BCS	-0.076	0.097				
Nodal status	Negative						
	Positive	**0.691**	0.102				
Adjuvant CT	No						
	Yes	-0.118	0.114				
Tamoxifen	No						
	Yes	-0.084	0.100				
Perioperative CT	No						
	Yes	-0.026	0.075				
		S_1		S_2		S_3	
Covariate	Category	B	SE	B	SE	B	SE
Tumor size	< 2 cm						
	2 - 5 cm	-0.069	0.252	**2.163**	1.031	-0.009	0.106
	> 5 cm	0.665	0.398	**3.478**	1.103	0.095	0.169
AgeC		**0.277**	0.121	0.143	0.238	-0.002	0.048
AgeC2		**0.382**	0.097	-0.206	0.204	-0.013	0.040
Stage				1.022	1.043	**4.759**	0.269

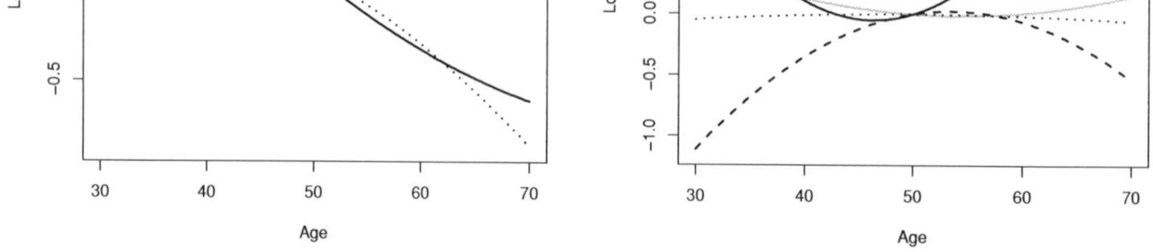

Figure 10.3 *Left: Effect of age on the hazards for LR, DM and LR+DM. Right: Effect of age on the hazard for death in the model without stage (right hand side of Table 10.3) and in the different stages*

10.3 Dynamic prediction models for breast cancer

The Cox models obtained in Section 10.3 could be used to define a multi-state model. Within such a multi-state model dynamic predictions can be obtained as discussed in Sections 9.2 and 9.3. This is quite cumbersome and could be prone to biased predictions caused by violation of the model assumptions. Here the focus is the landmark approach because of its robustness and the ease of obtaining models and predictions. At each landmark time-point, the current stage is an important predictor together with age, the tumor and treatment related covariates and the moment of prediction. As observed in Table 10.5, there might be important interaction between the stage and these predictors. For this reason, separate landmark models are developed for the different stages. As can be seen from Figure A.8, the number of patients in stage S_2 is quite small during the whole follow-up. Nevertheless, this is considered as a separate group because a preliminary analysis showed that S_2 behaves very much like S_1 in the early part of the follow-up and very much like S_3 later in the follow-up.

The endpoints to be considered are overall survival (OS) for patients in all stages and disease-free survival (DFS) for those in stage S_1. For patients in stage S_2 metastasis-free survival might be of some hypothetical interest, but this will not be pursued. Landmark time points are selected from $t_{\mathrm{LM}} = 0$ up to $t_{\mathrm{LM}} = 8$ years with steps of $\Delta_{t_{\mathrm{LM}}} = 1/12$. The width of the prediction window is taken to be $w = 5$.

A preliminary landmark analysis with the covariates as main effects showed that only type of surgery, tumor size, nodal status and age had significant effects for any of the four landmark models, so adjuvant chemotherapy, perioperative chemotherapy, and tamoxifen were no longer considered. Subsequently, after main effects of type of surgery, tumor size, nodal status and age were entered, interactions between t_{LM} and the covariates were added in a forward selection procedure, using $P < 0.05$ as entry criterion. Table 10.6 shows the results for the ipl* model, with $\theta(s) = \gamma_1(s/8) + \gamma_2(s/8)^2$. Higher tumor size and positive nodal status all imply bad prognosis for overall survival and disease-free survival, irrespective of the stage. In contrast, the effect of type of surgery appears to be different for overall survival from stage 2, compared to overall survival from the other stages and compared to DFS from stage 1. This could be explained by the fact that the type of surgery is primarily focused on local control; after a local recurrence has occurred (stage 2), the effect of the primary treatment can change as a result. Also the effect of age seems to be different from stage 2, compared to the other landmark models, although this is hard to judge because of the presence of interactions with the landmark time points in some of the models. To complete the picture, Figure 10.4 shows the baseline cumulative hazards of the four models considered. No attempt has been made to smooth the baseline cumulative hazards.

Figure 10.5 shows dynamic fixed width ($w = 5$ years) prediction probabilities of death from stages 1, 2 and 3, and of recurrence and/or death from stage 1, obtained from the landmark models of Table 10.6. Three patients were chosen, with ages 30,

Table 10.6 *Landmark Cox regression models for overall survival from the three stages and for disease-free survival (DFS) from stage 1*

		Overall survival						DFS	
		Stage 1		Stage 2		Stage 3		Stage 1	
Covariate	Category	B	SE	B	SE	B	SE	B	SE
Type of	Mast, RT-								
surgery	Mast, RT+	0.483	0.132	-0.754	0.513	0.151	0.182	0.099	0.105
	BCS	-0.014	0.120	0.348	0.463	-0.235	0.153	0.030	0.090
Tumor size	< 2 cm								
	2 - 5 cm	0.270	0.117	0.213	0.334	0.316	0.363	0.259	0.079
	> 5 cm	0.800	0.191	2.198	0.470	1.014	0.530	0.718	0.147
Tumor size * s	2 - 5 cm					-0.083	0.064		
	> 5 cm					-0.248	0.108		
Nodal status	Negative								
	Positive	1.036	0.136	0.443	0.331	0.771	0.134	0.461	0.095
Nodal status * s	Positive	-0.086	0.039					-0.055	0.023
AgeC		-0.140	0.063	0.763	0.342	-0.068	0.063	-0.115	0.048
AgeC2		0.107	0.054	0.639	0.213	-0.005	0.057	0.068	0.042
AgeC * s		0.089	0.021	-0.172	0.096			0.024	0.012
AgeC2 * s		-0.006	0.017	-0.189	0.063			0.013	0.011
$\theta(s)$	$s/8$	-5.016	0.408	-8.607	1.742	-5.332	0.912	-3.919	0.228
	$(s/8)^2$	0.626	0.377	3.563	1.452	1.627	0.621	0.889	0.189

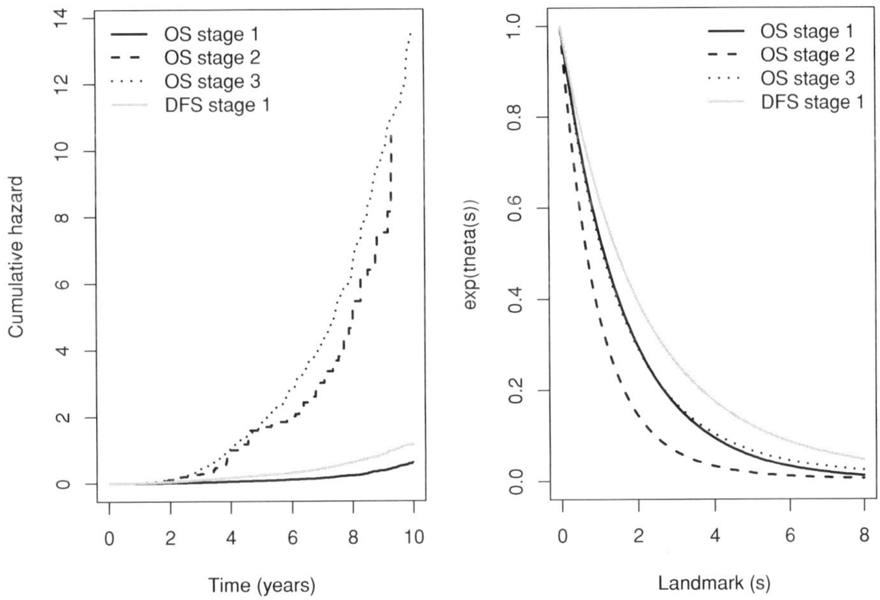

Figure 10.4 *Baseline cumulative hazards and landmark effects in the proportional baselines landmark supermodels*

50 and 70, each with relatively good prognosis: tumor size < 2 cm, negative nodal status, treated with mastectomy plus radiotherapy. From stage 1 and 2, the predicted probabilities of death and of recurrence and/or death are higher for the patient aged 70, compared to the patient aged 50. The patient aged 30 however, has a 5-year dynamic death probability which starts out as the worst of the three ages, but which decreases more rapidly and ends up as the best of the three ages. This phenomenon is probably due to a selection effect caused by the fact that recurrences are a mixture of recurrences caused by aggressive disease and by inadequate local treatment. It is well known that in breast cancer, young patients often present with aggressive disease (possibly because of genetic factors), see also Figure 10.3; those young patients with aggressive disease will have high progression and death rates, but the other young patients could be expected to live longer than their older counterparts. As a result, those that managed to survive the first couple of years have a better prognosis. It is evident from Figure 10.5 that all the 5-year dynamic prediction probabilities decrease as the time of prediction increases and that overall higher stages imply higher 5-year dynamic prediction death probabilities. For stage 3, the predicted 5-year death probabilities decrease to about 0.4, while for stages 1 and 2, they decrease to below 0.1.

Figure 10.6 shows similar plots, this time for three patients, aged 30, 50 and 70, with much worse prognosis, namely with tumor size > 5 cm and positive nodal sta-

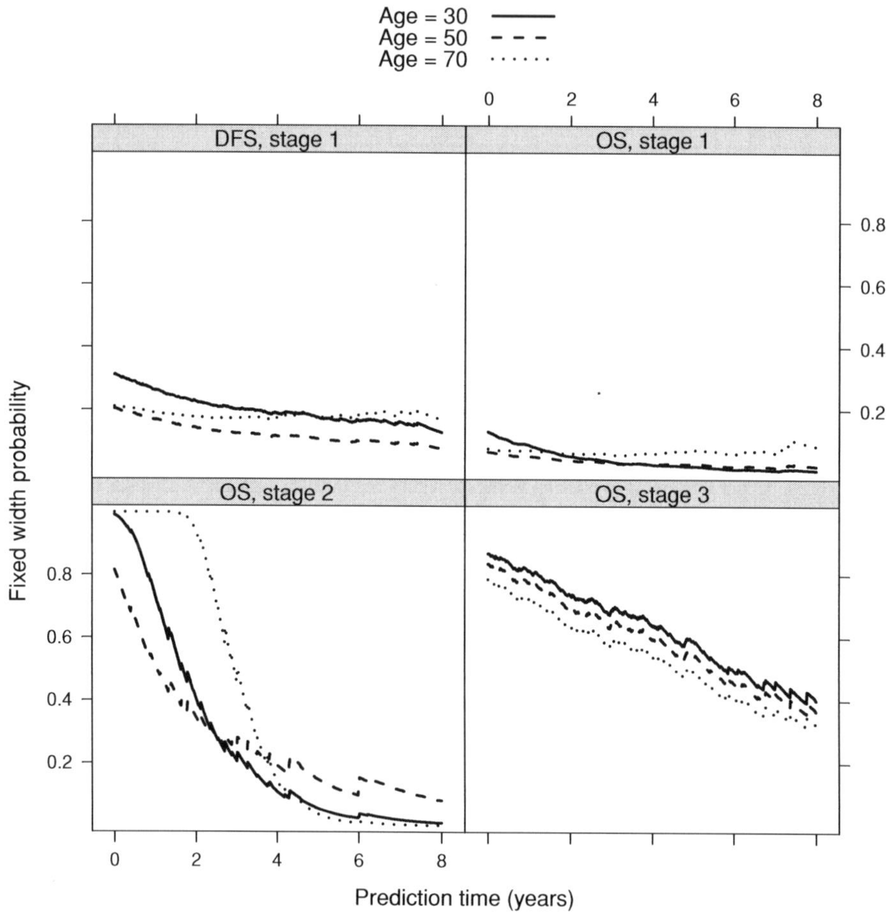

Figure 10.5 *Landmark dynamic fixed width ($w = 5$) prediction probabilities of death (OS) from stages 1, 2 and 3, and of recurrence and/or death (DFS) from stage 1, for a patient with relatively good prognosis*

tus, also treated with mastectomy plus radiotherapy. Generally, the dynamic death predictions are considerably higher in the beginning than those of the good prognosis patients, but at the end they are quite comparable.

10.4 Dynamic assessment of "cure"

In order to discuss the (im)possibilities of the assessment of cure, the upper two panels of Figure 10.5 are shown again on a probability scale from 0 to 0.3 in Figure 10.7. Very strictly speaking, the assessment "the patient is cured" could be translated into more statistical terms as "the probability of any disease related event is equal to zero." For the breast cancer example the disease related events are i) local recurrence, ii) distant metastasis or iii) death caused by the disease.

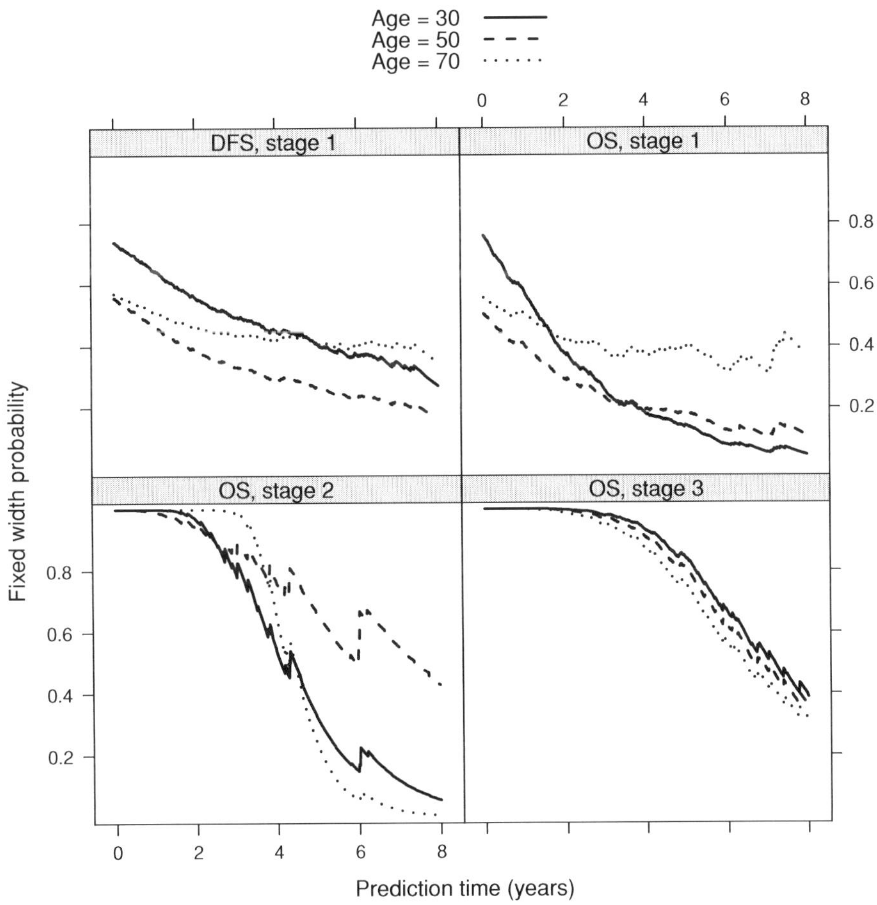

Figure 10.6 *Landmark dynamic fixed width ($w = 5$) prediction probabilities of death from stages 1, 2 and 3, and of recurrence and/or death from stage 1, for a patient with relatively bad prognosis*

There are a few observations to be made:

1. It is very hard to establish the cause of death for all patients in a large study and that piece of information is not available in the data analyzed in this book.
2. The strict definition involves the whole future life of a patient. To verify the statement, data with very long follow-up would be needed. Those are not available.
3. Absolute certainty can never be given. It is wiser to say that a probability is very small than to say that it is zero.

Therefore, a practical *statistical definition of cure* could be given as

Cured =   Probability of any disease related event within the next w years $\leq \varepsilon$

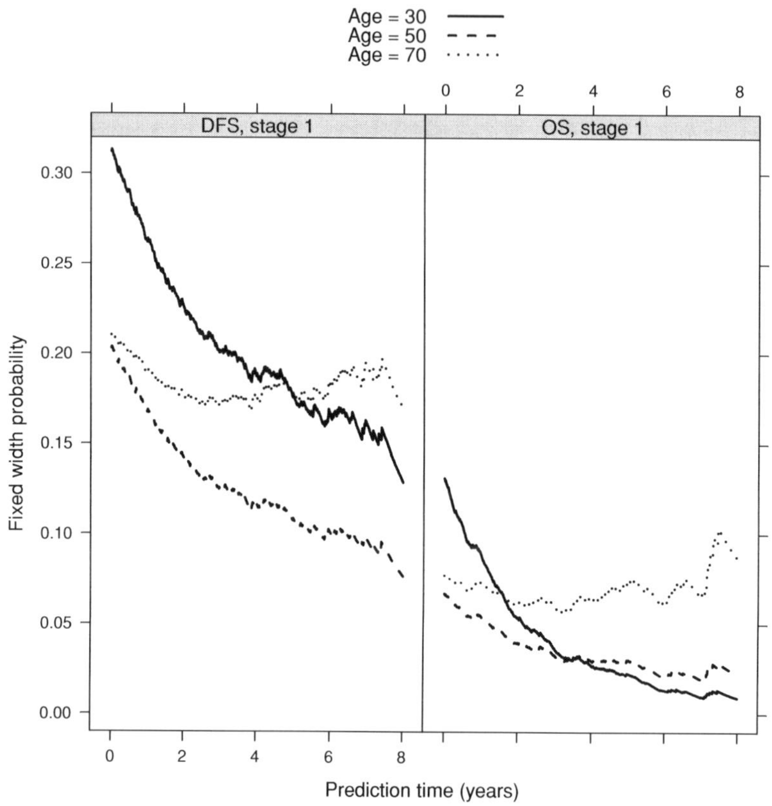

Figure 10.7 *Landmark dynamic fixed width ($w = 5$) prediction probabilities of death and of recurrence and/or death from stage 1, for a patient with relatively good prognosis*

Presumably, it would be reassuring enough for the patient if such a statement could be made for $w = 10$ and $\varepsilon = 1\%$. However, the majority of data sets will not allow such a wide window. In the breast cancer data considered here, a more realistic value is $w = 5$. Graphs like in Figures 10.5, 10.6 and 10.7 allow the assessment of similar predictive probabilities with $w = 5$ during the first eight years of the follow-up. The value of ε need not to be fixed beforehand. If a value has to be specified, a value of $\varepsilon = 5\%$ would be more realistic. For young patients the probability of death from other causes within the next 5 years is negligible. Hence, a curve like the one in the left hand panel of Figure 10.7 could be used to assess the "cured" status for younger patients. For the young patient shown in the curve the probability of no disease related event within the next 5 years falls sharply, but does not come near to an acceptable 5% level. For a Dutch patient of 70 years the probability of death from other causes within the next 5 years is in the order 15-20% (source: Dutch Central Bureau of Statistics, `statline.cbs.nl`) and the graph in the left panel of Figure 10.7 does not directly answer the "Am I cured?" question.

The definition above might be a bit too pessimistic in the sense that recurrence and metastasis can be treated and that the real question for the patient is not whether she is cured but whether she will live for another w years. That leads to the definition of

> *Survival Guarantee* =
> Probability of death within the next w years $\leq \varepsilon$

Those probabilities can be read from the right-hand side of Figure 10.7. For the patients of age $=$ 30 years and age $=$ 50 years with a good prognosis shown in Figure 10.5 these probabilities get reassuringly low after 4 years of follow-up, and even for the patients of the same age with a bad prognosis shown in Figure 10.6 those probabilities drop dramatically *if* the patient stays disease free.

For the older patients even these probabilities might be too pessimistic because they are inflated by the competing risk "death from other causes." That leads to the most optimistic definition of

> *Relative Survival Guarantee* =
> Probability of death within the next w years $\leq$ population mortality + ε

The population mortality could be obtained from external sources like the Dutch Central Bureau of Statistics. They show that for the 70 years old patient with good prognosis the predicted mortality as shown in Figure 10.7 is of the same order as the population mortality. For the patient of the same age with a bad prognosis, this is not true. Moreover staying disease free does not decrease mortality risk as quickly as for younger patients, but this might be explained by the increasing mortality within the window due to increasing age during follow-up.

A different approach to the same question could be obtained through the relative survival approach of Hakulinen & Tenkanen (1987), further developed among others by Estève et al. (1990) and Dickman et al. (2004). Such models write the hazards within the patients as

$$h(t|x) = h_{\text{disease}}(t|x) + h_{\text{population}}(t|\text{age}, \text{sex}) .$$

The population hazard is handled as an additive offset, while the disease specific hazard is modeled by a multiplicative Cox-type model. In the disease specific part the general covariate x can also include age and sex, if needed. Such models are a bit hard to fit, because of the condition that $h_{\text{disease}}(t|x) \geq 0$. The interesting paper by Janssen-Heijnen et al. (2007) gives conditional relative survival probabilities for different types of cancer, depending on the time since diagnosis. This is very close to the landmarking approach advocated in this chapter. However, building general landmark models around the relative survival model allowing general covariates in the disease specific part and not just age and sex, would require the development of new software. There is no direct need for such software, because the approach of this chapter can answer the question of "relative survival guarantee" within the setting of existing software.

10.5 Additional remarks

An interesting alternative for the modeling of chronic disease data is modeling through frailty models. Frailty models are a very useful tool in modeling recurrent events (see Hougaard 2000, Chapter 9), because they quantify the correlation within individuals and allow easy predictions by updating the frailty distribution given the history of a patient. Something similar can be done in chronic disease models. It is all a bit more complicated because the "consecutive events" are not copies of the same event as in "recurrent events" and patients with "consecutive events" might be rare. Interesting developments in this direction can be found in Rondeau et al. (2007) and Putter & van Houwelingen (2011). Fitting such models can lead to new insights in the data structure but it is not clear yet how easy it will be to obtain dynamic predictions if a more-dimensional frailty is needed and how other time-dependent information could be included in the prediction model.

Part IV

Dynamic prognostic models for survival data using genomic data

Chapter 11

Penalized Cox models

11.1 Introduction

In 1999, the Science paper by Golub et al. (1999) started a new era in the world of clinical diagnostics and prognosis by the introduction of high throughput data measured through the use of *microarrays*. These devices enable measuring tens of thousands of variables in one single experiment. Originally, micro-arrays were mainly used to measure *gene expression* levels in tissue or serum, but presently they are also used for measuring DNA content and DNA methylation. In the meantime new technology already has entered in the form of the so-called *proteomic data* produced by mass-spectrometry based techniques like MALDI-TOF. The pioneering paper breaking the ground for clinical applications of proteomics is the one by Petricoin et al. (2002). The focus in this part of the book will be on gene expression data, but the statistical methodology carries over to all type of high-dimensional data.

In general terms the situation can be described as having a large set of p covariates $X_1,...,X_p$ that can all be used to predict the survival outcome of interest. The number p is of the same order or even larger than the number of individuals in the data set. That makes it useless to fit a prognostic model like the Cox model because of the extreme overfitting. The regression parameters will explode, the model will fit the data perfectly, but prediction based on the model will be very poor. Some form of *tuning* of the model is needed. A short list of tuning methods is given below.

Method	Tuning parameter
Univariate selection	# "top genes"
Forward stepwise selection	# selected genes
Principal components regression	# principal components
Supervised principal components	# top genes, # principal components
Partial least squares	# components
Ridge regression	weight of the quadratic penalty
Lasso regression	weight of the absolute value penalty

The prognostic performance of these methods in cancer data sets has been compared in an extensive study by Bøvelstad et al. (2007). The classic approach of forward stepwise selection behaves quite poorly. It is too “greedy,” because it selects the most significant predictor in each step. Selection on univariate significance is more robust and easier to carry out. *Principal components regression* is a well-established method that takes the first k principal components of the set of predictors. It is quite stable, but might miss relevant information. *Supervised principal components regression* is a hybrid introduced by Bair & Tibshirani (2004) and Bair et al. (2006). Partial least squares (PLS) is an algorithmic method which is popular and successful in chemometrics. For details on the application of PLS in Cox regression see Nygård et al. (2008). Finally, ridge regression and lasso are both variants of the penalized likelihood approach.

Prognostic performance of each method can be controlled by optimal choice of the tuning parameter(s). To do so, some form of cross-validation is needed. Bøvelstad et al. (2007) consider “tuned” versions of Cox regression and use the cross-validated partial log-likelihood CVPL of Section 3.6 to compare different methods. Their conclusion is that ridge regression performs best among all methods and that lasso regression is best among the methods using a limited number of predictors. Both forms of penalized Cox regression will be discussed in the next section.

11.2 Ridge and lasso

Ridge

Ridge regression was introduced by Hoerl & Kennard (1970) as a modification of ordinary least squares (OLS) regression for continuous outcomes. Their purpose was to prevent degeneracy due to multi-collinearity of the predictors/covariates. Variants for logistic regression and Cox regression were introduced by le Cessie & van Houwelingen (1992) and Verweij & van Houwelingen (1994), respectively. An application for high-dimensional genomic data in survival data can be found in van Houwelingen et al. (2006).

Following the notation of Section 2.3, the log-likelihood for the ridge version of the Cox model is defined as

$$l_{\text{ridge}}(h_0,\beta) = l(h_0,\beta) - \frac{1}{2}\lambda \sum_{j=1}^{p} \beta_j^2 = l(h_0,\beta) - \frac{1}{2}\lambda \beta^\top \beta \,,$$

with $\lambda \geq 0$.

Adding a penalty on the regression parameters does not affect the maximization of the log-likelihood with respect to h_0 given the regression parameters. So, the baseline-hazard is still estimated by the Breslow estimator and the regression parameters are obtained by maximizing the *ridge partial log-likelihood*

$$\text{pl}_{\text{ridge}}(\beta) = \sum_{i=1}^{n} d_i \cdot \ln\left(\frac{\exp(x_i^\top \beta)}{\sum_{j\in R(t_i)} \exp(x_j^\top \beta)}\right) - \frac{1}{2}\lambda \beta^\top \beta \,.$$

The first and second derivative of the ridge partial log-likelihood are given by

$$\frac{\partial \mathrm{pl}_{\mathrm{ridge}}(\beta)}{\partial \beta} = \frac{\partial \mathrm{pl}(\beta)}{\partial \beta} - \lambda \beta = \sum_i^n d_i(x_i - \bar{x}_i(\beta)) - \lambda \beta \ ,$$

and

$$I_{\mathrm{pl}_{\mathrm{ridge}}}(\beta) = -\frac{\partial^2 \mathrm{pl}_{\mathrm{ridge}}(\beta)}{\partial \beta^2} = I_{\mathrm{pl}}(\beta) + \lambda I_p \ ,$$

with I_p the $p * p$ identity matrix.

Adding the quadratic penalty ensures the existence of a unique maximizer of the ridge partial log-likelihood. Computing the ridge-Cox estimator is as easy as computing the ordinary Cox estimator. The log-likelihood is concave and the optimum can be found by Newton-Raphson. It needs only a simple adjustment to turn Cox regression software into ridge-Cox software.

Notice that the ridge partial log-likelihood is not invariant under rescaling of the predictors. If the predictors are all on a similar scale the quadratic penalty for the original β_j's is OK. If not, it is recommended to standardize the x's by using $x^* = (x - \bar{x})/\mathrm{sd}(x)$, as done in the original work of Hoerl and Kennard.

Lasso

Lasso regression was introduced in Tibshirani (1996) and adapted for Cox regression in Tibshirani (1997). The difference with ridge regression is that the quadratic penalty is replaced by the absolute value penalty leading to

$$\mathrm{pl}_{\mathrm{lasso}}(\beta) = \sum_{i=1}^n d_i \cdot \ln\left(\frac{\exp(x_i^\top \beta)}{\sum_{j \in R(t_i)} \exp(x_j^\top \beta)}\right) - \lambda \sum_{j=1}^p |\beta_j| \ .$$

The lasso partial log-likelihood is concave as well, but the maximum cannot be found by Newton-Raphson because the first and second derivative of the lasso partial log-likelihood are only defined if $\beta_j \neq 0$. The first derivative is given by

$$\frac{\partial \mathrm{pl}_{\mathrm{lasso}}(\beta)}{\partial \beta} = \frac{\partial \mathrm{pl}(\beta)}{\partial \beta} - \lambda \mathrm{sign}(\beta) = \sum_i^n d_i(x_i - \bar{x}_i(\beta)) - \lambda \mathrm{sign}(\beta) \ ,$$

where $\mathrm{sign}(\beta_i) = 1$ if $\beta_i > 0$ and $\mathrm{sign}(\beta_i) = -1$ if $\beta_i < 0$ and not defined if $\beta_i = 0$. Where it exists, the negative second derivative of the lasso partial log-likelihood equals $I_{\mathrm{pl}}(\beta)$.

Although the partial log-likelihood looks very similar, the lasso estimators behave quite differently. Many regression coefficients are estimated to be zero. This might help interpreting the predictor. The price is that computing time to obtain a lasso estimator depends strongly on λ, in contrast to the ridge estimator. See Figure 3.11 of Hastie et al. (2009) for a geometrical comparison of ridge and lasso and Goeman (2010) for computational aspects of lasso-Cox.

Obtaining the optimal weight λ_{opt}

In van Houwelingen et al. (2006), cross-validated partial log-likelihood is advocated as the way to obtain λ_{opt}. This objective is also used in Bøvelstad et al. (2007) and Goeman (2010) and incorporated in Goeman's **penalized** R package. The idea is that λ_{opt} should optimize the internal prognostic performance of the resulting estimator. Cross-validation is essential in doing this. Leave-one-out cross-validation is attractive because it stays closest to validation on new data, but it might be time-consuming especially for lasso-Cox. An alternative could be k-fold cross-validation ($k = n$ for leave-one-out cross-validation) as applied in Bøvelstad et al. (2007) and incorporated in the **penalized** package. CVPL as a performance measure could be replaced by other prediction measures as discussed in Sections 3.5 and 3.6. However, concordance measures like Harrell's C-index are not appropriate because they only measure the discrimination and not the calibration.

It is common practice to evaluate the internal performance of the procedure at λ_{opt} by using the same cross-validation "folds" and cross-validation based regression parameters. This is slightly optimistically biased because *all* data are used to obtain λ_{opt}. Strictly speaking double cross-validation would be needed that uses different values of λ_{opt} for each cross-validation fold, but that is getting too demanding on computing time. The difference between single and double cross-validation will depend on how smoothly the cross-validated performance depends on the penalty weight λ. Generally, the CVPL plot for ridge is much smoother than the one for lasso, which can have many local modes. It should be noticed that even in single cross-validation the set of covariates with non-zero coefficients selected by lasso can vary between cross-validation folds.

11.3 Application to Data Set 3

This data set contains the clinical and genomic data on 295 patients as reported in van de Vijver et al. (2002). The data was reanalyzed by Van Houwelingen in cooperation with the statisticians of the NKI and published in Statistics in Medicine (van Houwelingen et al. 2006). Basic statistics are given in Section A.3.

For the time being the clinical covariates are ignored and attention is focused on the gene expression data. The gene expression covariates $X_1, \ldots, X_{4919}$ are natural logarithms of the standardized ratio, which is the ratio of the expression of an individual and the mean expression in the whole group. This would imply that $\overline{\exp(X_j)} = 1$ for all genes. That is not quite true in the data because of patient selection after standardization, but most $\overline{\exp(X_j)}$'s are close to 1. This also implies that mean and variance of the X_j's themselves satisfy $\overline{X}_j + 0.5\text{var}(X_j) \approx 0$. This relation can be seen in the data as well. The means of most X_j's are slightly below 0, while the typical standard deviation is about 0.2. Although there is some variation in standard deviation, the X_j's are not standardized for the sake of an easy interpretation.

Principal component analysis of the X_j's showed that the first principal component explains 11% of the total variation, while the first 10 principal components explain about 44%. This is in line with the observation that the average absolute correlation is about 0.14 and that correlations above 0.4 are very rare. The scree-plot of the eigenvalues is quite flat. This all suggests that principal components regression might ignore relevant information in higher components. For that reason, principal component regression is not pursued further. Attention is focused on ridge and lasso variants of the Cox model.

Information on the separate univariate regression coefficients is given in Figure A.4, which shows a funnel plot popular in meta-analysis. Although a large percentage (38%) is significant at the 5% level, the plot also shows that there are no very outspoken genes with a reliable large effect.

Ridge and lasso fits

Both ridge and lasso fits were obtained. Plots of the cross-validated partial log-likelihood as a function of the tuning parameter λ for both methods are given in Figure 11.1. For lasso also the number of non-zero coefficients are given. The optimal values λ_{opt} and the corresponding CVPL's are $\lambda_{\text{opt}} = 459$, $\text{CVPL}(\lambda_{\text{opt}}) = -476.22$ for ridge and $\lambda_{\text{opt}} = 7.70$, $\text{CVPL}(\lambda_{\text{opt}}) = -479.49$ for lasso with 16 non-zero coefficients. Notice that the CVPL plot for ridge is much smoother than the one for lasso. The explanation is that a small change in the tuning parameter λ might lead to a different selection of genes and an abrupt (not differentiable) change in prediction. To get some further insight in the models, the histogram of the prognostic indices and their means and standard deviations for the optimal choices of λ are given in Figure 11.2. Notice that ridge yields a larger standard deviation.

The predicted survival curves for selected percentiles are shown in Figure 11.3. Both models look very similar. However, the plots do not tell whether the models produce the same predicted survival curves for each individual. The "agreement" between the two models can be inferred from the correlation between the two prognostic indices. The correlation is quite high ($r = 0.904$), but this assessment may be optimistically biased because both prognostic indices fit the same data. Therefore, it might be better to explore the agreement after cross-validation.

Internal validation based on cross-validation

As described in Section 3.6 a cross-validation based predicted survival curve for individual i could be obtained from

$$\hat{S}_{\text{CV},i}(t) = \hat{S}_{(-i)}(t|x_i) = \exp\left(-\hat{H}_{0,(-i)}(t) \cdot \exp(x_i^\top \hat{\beta}_{(-i)})\right) ,$$

and a cross-validation based prognostic index from

$$\text{PI}_{\text{CV},i} = \text{PI}_{i,(-i)} = x_i^\top \hat{\beta}_{(-i)} .$$

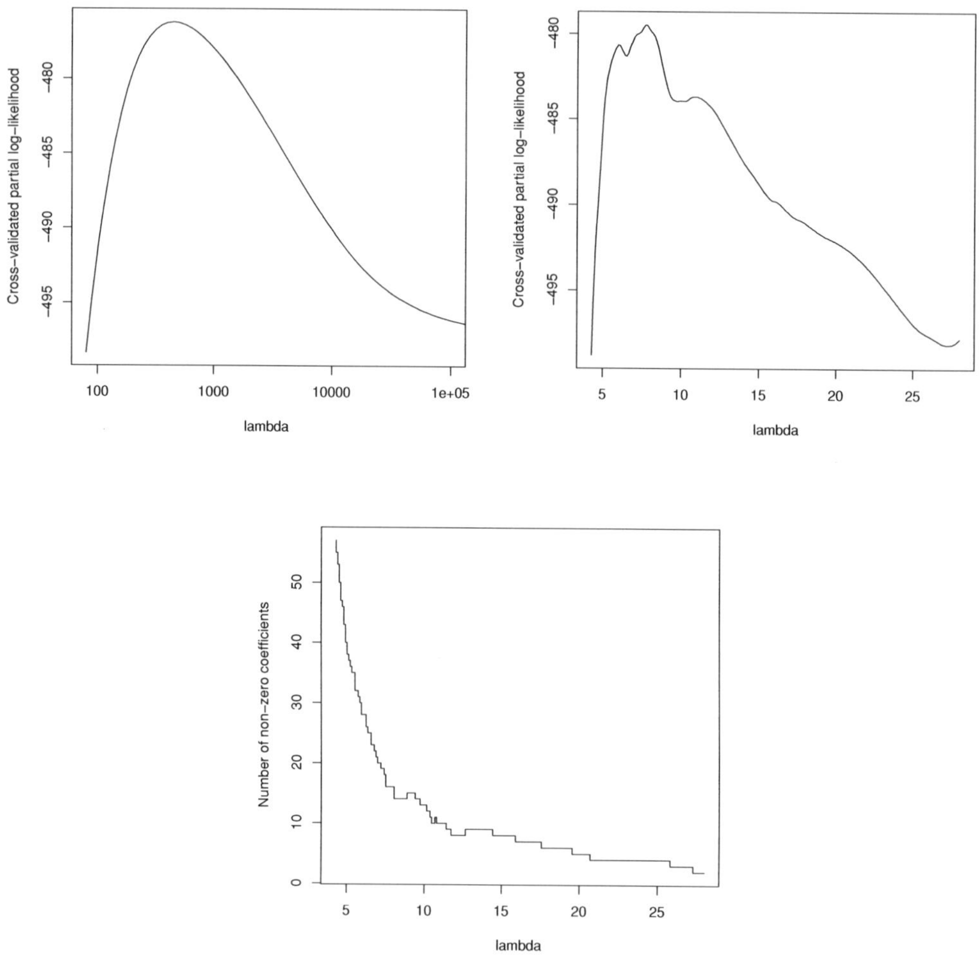

Figure 11.1 *CVPL for ridge (left) and lasso (right); below is the number of non-zero coefficients of the lasso*

The predicted survival curves are invariant under additive transformation of the X's, but the cross-validated prognostic indices are not. This could be partly remedied by centering all X's beforehand. A slightly more elegant solution is to define (a centered version of)

$$\mathrm{PI}_{\mathrm{CV},i} = \ln(-\ln(\hat{S}_{\mathrm{CV},i}(t_0))) ,$$

for a suitable choice of t_0. This is invariant under scaling of the covariates, but it intrinsically depends on t_0 because the cross-validation based predicted survival curves do not fully satisfy the proportional hazards assumption, although the violation only shows up at the end of follow-up. Following Goeman (2010),

$$\mathrm{PI}_{\mathrm{CV},i} = \ln(-\ln(\hat{S}_{\mathrm{CV},i}(5)))$$

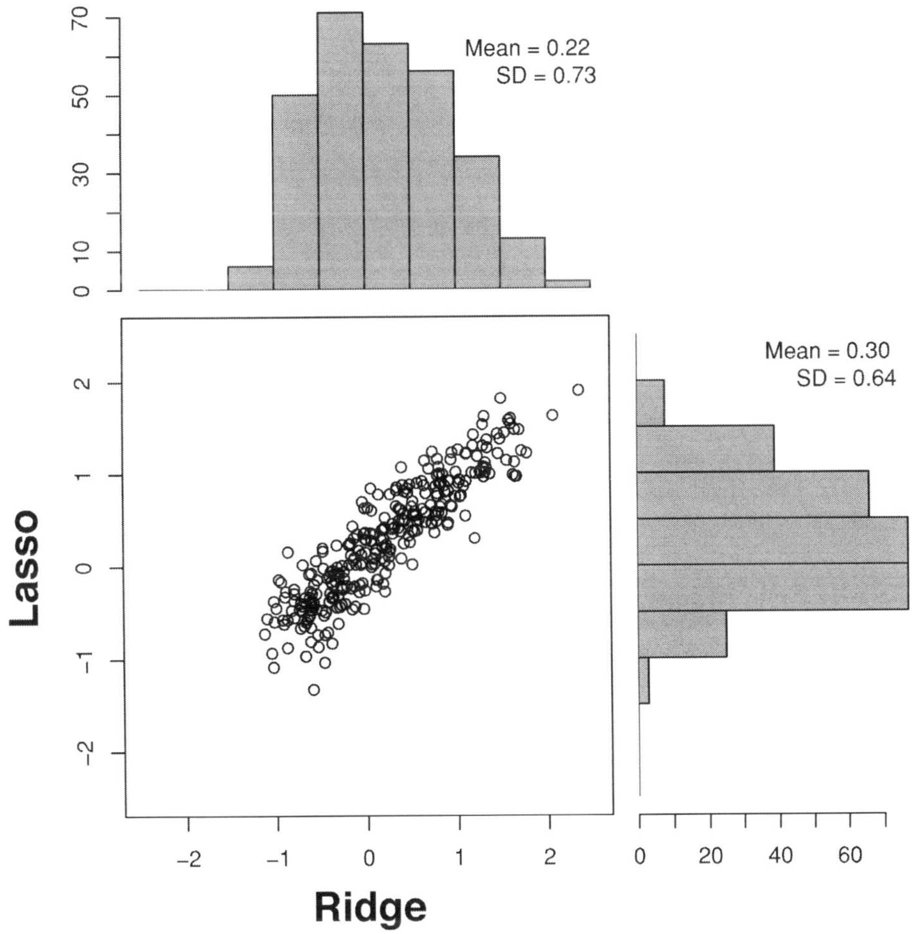

Figure 11.2 *Scatterplot of the prognostic indices for ridge and lasso. On the top and to the right are histograms of the prognostic indices for ridge and lasso, respectively*

after centering, is taken as the cross-validated prognostic index that is used for model validation and model comparison.

A visual check of the model validity is to compare the predicted survival curves with Kaplan-Meiers in subgroups based on PI_{CV}. The Kaplan-Meier curves for 4 groups of equal size are shown in Figure 11.4. Again, ridge and lasso give very similar graphs, that agree quite well with the curves in Figure 11.3. Both show some convergence of the survival curves later in the follow-up. The correlation between the two cross-validated prognostic indices is $r = 0.894$. Their standard deviations are $\text{SD}_{\text{ridge}} = 0.700$, $\text{SD}_{\text{lasso}} = 0.633$. The agreement can also be seen from the cross-table of the two quartile groupings shown in Table 11.1. Although the correlation is quite high, the agreement is far from perfect.

The cross-validated prognostic indices can also be used for the internal calibration as discussed in Section 4.2. Cox regression on the cross-validated indices yielded the interesting results of Table 11.2. Actually, the conclusion is quite strik-

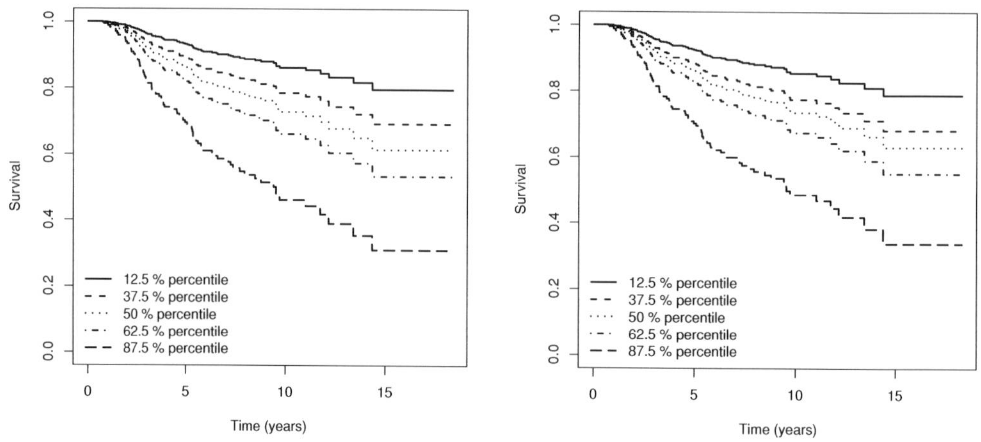

Figure 11.3 *Survival curves for ridge (left) and lasso (right)*

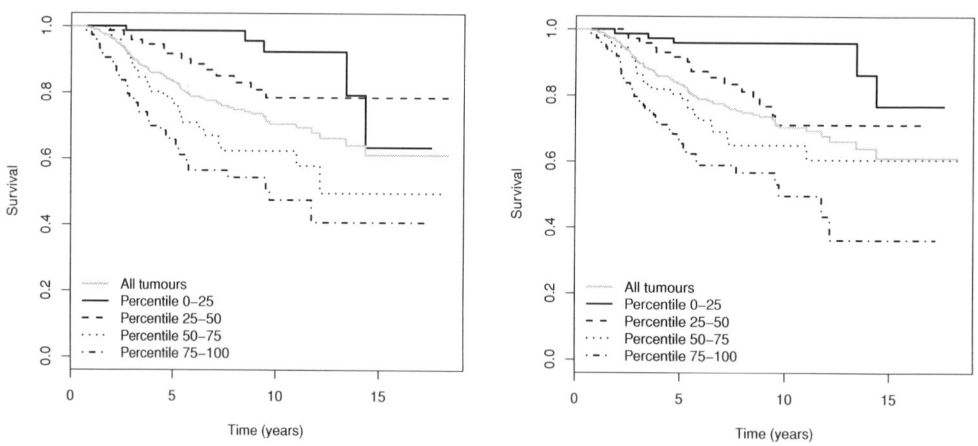

Figure 11.4 *Cross-validated Kaplan-Meier curves for ridge (left) and lasso (right)*

Table 11.1 *Crosstable of ridge versus lasso quartiles*

	Lasso			
Ridge	0-25	25-50	50-75	75-100
0-25	58	15	1	0
25-50	15	39	18	2
50-75	1	19	38	15
75-100	0	1	16	57

Table 11.2 *Cox regression on cross-validated prognostic indices*

Prognostic indices included	Ridge B	Lasso B	Model χ^2
Ridge	1.000		40.304
Lasso		0.998	33.053
Both	1.022	-0.026	40.309

ing. It could be phrased as: *The lasso predictor contains 80% of the information of the ridge predictor and nothing beyond that.* Given the ridge predictor, the lasso predictor is redundant. The difference in model χ^2 is in line with the difference in CVPL.

The conclusion must be that although the two fitted models are very similar, the predictive performance of the lasso is worse than that of the ridge estimator. Therefore, only ridge will be considered for further analysis. This has also practical advantages, because ridge is easier to compute. However, this does not mean that lasso is considered to be inferior. Most of the further analyses could be done for lasso as well. See also the discussion on *Gene finding versus prognostics* in Section 11.5.

11.4 Adding clinical predictors

The analysis so far has ignored the existing clinical information. For the data set in the application the clinical information and its univariate impact on survival is summarized in Table A.3. The question is how to combine both genomic information denoted by X and clinical information denoted by Z into a single prediction model. The approach taken by Bøvelstad et al. (2009) is to consider a Cox model containing both genomic covariates X and clinical covariates Z, which can be written as

$$h(t|X,Z) = h_0(t)\exp(X^\top\beta + Z^\top\gamma) .$$

The clinical covariates are supposed to be well-established and are all included in the model. The high-dimensional genomic part needs to be regularized. This can be achieved by adding a penalty on β leading to variants of ridge regression or lasso regression. Goeman's **penalized** package allows such models in which not all regression coefficients are penalized. The research question of Bøvelstad et al. (2009) was whether genomic information can help to improve the predictions and which approach is best able to do so. For the data considered here, their conclusion was that only ridge regression leads to improved predictions.

The disadvantage of using the model above is that the correlation between clinical and genomic covariates can lead to dramatic changes in both β and γ compared with the models with only clinical or genomic covariates. For example, adding clinical covariates can completely change the selection of genomic covariates made by lasso regression. The implication is that the prediction model depends completely on the selection of clinical covariates and cannot easily be transferred from one data set to another.

The approach taken by van Houwelingen et al. (2006) is in the spirit of the combination of models and the super learner of van der Laan et al. (2007) briefly touched upon in Section 4.5. So, instead of considering the model above that includes both types of information from the start, two models are defined

- Clinical model: $h(t|Z) = h_0(t)\exp(Z^\top \gamma)$ leading to $\mathrm{PI}_{\mathrm{clin}}(Z)$,
- Genomic model: $h(t|X) = h_0(t)\exp(X^\top \beta)$ leading to $\mathrm{PI}_{\mathrm{gen}}(X)$.

The models are then combined into

- Super model: $h(t|\mathrm{PI}) = h_0(t)\exp(\alpha_1 \mathrm{PI}_{\mathrm{clin}}(Z) + \alpha_2 \mathrm{PI}_{\mathrm{gen}}(X))$.

To prevent over-fitting, the parameters in the supermodel are fitted using cross-validated versions of $\mathrm{PI}_{\mathrm{clin}}(Z)$ and $\mathrm{PI}_{\mathrm{gen}}(X)$. This program is easily carried out and gives an unbiased view on the contribution of the two sources of information to the prediction model.

For the clinical model all covariates of Table A.3 are used in a simple Cox model. The prognostic index is again computed as $\mathrm{PI}_{\mathrm{CV},i} = \ln(-\ln(\hat{S}_{\mathrm{CV},i}(5)))$ after centering. Its standard deviation equals 1.133. For the genomic model, the results of the "optimal" ridge regression are used. The correlation of the two prognostic indices equals $r = 0.652$. The results from fitting Cox models on the cross-validated prognostic indices is given in Table 11.3. The first observation to be made is that the

Table 11.3 *Super model Cox regression*

Prognostic indices included	Clinical α_1	Genetic α_2	Model χ^2
Clinical $\mathrm{PI}_{\mathrm{clin,CV}}$	0.737		43.750
Genomic $\mathrm{PI}_{\mathrm{gen,CV}}$		1.000	40.304
Both $\mathrm{PI}_{\mathrm{clin,CV}}$ and $\mathrm{PI}_{\mathrm{gen,CV}}$	0.495	0.582	52.369
Calibrated coefficients	0.495/0.737 = 0.672	0.582/1.000 = 0.582	

clinical prognostic index is poorly calibrated, as could be expected since the clinical model has 11 degrees of freedom. The genomic prognostic index is well calibrated as observed before. Next, it should be noticed that the calibrated clinical model performs slightly better than the genomic model. The super model performs better than the individual models, but not spectacularly so. Apparently, the information that is shared by the genomic source and the clinical one and that is responsible for the correlation of the cross-validated indices is much more relevant than the independent parts. Unfortunately it is impossible to disentangle joint and independent parts of $\mathrm{PI}_{\mathrm{clin,CV}}$ and $\mathrm{PI}_{\mathrm{gen,CV}}$.

Finally, the predictive performance of the models at diagnosis can be assessed by Kullback-Leibler or Brier scores for the models of Table 11.3. These are shown

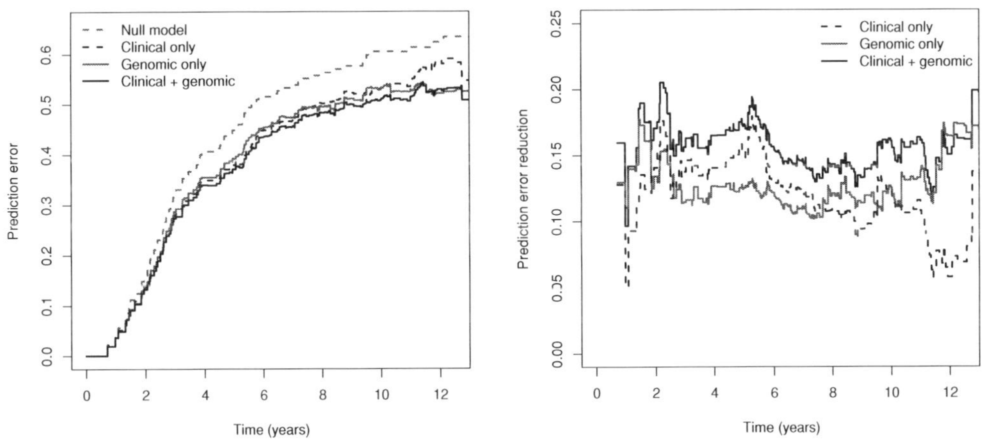

Figure 11.5 *Kullback-Leibler prediction error curves (left) and prediction error reduction curves (right) for the null model (Kaplan-Meier) and for the three models of Table 11.3*

in Figure 11.5. On the left the Kullback-Leibler prediction error curves are shown for the three Cox models of Table 11.3, as well as for the null model, given by the overall Kaplan-Meier estimate. The plot on the right shows the reduction in prediction error with respect to the null model for the three Cox models. Interestingly, the prediction error (reduction) curves seem to suggest that for short term prediction the clinical information is more important, while the genomic information reduces the prediction error in the long term.

11.5 Additional remarks

Computational simplification of ridge regression

Since the ridge estimator has to satisfy

$$\frac{\partial \mathrm{pl}_{ridge}(\beta)}{\partial \beta} = \sum_{i}^{n} d_i(x_i - \bar{x}_i(\beta)) - \lambda\beta = 0 \, ,$$

it is evident that β has to lie in the linear space spanned by the x_i's, the patient covariate vectors. That implies that $\beta = X^\top \gamma$, where X is the $n \times p$ design matrix with the covariates in the columns. The regression part of the model could be reparametrized as $\tilde{X}\gamma$, with $\tilde{X} = XX^\top$ the $n \times n$ design matrix $\tilde{X} = XX^\top$, γ the n-dimensional parameter vector and penalty $\frac{1}{2}\lambda\gamma^\top\tilde{X}\gamma$. This greatly reduces the complexity of the problem and the computing time needed to obtain the ridge estimate.

Built-in calibration

At first sight it is striking how well the optimal ridge and lasso estimator are calibrated. However, this is no coincidence. The local effect of increasing the penalty is shrinkage of the prognostic index towards the mean. Hence, the optimal λ will automatically produce an estimated prognostic index that is well calibrated under CVPL as described in Section 4.2 and hence also under Cox regression on the cross-validated prognostic index as shown in Figure 4.1.

Gene finding versus prognostic modeling; global test

Finding a good prognostic model is a different game from finding the biologically most relevant genes involved in the development of the tumor. For the latter purpose often gene lists are produced consisting of all genes at an appropriate level of significance controlling the False Discovery Rate by the approach of Benjamini & Hochberg (1995). Those gene lists are then compared with gene sets like the Gene Ontology pathways (Ashburner et al. 2000). It is a common misunderstanding that the gene selection produced by lasso is of a similar nature. That is not true. The advantage of lasso over ridge is that predictions can be obtained by a limited number of genes. Why the particular subset is selected is hard to understand unless all genes are independent, but that is very uncommon.

There is some link between obtaining list of significant genes and prognostic modeling. In Goeman et al. (2004, 2005) a global test for the null hypothesis that none of the genes have an effect is derived as a locally optimal score test in a regression model like the Cox model. The global test can also be used as a screening tool for predictive modeling. If the test fails to be significant, it is no use to start modeling, see van Houwelingen et al. (2006). See Goeman et al. (2006) for more theoretical background on the global test.

Bayesian interpretation

The penalized partial log-likelihoods of Section 11.2 can be interpreted as posterior likelihoods in a Bayesian model with independent normals or independent double exponentials as priors for the regression coefficients, respectively. However, the estimates obtained through penalization can hardly be seen as Bayesian estimates. A proper Bayesian analysis would use mean or median for a point estimate and would also produce credibility intervals for the parameters. The penalized estimate corresponds with the mode of the posterior distribution, which is hardly used by Bayesians. This is particularly a problem for lasso, where the mode and the mean may be quite different.

An example of a proper Bayesian approach to model selection is the so called spike-and-slab model that use a mixture of point mass at zero and a continuous density as prior for the regression coefficients. Predictors are selected on the basis of the posterior probability that a parameter is non-zero as described in Mitchell & Beauchamp (1988).

Confidence intervals

Penalized estimators reduce the variance of the estimated regression coefficients at the price on introducing a bias that could be quite substantial. For such biased estimators it is very hard, if not impossible, to obtain proper confidence intervals. In Verweij & van Houwelingen (1994) confidence intervals are given based on normal approximations of the posterior distribution induced by the penalty. These should be used with care and only for ridge estimators. The normal approximation will not hold for the lasso. Bootstrap and similar methods can be used to obtain confidence interval for the performance of the models but not for the regression parameters of the model.

Chapter 12

Dynamic prediction based on genomic data

12.1 Testing the proportional hazards assumption

The presentation in Chapter 11 is the usual statistical approach for the prediction of survival based on high-dimensional data. The validity of the proportional hazards (PH) assumption underlying the Cox model is hardly ever discussed. The validity of the PH assumption is of no concern if the prognostic index is only used to group patients in different risk categories, and the survival in each group is estimated separately by the Kaplan-Meier curves as shown in Figure 11.4. However, for model based predictions as in Figure 11.3 the PH assumption is crucial. An easy way to check the validity of the model is the two-stage procedure sketched in Section 6.3. In that approach, the prognostic index is obtained in the first stage and in the second stage a model is fitted with the prognostic index as the only predictor allowing a time-varying effect. A slightly more robust approach is to use a cross-validated version of the prognostic index as introduced in Section 11.3 and used in Section 11.4. This can be done for the genomic ridge regression predictor and the clinical predictor separately and for the super learner predictor that is defined in Table 11.3. The results are shown in Table 12.1. Evidently, both predictors and the super learner show a decreasing effect over time. Allowing such an effect might help to improve

Table 12.1 *Regression coefficients (standard errors) in the two-stage Cox regression approach*

Predictor	Model	Constant B (SE)	$\ln(1+t)$ B (SE)	Model χ^2
$PI_{clin,CV}$	Time-fixed	0.737 (0.117)		43.750
	Time-varying	1.748 (0.412)	-0.625 (0.239)	50.397
$PI_{gen,CV}$	Time-fixed	1.000 (0.156)		40.304
	Time-varying	2.220 (0.573)	-0.742 (0.334)	45.355
Super learner				
$(0.495 * PI_{clin,CV}+$	Time-fixed	1.000 (0.144)		52.369
$0.582 * PI_{gen,CV})$	Time-varying	2.391 (0.554)	-0.847 (0.318)	59.750

the predictive performance. One way is using the time-varying effect model. As argued in Section 7.2, simple robust dynamic predictions could be obtained by the landmark approach. This will be further explored in the next section.

12.2 Landmark predictions

To get an impression of the possibilities of obtaining sliding landmark predictions for this type of data, landmark data sets are created at $t_{\text{LM}} = 0, 1, 2, 3, 4$ and 5 years, with a prediction window of width $w = 5$ years. Some of the results for landmark predictions at each t_{LM} are given in Table 12.2. They confirm the decreasing effect of both predictors, the observation that the clinical predictor performs slightly better than the genomic predictor and the fact that the combined predictor (super learner) is not really much better than the clinical one. The results of a multivariate model are not shown in the table; they are virtually identical to those of the super learner. The results of Table 12.2 can be generalized in different ways. First of all landmark super models can be obtained using the two cross-validated predictors and the super learner, a prediction window $w = 5$ and landmarking time-points running from $t_{LM} = 0$ to $t_{LM} = 7$, with 0.1 years distance. Table 12.3 shows the results of a proportional baselines (ipl*) landmark supermodel, with $\theta(s) = \eta_1(s/7) + \eta_2(s/7)^2$ in Equation (7.3). In the column with "Fixed" the results of the model without interactions with t_{LM} are reported, while in the columns "Landmark-dependent" results are shown with linear interactions with t_{LM}. The super learner uses the fixed combination $0.495 * \text{PI}_{\text{clin,CV}} + 0.582 * \text{PI}_{\text{gen,CV}}$. Instead a landmark supermodel can be fitted with the clinical and genomic cross-validated prognostic indices as two covariates. The results of these models in terms of predictive accuracy appeared to be no better than those of the super learner, so they are not shown here. Figure 12.1 shows the dynamic predictions of death within the fixed width window of $w = 5$ years, for four individuals. These were chosen so as to represent the 25% and 75% quantiles of the clinical (-0.85 and 0.81, respectively) and the genomic (-0.55 and 0.55, respectively) cross-validated prognostic indices in the data. The correspond-

Table 12.2 *Univariate Cox regression on $PI_{clin,CV}$ (clinical) and $PI_{gen,CV}$ (genomic) in different landmark data sets; the regression coefficients (B) and model χ^2 are given for the clinical, genomic and super learner models*

			Clinical		Genomic		Super learner	
t_{LM}	At risk	Events	B	χ^2	B	χ^2	B	χ^2
0	295	48	0.916	41.934	1.179	33.869	1.222	47.093
1	292	58	0.837	42.260	1.125	38.069	1.128	49.675
2	281	52	0.766	32.026	1.006	27.568	1.015	36.864
3	260	39	0.690	19.438	0.965	19.546	0.940	24.046
4	246	29	0.598	10.812	0.787	9.510	0.791	12.653
5	232	26	0.606	9.471	0.770	8.067	0.795	11.040

Table 12.3 *Estimated regression coefficients (standard errors) for the proportional baselines (ipl*) landmark super models without (fixed) and with (landmark-dependent) linear landmark interactions. Estimates of* η_1 *and* η_2 *for the baselines* $\theta(s) = \eta_1(s/7) + \eta_2(s/7)^2$ *are not shown*

Model	Time	Clinical	Genomic	Super learner
Fixed	1	0.741 (0.122)	0.946 (0.155)	0.970 (0.138)
Landmark-	1	0.891 (0.155)	1.191 (0.209)	1.195 (0.181)
dependent	$s/7$	-0.412 (0.289)	-0.644 (0.368)	-0.595 (0.311)

ing quantiles of the super learner in the data for these four individuals were 23% (low risk clinical, low risk genomic), 47% (low risk clinical, high risk genomic), 54% (high risk clinical, low risk genomic) and 76% (high risk clinical, high risk genomic). In the plot of the clinical model, no distinction is made between genomic high/low risk, and similarly for the genomic model, where no distinction is made between clinical high/low risk. Hence only two curves are shown for the landmark fixed model (no linear interactions with the landmark time points) and two for the landmark dependent model (including linear interactions with the landmark time points). Especially for the low risk patient, the landmark fixed and landmark dependent models give different predictions. In the plot of the super learner, four patients are shown, with little difference between the discordant (high clinical risk, low genomic risk, and low clinical risk, high genomic risk) patients.

The moving Kullback-Leiber dynamic prediction error curves for the separate models and the combined model are shown in Figure 12.2. The corresponding dynamic prediction error reduction curves are shown in Figure 12.3. It is obvious that combining the predictors improves the prediction. With respect to the choice of landmark fixed or landmark dependent models, the improvement in allowing the models to be landmark dependent is less clear in this data set. The overall conclusion seems to be that inclusion of linear landmark interactions in the model gives a slight improvement in terms of predictive accuracy in the beginning of the follow-up and does not do too badly at the end of the follow-up.

The next extension is that the genomic predictor could be improved by the landmark analogue of the rank = 1 reduced rank model of Section 6.3. This would require a penalized version of the landmark supermodels. It is not hard to write down formulas for such models, but there is no software for fitting them yet. Moreover, as shown in Section 6.3, the improvement of the rank = 1 model over the two-stage model is usually very modest.

The challenge is to develop something like the general reduced rank model with different predictors depending on the moment of prediction t_{LM}. In general terms this could be achieved by a model with

$$\beta_{t_{\mathrm{LM}}}(s) = \sum_{j=1}^{m_b} \gamma_j f_j(s) ,$$

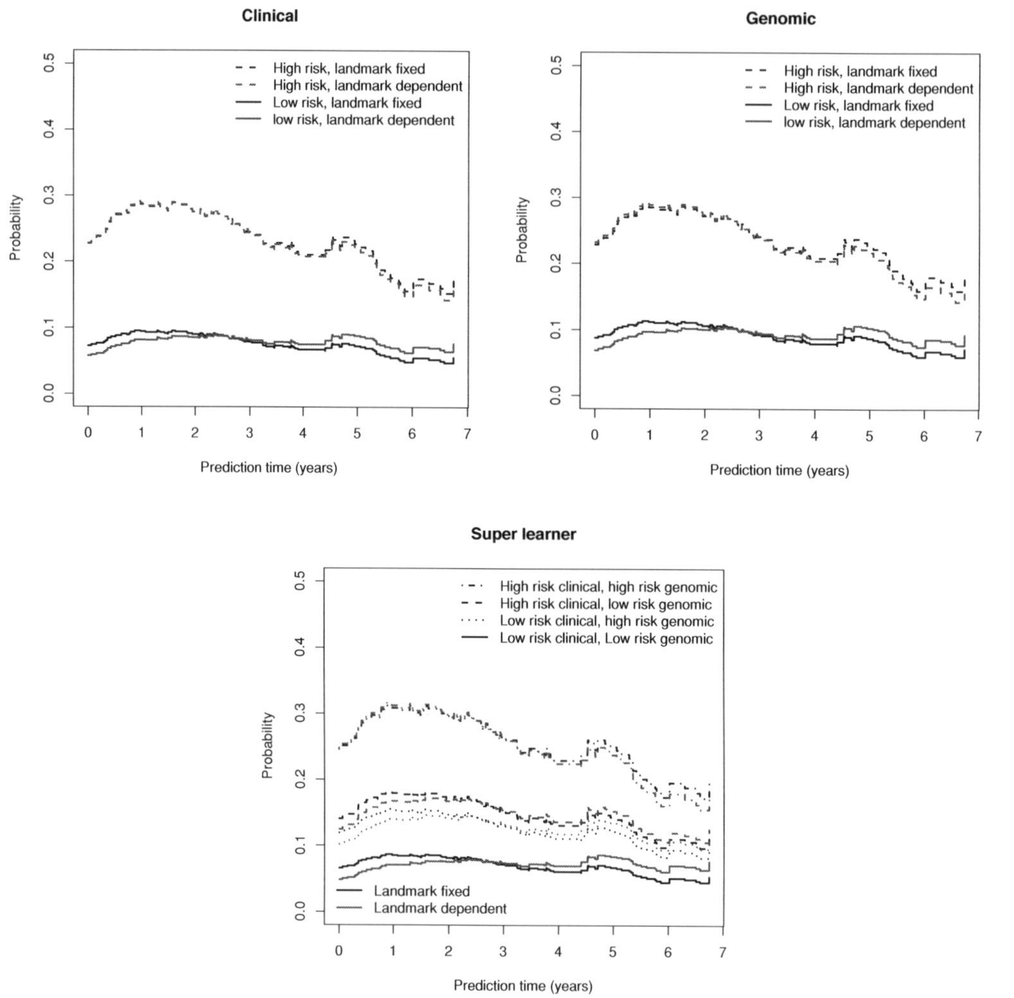

Figure 12.1 *Dynamic fixed width predictions without (black) and with (grey) landmark interactions*

as discussed in Section 7.2. In this model each of the γ_j's is of dimension p. Extension of ridge regression would require the definition of a suitable penalty on the set of all γ_j's, combining the traditional ridge regression with the proposals for regularization over time that can be found in Verweij & van Houwelingen (1995) and Kauermann & Berger (2003). Unfortunately, there are no generally accepted proposals for this problem yet.

In the absence of general theory, it will be explored more informally whether the results of Table 12.2 could be improved by allowing more than a single predictor. For the time being the focus is on the genomic data. For each of the landmark data sets at $t_{\mathrm{LM}} = 0, 1, 2, 3, 4$ and 5 years, a ridge regression estimate was obtained and the cross-validated predictor for that landmark data set, defined as $\mathrm{PI}_{\mathrm{gen,CV},t_{\mathrm{LM}}} = \ln(-\ln(\hat{S}(t_{\mathrm{LM}}+5|t_{\mathrm{LM}},x))$. This predictor is well-defined for every

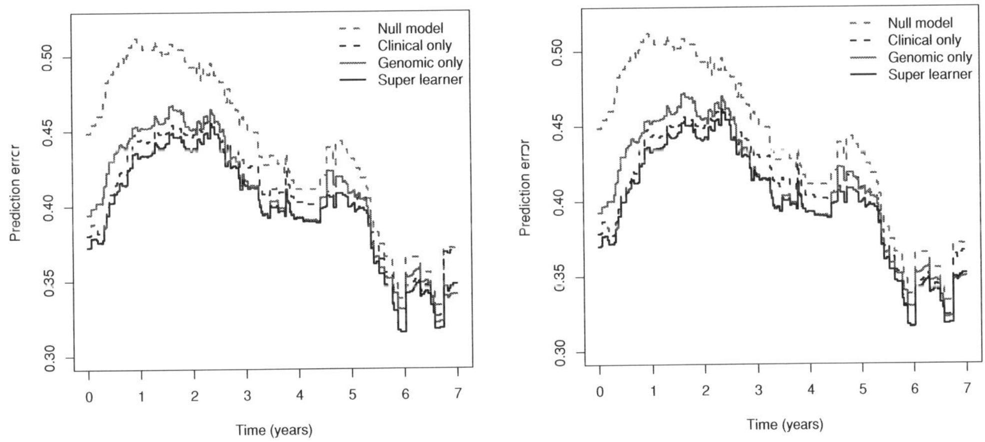

Figure 12.2 *Kullback-Leibler dynamic (fixed width $w = 5$) prediction error curves for the landmark supermodels without (left) and with (right) landmark interactions*

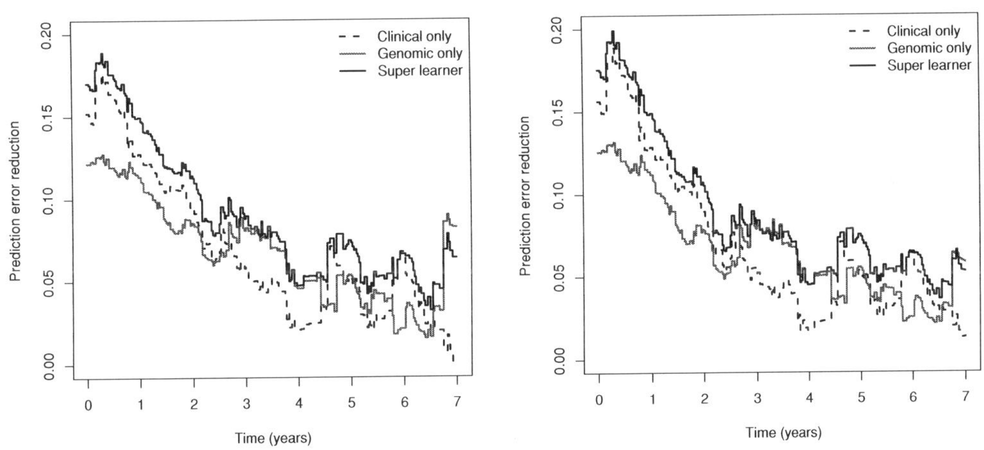

Figure 12.3 *Kullback-Leibler dynamic (fixed width $w = 5$) prediction error reduction curves for the landmark supermodels without (left) and with (right) landmark interactions*

landmark time-point. The crucial question is how correlated the different predictors are. Improved prediction can only be expected if these correlations are not too high. Standard deviations and correlations are shown in Table 12.4. The increasing standard deviations in this table and the decreasing correlations with $PI_{gen,CV}$ are hopeful indications that ridge regression per landmark might be profitable. The model χ^2 of Cox regressions with the landmark specific predictors are shown in Table 12.5. The interesting finding is that a calibrated version of $PI_{gen,CV}$ does well at $t_{LM} = 0, 1, 2$, but is outperformed by the "local" ridge regression for $t_{LM} = 3$

Table 12.4 *Standard deviations and correlations of cross-validated landmark specific genomic ridge predictors*

Predictor	SD	$\mathrm{PI}_{\mathrm{gen,CV}}$	$\mathrm{PI}_{\mathrm{gen,CV},t_{\mathrm{LM}}}$				
			$t_{\mathrm{LM}}=0$	$t_{\mathrm{LM}}=1$	$t_{\mathrm{LM}}=2$	$t_{\mathrm{LM}}=3$	$t_{\mathrm{LM}}=4$
$\mathrm{PI}_{\mathrm{gen,CV}}$	0.69						
$\mathrm{PI}_{\mathrm{gen,CV},0}$	0.70	0.947					
$\mathrm{PI}_{\mathrm{gen,CV},1}$	0.75	0.985	0.962				
$\mathrm{PI}_{\mathrm{gen,CV},2}$	0.63	0.974	0.926	0.982			
$\mathrm{PI}_{\mathrm{gen,CV},3}$	0.86	0.870	0.733	0.855	0.865		
$\mathrm{PI}_{\mathrm{gen,CV},4}$	0.81	0.737	0.495	0.670	0.708	0.839	
$\mathrm{PI}_{\mathrm{gen,CV},5}$	0.84	0.632	0.350	0.556	0.608	0.792	0.947

Table 12.5 *Comparison of model χ^2 for different approaches, using the genomic data*

Predictor	$t_{\mathrm{LM}}=0$	$t_{\mathrm{LM}}=1$	$t_{\mathrm{LM}}=2$	$t_{\mathrm{LM}}=3$	$t_{\mathrm{LM}}=4$	$t_{\mathrm{LM}}=5$
$\mathrm{PI}_{\mathrm{gen,CV}}$	33.869	38.069	27.568	19.546	9.510	8.067
$\mathrm{PI}_{\mathrm{gen,CV},t_{\mathrm{LM}}}$	34.758	36.101	23.596	25.968	19.163	19.120

and further. Another interesting observation is that at $t_{\mathrm{LM}}=5$ the prediction can be slightly improved (model $\chi^2=20.947$) by using the predictor obtained at $t_{\mathrm{LM}}=4$ instead of using the one defined at $t_{\mathrm{LM}}=5$.

A similar analysis can be carried out for the clinical data. The results are shown in Table 12.6. An explanation for the very poor performance of the landmark mod-

Table 12.6 *Comparison of model χ^2 for different approaches, using the clinical data*

Predictor	$t_{\mathrm{LM}}=0$	$t_{\mathrm{LM}}=1$	$t_{\mathrm{LM}}=2$	$t_{\mathrm{LM}}=3$	$t_{\mathrm{LM}}=4$	$t_{\mathrm{LM}}=5$
$\mathrm{PI}_{\mathrm{clin,CV}}$	41.934	42.260	32.026	19.438	10.812	9.472
$\mathrm{PI}_{\mathrm{clin,CV},t_{\mathrm{LM}}}$	28.117	38.703	27.930	3.915	1.133	0.160

els at later time-points can be found in the group of 75 patients with well differentiated histological grade. Only four of them die, at $t=2.5, 5.8, 13.4$ and 14.4 years, respectively. In a prediction window with only one event in this group of patients, a cross-validated predictor will not be able to predict this event. This could be remedied by assuming smooth dependence of the model on the landmark time-point as in the landmark super models. But even then, models with 11 degrees of freedom in data with 79 events will be prone to overfitting. Using the calibrated cross-validated predictor coming from a simple Cox model on the total data set is quite robust, although it might overlook relevant clinical information later on in the follow-up. Combining landmark specific genomic and global clinical predictor leads to the results of Table 12.7. It shows that using the landmark specific genomic predictor can substantially improve the predictive performance.

Table 12.7 *Comparison of model χ^2 for different approaches, combining the landmark specific genomic predictor with the global clinical predictor*

Predictors	$t_{LM} = 0$	$t_{LM} = 1$	$t_{LM} = 2$	$t_{LM} = 3$	$t_{LM} = 4$	$t_{LM} = 5$
$PI_{clin,CV}$ + $PI_{gen,CV}$	47.289	49.683	36.912	24.085	12.662	11.062
$PI_{clin,CV}$ + $PI_{gen,CV,t_{LM}}$	47.869	48.877	34.770	31.006	22.068	24.113

The conclusion of this explorative analysis is that there is more useful information hidden in the genomic data than the prognostic index coming from a ridge regression model on the whole data sets. However, there are no clear-cut strategies how to build robust dynamic predictors. Predictive models based on an explorative analysis as described above is doomed to be optimistically biased.

A possible approach that comes to mind is going back to the ideas of Blackstone et al. (1986). The model described there is not easy to fit, but the idea of considering different phases "early," "middle" and "late" in the follow-up and having phase specific models is attractive, if there is consensus about such phases. Simple Cox models could be built for each phase and those models could be combined into dynamic models using the landmark version of the super learner as described in Section 11.4. If the phases are to be derived from the data, overfitting is again inevitable.

Such an approach will not be further explored here. Reasons are that the findings will be quite speculative on this small data set. Moreover, truly dynamic prediction will also involve clinical follow-up data as in Chapter 10. The really interesting question is whether genomic data obtained from the tumor will be relevant later in the follow-up for patients in the different stages defined in Chapter 10. Unfortunately, clinical follow-up is not available in Data Set 3.

12.3 Additional remarks

The Cox model for genomic data as used in this chapter and Chapter 11 uses a simple additive model to capture the effect of the high-dimensional genomic information. This chapter can be seen as an attempt to relax the proportional hazards assumption, keeping the simple additive structure. The big challenge is to go beyond additivity by adding interactions or considering regression tree models and to include more biological information like pathways, GO-terms and the like. The experience so far is that such extensions can lead to interesting biological insights but are of limited use in building better prediction models. More biological insight is needed that has to come from the biological and medical side and not from the statistical side. Statistics can tests hypotheses and assess performances but can never replace biological and medical research.

Part V

Appendices

Appendix A

Data sets

There are six data sets that are used throughout the book. Since some of the data sets are used more than once, it is convenient to give the relevant information in this appendix, as an internal reference. For each data set references will be given to the clinical studies from which they arise and the methodological studies in which they have been used before. Basic information will be provided on the survival and censoring distributions and the cumulative incidence functions of competing risks (if present). Furthermore, a description will be given of the main risk factors in the data set, their descriptive statistics and their univariate effects on the outcomes of interest.

A.1 Data Set 1: Advanced ovarian cancer

The data originate from two clinical trials comparing different combination chemotherapies that were carried out in The Netherlands around 1980. For details see Neijt et al. (1984) and Neijt et al. (1987). The outcome of interest is overall survival, that is the time from diagnosis until death from any cause. Actually, the difference in survival between the chemotherapy regimens is very limited.

The data from the two trials were combined into one data set for the purpose of prognostic modeling. The first analysis aimed at prognostic modeling was published as van Houwelingen et al. (1989). In this analysis the slightly inferior HEXA-CAF arm was excluded. Later on, this data set was used in the PhD research of Pierre Verweij (Verweij & van Houwelingen 1993, 1994, 1995, Verweij et al. 1998) and Aris Perperoglou (Perperoglou, le Cessie & van Houwelingen 2006*a*,*b*).

There are 358 patients in the data set. The survival function and the censoring function are shown in Figure A.1. The survival function shows that advanced ovarian cancer has a very poor prognosis with a long term survival of about 25%. The censoring function shows that the data set contains the information available directly after the last patient completed four years of follow-up.

The clinical risk factors and their univariate effects on survival as obtained in a simple Cox model are shown in Table A.1. The FIGO-index (FIGO) is a staging system for ovarian cancer. Advanced ovarian cancer comprises the stages III and IV. It is clear that stage IV patients have a worse prognosis. It is very hard to remove the

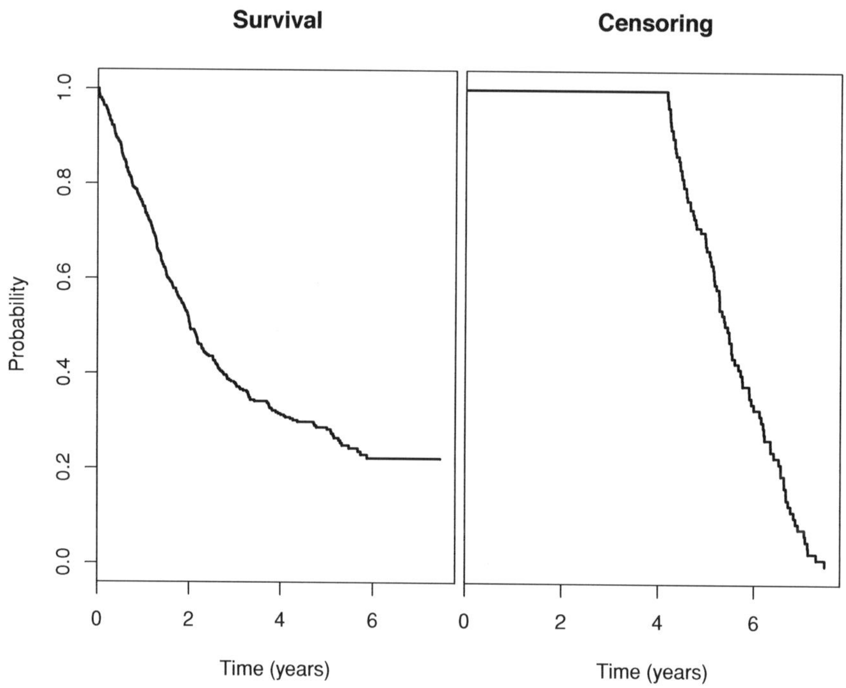

Figure A.1 *Survival and censoring functions for Data Set 1*

tumor completely. The diameter of the residual tumor after surgery is an important prognostic factor. The Karnofsky-index is a measure that indicates how well the patients are at the moment of diagnosis. A score of 100 is an indication of no physical limitations. The lower the score, the more the patient is affected by the cancer and the worse the prognosis. Broders' grade is a histological grading system for the tumor cells. Unknown could be missing at random but also an indication of the severity of the disease. Broders=1 looks like being favorable. Ascites stands for abdominal fluid. Again unknown is not just missing at random, but could be an indication of emergency surgery.

A.2 Data Set 2: Chronic Myeloid Leukemia (CML)

This data set originates from the Benelux CML study that was reported in Kluin-Nelemans et al. (1998). The outcome of interest is overall survival. As in the first data set there is little difference in survival between the treatment arms and therefore treatment is not considered as a risk factor. In this data set there are clinical risk factors measured at diagnosis that can be used for individual prognosis. Moreover, the white blood cell count (WBC) of each patient is recorded frequently during the follow-up. It is of interest to see how this biomarker can be used for dynamic prognosis.

Table A.1 *The clinical risk factors and their univariate effects on survival in Data Set 1. Shown are the regression coefficients (B) and their standard errors (SE) in separate Cox models for each risk factor*

Covariate	Category	Frequency	B	SE
FIGO	III	262		
	IV	96	0.706	0.132
Diameter	Microscopic	29		
	< 1 cm	67	0.424	0.319
	1-2 cm	49	0.934	0.322
	2-5 cm	68	1.023	0.310
	> 5 cm	145	1.234	0.293
Karnofsky	≤ 60	20	1.168	0.253
	70	46	0.811	0.188
	80	47	0.314	0.196
	90	108	0.070	0.155
	100	137		
Broders	1	42		
	2	89	0.594	0.242
	3	127	0.663	0.233
	4	49	0.443	0.269
	Unknown	51	0.813	0.262
Ascites	Absent	94		
	Present	212	0.496	0.152
	Unknown	52	0.565	0.205

The data set is used in the PhD research of Mark de Bruijne (de Bruijne 2001, de Bruijne et al. 2001) and re-analyzed in Van Houwelingen's paper on landmarking (van Houwelingen 2007).

The data set contains 190 patients. There were 195 patients in the trial. However, five patients were dropped that had no WBC measurement at all. The survival function and the censoring function are shown in Figure A.2. The survival function shows that Chronic Myeloid Leukemia has a very poor prognosis with eight year survival of about 20%. The censoring function shows that the follow-up in this data set ranges from three to eight years.

The risk factors at diagnosis are age and the Sokal score, which is a clinical index based upon spleen size, percentage of circulating blasts, platelet and age at diagnosis. Although age is used in the Sokal score, the correlation between age and Sokal is only 0.28. The distribution of risk factors and their univariate effects in a simple Cox model are given in Table A.2. It is clear that both age and Sokal are of prognostic importance.

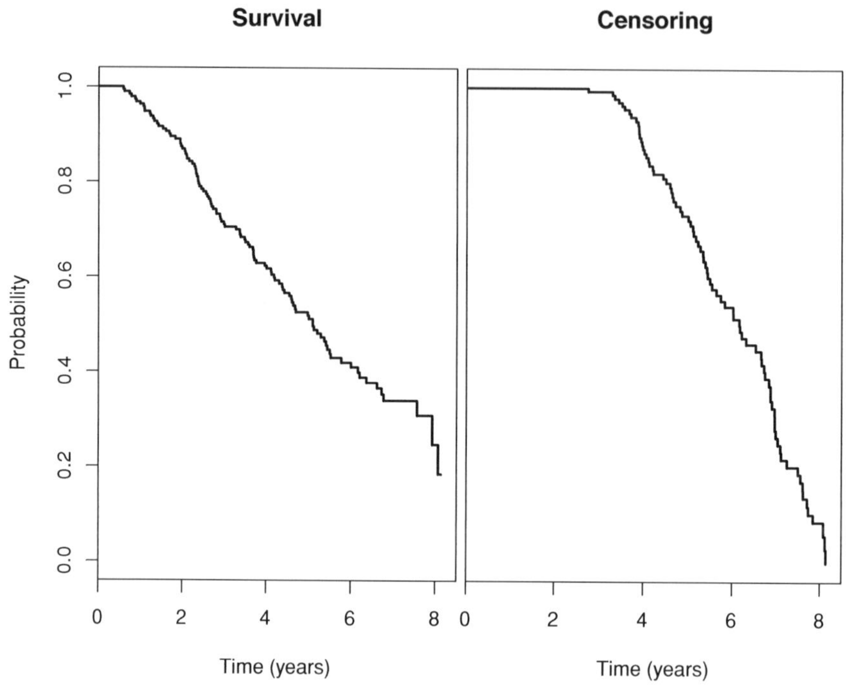

Figure A.2 *Survival and censoring functions for Data Set 2*

Table A.2 *The clinical risk factors and their effects on survival in Data Set 2; shown are the regression coefficients (B) and their standard errors (SE) in separate Cox models for each risk factor*

Covariate	Min	Max	Mean	SD	B	SE
Age	19.90	84.20	53.41	13.14	0.022	0.008
Sokal	0.32	4.44	1.07	0.51	0.626	0.163

The WBC is a very skewed variable. Therefore, it was transformed and centered by defining LWBC $= {}^{10}\log(\text{WBC}) - 0.95$. This is the biomarker as in de Bruijne et al. (2001). Its mean value over all observations is zero and its standard deviation equals 0.33. Roughly speaking WBC is measured monthly. The largest number of observations is about 100 for patients that had a follow-up of 8 years. A complication in the data is that for some patients measurement of WBC stopped well before the end of the follow-up. This is further discussed in Chapter 8. To assess the effect of WBC on survival, a Cox model was fitted with the last observed LWBC-value as (single) time-dependent covariate. A value LWBC $= 0$ was imputed before the first WBC measurement. The obtained regression coefficient is B=1.248 with SE=0.255 showing a substantial effect of increased WBC on survival.

A.3 Data Set 3: Breast Cancer I (NKI)

This data set contains data on the overall survival of breast cancer patients as collected in the Dutch Cancer Institute (NKI) in Amsterdam. This data set became very well-known because it was used in one of the first successful studies that related the survival of breast cancer to gene expression. The findings of this study were reported in two highly cited and highly influential papers in 2002 (van't Veer et al. 2002, van de Vijver et al. 2002).

This data set contains the clinical and genomic data of 295 patients as reported in van de Vijver et al. (2002). The data was reanalyzed by van Houwelingen in cooperation with the statisticians of the NKI and published in Statistics in Medicine (van Houwelingen et al. 2006). The survival and censoring functions of this data set are shown in Figure A.3. The survival curve appears to stabilize at a

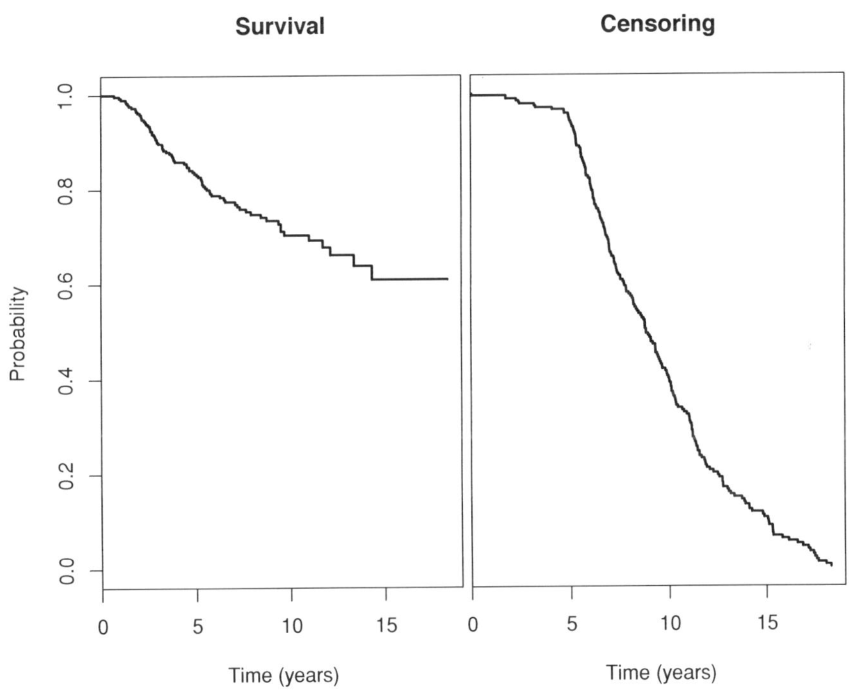

Figure A.3 *Survival and censoring functions for Data Set 3*

long term survival rate of about 60%. The censoring curve shows that the median follow-up in the data set is about 9 years.

The information about the clinical risk factors available after surgery is given in Table A.3. Of the categorical covariates histological grade and vascular invasion appear to have a significant univariate effect. For the continuous covariates the univariate Cox regression coefficients are given in Table A.3 as well. No attempt is made at this stage to optimize the scaling of these covariates. A simple model is ap-

Table A.3 *The clinical risk factors and their effects on survival in Data Set 3; shown are the regression coefficients (B) and their standard errors (SE) in separate Cox models for each risk factor*

Covariate	Category	Frequency	B	SE
Chemotherapy	No	185		
	Yes	110	-0.235	0.240
Hormonal therapy	No	255		
	Yes	40	-0.502	0.426
Type of surgery	Excision	161		
	Mastectomy	134	0.185	0.225
Histological grade	Intermediate	101		
	Poorly differentiated	119	0.789	0.248
	Well differentiated	75	-1.536	0.540
Vascular invasion	-	185		
	+	80	0.682	0.234
	+/-	30	-0.398	0.474

Covariate	Min	Max	Mean	SD	B	SE
Diameter	2	50	22.54	8.86	0.037	0.011
Number of positive nodes	0	13	1.38	2.19	0.064	0.046
Age (years)	26	53	43.98	5.48	-0.058	0.020
Estrogen level	-1.591	0.596	-0.260	0.567	-1.000	0.183

plied with a linear effect of each continuous covariate. Apparently tumor diameter, age of the patient and estrogen level have a significant univariate effect.

In van Houwelingen et al. (2006) the information on (tumor) gene expression of 4919 genes is used. This is a selection of those genes for which reliable expression is available. To obtain an impression of the association of gene expression with overall survival the univariate regression coefficients (B) and their standard errors (SE) are computed. They are shown in the funnel plot of Figure A.4. The curved lines in that plot correspond with the upper and lower cut-off critical values of $\pm 2 * \mathrm{SE}$, corresponding to a significance level of about 5%. Apparently there are no genes with a very outspoken effect. A total of 1876 genes (38%) fall outside the funnel.

A.4 Data Set 4: Gastric Cancer

This data set originates from the Dutch Gastric Cancer Trial, a randomized clinical trial comparing two surgical techniques for gastric cancer patients. The techniques differ with respect to the amount of tumor tissue resected. At the time the trial was planned, the so-called D1-dissection (conventional lymph node dissection) was considered the standard in Western Europe and the U.S., while the other,

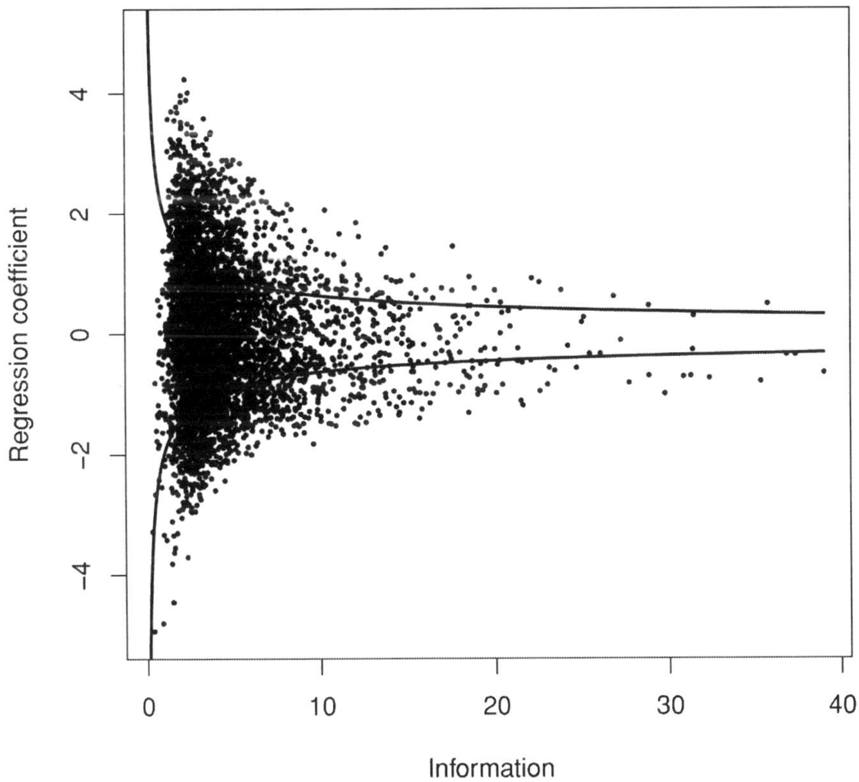

Figure A.4 *Funnel plot of the univariate Cox regression coefficients of gene expression in Data Set 3; information*$=1/SE^2$. *Lines correspond with* $\pm 2 * SE$

D2-dissection (extended lymph node dissection) was the standard in Japan. For the complete trial, 1078 patients were randomized (539 D1, 539 D2), of which 996 were eligible. Of these, 711 patients were resected with curative intent (381 D1, 330 D2), and this subset was the subject of the major clinical publications of the trial (Bonenkamp et al. 1999, Hartgrink et al. 2004). Fewer patients remained in the D2 group compared to the D1 group, because of remnant tumor or peritoneal metastases. The same data were used in Putter et al. (2005) and van Houwelingen (2007), while data on all 1078 patients were used in van Houwelingen et al. (2005). The primary endpoint of the trial was overall survival.

Figure A.5 shows the censoring distribution (median follow-up is nine years) and the Kaplan-Meier overall survival curve for both treatment arms. From the survival curves it can be seen that overall survival for the more radical D2-arm is lower than for the D1-arm over the first four years (due to higher post-operative mortality). However, after four years overall survival is higher for the D2-arm than for the D1-arm, most probably due to lower cancer recurrence rates. Over the whole

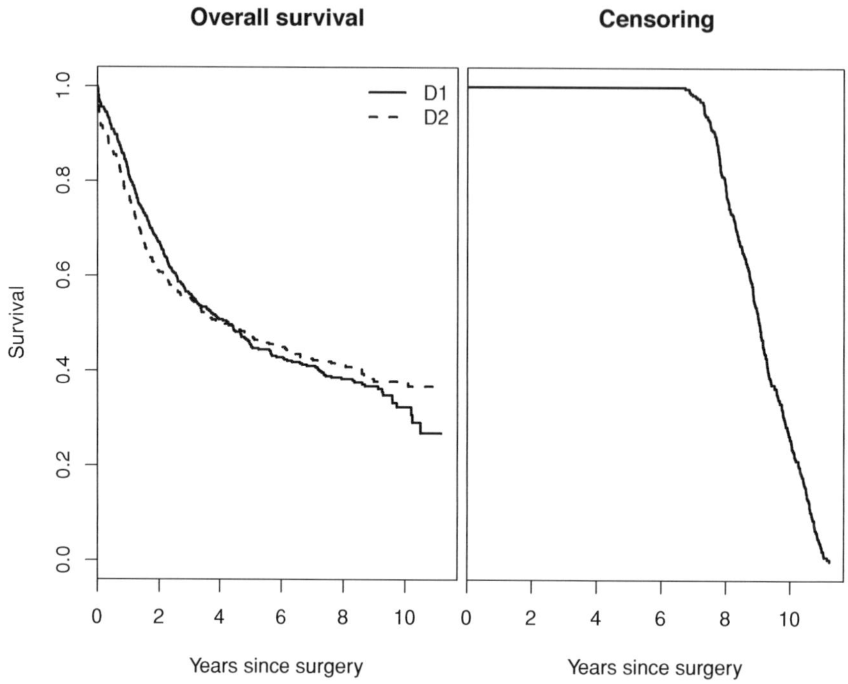

Figure A.5 *Survival and censoring functions for Data Set 4*

follow-up period, the log-rank test comparing D1 and D2 resulted in a P-value of 0.71, and the estimated regression coefficient of D2 with respect to D1 from a Cox proportional hazards model was -0.033 (SE = 0.095).

The initial advantage for D1 changing to an advantage for D2 later during follow-up results in a time-varying treatment effect and a violation of the proportional hazards assumption, as described in Putter et al. (2005). Such time-dependence can be tested formally by adding interaction terms of treatment with $f(t)$, where f is a given function of time, and t is time since surgery. The interaction of treatment and time, expressed as $f(t) = \log(t+1)$, with time in months, had a significant regression coefficient of -0.273 (SE $= 0.084$, P $= 0.001$). If this model were correct, the time-varying hazard ratio would follow a pattern like that shown in Figure A.6. Dashed curves are pointwise 95% confidence intervals of the hazard ratio.

The clinical covariates of this data set and their univariate effects on overall survival are shown in Table A.4. Higher T-stage, older age, total resection, CMA as tumor location, residual tumor and positive lymph node involvement are all significantly associated with higher death rates.

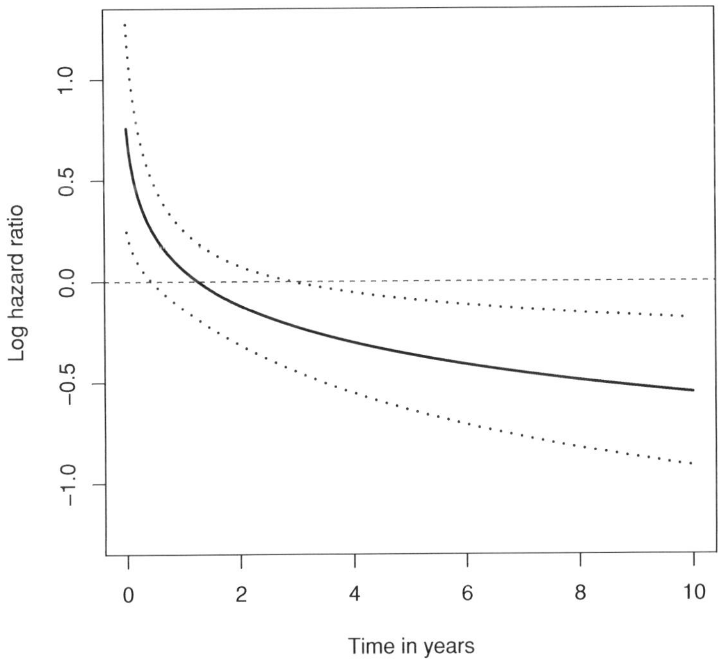

Figure A.6 *The estimated log hazard ratio with 95 per cent confidence intervals based on Cox regression with treatment as time-varying effect for Data Set 4*

A.5 Data Set 5: Breast Cancer II (EORTC)

This data set originates from a clinical trial in breast cancer patients, conducted by the European Organization for Research and Treatment of Cancer (EORTC trial 10854). The objective of the trial was to study whether a short intensive course of perioperative chemotherapy yields better therapeutic results than surgery alone. The trial included patients with early breast cancer, who underwent either radical mastectomy or breast conserving therapy before being randomized. The trial consisted of 2795 patients, randomized to either perioperative chemotherapy or no perioperative chemotherapy. Results of the trial were reported in Clahsen et al. (1996), van der Hage et al. (2001). The data have been studied by multi-state modeling in Putter et al. (2006).

This book uses a subset of all 2687 patients with complete information on the covariates detailed below. Figure A.7 shows survival and censoring functions for this subset. Median follow-up is 10.8 years. Notice that very few patients die within the first year. This might be due to the selection of patients with complete covariate information. As in the ovarian cancer patients of Data Set 1, missing information at $t = 0$ might be related to the condition of the patients and therefore to their survival.

Table A.4 *The clinical risk factors and their univariate effects on overall survival in Data Set 4; C = Cardia (proximal), M = Middle, A = Antrum (distal), R0 = no residual tumor, R1 = residual tumor; shown are the regression coefficients (B) and their standard errors (SE) in separate Cox models for each risk factor*

Covariate	Category	Frequency	B	SE
Randomized treatment	D1	380		
	D2	331	-0.033	0.095
Gender	Male	401		
	Female	310	-0.148	0.096
Age	$\leq$ 65 years	346		
	$>$ 65 years	365	0.549	0.096
T-stage	T1	188		
	T2	333	0.872	0.138
	T3	188	1.543	0.145
	Unknown	2		
Type of resection	Total	241		
	Partial	470	-0.617	0.096
Tumor location	CMA	69		
	C	74	-0.639	0.188
	M or A	568	-0.952	0.140
Residual tumor	R0	639		
	R1	72	1.266	0.135
Lymph node involvement	Negative	315		
	Positive	386	1.100	0.104

This might cause some bias for predictions at an early stage, but will fade out for predictions later on.

The interesting aspect of these data is the occurrence of the intermediate events local recurrence (LR) and distant metastasis (DM). Figure A.8 shows a plot of the proportions of patients with a history of local recurrence and distant metastasis, relative to the number of patients at risk, over time (there were 261 local recurrences, 821 distant metastases, and 122 patients experienced both local recurrence and distant metastasis (LR+DM)). The proportions are stacked; the distance between two adjacent curves represents the proportion of patients at risk with the event. The proportion of patients at risk with a history of local recurrence, irrespective of distant metastasis, may be read off by adding the proportions of patients at risk with LR only and of patients at risk with LR+DM, and similarly the proportion of patients at risk with a history of distant metastasis, irrespective of local recurrence can be read off as well. After 12 years the behavior starts to get somewhat erratic, due to the smaller number of patients at risk. It can be seen from Figure A.8 that both local recurrence and distant metastasis occur over the whole follow-up period.

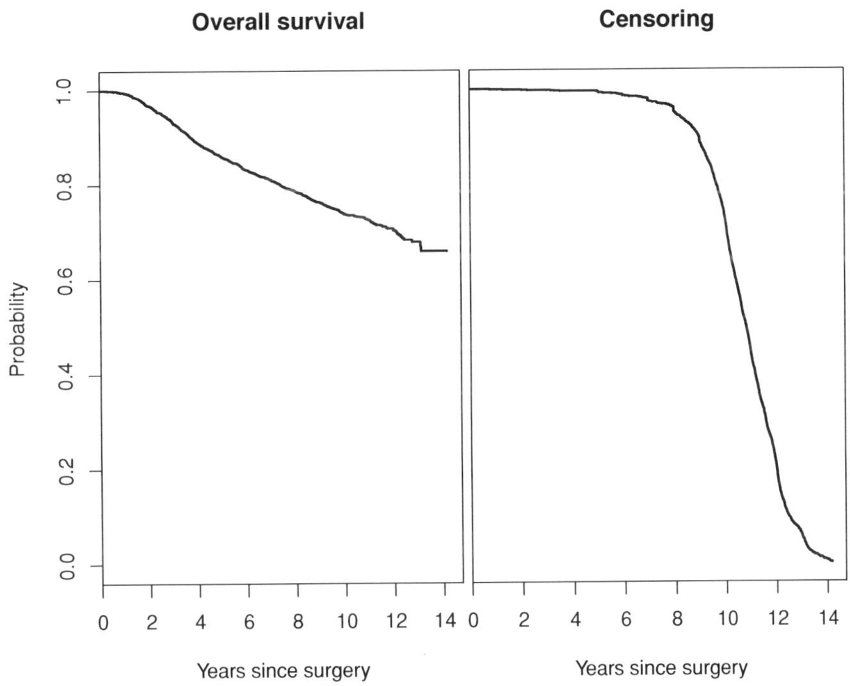

Figure A.7 *Survival and censoring functions for Data Set 5*

The occurrence of these intermediate events influences survival. This can be seen by defining the time-dependent covariates $Z_{\mathrm{LR}}(t) = \mathbf{1}\{t > \text{time of LR}\}$ and $Z_{\mathrm{DM}}(t) = \mathbf{1}\{t > \text{time of DM}\}$. The estimated univariate regression coefficients (SE) of $Z_{\mathrm{LR}}(t)$ and $Z_{\mathrm{DM}}(t)$ are highly significant; 1.842 (0.084) and 3.952 (0.104), respectively. In terms of these time-dependent covariates, what Figure A.8 shows is the (time-varying) mean of $Z_{\mathrm{LR}}(t)$ and $Z_{\mathrm{DM}}(t)$ among patients at risk at time t.

Table A.5 shows the distribution of clinical risk factors in the data and their univariate effects on survival. As expected, larger tumor size and positive nodal status are associated with higher death rates. The positive regression coefficients for use of tamoxifen and adjuvant chemotherapy are most probably caused by indication, i.e., patients with higher risk receive tamoxifen and/or adjuvant chemotherapy.

A.6 Data Set 6: Acute Lymphatic Leukemia (ALL)

This data set originates from the EBMT (the European Group for Blood and Marrow Transplantation, http://www.ebmt.org/) registry. We consider all 2297 Acute Lymphoid Leukemia (ALL) patients who had an allogeneic bone marrow transplantation from an HLA-identical sibling donor between 1985 and 1998. The data were extracted from the EBMT database in 2004. All patients were transplanted in first complete remission. Events recorded during the follow-up of these patients are: Acute Graft versus Host Disease (AGvHD), Platelet Recovery (PR, the recovery of

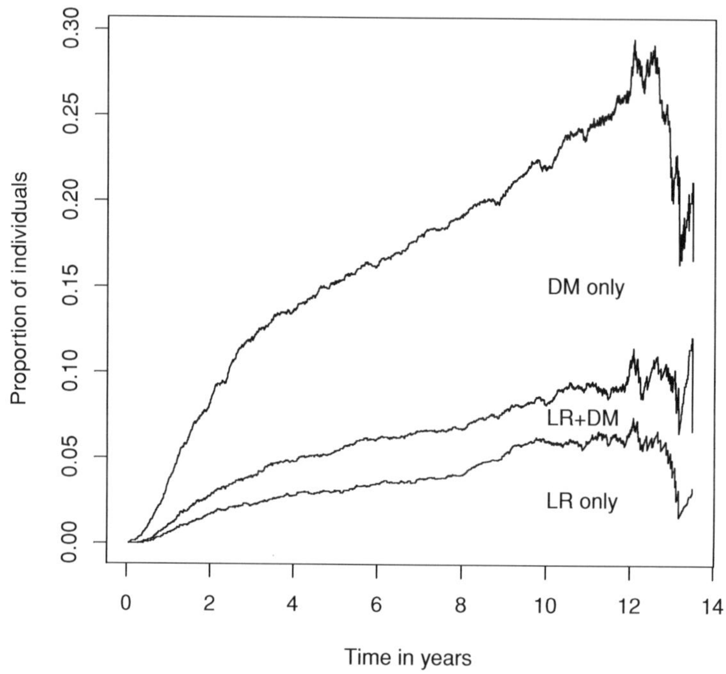

Figure A.8 *Estimated proportions of patients at risk with a history of local recurrence and distant metastasis for Data Set 5; LR = local recurrence, DM = distant metastasis*

platelet counts to normal level), Relapse and Death. AGvHD has been defined as a GvHD of grade 2 or higher, appearing before 100 days post-transplant. We consider both relapse and death as endpoints, and our main outcome is the first occurrence of relapse or death, denoted in short relapse-free survival (RFS). We do not use data on survival after relapse, because this information is not always reliable. The data have been used in Fiocco et al. (2008) and van Houwelingen & Putter (2008).

Figure A.9 shows a Kaplan-Meier plot of relapse-free survival. The lower curve of the left plot is relapse-free survival, and the upper curve distinguishes between relapse and death. The distance between $y = 1$ and the upper curve is the cumulative incidence of relapse, and the distance between the upper and the lower curve is the cumulative incidence of death before relapse. The plot on the right shows the censoring distribution; it has a different form from those of the clinical trials. Median follow-up is 6.6 years.

Figure A.10 shows, in a similar stacked plot as Figure A.8, the proportions of patients at risk with a history of PR only, with PR and AGvHD, and with AGvHD only. There were 1201 platelet recoveries, 1117 AGvHD's, and 643 patients experienced both of these events. The dynamics of these intermediate events is quite different from that of local recurrence and distant metastasis in Data Set 5. Both

Table A.5 *The clinical risk factors and their effects on survival in Data Set 5; chemo = chemotherapy, RT = radiotherapy; shown are the regression coefficients (B) and their standard errors (SE) in separate Cox models for each risk factor*

Covariate	Category	Frequency	B	SE
Type of surgery	Mastectomy with RT	636		
	Mastectomy without RT	543	-0.154	0.103
	Breast conserving	1508	-0.553	0.086
Tumor size	< 2 cm	808		
	2 - 5 cm	1723	0.531	0.093
	> 5 cm	156	1.201	0.143
Nodal status	Node negative	1425		
	Node positive	1262	0.835	0.077
Age	≤ 50	1074		
	> 50	1613	0.132	0.077
Adjuvant chemo	No	2195		
	Yes	492	0.173	0.092
Tamoxifen	No	1937		
	Yes	750	0.471	0.077
Perioperative chemo	No	1343		
	Yes	1344	-0.129	0.074

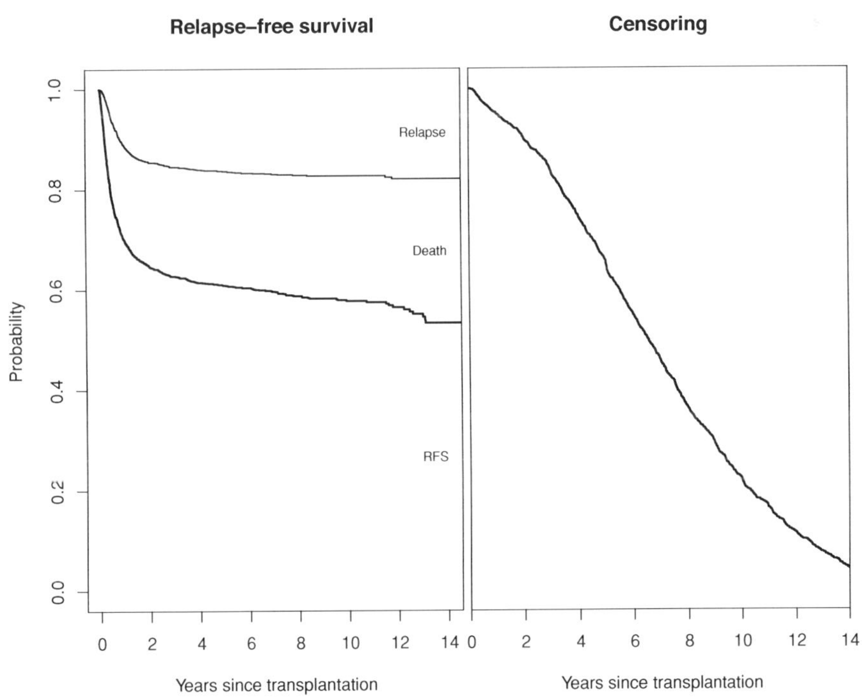

Figure A.9 *Survival and censoring functions for Data Set 6*

PR and AGvHD occur much earlier (AGvHD within 100 days post-transplant by definition). The univariate regression coefficients (SE) of $Z_{PR}(t)$ and $Z_{AGvHD}(t)$, defined in the same way as $Z_{LR}(t)$ and $Z_{DM}(t)$ in Data Set 5, are estimated as -0.341 (0.069) and 1.487 (0.083), respectively, indicating significant beneficial and harmful effects of the occurrence of PR and AGvHD, respectively, on relapse-free survival.

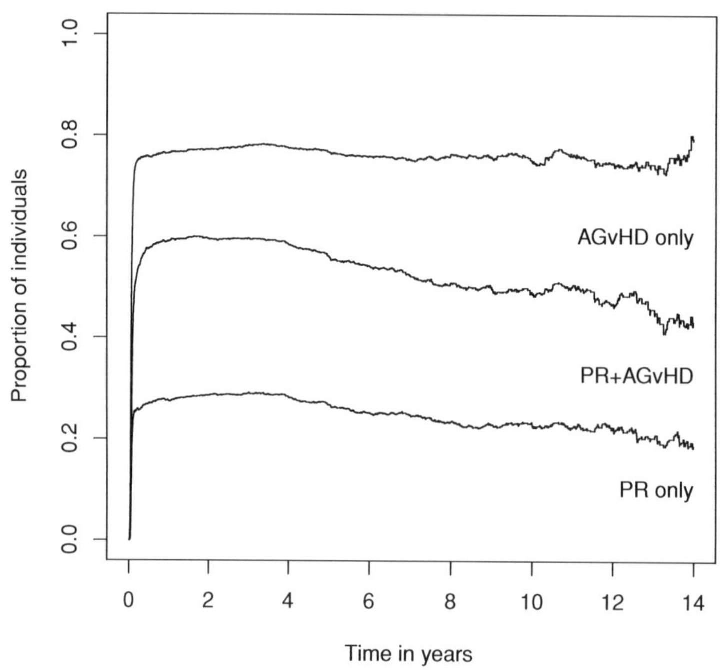

Figure A.10 *Estimated proportions of patients at risk with a history of platelet recovery and acute GvHD for Data Set 6*

Prognostic information at time of transplant are: donor recipient gender mismatch, T-cell depletion (TCD), year of transplant and age at transplant. Their frequencies and univariate effects on relapse-free survival are shown in Table A.6. Use of T-cell depletion, year of transplantation and age at transplant have clear effects on relapse-free survival.

Table A.6 *Prognostic factors and their univariate effects on relapse-free survival in Data Set 6; TCD = T-cell depletion. Shown are the regression coefficients (B) and their standard errors (SE) in separate Cox models for each risk factor*

Covariate	Category	Frequency	B	SE
Donor recipient match	No gender mismatch	1734		
	Gender mismatch	545	0.131	0.076
GvHD prevention	No TCD	1730		
	TCD	549	0.251	0.074
Year of transplantation	1985-1989	634		
	1990-1994	896	-0.304	0.079
	1995-1998	749	-0.313	0.086
Age at transplant	≤ 20	551		
	20-40	1213	0.317	0.088
	> 40	515	0.494	0.100

Appendix B

Software and website

Throughout this book, the programming language R (R Development Core Team 2010) has been used extensively for the analyses. The major advantage of R is the fact that it is freely available under the GNU license, and as a consequence that it has a large community of users that actively contribute by making available their software through so called R packages. R packages are collections of routines grouped together using a uniform packaging system that contains documentation of the functions in the packages and that ensures a minimal degree of internal consistency of the software provided. These R packages can be used for specific, often highly specialized purposes. All the analyses in this book were performed in R, and the code of the analyses, as well as most of the data sets used can be found on the book website `www.msbi.nl\DynamicPrediction`. Many analyses have been performed using a number of such R packages, and it is fair to say that many more appropriate R packages have *not* been used. For some of the methods in this book, no satisfactory implementation was available at the time of writing. For this purpose, a set of R functions for dynamic prediction in survival analyses have been gathered in the **dynpred** package.

The **dynpred** package also contains the data sets for which the authors have obtained permission to distribute freely. As a matter of courtesy towards the clinical researchers involved, reader who use these data are requested to cite the clinical papers from which these data originate. References can be found in Appendix A and at the bottom of the help pages in the **dynpred** package.

In this appendix, a summary is given of the R packages that were used or that could have been used in this book. The first section is about existing packages used in the book, the second about the new **dynpred** package. Each of these packages is available from the Comprehensive R Archive Network (CRAN, `cran.r-project.org/`), unless explicitly stated. It is important to realize that everything that is said here about these packages is necessarily only true at the time of writing; R packages continually evolve, while the content of this book is of course static. Please check the book website for major changes with respect to R packages, relevant for the content of this book. In this respect it is worthwile to mention the CRAN task view on survival analysis (`http://cran.r-project.org/web/views/Survival.html`), maintained by Arthur Allignol and Aurélien Latouche, which contains a concise summary of R packages for survival analysis.

B.1 R packages used

The ***survival*** *package*

The **survival** package (Therneau & original Splus-¿R port by Thomas Lumley 2010) is one of the oldest R packages for survival analysis, and it is certainly the most widely used and most versatile R package for survival analysis. Most, if not all, of the more specialized R packages to be discussed later in this chapter build on the **survival** package. Originally written for S-PLUS by Terry Therneau, it has been ported to R by Thomas Lumley.

The **survival** package contains all the basic methods for survival analysis and much more. Non-parametric methods include the Kaplan-Meier estimate of Chapter 1 with standard errors (both Aalen and Greenwood), and the Nelson-Aalen estimator of Chapter 2 with standard errors. Both can be obtained using the function `survfit()`. The Cox model introduced in Chapter 2 is implemented through the `coxph()` function. Stratified Cox regression is possible through the addition of a `strata` term in the `coxph()` formula. Time-dependent covariates in the Cox model can be fitted by extending the data structure as described in Section 2.5. All these methods are implemented both for right censored and left truncated data (Chapter 2). Survival functions and associated standard errors for given covariate values based on the Cox model are implemented through the function `survfit()`. Of the alternative survival models of Section 2.6, the accelerated failure time models are available through the function `survreg()`, which performs parametric regression for survival data. The function `survreg()` is used for some of the calibration methods of Chapter 4. Both martingale and Schoenfeld residuals (Section 2.6) can be obtained using `residuals.coxph()`.

In Chapter 6 Cox proportional hazards models with time-varying covariate effects are discussed. When these time-varying covariate effects are characterized through one or more pre-specified functions of time, these models can be fitted, as in the case of time-dependent covariates, by extending the data structure as in Section 2.5. The function `survSplit()` helps in creating such extended data sets. Other statistical programs, like SPSS and SAS, work in a different way by internally adjusting the Cox partial likelihood. In the end, the resulting partial likelihoods are the same, but the internal calculations of SPSS and SAS may well be quicker and do not require possibly large data sets to be stored. In the extended data set, when the pre-specified functions are continuous, an individual with observed time t needs one data row for every event time point $\leq t$. For the ovarian cancer data set, consisting of 358 individuals, the extended data set already has 59385 rows. For functions able to fit Cox proportional hazards models with time-varying effects, bypassing the construction of extended data sets, see the `coxvc()` function in B.1.

The **survival** package also fits shared frailty models (Gaussian, t or gamma distributions) through the addition of a `frailty` term in the `coxph()` formula. When the frailties are individual frailties, the gamma frailty model coincides with the Burr model. Similarly, adding a `cluster` term in the `coxph()` formula gives adjusted standard errors of the regression coefficients using sandwich estimators (Lin & Wei 1989). This feature is extensively used in the landmark super models appearing in Chapters 7-10 and 12.

Non-proportional hazards models

The cure models of Section 6.2 were fitted using the **semicure** software by Paul Peng, available from `http://www.math.mun.ca/~ypeng/research/semicure/`. It is not a true R package, but the functions within the software can be used with minor changes. See the book website for details.

The **coxvc** package (Perperoglou 2005) is available from the book website. It contains functions for the reduced rank models discussed in Section 6.3. Input are the data, covariates, the desired rank, and the time functions which are specified through a matrix of their values evaluated at the time points in the data. By specifying maximal rank, the equivalent of a full model where each covariate has different time effects is fitted. This gives an alternative to the extended data structure mentioned in B.1.

Competing risks and multi-state models

For data preparation and for dynamic prediction in non-parametric and semi-parametric multi-state models, the **mstate** (de Wreede et al. 2010) was used.

Ridge and lasso regression for survival data

For ridge and lasso regression used in Chapters 11 and 12, the **penalized** package (Goeman 2010) was used. The **penalized** package performs both ridge and lasso regression (and the combination) and selects optimal λ with cross-validated partial log-likelihood as optimality criterion. The values of the resulting prognostic index and cross-validated survival curves for the individuals may be obtained.

B.2 The dynpred package

With just a few exceptions, existing packages in R proved sufficient for estimation and for prediction from baseline. For *dynamic* prediction, however, not so much is readily available. For that reason, a number of functions have been gathered in the **dynpred** package, available from CRAN and from the book website. In the remainder of this section, a list is given of data and functions available in **dynpred** at the time of writing. This list is not intended as a complete documentation of the package; the documentation is part of the **dynpred** package and is subject to changes in the future. The book website contains R scripts in which these functions are used.

The **dynpred** package contains

Data:

- `ova`: Advanced ovarian cancer data (Section A.1).
- `wbc1`: Clinical and follow-up data from the Benelux CML study (Section A.2).
- `wbc2`: White blood cell count (WBC) data from the Benelux CML study (Section A.2).
- `nki`: Clinical and follow-up data from the NKI breast cancer data (Section A.3).
- `nkigen`: The genomic data from the NKI breast cancer data (Section A.3).
- `ALL`: Acute lymphoid leukemia (ALL) from the European Group for Bone and Marrow Transplantation (Section A.6).

Functions:

- `Fwindow()`: Calculates dynamic probabilities of dying within a fixed window, see Section 1.2.
- `toleranceplot()`: Produces tolerance plot, as described in Section 3.2.
- `scatterplot()`: Produces a scatter plot, as described in Section 3.2.
- `AUC()`: Produces an AUC(t) curve, as described in Section 3.3; it also calculates Harrell's C-index through formula (3.3).
- `AUCw()`: Produces a dynamic version of `AUC()` and of the C-index, as described in Section 3.3.
- `cindex()`: Calculates Harrell's C-index directly by counting concordant and discordant pairs.
- `CVcindex()`: Calculates a cross-validated version of the C-index, see Section 3.5.
- `CVPL()`: Calculates cross-validated log-partial likelihood (with shrinkage), see Section 3.5.
- `pe()`: Basic function that calculates Brier of Kullback-Leibler prediction error curves, given survival data, and given estimated survival and censoring curves for each individual in the data.
- `pecox()`: This function is a wrapper for `pe()`; it takes survival data and formulas for Cox models for survival and censoring. From this it calculates survival and censoring curves for each individual in the data and subsequently calls `pe()`.
- `pew()`: Dynamic version of `pe()`. The arguments are the same as for `pe()`, with the addition of the window width. It calculates dynamic fixed width Brier or Kullback-Leibler prediction error curves using (3.6).
- `pewcox()`: Dynamic version of `pecox()`, used as a wrapper for `pew()`.
- `cutLM()`: Constructs a landmark data set, given original data, a landmark

time point and a horizon at which administrative censoring is enforced. Time-dependent covariates may be specified, in which case the value at the landmark time point is included in the resulting data set.

- `predLM()`: Function for landmark prediction. It takes a landmark super model, covariate values and a sequence of time points at which prediction is required. A data frame including predictions is returned.

B.3 Additional remarks

Some R packages not used

R packages for prediction error

The **risksetROC** package (Heagerty & packaging by Paramita Saha 2011) contains functions for time-dependent ROC curves; its function `risksetROC()` returns an ROC curve, while `risksetAUC()` gives the C index. Brier prediction error curves may be obtained from the **pec** package (Gerds 2009). Unfortunately, it does not calculate Kullback-Leibler prediction error curves, but it does have possibilities of dealing with overfitting.

Competing risks and multi-state models

There has been a recent impetus with respect to software for competing risks and multi-state models. Non-parametric cumulative incidence functions can be estimated using a range of packages, such as **cmprsk** (Gray 2010) and **prodlim** (Gerds 2010). Cumulative incidence functions given covariate values within the context of Cox models for the cause-specific hazards may be obtained using **mstate** (de Wreede et al. 2010). The oldest R package for multi-state models is the **msm** package (Jackson 2011) of Chris Jackson. It is based on (piecewise) exponential models for the transition hazards, and can be used for dynamic prediction within the context of these models. For dynamic prediction in non-parametric and semi-parametric models, the **etm** (Allignol et al. 2011) can be used for non-parametric models and **mstate** for both non- and semi-parametric models. For a wider overview, readers are referred to a recent special issue of the Journal of Statistical Software (Putter 2011) on software for competing risks and multi-state models.

Miscellaneous

An R program to fit the Relaxed Burr model (not part of **dynpred**) is available from the book website. Code to fit the relaxed Burr models in Section 6.2 can also be found at the book website.

We have not discussed estimation of the variance of landmark dynamic predictions obtained from landmark super models in the book. This would entail obtaining the variance of the baseline hazard, and combining that with the uncertainty of the estimated regression coefficients. See Sections 2.7 and 6.4 for more details. Variances of landmark dynamic predictions are not yet part of **dynpred**.

Data not contained in ***dynpred***

The data of Sections A.1-A.3 and A.6 are all available from **dynpred**. Two data sets that were used in this book are not available in **dynpred** nor directly from the book website. These are the Dutch Gastric Cancer data of Section A.4 and the EORTC 10854 Breast Cancer data of Section A.5. The latter data set *is* available from the EORTC, after an agreement has been signed between the user and EORTC, guaranteeing confidentiality and privacy of the patients involved. See the book website for details and a link to the EORTC website hosting the data.

References

Aalen, O. O. (1975), Statistical Inference for a Family of Counting Processes, PhD thesis, University of California, Berkeley.

Aalen, O. O. (1980), A model for non-parametric regression analysis of life times, *in* W. Klonecki, A. Kozek & J. Rosinski, eds, 'Mathematical Statistics and Probability Theory', Vol. 2 of *Lecture Notes in Statistics*, Springer, New York, pp. 1–25.

Aalen, O. O. (1989), 'A linear-regression model for the analysis of life times', *Statistics in Medicine* **8**, 907–925.

Aalen, O. O. (1994), 'Effects of frailty in survival analysis', *Statistical Methods in Medical Research* **3**, 227–243.

Aalen, O. O., Borgan, Ø. & Gjessing, H. K. (2008), *Survival and Event History Analysis: A Process Point of View*, Statistics for Biology and Health, Springer, New York.

Aalen, O. O. & Johansen, S. (1978), 'An empirical transition matrix for nonhomogeneous Markov chains based on censored observations', *Scandinavian Journal of Statistics* **5**, 141–150.

Abbring, J. H. & van den Berg, G. J. (2003), 'The identifiability of the mixed proportional hazards competing risks model', *Journal of the Royal Statistical Society - Series B* **65**, 701–710.

Akaike, H. (1974), 'A new look at the statistical model identification', *IEEE Transactions on Automatic Control* **19**, 716–723.

Allignol, A., Schumacher, M. & Beyersmann, J. (2011), 'Empirical transition matrix of multi-state models: The **etm** package', *Journal of Statistical Software* **38**(4), 1–15.

Altman, D. G. & Royston, P. (2000), 'What do we mean by validating a prognostic model?', *Statistics in Medicine* **19**, 453–473.

Andersen, P. K. (1991), 'Survival analysis 1982–1991: The second decade of the proportional hazards regression', *Statistics in Medicine* **10**, 1931–1941.

Andersen, P. K. & Borgan, Ø. (1985), 'Counting process models for life history data: A review (with discussion)', *Scandinavian Journal of Statistics* **12**, 97–158.

Andersen, P. K., Borgan, Ø., Gill, R. D. & Keiding, N. (1988), 'Censoring, truncation and filtering in statistical models based on counting processes', *Contemporary Mathematics* **80**, 19–60.

Andersen, P. K., Borgan, Ø., Gill, R. D. & Keiding, N. (1993), *Statistical Models Based on Counting Processes*, Springer-Verlag.

Andersen, P. K. & Gill, R. D. (1982), 'Cox's regression model for counting processes: A large sample study', *Annals of Statistics* **10**, 1100–1120.

Andersen, P. K. & Keiding, N. (2002), 'Multi-state models for event history analysis', *Statistical Methods in Medical Research* **11**, 91–115.

Andersen, P. K. & Klein, J. P. (2007), 'Regression analysis for multistate models based on a pseudo-value approach, with applications to bone marrow transplantation studies', *Scandinavian Journal of Statistics* **34**, 3–16.

Andersen, P. K., Klein, J. P., Kundsen, K. M. & Tabanera y Palacios, R. (1997), 'Estimation of variance in Cox's regression model with shared gamma frailties', *Biometrics* **53**, 1475–1484.

Andersen, P. K., Klein, J. P. & Rosthøj, S. (2003), 'Generalised linear models for correlated pseudo-observations, with applications to multi-state models', *Biometrika* **90**, 15–27.

Andersen, P. K. & Liestøl, K. (2003), 'Attenuation caused by infrequently updated covariates in survival analysis', *Biostatistics* **4**, 633–649.

Anderson, J. R., Cain, K. C. & Gelber, R. D. (1983), 'Analysis of survival by tumor response', *Journal of Clinical Oncology* **1**, 710–719.

Anderson, T. W. (1951), 'Estimating linear restrictions on regression coefficients for multivariate normal distribution', *Annals of Mathematical Statistics* **22**, 327–351.

Ashburner, M., Ball, C. A., Blake, J. A., Botstein, D., Butler, H., Cherry, J. M., Davis, A. P., Dolinski, K., Dwight, S. S., Eppig, J. T., Harris, M. A., Hill, D. P., Issel-Tarver, L., Kasarskis, A., Lewis, S., Matese, J. C., Richardson, J. E., Ringwald, M., Rubin, G. M. & Sherlock, G. (2000), 'Gene ontology: tool for the unification of biology', *Nature Genetics* **25**, 25–29.

Bair, E., Hastie, T., Paul, D. & Tibshirani, R. (2006), 'Prediction by supervised principal components', *Journal of the American Statistical Association* **101**, 119–137.

Bair, E. & Tibshirani, R. (2004), 'Semi-supervised methods to predict patient survival from gene expression data', *PLOS Biology* **2**, 511–522.

Barker, P. & Henderson, R. (2004), 'Modelling converging hazards in survival analysis', *Lifetime Data Analysis* **10**, 263–281.

Barlow, R. E. & Proschan, F. (1975), *Statistical Theory of Reliability and Life Testing: Probability Models*, New York: Holt.

Benjamini, Y. & Hochberg, Y. (1995), 'Controlling the false discovery rate: A practical and powerful approach to multiple testing', *Journal of the Royal Statistical Society - Series B* **57**, 289–300.

Bennett, S. (1983*a*), 'Analysis of survival data by the proportional odds model', *Statistics in Medicine* **2**, 273–277.

Bennett, S. (1983*b*), 'Log-logistic regression models for survival data', *Journal of the Royal Statistical Society - Series C (Applied Statistics)* **32**, 165–171.

Blackstone, E. H., Naftel, D. C. & Turner, M. E. (1986), 'The decomposition of time-varying hazard into phases, each incorporating a separate stream of concomitant information', *Journal of the American Statistical Association* **81**, 615–624.

Bonenkamp, J. J., Hermans, J., Sasako, M. & van de Velde, C. J. H. (1999), 'Extended lymph-node dissection for gastric cancer', *New England Journal of Medicine* **340**, 908–914.

Bøvelstad, H. M., Nygård, S. & Borgan, Ø. (2009), 'Survival prediction from clinico-genomic models - a comparative study', *BMC Bioinformatics* **10**, art. no. 413.

Bøvelstad, H. M., Nygård, S., Størvold, H. L., Aldrin, M., Borgan, Ø., Frigessi, A. & Lingjærde, O. C. (2007), 'Predicting survival from microarray data — a comparative study', *Bioinformatics* **23**, 2080–2087.

Brant, L. J., Sheng, S. L., Morrell, C. H., Verbeke, G. N., Lesaffre, E. & Carter, H. B. (2003), 'Screening for prostate cancer by using random-effects models', *Journal of the Royal Statistical Society - Series A* **166**, 51–62.

Breslow, N. E. (1972), 'Discussion of Professor Cox's paper', *Journal of the Royal Statistical Society - Series B* **34**, 216–217.

Breslow, N. E. (1974), 'Covariance analysis of censored survival data', *Biometrics* **30**, 89–99.

Burr, I. W. (1942), 'Cumulative frequency functions', *Annals of Mathematical Statistics* **13**, 215–232.

Carroll, R. J., Rupert, D. & Stefanski, L. A. (2006), *Measurement Error in Nonlinear Models: A Modern Perspective*, 2nd edn, Chapman & Hall/CRC, Boca Raton.

Clahsen, P. C., van de Velde, C. J. H., Julien, J.-P., Floiras, J.-L., Delozier, T., Mignolet, F. Y., Sahmoud, T. M. & cooperating investigators. (1996), 'Improved local control and disease-free survival after perioperative chemotherapy for early-stage breast cancer', *Journal of Clinical Oncology* **14**, 745–753.

Commenges, D. (1999), 'Multi-state models in epidemiology', *Lifetime Data Analysis* **5**, 315–327.

Cook, R. J. & Lawless, J. F. (2007), *The Statistical Analysis of Recurrent Events*, Springer, New York.

Copas, J. B. (1983), 'Regression, prediction and shrinkage', *Journal of the Royal Statistical Society - Series B* **45**, 311–354.

Copas, J. B. (1987), 'Cross-validation shrinkage of regression predictors', *Journal of the Royal Statistical Society - Series B* **49**, 175–183.

Cortese, G. & Andersen, P. K. (2010), 'Competing risks and time-dependent covariates', *Biometrical Journal* **52**, 138–158.

Cox, D. R. (1958), 'Two further applications of a model for binary regression', *Biometrika* **45**, 562–565.

Cox, D. R. (1972), 'Regression models and life-tables', *Journal of the Royal Statistical Society - Series B* **34**, 187–220.

Dabrowska, D. M., Sun, G. W. & Horowitz, M. M. (1994), 'Cox regression in a Markov renewal model: and application to the analysis of bone-marrow transplant data', *Journal of the American Statistical Association* **89**, 867–877.

de Bruijne, M. H. J. (2001), Survival Prediction using Repeated Follow-up Measurements, PhD thesis, Leiden University.

de Bruijne, M. H. J., le Cessie, S., Kluin-Nelemans, H. C. & van Houwelingen, H. C. (2001), 'On the use of Cox regression in the presence of an irregularly observed time-dependent covariate', *Statistics in Medicine* **20**, 3817–3829.

De Gruttola, V. & Tu, X. M. (1994), 'Modelling progression of CD4-lymphocyte count and its relationship to survival time', *Biometrics* **50**, 1003–1014.

de Wreede, L. C., Fiocco, M. & Putter, H. (2010), 'The **mstate** package for estimation and prediction in non- and semi-parametric multi-state and competing risks models', *Computer Methods and Programs in Biomedicine* **99**, 261–274.

Di Serio, C. (1997), 'The protective impact of a covariate on competing failures with an example from a bone marrow transplantation study', *Lifetime Data Analysis* **3**, 99–122.

Dickman, P. W., Sloggett, A., Hills, M. & Hakulinen, T. (2004), 'Regression models for relative survival', *Statistics in Medicine* **23**, 51–64.

Duchateau, L. & Janssen, P. (2008), *The Frailty Model*, Springer, New York.

Eilers, P. H. C. & Marx, B. D. (1996), 'Flexible Smoothing with B-splines and Penalties', *Statistical Science* **11**, 89–121.

Elbers, C. & Ridder, G. (1982), 'True and spurious duration dependence: The identifiability of the proportional hazard model', *The Review of Economic Studies* **XLIX**, 403–409.

Estève, J., Benhamou, E., Croasdale, M. & Raymond, L. (1990), 'Relative survival and the estimation of net survival - Elements for further discussion', *Statistics in Medicine* **9**, 529–538.

Feller, W. (1950), *An Introduction to Probability Theory and its Applications*, Vol. 1, Wiley, Chichester.

Fieuws, S., Verbeke, G., Maes, B. & Vanrenterghem, Y. (2008), 'Predicting renal graft failure using multivariate longitudinal profiles', *Biostatistics* **9**, 419–431.

Fine, J. P. & Gray, R. J. (1999), 'A proportional hazards model for the subdistribution of a competing risk', *Journal of the American Statistical Association* **94**, 496–509.

Fiocco, M., Putter, H. & van Houwelingen, H. C. (2008), 'Reduced-rank proportional hazards regression and simulation-based prediction for multi-state models', *Statistics in Medicine* **27**, 4340–4358.

Fiocco, M., Putter, H. & van Houwelingen, J. C. (2005), 'Reduced rank proportional hazards model for competing risks', *Biostatistics* **6**, 465–478.

Gerds, T. A. (2009), ***pec**: Prediction Error Curves for Survival Models.* R package version 1.1.1, URL: `http://CRAN.R-project.org/package=pec`.

Gerds, T. A. (2010), ***prodlim**: Product Limit Estimation.* R package version 1.1.3, URL: `http://CRAN.R-project.org/package=prodlim`.

Gerds, T. A. & Schumacher, M. (2006), 'Consistent estimation of the expected brier score in general survival models with right-censored event times', *Biometrical Journal* **48**, 1029–1040.

Gifi, A. (1990), *Nonlinear Multivariate Analysis*, Wiley, Chichester.

Goeman, J. J. (2010), 'L_1 penalized estimation in the Cox proportional hazards model', *Biometrical Journal* **52**, 70–84.

Goeman, J. J., Oosting, J., Cleton-Jansen, A.-M., Anninga, J. K. & van Houwelingen, H. C. (2005), 'Testing association of a pathway with survival using gene expression data', *Bioinformatics* **21**, 1950–1957.

Goeman, J. J., van de Geer, S. A., de Kort, F. & van Houwelingen, H. C. (2004), 'A global test for groups of genes: testing association with a clinical outcome', *Bioinformatics* **20**, 93–99.

Goeman, J. J., van de Geer, S. A. & van Houwelingen, H. C. (2006), 'Testing against a high dimensional alternative', *Journal of the Royal Statistical Society - Series B* **68**, 477–493.

Golub, T. R., Slonim, D. K., Tamayo, P., Huard, C., Gaasenbeek, M., Mesirov, J. P., Coller, H., Loh, M. L., Downing, J. R., Caligiuri, M. A., Bloomfeld, C. D. & Lander, E. S. (1999), 'Molecular classification of cancer: class discovery and class prediction by gene expression monitoring', *Science* **286**, 531–537.

Graf, E., Schmoor, C., Sauerbrei, W. & Schumacher, M. (1999), 'Assessment and comparison of prognostic classification schemes for survival data', *Statistics in Medicine* **18**, 2529–2545.

Grambsch, P. M. & Therneau, T. M. (1994), 'Proportional hazards tests and diagnostics based on weighted residuals', *Biometrika* **81**, 515–526.

Grambsch, P. M. & Therneau, T. M. (1995), 'Diagnostic plots to reveal functional form for covariates in multiplicative intensity models', *Biometrics* **51**, 1469–1482.

Gran, J. M., Røysland, K., Wolbers, M., Didelez, V., Sterne, J. A. C., Ledergerber, B., Furrer, H., von Wyl, V. & Aalen, O. O. (2010), 'A sequential Cox approach for estimating the causal effect of treatment in the presence of time-dependent confounding applied to data from the Swiss HIV Cohort Study', *Statistics in Medicine* **29**, 2757–2768.

Graunt, J. (1662), Natural and political observations made upon the bills of mortality, *in* 'The World of Mathematics', Vol. 3, Simon and Schuster, New York, pp. 1421–1436.

Gray, B. (2010), ***cmprsk**: Subdistribution Analysis of Competing Risks.* R package version 2.2-1, URL: `http://CRAN.R-project.org/package=cmprsk`.

Greenwood, M. (1926), The natural duration of cancer, *in* 'Reports on Public Health and Medical Subjects 33', London: His Majesty's Stationery Office, pp. 1–26.

Hakulinen, T. & Tenkanen, L. (1987), 'Regression analysis of relative survival rates', *Journal of the Royal Statistical Society - Series C (Applied Statistics)* **36**, 309–317.

Halley, E. (1693), An estimate of the degrees of the mortality of mankind, drawn from curious tables of the births and funerals at the city of Breslaw; With an attempt to ascertain the price of annuities upon lives, *in* 'The World of Mathematics', Vol. 3, Simon and Schuster, New York, pp. 1437–1447.

Hansen, B. E., Thorogood, J., Hermans, J., Ploeg, R. J., van Bockel, J. H. & van Houwelingen, J. C. (1994), 'Multistate modelling of liver transplantation data', *Statistics in Medicine* **13**, 2517–2529.

Harrell, F. E. (2001), *Regression Modeling Strategies: with Applications to Linear Models, Logistic Regression, and Survival Analysis*, Springer, New York.

Harrell, F. E., Lee, K. L. & Mark, D. B. (1996), 'Multivariable prognostic models: issues in developing models, evaluating assumptions and adequacy, and measuring and reducing errors', *Statistics in Medicine* **15**, 361–387.

Hartgrink, H. H., van de Velde, C., Putter, H., Bonenkamp, J., Klein Kranenbarg, E., Songun, I., Welvaart, K., van Krieken, J., Meijer, S., Plukker, J., van Elk, P., Obertop, H., Gouma, D., van Lanschot, J., Taat, C., de Graaf, P., von Meyenfeldt, M., Tilanus, H. & Sasako, M. (2004), 'Extended lymph node dissection for gastric cancer: Who may benefit? Final results of the randomized Dutch Gastric Cancer Group Trial', *Journal of Clinical Oncology* **22**, 2069–2077.

Hastie, T., Tibshirani, R. & Friedman, J. (2009), *The Elements of Statistical Learning: Data Mining, Inference, and Prediction*, 2nd edn, Springer, New York.

Heagerty, P. J. & packaging by Paramita Saha (2011), ***risksetROC**: Riskset ROC curve estimation from censored survival data.* R package version 1.0.3, URL: `http://CRAN.R-project.org/package=risksetROC`.

Heagerty, P. J. & Zheng, Y. (2005), 'Survival model predictive accuracy and ROC curves', *Biometrics* **61**, 92–105.

Henderson, R., Diggle, P. & Dobson, A. (2000), 'Joint modelling of longitudinal measurements and event time data', *Biostatistics* **1**, 465–480.

Henderson, R., Jones, M. & Stare, J. (2001), 'Accuracy of point predictions in survival analysis', *Statistics in Medicine* **20**, 3083–3096.

Hermans, J., Bonenkamp, J. J., Sasako, M. & van de Velde, C. J. H. (1999), 'Stage migration in gastric cancer: Its influence on survival rates', *Chirurgische Gastroenterologie* **15**, 249–252.

Hjort, N. L. (1992), 'On inference in parametric survival data models', *International Statistical Review* **60**, 355–387.

Hoerl, A. E. & Kennard, R. W. (1970), 'Ridge regression: biased estimation for nonorthogonal problems', *Technometrics* **1**, 55–67.

Hoeting, J. A., Madigan, D., Raftery, A. E. & Volinsky, C. T. (1999), 'Bayesian model averaging: A tutorial', *Statistical Science* **14**, 382–401.

Hogan, J. W. & Laird, N. M. (1997*a*), 'Mixture models for the joint distribution of repeated measures and event times', *Statistics in Medicine* **16**, 239–257.

Hogan, J. W. & Laird, N. M. (1997*b*), 'Model-based approaches to analysing incomplete longitudinal and failure time data', *Statistics in Medicine* **16**, 259–272.

Hosmer, D. W., Hosmer, T., le Cessie, S. & Lemeshow, S. (1997), 'A comparison of goodness-of-fit tests for the logistic regression model', *Statistics in Medicine* **16**, 965–980.

Hosmer, D. W. & Lemeshow, S. (1980), 'A goodness-of-fit test for the multiple logistic regression model', *Communications in Statistics A* **10**, 1043–1069.

Hougaard, P. (1999), 'Multi-state models: A review', *Lifetime Data Analysis* **5**, 239–264.

Hougaard, P. (2000), *Analysis of Multivariate Survival Data*, Springer, New York.

Jackson, C. H. (2011), 'Multi-state models for panel data: The **msm** package for R', *Journal of Statistical Software* **38**(8), 1–29.

Janssen-Heijnen, M. L., Houterman, S., Lemmens, E. V., Brenner, H., Steyerberg, E. W. & Coebergh, J. W. (2007), 'Prognosis for long-term survivors of cancer', *Annals of Oncology* **18**, 1408–1413.

Jewell, N. P. & Nielsen, J. P. (1993), 'A framework for consistent prediction rules based on markers', *Biometrika* **80**, 153–164.

Kalbfleisch, J. D. & Prentice, R. L. (2002), *The Statistical Analysis of Failure Time Data*, Wiley, New York.

Kaplan, E. & Meier, P. (1958), 'Nonparametric estimation from incomplete observations', *Journal of the American Statistical Association* **43**, 457–481.

Kauermann, G. & Berger, U. (2003), 'A smooth test in proportional hazard survival models using local partial likelihood fitting', *Lifetime Data Analysis* **9**, 373–393.

Keiding, N., Andersen, P. K. & Klein, J. P. (1997), 'The role of frailty models and accelerated failure time models in describing heterogeneity due to omitted covariates', *Statistics in Medicine* **16**, 215–224.

Keiding, N., Klein, J. P. & Horowitz, M. M. (2001), 'Multi-state models and outcome prediction in bone marrow transplantation', *Statistics in Medicine* **20**, 1871–1885.

Kent, J. T. & O'Quigley, J. (1988), 'Measures of dependence for censored survival data', *Biometrika* **75**, 525–534.

Klein, J. P. (1992), 'Semiparametric estimation of random effects using the Cox model based on the EM algorithm', *Biometrics* **48**, 795–806.

Klein, J. P. & Andersen, P. K. (2005), 'Regression modeling of competing risks data based on pseudovalues of the cumulative incidence function', *Biometrics* **61**, 223–229.

Klein, J. P., Keiding, N. & Copelan, E. A. (1994), 'Plotting summary predictions in multistate survival models: Probabilities of relapse and death in remission for bone-marrow transplantation patients', *Statistics in Medicine* **12**, 2315–2332.

Klein, J. P. & Moeschberger, M. L. (2003), *Survival Analysis: Techniques for Censored and Truncated Data*, Statistics for Biology and Health, 2nd edn, Springer, New York.

Klein, J. P. & Shu, Y. Y. (2002), 'Multi-state models for bone marrow transplantation studies', *Statistical Methods in Medical Research* **11**, 117–139.

Kluin-Nelemans, J. C., Delannoy, A., Louwagie, A., le Cessie, S., Hermans, J., van der Burgh, J. F., Hagemeijer, A. M., van den Berghe, H. & Benelux CML Study Group (1998), 'Randomized study on hydroxyurea alone versus hydroxyurea combined with low-dose interferon-alpha 2b for chronic myeloid leukemia', *Blood* **91**, 2713–2721.

Korn, E. L. & Simon, R. (1990), 'Measures of explained variation for survival data', *Statistics in Medicine* **9**, 487–503.

Kortram, R. A., van Rooij, A. C. M., Lenstra, A. J. & Ridder, G. (1995), 'Constructive identification of the mixed proportional hazards model', *Statistica Neerlandica* **49**, 269–281.

Koziol, J. A. & Jia, Z. (2009), 'The concordance index C and the Mann-Whitney parameter Pr(X¿Y) with randomly censored data', *Biometrical Journal* **51**, 467–474.

Krzanowski, W. J. & Hand, D. J. (2009), *ROC Curves for Continuous Data*, CRC Press.

Kuk, A. Y. C. & Chen, C.-H. (1992), 'A mixture model combining logistic regression with proportional hazards regression', *Biometrika* **79**, 531–541.

le Cessie, S. & van Houwelingen, H. C. (1995), 'Testing the fit of a regression model via score tests in random effects models', *Biometrics* **51**, 600–614.

le Cessie, S. & van Houwelingen, J. C. (1991), 'A goodness-of-fit test for binary regression models, based on smoothing methods', *Biometrics* **47**, 1267–1282.

le Cessie, S. & van Houwelingen, J. C. (1992), 'Ridge estimators in logistic regression', *Journal of the Royal Statistical Society - Series C (Applied Statistics)* **41**, 191–201.

Liang, K. Y. & Zeger, S. L. (1986), 'Longitudinal data-analysis using generalized linear models', *Biometrika* **73**, 13–22.

Lin, D. Y. & Wei, L. J. (1989), 'The robust inference for the Cox proportional hazards model', *Journal of the American Statistical Association* **84**, 1074–1078.

Lin, D. Y. & Ying, Z. (1994), 'Semiparametric analysis of the additive risk model', *Biometrika* **81**, 61–71.

Martinussen, T. & Scheike, T. H. (2006), *Dynamic Regression Models for Survival Data*, Statistics for Biology and Health, Springer, New York.

McKeague, I. W. & Sasieni, P. D. (1994), 'A partly parametric additive risk model', *Biometrika* **81**, 501–514.

Mitchell, T. J. & Beauchamp, J. J. (1988), 'Bayesian variable selection in linear regression', *Journal of the American Statistical Association* **83**, 1023–1032.

Murphy, S. A., Rossini, A. J. & van der Vaart, A. W. (1997), 'Maximum likelihood estimation in the proportional odds model', *Journal of the American Statistical Association* **92**, 968–976.

Neijt, J. P., ten Bokkel Huinink, W. W., van der Burg, M. E., van Oosterom, A. T., Vriesendorp, R., Kooyman, C. D., van Lindert, A. C., Hamerlynck, J. V., van Lent, M. & van Houwelingen, J. C. (1984), 'Randomised trial comparing two combination chemotherapy regimens (Hexa-CAF vs CHAP-5) in advanced ovarian carcinoma', *Lancet* **2**, 594–600.

Neijt, J. P., ten Bokkel Huinink, W. W., van der Burg, M. E., van Oosterom, A. T., Willemse, P. H., Heintz, A. P., van Lent, M., Trimbos, J. B., Bouma, J. & Vermorken, J. B. (1987), 'Randomized trial comparing two combination chemotherapy regimens (CHAP-5 vs CP) in advanced ovarian carcinoma', *Journal of Clinical Oncology* **5**, 1157–1168.

Nelson, W. (1969), 'Hazard plotting for incomplete failure data', *Journal of Quality Technology* **1**, 27–52.

Nicolaie, M. A., van Houwelingen, H. C. & Putter, H. (2010), 'Vertical modeling: a pattern mixture approach for competing risks modeling', *Statistics in Medicine*

29, 1190–1205.

Nielsen, G., Gill, R. D., Andersen, P. K. & Sørensen, T. I. A. (1992), 'A counting process approach to maximum likelihood estimation in frailty models', *Scandinavian Journal of Statistics* **19**, 25–43.

Nygård, S., Borgan, Ø., Lingjærde, O. C. & Storvold, H. L. (2008), 'Partial least squares Cox regression for genome-wide data', *Lifetime Data Analysis* **14**, 179–195.

O'Quigley, J. & Flandre, P. (1994), 'Predictive capability of proportional hazards regression', *Proceedings of the National Academy of Sciences USA* **91**, 2310–2314.

O'Quigley, J., Xu, R. & Stare, J. (2005), 'Explained randomness in proportional hazards models', *Statistics in Medicine* **24**, 479–489.

Pawitan, Y. & Self, S. (1993), 'Modeling disease marker processes in AIDS', *Journal of the American Statistical Association* **88**, 719–726.

Peng, Y. (2003), 'Fitting semiparametric cure models', *Computational Statistics and Data Analysis* **41**, 481–490.

Perperoglou, A. (2005), ***coxvc**: Cox models with time varying effects of the covariates and Reduced Rank models.* R package version 1-1-1.

Perperoglou, A., Keramopoullos, A. & van Houwelingen, H. C. (2007), 'Approaches in modelling long-term survival: an application to breast cancer', *Statistics in Medicine* **26**, 2666–2685.

Perperoglou, A., le Cessie, S. & van Houwelingen, H. C. (2006*a*), 'A fast routine for fitting cox models with time varying effects of the covariates', *Computer Methods and Programs in Biomedicine* **81**, 154–161.

Perperoglou, A., le Cessie, S. & van Houwelingen, H. C. (2006*b*), 'Reduced-rank hazard regression for modelling non-proportional hazards', *Statistics in Medicine* **25**, 2831–2845.

Perperoglou, A., van Houwelingen, H. C. & Henderson, R. (2006), 'A relaxation of the gamma frailty (Burr) model', *Statistics in Medicine* **25**, 4253–4266.

Peto, R. (1972), 'Discussion of Professor Cox's paper', *Journal of the Royal Statistical Society - Series B* **34**, 205–207.

Petricoin, E. F., Ardekani, A. M., Hitt, B. A., Levine, P. J., Fusaro, V. A., Steinberg, S. M., Mills, G. B., Simone, C., Fishman, D. A., Kohn, E. C. & Liotta, L. A. (2002), 'Use of proteomic patterns in serum to identify ovarian cancer', *Lancet* **359**, 572–577.

Pettitt, A. N. (1982), 'Inference for the linear model using a likelihood based on ranks', *Journal of the Royal Statistical Society - Series B* **44**, 234–243.

Proust-Lima, C. & Taylor, J. M. G. (2009), 'Development and validation of a dynamic prognostic tool for prostate cancer recurrence using repeated measures of posttreatment PSA: a joint modeling approach', *Biostatistics* **10**, 535–549.

Putter, H. (2011), 'Special issue about competing risks and multi-state models', *Journal of Statistical Software* **38**(1), 1–4.

Putter, H., Fiocco, M. & Geskus, R. (2007), 'Tutorial in biostatistics: competing risks and multi-state models', *Statistics in Medicine* **26**, 2277–2432.

Putter, H., le Cessie, S. & Stijnen, T. (2010), 'Hans van Houwelingen, 40 years in biostatistics', *Biometrical Journal* **52**, 5–9.

Putter, H., Sasako, M., Hartgrink, H. H., van de Velde, C. J. H. & van Houwelingen, J. C. (2005), 'Long-term survival with non-proportional hazards: results from the Dutch Gastric Cancer Trial', *Statistics in Medicine* **24**, 2807–2821.

Putter, H., van der Hage, J., de Bock, G. H., Elgalta, R. & van de Velde, C. J. H. (2006), 'Estimation and prediction in a multi-state model for breast cancer', *Biometrical Journal* **48**, 366–380.

Putter, H. & van Houwelingen, J. C. (2011), Frailties in multi-state models: Are they identifiable? Do we need them? Submitted.

R Development Core Team (2010), *R: A Language and Environment for Statistical Computing*, R Foundation for Statistical Computing, Vienna, Austria. ISBN 3-900051-07-0, URL: `http://www.R-project.org/`.

Ripatti, S. & Palmgren, J. (2000), 'Estimation of multivariate frailty models using penalized partial likelihood', *Biometrics* **56**, 1016–1022.

Robbins, H. (1956), An empirical Bayes approach to statistics, *in* 'Proceedings of the 3rd Berkeley Symposium on Mathematical Statistics and Probabilitiy', Vol. 1, pp. 157–163.

Rondeau, V., Mathoulin-Pelissier, S., Jacqmin-Gadda, H., Brouste, V. & Soubeyran, P. (2007), 'Joint frailty models for recurring events and death using maximum penalized likelihood estimation: Application on cancer events', *Biostatistics* **8**, 708–721.

Royston, P. (2001), 'The lognormal distribution as a model for survival time in cancer, with an emphasis on prognostic factors', *Statistica Neerlandica* **55**, 89–104.

Royston, P. (2006), 'Explained variation for survival models', *The Stata Journal* **6**, 83–96.

Royston, P. & Parmar, M. K. B. (2002), 'Flexible parametric proportional-hazards and proportional-odds models for censored survival data, with application to prognostic modelling and estimation of treatment effects', *Statistics in Medicine* **21**, 2175–2197.

Royston, P. & Sauerbrei, W. (2004), 'A new measure of prognostic separation in survival data', *Statistics in Medicine* **23**, 723–748.

Royston, P. & Sauerbrei, W. (2008), *Multivariable Model-building: A Pragmatic Approach to Regression Analysis based on Fractional Polynomials for Modelling Continous Variables*, Wiley, Chichester.

Sauerbrei, W; Royston, P. (1999), 'Building multivariable prognostic and diagnostic models: transformation of the predictors by using fractional polynomials', *Journal of the Royal Statistical Society - Series A* **162**, 71–94.

Sauerbrei, W., Royston, P. & Look, M. (2007), 'A new proposal for multivariable modelling of time-varying effects in survival data based on fractional polynomial time-transformation', *Biometrical Journal* **49**, 453– 473.

Scheike, T. H. & Zhang, M. J. (2007), 'Direct modelling of regression effects for transition probabilities in multistate models', *Scandinavian Journal of Statistics* **34**, 17–32.

Scheike, T. H., Zhang, M. J. & Gerds, T. A. (2008), 'Predicting cumulative incidence probability by direct binomial regression', *Biometrika* **95**, 205–220.

Schemper, M. (2003), 'Predictive accuracy and explained variation', *Statistics in Medicine* **22**, 2299–2308.

Schemper, M. & Henderson, R. (2000), 'Predictive accuracy and explained variation in Cox regression', *Biometrics* **56**, 249–255.

Schemper, M. & Stare, J. (1996), 'Explained variation in survival analysis', *Statistics in Medicine* **15**, 1999–2012.

Schoenfeld, D. (1982), 'Partial residuals for the proportional hazards regression model', *Biometrika* **69**, 239–241.

Smits, J. M. A., Deng, M. C., Hummel, M., De Meester, J., Schoendube, F., Scheld, H. H., Persijn, G. G., Laufer, G., Van Houwelingen, H. C. & Comparative Outcome and Clinical Profiles in Transplantation (COCPIT) Study Group (2003), 'A prognostic model for predicting waiting-list mortality for a total national cohort of adult heart-transplant candidates', *Transplantation* **76**, 1185–1189.

Stare, J., Pohar Perme, M. & Henderson, R. (2011), 'A measure of explained variation for event history data', *Biometrics*. In press; DOI: 10.1111/j.1541-0420.2010.01526.x.

Steyerberg, E. W. (2009), *Clinical Prediction Models: A Practical Approach to Development, Validation, and Updating*, Springer, New York.

Steyerberg, E. W., Borsboom, G. J. J. M., van Houwelingen, H. C., Eijkemans, M. J. C. & Habbema, J. D. F. (2004), 'Validation and updating of predictive logistic regression models: a study on sample size and shrinkage', *Statistics in Medicine* **23**, 2567–2586.

Steyerberg, E. W., Eijkemans, M. J. C. & Habbema, J. D. F. (2001), 'Application of shrinkage techniques in logistic regression analysis: a case study', *Statistica Neerlandica* **55**, 76–88.

Steyerberg, E. W., Eijkemans, M. J., Houwelingen, J. C. V., Lee, K. L. & Habbema, J. D. (2000), 'Prognostic models based on literature and individual patient data in logistic regression analysis', *Statistics in Medicine* **19**, 141–160.

Struthers, C. A. & Kalbfleisch, J. D. (1986), 'Misspecified proportional hazard models', *Biometrika* **73**, 363–369.

The Benelux CML Study Group (1998), 'Randomized study on hydroxyurea alone versus hydroxyurea combined with low-dose interferon-α2b for chronic myeloid leukemia', *Blood* **91**, 2173–2721.

Therneau, T., Grambsch, P. & Pankratz, V. S. (2003), 'Penalized survival models and frailty', *Journal of Computational and Graphical Statistics* **12**, 156–175.

Therneau, T. M. & Grambsch, P. M. (2000), *Modeling Survival Data: Extending the Cox Model*, Springer, New York.

Therneau, T. & original Splus-¿R port by Thomas Lumley (2010), ***survival**: Survival analysis, including penalised likelihood.* R package version 2.36-2, URL: `http://CRAN.R-project.org/package=survival`.

Thorogood, J., van Houwelingen, J. C., Persijn, G. G., Zandvoort, F. A., Schreuder, T. G. M. & van Rood, J. J. (1991), 'Prognostic indices to predict survival of first and second renal allografts', *Transplantation* **52**, 831–836.

Tibshirani, R. (1996), 'Regression shrinkage and selection via the lasso', *Journal of the Royal Statistical Society - Series B* **28**, 267–288.

Tibshirani, R. (1997), 'The lasso method for variable selection in the Cox model', *Statistics in Medicine* **16**, 385–395.

Tsiatis, A. A. (1981), 'A large sample study of Cox's regression model', *Annals of Statistics* **9**, 93–108.

Tsiatis, A. A., DeGruttola, V. & Wulfsohn, M. S. (1995), 'Modeling the relationship of survival to longitudinal data measured with error. Applications to survival and CD4 counts in patients with AIDS', *Journal of the American Statistical Association* **90**, 27–37.

van de Vijver, M. J., He, Y. D., van 't Veer, L. J., Dai, H., Hart, A. A. M., Voskuil, D. W., Schreiber, G. J., Peterse, J. L., Roberts, C., Marton, M. J., Parrish, M., Atsma, D., Witteveen, A., Glas, A., Delahaye, L., van der Velde, T., Bartelink, H., Rodenhuis, S., Rutgers, E. T., Friend, S. H. & Bernards, R. (2002), 'A gene-expression signature as a predictor of survival in breast cancer', *New England Journal of Medicine* **347**, 1999–2009.

van der Hage, J. A., van de Velde, C. J. H., Julien, J.-P., Floiras, J.-L., Delozier, T., Vandervelden, C., Duchateau, L. & cooperating investigators (2001), 'Improved survival after one course of perioperative chemotherapy in early breast cancer patients: long-term results from the European Organization for Research and Treatment of Cancer (EORTC) trial 10854', *European Journal of Cancer* **37**, 2184–2193.

van der Laan, M. J., Polley, E. C. & Hubbard, A. E. (2007), 'Super learner', *Statistical Applications in Genetics and Molecular Biology* **6**. Article 25.

van Houwelingen, H. C. (2000), 'Validation, calibration, revision and combination of prognostic survival models', *Statistics in Medicine* **19**, 3401–3415.

van Houwelingen, H. C. (2007), 'Dynamic prediction by landmarking in event history analysis', *Scandinavian Journal of Statistics* **34**, 70–85.

van Houwelingen, H. C., Arends, L. R. & Stijnen, T. (2002), 'Advanced methods in meta-analysis: multivariate approach and meta-regression', *Statistics in Medicine* **21**, 589–624.

van Houwelingen, H. C., Bruinsma, T., Hart, A. A. M., van't Veer, L. J. & Wessels, L. F. A. (2006), 'Cross-validated Cox regression on microarray gene expression data', *Statistics in Medicine* **25**, 3201–3216.

van Houwelingen, H. C. & Putter, H. (2008), 'Dynamic predicting by landmarking as an alternative for multi-state modeling: an application to acute lymphoid leukemia data', *Lifetime Data Analysis* **14**, 447–463.

van Houwelingen, H. C. & Thorogood, J. (1995), 'Construction, validation and updating of a prognostic model for kidney graft survival', *Statistics in Medicine* **14**, 1999–2008.

van Houwelingen, H. C., van de Velde, C. J. H. & Stijnen, T. (2005), 'Interim analysis on survival data: Its potential bias and how to repair it', *Statistics in Medicine* **24**, 2823–2835.

van Houwelingen, J. C. (1976), 'Monotone empirical Bayes tests for continuous one-parameter exponential family', *Annals of Statistics* **4**, 981–989.

van Houwelingen, J. C. & le Cessie, S. (1990), 'Predictive value of statistical models', *Statistics in Medicine* **9**, 1303–1325.

van Houwelingen, J. C., ten Bokkel Huinink, W. W., van der Burg, M. E., van Oosterom, A. T. & Neijt, J. P. (1989), 'Predictability of the survival of patients with advanced ovarian cancer', *Journal of Clinical Oncology* **7**, 769–773.

van Nes, J. G. H., Putter, H., van Hezewijk, M., Hille, E. T. M., Bartelink, H., Collette, L. & van de Velde, C. J. H. (2010), 'Tailored follow-up for early breast cancer patients: a prognostic index that predicts locoregional recurrence', *European Journal of Surgical Oncology* **36**, 617–624.

van't Veer, L. J., Dai, H. Y., van de Vijver, M. J., He, Y. D. D., Hart, A. A. M., Mao, M., Peterse, H. L., van der Kooy, K., Marton, M. J., Witteveen, A. T., Schreiber, G. J., Kerkhoven, R. M., Roberts, C., Linsley, P. S., Bernards, R. & Friend, S. H. (2002), 'Gene expression profiling predicts clinical outcome of breast cancer', *Nature* **415**, 530–536.

Varma, S. & Simon, R. (2006), 'Bias in error estimation when using cross-validation for model selection', *BMC Bioinformatics* **7**, art. no. 91.

Verweij, P. J. M. & van Houwelingen, H. C. (1993), 'Cross-validation in survival analysis', *Statistics in Medicine* **12**, 2305–2314.

Verweij, P. J. M. & van Houwelingen, H. C. (1994), 'Penalized likelihood in Cox regression', *Statistics in Medicine* **13**, 2427–2436.

Verweij, P. J. M. & van Houwelingen, H. C. (1995), 'Time-dependent effects of fixed covariates in Cox regression', *Biometrics* **51**, 1550–1556.

Verweij, P. J. M., van Houwelingen, H. C. & Stijnen, T. (1998), 'A goodness-of-fit test for Cox's proportional hazards model based on martingale residuals', *Biometrics* **54**, 1517–1526.

Wulfsohn, M. S. & Tsiatis, A. A. (1997), 'A joint model for survival and longitudinal data measured with error', *Biometrics* **53**, 330–339.

Xu, R. & O'Quigley, J. (2000), 'Estimating averate regression effect under non-proportional hazards', *Biostatistics* **1**, 423–439.

Zheng, Y. Y. & Heagerty, P. J. (2005), 'Partly conditional survival models for longitudinal data', *Biometrics* **61**, 379–391.

Index

An environmentally friendly book printed and bound in England by www.printondemand-worldwide.com

PEFC Certified

This product is from sustainably managed forests and controlled sources

PEFC

PEFC/16-33-415

www.pefc.org

FSC

www.fsc.org

MIX

Paper from responsible sources

FSC® C004959

This book is made entirely of sustainable materials; FSC paper for the cover and PEFC paper for the text pages.

#0118 - 070114 - C0 - 254/178/13 [15] - CB

The Science of Renewable Energy

The Science of Renewable Energy

Frank Spellman and Revonna Bieber

CRC Press
Taylor & Francis Group
Boca Raton London New York

CRC Press is an imprint of the
Taylor & Francis Group, an **informa** business

CRC Press
Taylor & Francis Group
6000 Broken Sound Parkway NW, Suite 300
Boca Raton, FL 33487-2742

Printed in the United States of America on acid-free paper
10 9 8 7 6 5 4 3 2 1

International Standard Book Number: 978-1-4398-2502-0 (Hardback)

Visit the Taylor & Francis Web site at
http://www.taylorandfrancis.com

and the CRC Press Web site at
http://www.crcpress.com

For Lyn Hunter

Contents

Preface

Studying renewable energy without science is analogous to attempting to cook without being able to read a recipe or to measure ingredients. Moreover, studying renewable energy through science affords us the opportunity not only to grasp the pressing need for renewable energy but also to evaluate and define the complicated task involved in developing viable energy sources for the future. For future energy needs, there is no question or doubt that we are facing an international crisis of unprecedented proportion. We are headed for a train wreck—the end of affordable and accessible petroleum products and the resulting economic and political impact. Keep in mind that we need to replace liquid hydrocarbon fuels because the wells are almost dry, but also remember that we must clean up the environment as well. Production of the reliable, renewable energy that is so essential to us should not destroy or further pollute our fragile environment.

Even though peak oil is here and hydrocarbon energy resources will soon be strained to the limit, and even though we recognize the disastrous, oncoming train wreck of world economies collapsing, we need to follow Adrienne Rich's* perfect words and "dive into the wreck." Simply, a substantial and reliable supply of renewable energy is the wreck preventer we need; simply by pursuing emerging technologies that include making oil or energy from our usual discards—garbage, algae, plastics, sewage, agricultural and forestry biomass and waste—to power fuel cells and full-sized, plug-in diesel hybrid SUVs, we can minimize or even avoid the damage this approaching train wreck is certain to cause if we continue on our present course. The development of these emerging technologies must be accompanied by advancements in solar, wind, and ocean technologies.

Many of us have come to realize that we pay a price (sometimes a high price) for what we call the "good life." Consumption of the world's resources makes us all at least partially responsible for the shortages and the resulting environmental pollution. It is a matter of proven record that pollution and its ramifications are the inevitable byproducts of a good life maintained by fossil fuels. Fuel shortages and pollution due to the use of fossil fuel cannot be attributed to any one individual. To reduce fuel shortages and its harmful effects, we must band together as an informed, knowledgeable electorate and pressure our elected decision-makers to manage the problem now, for the future. It is imperative that we replace fossil fuels with clean and efficient renewable energy. We must quiet that vocal minority of doomsayers and naysayers who claim that the transition from our current fossil fuel–based economy to one that is based entirely on renewable energy would be too costly and impractical. This text was not written to silence these doubting Thomases, but instead to make their arguments hollow, and at best, moot. Even as individuals we can have an impact on the quest to develop renewable energy sources, to replace dwindling supplies of crude oil, and to preserve and repair the environment. The concepts and guidelines provided in this text will equip the reader with the critical scientific rationale required to make decisions directly affecting our quest for renewable energy sources—decisions that not only make sense but also work to preserve the environment around us.

Throughout this text, we present common-sense approaches and practical (and sometimes poetic) examples. Because this is a science text, we have adhered to scientific principles, models, and observations; however, you need not be a scientist to understand the principles and concepts we present—we go easy on the hard math and science and present the material in a user-friendly manner. You do need an open mind, though, as well as a love of the challenge of wading through the muck, an ability to decipher problems, and the patience to answer questions relevant to each topic presented. We weave real-life situations throughout the fabric of this text and present them in straightforward, plain English to equip you with the facts, knowledge, and information necessary to understand the complex issues and to make your own informed decisions.

* Rich, A., *Diving into the Wreck*, Norton, New York, 1994.

As a companion text to *The Science of Water*, *The Science of Air*, and *The Science of Environmental Pollution*, *The Science of Renewable Energy* follows the same proven format used in the other best-selling texts. Like its forerunners, this text is not an answer book. Instead, it is designed to stimulate thought. Although possible solutions to specific renewable energy questions are provided, we also point out the disadvantages and hurdles that must be overcome to make renewable energy a viable alternative to fossil fuels. Our goal is to provide a framework of principles on which to build an understanding of the complexities of substituting renewables for fossil fuels.

The Science of Renewable Energy is designed for a wide range of diverse student backgrounds. The text focuses on renewable energy derived from natural processes that replenish constantly: the sun, wind, and water (tides and waves); geothermal resources; biomass; hydroelectricity; biofuels; and hydrogen fuel cells. We explore ways to derive and process energy while limiting or abolishing pollution of the atmosphere, of surface and groundwater, and of soil (the three environmental media), all critical to our very survival. Because renewable energy and pollution prevention are real-world problems, it logically follows that we can solve these problems by using real-world methods—that's what *The Science of Renewable Energy* is all about.

Frank R. Spellman
Revonna M. Bieber

Authors

Frank R. Spellman, PhD, is a retired assistant professor of environmental health at Old Dominion University, Norfolk, Virginia, and author of more than 70 books covering topics ranging from concentrated animal feeding operations (CAFOs) to all areas of environmental science and occupational health. Many of his texts are readily available online, and several have been adopted for classroom use at major universities throughout the United States, Canada, Europe, and Russia; two have been translated into Spanish for South American markets. Dr. Spellman has been cited in more than 400 publications. He serves as a professional expert witness for three law groups and as an incident/accident investigator for the U.S. Department of Justice and a northern Virginia law firm. In addition, he consults on homeland security vulnerability assessments for critical infrastructures including water/wastewater facilities nationwide and conducts pre-Occupational Safety and Health Administration (OSHA)/Environmental Protection Agency (EPA) audits throughout the country. Dr. Spellman receives frequent requests to co-author with well-recognized experts in several scientific fields; for example, he is a contributing author for the prestigious text *The Engineering Handbook*, 2nd ed. (CRC Press). Dr. Spellman lectures on sewage treatment, water treatment, and homeland security and health and safety topics throughout the country and teaches water/wastewater operator short courses at Virginia Tech (Blacksburg, Virginia). He holds a BA in public administration, a BS in business management, an MBA, and an MS and PhD in environmental engineering.

Revonna M Bieber is currently working as a staff industrial hygienist in Norfolk, Virginia. She holds an AAS in radiography, a BS in biology, and a MS and MPH in environmental health and safety. She has worked with Dr. Spellman on various publications.

"We won't know we have too many cars—until we have too many cars."

—Eugene Odum, U.S. ecologist (2001)

The eminent ecologist, Eugene Odum, is probably correct in his assertion about knowing when we have too many cars only after that day actually arrives. A more pressing concern is if we acknowledge that a liquid fuel crisis exists only after the fuel taps run dry.

—Frank R. Spellman (2011)

1 Post-Oil Energy

Oil creates the illusion of a completely changed life, life without work, life for free....The concept of oil expresses perfectly the eternal human dream of wealth achieved through lucky accident....In this sense, oil is a fairy tale and like every fairy tale a bit of a lie.

—Ryszard Kapuscinski

Massive changes in the existence of humanity are imminent. The oil shortage? Blame it all on the Greenies; it is their entire fault. There's something guttural, something personal, about the price of gas.

—Christopher Steiner

THE GATHERING STORM

As we begin writing this book, several significant events and situations have presented themselves that warrant our attention. The devastating earthquake in impoverished Haiti tops the list. It is a Third World nation with a disastrous economy compounded by a disaster event of historical proportions. Second on our list of significant events or situations is the current state of the U.S. economy. The U.S. unemployment rate is at 9 to 10+%, with increasing numbers of workers losing their jobs daily. For those still working in factory jobs, their future in these same types of jobs is threatened and, in some cases, doomed—these jobs are being done away with or outsourced overseas. Many of these are jobs are unlikely to return to the United States. Third on our list is the nominal world population clock that just posted a world population figure of 6,825,276,728 and counting—pointing to the increasing need for more resources, including energy.

Wind turbines at summit of Tehachapi Pass, California. (Photo by Frank R. Spellman.)

How are these significant events related? What do they have to do with energy? What do they have to do with the pending oil crisis? Finally, what do they have to do with renewable energy? We discuss the interrelationships between oil and renewable energy and the significance of these three events in the following.

In regard to Haiti, it is the poorest country in the Americas. Not only is it impoverished but it also is the least developed. The recent earthquake only exacerbated an already desperate situation in Haiti. Moreover, Haiti has consistently ranked among the most corrupt countries in the world. And, with the exception of a few agricultural crops, it has no exports. Haiti has only survived because of the foreign aid it receives, mostly from the United States and lesser amounts from other countries.

Haiti's energy supplies must be imported; it has no indigenous supply of oil-derived energy. Haiti does have extensive exposure to sunlight and wind, as well as tide and wave energy. These sources can be employed in the development and production of renewable energy, all of which would lessen the need and expense for Haiti to import expensive hydrocarbon fuel products such as gasoline. Development of renewable energy sources in Haiti could go a long way in helping it reduce its dependence on worldwide generosity and loans. Certainly, economic development could benefit if it were tied to a decrease in corruption, a shoring up of a stable political footing, and an increase in human health and medical care.

The second event or situation that is interrelated with petroleum products and renewable energy production and the current and future state of the U.S. economy is the lack of viable jobs and lackluster progress in technological innovation. For some time, the United States has been the leader in technological development throughout the world; however, recent reports have pointed out that from 1999 to the present, the U.S. population grew by 32 million, or 1%, yet we had zero net job creation. Why? Take your pick; there are several reasons for our current economic condition. To begin with, we develop technology or technological advances in products such as computers, computer software, automobiles, farm machinery, furniture, and textiles, among others. But, the vast majority of clothing, cell phones, computers, and televisions and a large percentage of our construction products (e.g., Chinese drywall) are produced in a foreign country. After we have spent large amounts of capital and time and volumes of expertise developing U.S. innovations, these are simply copied by foreign industries and produced at about nine times lower cost than if they were produced in the United States. These foreign producers have much lower labor costs and no regulatory pressure, and many receive financial assistance from their government. When our technological innovations leave the country, it creates an uneven playing field that simply does not teeter-totter in a direction favorable to us.

In addition to labor, regulatory, and other related costs, the United States is not self-sustaining in oil production anymore. Our dependence on foreign energy sources is not only humiliating and shameful but also quite costly to all of us in so many different ways. For example, we have failed to grow jobs because we are no longer self-sufficient as a nation, with the exception of producing agricultural products for the country's needs.

Complicating any economic recovery we may achieve (at this writing, this seems doubtful and may be wishful thinking) is not only our dependence on foreign oil but also the price of oil. One thing is certain—if and when economic conditions turn around and begin to climb in the positive direction, the price of a gallon of gasoline ($2.78 and similar prices for diesel fuel at this writing) will simply increase, and increase, and increase again. The foreign energy producers are just sitting back waiting for us to recover, so they can turn the price increase lever one notch higher … and higher …, etc. These increases in the price of gasoline and diesel fuel will quickly put a lid on any economic recovery; we will simply descend that slippery oil-soaked slope back to our current condition of stagnant malaise, or maybe worse.

The third situation we mentioned is the continuing population growth and its impact not only on future fuel requirements but on many other resources, as well. Consider, for example, that at this very moment the nominal world population clock reads 6,825,276,728 and counting. Population growth on this globe is something that directly affects energy availability, use, and the pollution effects associated with using hydrocarbon fuels. The U.S. Census Bureau (2008) predicts that the world's population in 2030 will be almost double that of 1980. Savinar (2010) predicts that oil production in 2030 will have declined back to 1980 levels while worldwide demand for oil will significantly outpace production. Bartlett (2004) claimed that the rate of oil production per capita is falling, and that the decline has gone undiscussed because a politically incorrect form of population control may be implied by mitigation.

It is important to point out that one factor that has helped to somewhat ameliorate the effect of population growth on demand is the decline of the population growth rate since the 1970s, although this is offset to a degree by increasing average longevity in developed nations. In 1970, the population grew at a rate of 2.1%. By 2007, the growth rate had declined to 1.167% (CIA, 2008), but the fact remains that the world's population is still growing and is continuing to put pressure on our limited energy supplies.

As mentioned, energy is only one of several issues related to an increasing human population. The basic requirements of any population include air, water, food, shelter, and clothing. All of these human needs are tied directly or indirectly to energy.

In regard to the air we breathe, the air itself is ubiquitous and fills a vital need naturally for each of us; however, the air we breathe has been affected or tainted by energy use—namely, the air pollution that commonly accompanies the use of some forms of energy generation or use. At the present time, there is continuing debate on hydrocarbon energy use and its impact on global climate change. Although it is likely that we are undergoing global climate change—a general warming trend—it is not that apparent or proven that air pollution caused by energy use is the main causal factor of climate change. There can be little doubt, however, that human-caused (anthropogenic) pollution is a contributor to the current warming trend. Still, some critics argue that, in a discussion about oil, oil resources, peak oil supplies, and oil shortages, air and oil do not mix, and the use of hydrocarbons is not a primary cause of air pollution. They say there is no connection. In countering this argument, we suggest that these critics drive an automobile from any out-of-town location into downtown Los Angeles on one of those especially warm days in August or September, park the car, step out of the car, and take a deep breath of Los Angeles Basin air. We think that when this occurs the critic will get a mouthful and lungs full of more than just air and will readily come to realize that oil-derived vapors and air do make a noxious mix, making them intolerable or at least uncomfortable to inhale.

Along with good, clean air being one of those substances that we simply cannot live without is our need for water. In addition to its nutritional benefits, water used to be the primary producer of energy in the early days of humans' existence. Water is still used in hydropower applications and as such is a renewable energy source (discussed later in the text) and a major contributor to the generation of electricity. The problem with water, as with hydrocarbons, is that it is a finite natural resource. We have on Earth today the same amount of water that we have always had. We can't add to or delete from our present total water supply to any significant degree. The problem with our potable water is location; it is sparsely located or basically not available in many parts of the globe. Water is heavy and cumbersome, takes up a lot of space, and is not easily transported from location to location. Much of the freshwater of the globe is polluted, not exclusively by energy use but instead by the old mantra: We do not want it anymore so throw it away. Often the receivers of our throwaways are and have been our ponds, lakes, streams, rivers, and oceans. Beyond the use of water for the production of hydropower energy, the interface between energy and water is often ugly, unsightly, and unhealthy and results in ponds, lakes, streams, and rivers polluted by hydrocarbon-derived products such as pesticides, herbicides, rodenticides, fertilizers, and so forth. It seems rather odd that we so carelessly pollute our air and water supplies, the two substances we absolutely cannot live without.

Water also interfaces with energy use and its production in that it is used as an industrial cooling medium for machinery used in the refining, production, or use of hydrocarbon products, and it is transported by truck, tanker car, or pipeline or distributed via irrigation networks through power derived (e.g., by pumps) from the use hydrocarbon fuels such as gasoline and diesel fuel.

The connection between energy use and food has always existed and always will. Since time immemorial to present, we have expended one form of energy or another to hunt, gather, sow, and harvest our foodstuffs. In the current era, the mechanized farm machinery we are familiar with could not function without oil-derived fuels and lubricants. Moreover, many of the chemicals used to fertilize and protect crops from insects and disease are oil derived. Simply, ours is an oil-based agriculture, and it is this oil-based agriculture that is driving the world's exploding population growth. As oil production increased, so did food production. As food production increased, so did population. In step with population growth, demand for food increased, which escalated the demand for oil. Today, almost all the energy used by the agricultural industry to produce, store, and transport food comes from oil. As mentioned, from tractor to combine to harvester to irrigation pumps, the energy required for storage and the trucking, railing, and flying of these foods to market depends on oil-derived products. Because supplies of oil-derived products such as diesel fuel and gas are essential to modern agriculture techniques, a fall in global oil supplies could cause spiking food prices and unprecedented famine in the coming decades (Goodchild, 2007). Pfeiffer (2004) contends that current population levels are unsustainable and that to achieve a sustainable economy and avert disaster the U.S. population would have to be reduced by at least one third and the world population by two thirds.

DID YOU KNOW?

The largest consumer of fossil fuels in modern agriculture is ammonia production (for fertilizer) via the Haber process, which is essential to high-yielding intensive agriculture. The specific fossil fuel input to fertilizer production is primarily natural gas, to provide hydrogen via steam reforming.

The bottom line: Global population growth is only sustainable through our dependency on the infrastructure provided to us by oil. Whenever we make this statement to our college-level environmental students some buy our conclusion without too much thought and others give us that surprised or doubtful look. We discovered early in our environmental lecture presentations that many students and others just do not know that much about the importance of oil and its impact on our present-day society. In these lectures, we point out that many of the more than 500,000 products we are all accustomed to using every day are manufactured using oil as a material or ingredient. To illustrate our point, we discuss Table 1.1.

PEAK OIL

According to the U.S. Department of Energy (2007), the *peak oil theory* states that the world's oil production rate will reach a maximum rate and then enter terminal decline. This concept is based on the observed production rates of individual oil wells, and the combined production rate of a field or related oils wells. This is not to say that the world is running out of oil but it does signal the end of abundant and cheap oil on which all industrial nations depend (Campbell and Laherrere, 1998). Peak oil theory, also known as Hubbert peak theory, concerns the long-term rate of conventional oil (and other fossil fuel) extraction and depletion. Hubbert proposed, in a 1956 paper he presented at a meeting of the American Petroleum Institute, that oil production in the United States would peak between 1965 and 1970 (Hubbert, 1956). U.S. oil production peaked in 1971 (Brandt, 2007) and has been decreasing since then. Hubbert's theory is subject to continued discussion because of the potential effects of lowered oil production and because of the ongoing debate over aspects of energy policy. Opinions on the effect of passing Hubbert's peak range from faith that the market economy will produce a solution, assuming major investments in alternatives will occur before a crisis, to

TABLE 1.1
Products Manufactured Using Oil or Oil Byproducts

Plastic cups	Water skis	Plastic forks	Dresses
Transparent tape	Lens cleaner	Lip balm	Balloons
Water pipes	Rubber boots	Panty hose	Shoes
Hair dryers	Oil filters	Bandages	Pens
Upholstery	Insulation	Running shoes	Bubble gum
Vaporizers	Trash bags	Chewing gum	Glue
Lipstick	Floor wax	Footballs	Basketballs
Caulking	Shoe polishes	Hair coloring	Crayons
Umbrellas	Dice	Golf balls	Mops
Computers	Fertilizers	Rubber bands	Tires
Petroleum jelly	Volley balls	Hearing aids	Perfumes
Nail polish	Synthetic clothing	House paint	Carpet
Loudspeakers	Movie film	Combs/brushes	Stereos
Electrician's tape	Air conditioners	Tape recorders	Car bodies
Electric blankets	Toothpaste	Denture adhesives	Floor wax
Motorcycle helmets	Fishing rods	Fishing lures	Antiseptics
Shower curtains	Roofing paper	Heart valves	Ice chests
Artificial turf	Paint brushes	Rubber cement	Toilet seats
Boats	Cold cream	Contacts	Telephones
Ammonia	Candles	Bras	Cameras
Fiberglass	Solvents	Purses	Cortisone
Car batteries	Eyeglasses	Parachutes	Plastic wrap
Insect repellent	Deodorant	Antihistamine	Compact discs
Inks of all types	Plywood adhesive	Hair curlers	Plastic chairs
Dishwasher liquid	Refrigerator seals	Thermos bottle	Plastic toys
Shirts (non-cotton)			

prediction of doomsday scenarios of a global economy unable to meet its energy needs (Gwyn, 2004). This thesis postulates that the price of oil at first will escalate and then retreat as other types of fuel and energy sources are developed and used (CERA, 2006). Pessimistic predictions of future oil production operate on the thesis that the peak has already occurred (Deffeyes, 2007), oil production is on the peak, or it will occur shortly (Koppelaar, 2006). Hubbert's prediction remained accurate even after discovery of the Prudhoe Bay oil field (after 1971), with its great volume of oil still not enough to bring U.S. oil production out of a long-term decline (USDOE, 2007).

Peak global oil production and its subsequent decline (based on optimistic assessments) is forecast to begin by about 2020 or later. Keep in mind that the exact date of world oil peaking is a best guess; the actual date cannot be known with any certainty, complicating the decision-making process. The decision-making process refers to deciding what or which alternative sources of energy to pursue to prevent the economic and social impacts of oil peaking from being chaotic. A basic problem in predicting oil peaking is uncertain and politically biased oil reserve claims from many oil-producing countries. In some instances, the veracity of such claims may be questionable (Hirsch, 2005).

If we accept predictions about peaking oil, the possible effects and consequences of peaking oil cannot be painted and displayed without using a very wide brush. Few would argue against the fact that the widespread use of fossil fuels has been one of the most important agents of economic growth and prosperity since the industrial revolution. Fossil fuels have allowed humans to participate in the consumption of energy at a greater rate than it is being replaced. Some believe that mankind is in for a momentous paradigm shift when oil production decreases. This shift will cause humans and our modern technological society to change drastically. The actual impact of peak oil will depend heavily on the rate of decline and the development and adoption of effective alternatives. If alternatives are not forthcoming, the products produced with oil (including fertilizers, detergents, adhesives, solvents, and most plastics) would become scarce and expensive. In 2005, thc U.S Department of Energy published a report (known as the Hirsch Report) with a stark warning: "The peaking of world oil production presents the U.S. and the world with an unprecedented risk management problem. As peaking is approached, liquid fuel prices and price volatility will increase dramatically, and, without timely mitigation, the economic, social, and political costs will be unprecedented. Viable mitigation options exist on both the supply and demand sides, but to have substantial impact, they must be initiated more than a decade in advance of peaking" (USDOE, 2005).

Hirsch Report

After careful analysis, the Hirsch Report came to a number of conclusions and three possible scenarios:

Conclusions

1. *World oil peaking is going to happen.* Some forecasters predict it will occur within a decade, others later; for example, Simmons (2006) predicted global peak oil production from 2007 to 2009, Deffeyes (2008) predicted global peaking before 2009, and Laherrere (1998) predicted global peaking sometime between 2010 and 2020.
2. *Oil peaking could cost economies dearly*, particularly that of the United States.
3. *Oil peaking presents a unique challenge.* Previous transitions were gradual and evolutionary; oil peaking will be abrupt and revolutionary.
4. *The real problem is liquid fuels for transportation.* Motor vehicles, aircraft, trains, and ships have no readily available substitute.
5. *Mitigation efforts will require an intense effort over decades.*
6. *Both supply and demand will require attention.* Higher efficiency can reduce demand, but large amounts of substitute fuels must be produced.
7. *It is a matter of risk management.* Early mitigation will be less damaging than delayed mitigation.
8. *Government intervention will be required*; otherwise, the economic and social implications would be chaotic.
9. *Economic upheaval is not inevitable.* Without mitigation, peaking will cause major upheaval, but given enough lead time, the problems are solvable.
10. *More information is needed.* Effective action requires better understanding of a number of issues.

Scenarios

1. Waiting until world oil production peaks before taking crash program action leaves the world with a significant liquid fuel deficit for more than two decades.
2. Initiating a mitigation crash program ten years before world oil peaking helps considerably but still leaves a liquid fuels shortfall roughly a decade after the time that oil would have peaked.
3. Initiating a mitigation crash program 20 years before peaking appears to offer the possibility of avoiding a world liquid-fuels shortfall for the forecast period.

A MALTHUSIAN CATASTROPHE?

Obviously, the peaking of oil presents not only a number of questions but also a stark reality. Questions include: How will shortages and prices of oil affect you and your family? What about the effects on your community? How will oil shortages and prices affect industry and the economy?

The stark reality is that the decline of oil production is not temporary. For those sitting back and waiting for some White Knight to come charging into the country on his great steed and to then suddenly sink his great platinum-clad spear into the ground to free a gigantic black stream of crude oil that shoots upward and forms a giant shaft reaching toward the clouds, it just ain't going to happen; literally, that is a pipedream of mythical proportions.

The day is fast approaching when the inability to meet demand will raise the costs of fuel. Concurrently, every sector of the economy will be affected in some way, whether it be a shortage of new materials for manufactured goods, such as those listed in Table 1.1, or increasing costs associated with transporting those goods to market, storing them in climate-controlled facilities, purchasing the machinery used in the construction of those facilities (or structures of any kind), or fertilizers and pesticides used to maintain production yields in agriculture, among other dependencies too numerous to list here. That is only the beginning of the impact that such a situation will have on our society, as the resulting domino effect will generate enormous price increases across the board for all products and services in existence, making a significant number of them economically impractical. And, while the lack of oil supplies will affect the manufacture or availability of all of these products and services, it is food that is likely to be affected the most. An adequate food supply is essential not only to maintaining world peace and preventing famine but also to ensuring the health and wellbeing of the populace—a populace that is still increasing in size and will likely continue to grow even though the liquid fuel spigots are running dry, for good. This deadly doom and gloom scenario has been described, in somewhat related terms and meaning, as a Malthusian catastrophe. For our purposes here, this term denotes a forced return to subsistence level conditions once energy use has outpaced energy production. The tragic irony is that if this occurs and we are not prepared, for the first time in history Third World subsistence economies will have the one-up on many of the rest of us who, in our lifetimes, have never planted, tended, or harvested any type of crop.

Blame or Tame?

Whenever anything goes wrong on this globe of ours, it is human tendency for someone, anyone, or all of us to point a long finger of blame at someone or anyone beside ourselves, of course. This finger pointing has certainly been seen with regard to energy shortages and energy-derived pollution problems. In particular, many critics point the finger of blame for our current and pending shortage of liquid fuels at the environmentalists; however, the U.S. Environmental Protection Agency (USEPA, 1974) suggested that a finger of blame pointed at the environmental movement is misdirected. Although it is true that environmental regulations have caused some increased energy demand and have restricted supply to some extent, it is important to recognize that other factors have been considerably more significant. As mentioned, these include rapidly escalating demand for energy, energy pricing policies, oil import quotas, lack of incentives to invest in domestic energy facilities, and depletion of domestic oil and gas reserves.

We all realize that the price of energy has increased. The summer of 2008 with its $4+ gas prices got a lot of people's attention. The higher prices for heating oil, natural gas, and other oil-derived products have also attracted our attention. The problem is we tend to blame the higher costs of petroleum-based products on those who sell the products; we think they are greedy, and that is about the extent of our concern. What we don't understand is that it is the policymakers who have failed to understand the concept of exponential growth and have failed to formulate and pass a long-term energy strategy that is sensible. The result of these failures, and others, is the creation of an escalating energy crisis as time continues to run out. Unfortunately, most people are either oblivious to this or believe that some new source of energy will magically appear—again, some White Knight will ride into town with the grand solution.

We always need to blame someone or anyone for things we do not like; however, blame is never the solution to any problem. Instead, we need to get a hold of the issue and tame it. We need to think it through. Before doing that, however, we need to know the extent of the issue—the liquid fuel crisis.

WEATHERING THE STORM

With the energy crisis pending, we have three choices: We can mine for more oil, we can aggressively pursue energy conservation, or we can develop renewable energies. Because of limited fossil fuel resources, increased demand, and environmental impacts, renewable sources of energy are needed. Renewable energy is a form of energy capable of being constantly regenerated by natural processes at reasonable rates. Renewable energy technologies turn these fuels into usable forms of energy—most often electricity, but also heat, chemicals, or mechanical power. Some possibilities include:

- Biomass
- Wind energy
- Hydropower
- Tidal energy
- Wave energy
- Solar energy
- Geothermal power
- Hydrogen fuel cells

ALTERNATIVE AND RENEWABLE ENERGY*

The worldwide use of liquid fossil fuels and their decreasing availability, along with the politics involved and other economic forces, are pushing for substitute, alternative, and renewable fuel sources. This is the case, of course, because of the current and future economic problems that \$4+/gal gasoline have generated (especially in the United States) and because of the perceived crisis regarding high carbon dioxide emissions, the major contributing factor of global climate change.

Before proceeding with an introductory discussion of alternative and renewable energy sources, it is important to make a clear distinction between the two terms, the current buzzwords. *Alternative energy* is an umbrella term that refers to any source of usable energy intended to replace fuel sources without the undesired consequences of the replaced fuels. The use of the term "alternative" presupposes the existence of undesirable types of energy production (for many people, fossil fuel has joined that endless list of four-letter words). Today, alternative energy is considered to be fueled energy that is an alternative to fossil fuels and that does not use up natural resources or harm the environment. Examples of "alternative" fuels throughout history, though, have included petroleum as an alternative to whale oil, coal as an alternative to wood, alcohol as an alternative to fossil fuels, and coal gasification as an alternative to petroleum. The key point in understanding the concept of alternative energy is that these fuels need not necessarily be renewable.

Renewable energy is energy generated from natural resources—sunlight, wind, water (hydro), ocean thermal, wave and tide action, biomass, geothermal heat—which are naturally replenished (and thus renewable). Renewable energy resources are virtually inexhaustible, as they are replenished at the same rate as they are used, but they are limited in the amount of energy that is available per unit time. If we have not come full circle in our cycling from renewable to nonrenewable back to renewable we are getting close to that. Consider, for example, that in 1850, about 90% of the energy consumed in the United States was from renewable energy resources (e.g., hydropower, wind, wood). Now, however, the United States is heavily reliant on nonrenewable fossil fuels: natural gas, oil, and coal. In 2008, about 7% of all energy consumed and about 8.5% of total electricity production was from renewable energy resources (see Table 1.2).

* Some information in this section is from U.S. Energy Information Association, *Renewable Energy Trends*, 2004 ed., U.S. Department of Energy, Washington, D.C., 2005 (http:www.eia.doe.gov/cneaf/solar.renewables/page/trends/rentrends04.html); U.S. Energy Information Association, *How Much Renewable Energy Do We Use?*, U.S. Department of Energy, Washington, D.C., 2010 (http://tonto.eia.doe.gov/energy_in_brief/renewable_energy.cfm).

TABLE 1.2
U.S. Energy Consumption by Energy Source, 2008

Energy Source	Energy Consumption (quadrillion Btu)
Total	99.438
Fossil fuels (coal, natural gas, petroleum)	83.532
Electricity net imports	0.113
Nuclear electric power	8.427
Renewable	7.367
Biomass (total)	3.852
Biofuels	1.372
Waste	0.436
Wood and wood-derived fuels	2.044
Geothermal energy	0.360
Hydroelectric conventional	2.512
Solar thermal/PV energy	0.097
Wind energy	0.546

Source: EIA, *U.S. Energy Consumption by Energy Source*, Energy Information Administration, Washington, D.C., 2007 (www.eia.doe.gov/cneaf/solar.renewables/page/trends/table1.html).

Most renewable energy today is used for electricity generation, heat in industrial processes, heating and cooling buildings, and transportation fuels. Electricity producers (utilities, independent producers, and combined heat and power plants) consumed 51% of total U.S. renewable energy in 2007. Most of the rest of the remaining 49% was biomass consumed for industrial applications (principally papermaking) by plants producing only heat and steam. Biomass is also used for transportation fuels (ethanol) and to provide residential and commercial space heating. The largest share of renewable-energy-generated electricity comes from hydroelectric energy (71%), followed by biomass (16%), wind (9%), geothermal (4%), and solar (0.2%). Wind-generated electricity increased by almost 21% in 2007 over 2006, more than any other energy source. Its growth rate was followed closely by solar, which increased by over 19% in 2007 over 2006. Table 1.2 shows the primary forms of renewable energy available today: solar, wind, biomass, geothermal, and hydroelectric energy sources. Each of theses holds promise and poses challenges regarding future development.

The United States imports more than 50% of its oil. Replacing some of our petroleum with fuels provided from solar power or plant matter, for example, could save money and strengthen our energy security. Renewable energy is plentiful, and the technologies are improving all the time. There are many ways to use renewable energy. Our main focus should be on finding and developing a renewable source of liquid fuels (e.g., developed from biomass), because our economy runs on liquid fuels. Most of the nonliquid renewable energies will not (at the present time) provide power for airplanes and heavy trucks; that is, neither solar, wind, hydro, geothermal, wave, and tidal energy nor fuel cells can power the main transportation vehicles we use today. Note that we did not mention trains. Remember that today many trains are powered by diesel-power or diesel-electric systems. If necessary, trains could be retrofitted to steam power developed by burning coal and wood products (a step back into the past); however, we cannot do this for heavy trucks and airplanes. Most Americans still do not understand that we are running short of the fuels that we use every day, the fuels that made us what we are today. We built the world's greatest economy on oil that sold for \$3 to \$4 per barrel. That is no longer the case. To maintain our current level of living, the so-called good life, we must find and develop renewable energy sources to power and secure our future.

THE BOTTOM LINE: THE NEED FOR INNOVATION

Identifying and addressing the problems related to our dwindling supply of crude oil are the easy part of the fuel substitution equation; however, solving the equation is a different matter. What is needed to tackle this pending energy dilemma is good old-fashioned American innovative genius. History has shown that when humans put their minds and energies together to solve difficult problems, the problems eventually get solved. The solution to the energy crisis, in a word, is *innovation, innovation, innovation*! Consider the imaginative license we have taken in the following hypothetical case study that illustrates the kind of energy innovation we are talking about.

CASE STUDY

Our select team of renewable energy experts has been given a $250 million grant to do whatever it takes to develop a super battery or fuel cell to power the cars of the future. Accompanying specifications simply state that this battery or fuel cell must be large capacity (large ampere-hour capacity) and can be continuously recharged via light striking supersensitive solar sensors. To charge any rechargeable battery, a certain amount of current or electron flow is necessary. The solar super sensors we have invented will produce substantial current flow. These super sensors are wired from the battery itself to at least four positions on the roof of our experimental automobile, where they can constantly be exposed to daylight—any level of daylight. Okay, we know what you're thinking. You're thinking that it gets dark at night, so how are you going to continuously recharge the car's battery, fuel cell, or whatever else we choose to call it? Good question, but one with a simple answer. Two additional wires and ultra-super-super light sensors (that we also invented) will run from the battery to the car's headlamps, one to each. Thus, during nighttime driving, the light from the head lamps will be sensed by the ultra-super-super sensors and sent back to the battery. Now, we know electrical engineers out there in La La Land are scratching their heads and wondering where did we got such an unworkable idea. We are not finished with our explanation yet … hold onto your brain cells. An amplification or booster device (similar to a transistor that amplifies a signal but more powerful) will be located within the wiring between the battery and the external and headlamp cell connections. We are not talking about perpetual motion here, but perpetual energy. Is this scenario possible? Anything, absolutely anything, is possible. We are Americans; we think we can do anything if we put our minds to it and history has shown this to be true. Now is the time.

Let's say that we are able to create a power-plant arrangement consisting of a powerful battery or fuel cell with a built-in recharge system like the one just described, what's next? Another good question. What's next is the final ingredient in any innovative processes: education, training, and on-the-job training (OJT). To come up with innovation and research and development grants, we must have personnel who are highly educated and trained. Unfortunately, the United States is losing (or has already lost, in some areas) our edge in higher education. The United States will not regain its preeminence in science, technology, engineering, and mathematics until we shift our focus, money, and talent to educating our youth in these critical subjects.

DISCUSSION QUESTIONS

1. What is nonrenewable energy?
2. Explain the difference between alternative energy and renewable energy.
3. Is nuclear energy considered to be renewable energy?
4. What are the economic ramifications of running out of crude oil? Explain.

5. Will the crude oil crisis come about gradually or suddenly? Explain.
6. An economy based on renewable energy will help in our ongoing fight to reduce pollution. Explain.
7. Name 15 products produced by oil that we could get along without.
8. Summarize the Hirsch Report.
9. What is a Malthusian catastrophe?
10. Is the point made in the case study feasible? Explain.

REFERENCES AND RECOMMENDED READING

Bartlett, A.A. (2004). Thoughts on long-term energy supplies: scientists and the silent lie. *Physics Today*, 57(7), 53–57.

Brandt, A.R. (2007). Testing Hubbert. *Energy Policy*, 35(5), 3074–3088.

Campbell, C.J. (2004a). *The Essence of Oil and Gas Depletion*. Brentwood, U.K.: Multi-Science Publishing.

Campbell, C.J. (2004b). *The Coming Oil Crisis*. Brentwood, U.K.: Multi-Science Publishing.

Campbell, C.J. (2005). *Oil Crisis*. Brentwood, U.K.: Multi-Science Publishing.

Campbell, C.J. and Laherrere, J.H. (1998). The end of cheap oil. *Scientific American*, March, 80–86.

Cambridge Energy Research Associates. (2006). CERA says peak oil theory is faulty. *Post Carbon Institute Energy Bulletin*, November 14 (http://www.energybulletin.net/node/22381).

CIA. (2008). *The World Factbook*. Washington, D.C.: Central Intelligence Agency (http://www.cia.gov/library/publications/the-world-factbook/geos/X.html).

Deffeyes, K.S. (2008). *Hubbert's Peak: The Impending World Oil Shortage*. Princeton, NJ: Princeton University Press.

Deffeyes, K.S. (2010). *Current Events: Join Us As We Watch the Crisis Unfolding*, http://www.princeton.edu/hubbert/current-events.html.

Goodchild, P. (2007). *Peak Oil and Famine: Four Billion Deaths*, www.countercurrents.org/goodchild291007.htm.

Goodstein, D. (2005). *Out of Gas: The End of the Age of Oil*. New York: W.W. Norton.

Gwyn, R. (2004). *Demand for Oil Outstripping Supply*, http://www.commondreams.org/views04/0128-10.htm.

Hirsch, R.L. (2005). *The Mitigation of the Peaking of World Oil Production: Summary of an Analysis*. Uppsala, Sweden: Association for the Study of Peak Oil & Gas (http://www.peakoil.net/USDOE.html).

Hubbert, M.K. (1956). *Nuclear Energy and the Fossil Fuels Drilling and Production Practice*, http://www.hubbertpeak.com/hubbert/1956/1956.pdf.

Huber, P. (2005). *The Bottomless Well*. New York: Basic Books.

Koppelaar, R.H.E.M (2006). *World Oil Production & Peaking Outlook*. Amsterdam: Peak Oil Netherlands Foundation.

Leggett, J.K. (2005). *The Empty Tank: Oil, Gas, Hot Air, and the Coming Financial Catastrophe*. New York: Random House.

Pfeiffer, D.A. (2004). *Eating Fossil Fuels: Oil, Food, and the Coming Crises in Agriculture*. Gabriola Island, BC: New Society Publishers.

Rifkin, J. (2003). *The Hydrogen Economy: After Oil, Clean Energy from a Fuel-Cell-Driven Global Hydrogen Web*. New York: J.P. Tarcher.

Savinar, M. (2010). *Are We Running Out? I Thought There Was 40 Years of the Stuff Left*, http://www.lifeafter-theoilcrash.net.

Simmons, M. (2006). *Twilight in the Desert: The Coming Saudi Oil Shock and the World Economy*. New York: John Wiley & Sons.

Simon, J.I. (1998). *The Ultimate Resource*. Princeton, NJ: Princeton University Press.

Tertzakian, P. (2006). *A Thousand Barrels a Second*. New York: McGraw-Hill.

U.S. Census Bureau. (2008). *Total Midyear Population for the World: 1950–2050*, http://www.census.gov/compendia/statab/tables/09s1285.xls.

USDOE. (2001). *Renewable Energy: An Overview*. Washington, D.C.: U.S. Department of Energy.

USDOE. (2005). *Peaking of World Oil Production: Impacts, Mitigation, & Risk Management*. Washington, D.C.: U.S. Department of Energy.

USDOE. (2007). *Peak Oil—The Turning Point*. Washington, D.C.: U.S. Department of Energy.

USEPA. (1974). *EPA's Position on the Energy Crisis*. Washington, D.C.: U.S. Environmental Protection Agency (http://www.epa.gov/history/topics/energy/01.htm).

2 Energy Basics

Sunlight pours energy on the Earth, and the energy gets converted from one form to another, in an endless cycle of life, death, and renewal. Some of the sunlight got stored underground, which has provided us with a tremendous "savings account" of energy on which we can draw. Our civilization has developed a large thirst for the energy, as we've built billions and billions of machines large and small that all depend on fuel and electricity.

—Thom Hartmann

(The Last Hours of Ancient Sunlight: Revised and Updated: The Fate of the World and What We Can Do Before It's Too Late, 2004)

ENERGY

Energy can be defined in a number of ways. In the broad sense, energy means the capacity of something—a person, an animal, or a physical system (machine)—to do work and produce change. It can be used to describe someone doing energetic things such as running, talking, and acting in a lively and vigorous way. It is used in science to describe how much potential a physical system has to change. It also is used in economics to describe the part of the market where energy itself is harnessed and sold to consumers. For our purposes in this text, we simply define energy as something that can do work.

Two basic forms of energy are discussed in this text: kinetic and potential energy. Kinetic energy is energy at work or in motion; for example, a car in motion or a rotating shaft has kinetic energy. Potential energy is stored energy, like the energy stored in a coiled or stretched spring; an object stationed above a table has potential energy.

Chief Joseph Dam, Washington State. (Photo by Frank R. Spellman.)

According to the law of conservation of energy, energy cannot be made or destroyed but can be made to change forms. Moreover, when energy changes from one form to another, the amount of energy stays the same. Let's consider an example of the law of conservation of energy: Measure the potential energy of something, then change the energy from potential (stored) to kinetic (moving) and back again. Measure the energy again. The energy measured initially will be the same as that measured at the end; it will always be the same. One caveat to this explanation is that we now know that matter can be converted into energy through processes such as nuclear fission and nuclear fusion. The law of conservation of energy, therefore, has been modified or amplified to become the law of conservation of matter and energy.

Types of Energy

There are many types of energy; for example:

- Kinetic (motion) energy
- Water energy
- Potential energy
- Elastic energy
- Nuclear energy
- Chemical energy
- Sound energy
- Internal energy
- Heat/thermal energy
- Light (radiant) energy
- Electric energy

The electricity we use in our homes was probably generated by burning coal, by a nuclear reaction, or by a hydroelectric plant at a dam; the fuels for these processes are called *energy sources*. When we fill up a gas tank, the source might be petroleum or ethanol made by growing and processing corn (U.S. Energy Information Administration, 2009). Energy sources are divided into two groups—*renewable*, an energy source that can be easily replenished, and *nonrenewable*, an energy source that we are using up and cannot recreate (petroleum, for example, was formed millions of years ago from the remains of ancient sea plants and animals). In the United States, most of our energy comes from nonrenewable energy sources. Coal, petroleum, natural gas, propane, and uranium are nonrenewable energy sources. They are used to make electricity, to heat our homes, to move our cars, and to manufacture all kinds of products. Renewable and nonrenewable energy sources can be used to produce secondary energy sources, including electricity and hydrogen. Types of renewable energy sources include:

- Solar
- Hydro
- Wind
- Geothermal
- Ocean thermal energy conversion
- Tidal energy
- Hydrogen burning
- Biomass burning

Use of energy in the United States can be broken down into four major sectors of the economy (U.S. Energy Information Administration, 2008):

- *Commercial* (19%)—Buildings such as offices, malls, stores, schools, hospitals, hotels, warehouses, restaurants, places of worship, and more
- *Industrial* (31%)—Facilities and equipment used for manufacturing, agriculture, mining, and construction
- *Residential* (22%)—Homes and apartments
- *Transportation* (28%)—Vehicles that transport people or goods, such as cars, trucks, buses, motorcycles, trains, subways, aircraft, boats, barges, and even hot air balloons

Measuring Energy

Energy can be measured; that is, the amount of energy a thing has can be given a number. As for other kinds of measurements, there are measurement units. The units of measurement for measuring energy are used to make the numbers understandable and meaningful.

The SI unit for both energy and work is the joule (J) (as in "jewel," like a diamond or emerald). It is named after James Joule, who discovered that heat is a type of energy. One joule is equal to 1 newton meter. In terms of SI base units, 1 J is equal to 1 $kg \cdot m^2 \cdot s^{-2}$. The energy unit of measurement for electricity is the kilowatt-hour (kWh). One kWh is equivalent to 3,600,000 J (3600 kJ or 3.6 MJ).

Energy is usually expressed in British thermal units (Btu). One Btu is the amount of heat energy it takes to raise the temperature of 1 pound of water by 1°F at sea level:

1000 joules = 1 Btu
1000 joules = 1 kilojoule (kJ) = 1 Btu
1 therm = 100,000 Btu

Following are Btu conversion factors (U.S. Energy Information Administration, 2008):*

Electricity	1 kilowatt-hour (kWh) = 3412 Btu
Natural gas	1 cubic foot (ft^3) = 1028 Btu
	1 cubic foot (ft^3) = 0.01 therms
Motor gasoline	1 gallon = 124,000 Btu
Diesel fuel	1 gallon = 139,000 Btu
Heating oil	1 gallon = 139,000 Btu
Propane	1 gallon = 91,333 Btu
Wood	1 cord = 20,000,000 Btu

Electricity is a *secondary* energy source. The energy sources we use to make electricity can be renewable or nonrenewable, but electricity itself is neither renewable nor nonrenewable. Many of the producers of renewable energy primarily produce electricity; for example, hydropower can be converted to mechanical power via a waterwheel connected to a gear train to power a pump or other machine, but the primary purpose of hydropower is to generate electricity. Likewise, a windmill can also be connected to a gear train to power a pump or other machine, but the primary purpose of wind turbines is to produce electricity. Solar power can be used to heat water directly to produce steam which in turn powers a turbine to perform mechanical functions. Solar power can also be used to produce photovoltaic (PV) electrical energy. Even biomass liquid fuels can be used to power boilers that produce steam to power turbines connected to generators to produce electricity. In addition, waste off-gases or exhaust from various machines such as gas turbine generators can be used to heat boilers.

* These are weighted averages.

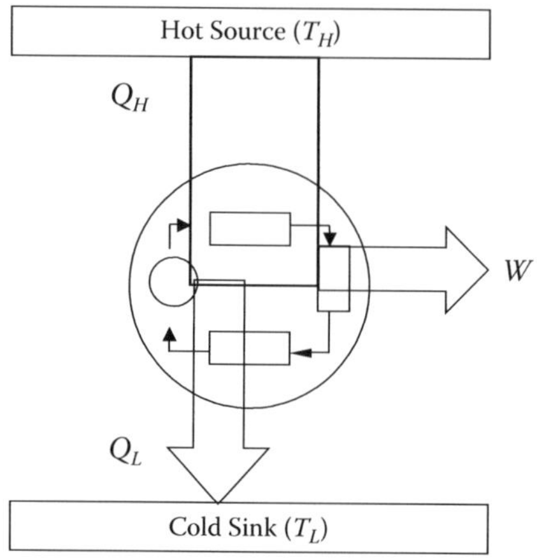

where:

Q_H = Heat energy taken from the high-temperature system
Q_L = Heat energy delivered to the cold-temperature system
W = Work
T_H = Absolute temperature of heat source
T_L = Absolute temperature of cold sink

FIGURE 2.1 Heat engine diagram.

The use of steam to power turbines to convert energy is accomplished by machines called *heat engines*. An automobile engine converts the chemical energy of gasoline into the mechanical energy of a piston and camshaft; the turbine in an electrical generating plant converts heat energy into mechanical work to run a generator which, in turn, produces electrical power. The two heat engines of interest to us in this text are the Rankine cycle (or vapor cycle) and the Stirling cycle (or gas cycle) heat engines (Hinrichs and Kleinbach, 2006).

Because electrical energy is the main product of heat engines powered by various renewable energy sources, it is important for general readers, students, and others to have some fundamental knowledge of heat engines and electricity, electrical circuitry, electrical power generation, and electrical power uses. Accordingly, prior to beginning a detailed discussion of each major renewable energy source—solar, wind, hydropower, bioenergy, geothermal, ocean, fuel cells—we first present a basic overview of heat engines and then an even more thorough discussion of basic electricity.

Heat Engines

A *heat engine* is a device that converts thermal energy to mechanical output. The thermal energy input is called *heat*, and the mechanical output is called *work*. Typically, heat engines run on a specific thermodynamic cycle. Heat engines can be open to the atmosphere (open cycle) or sealed and closed off to the outside (closed cycle). The driving agent of a heat engine is a temperature differential. That is, heat engines convert heat energy to mechanical work by exploiting the temperature gradient between a hot *source* and a cold *sink* (see Figure 2.1). Heat is transferred from the source, through the working body of the engine, to the sink, and in this process some of the heat is converted into work by exploiting the properties of a gas or liquid (the working substance). The lower the sink temperature or the higher the source temperature, the more work is available from the heat engine.

After doing work, the working substance can either be released into the environment or sent back to the heat source to start the cycle over. If the working substance is returned to its initial state (gas or liquid), there has been no change in its total energy. Consequently, from the first law

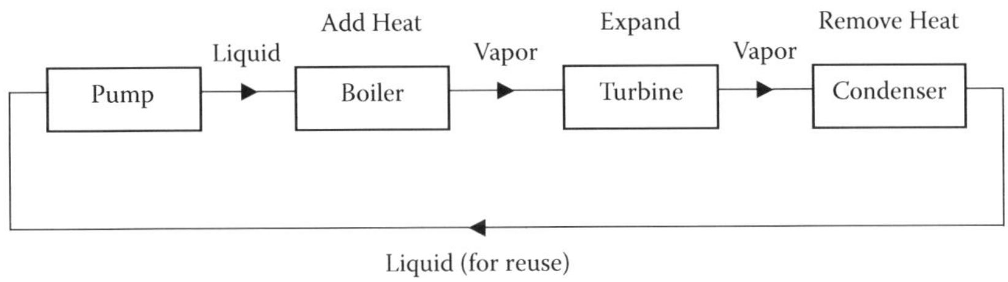

FIGURE 2.2 The four processes of the Rankine cycle.

of thermodynamics, the total energy of a system can be increased by doing work on it or by adding heat, and the total work done by the system is just equal to the heat added (i.e., heat in minus heat out).

The heat source may be direct solar radiation, geothermal steam, geothermal water, ocean water heated by the sun, combustion fuel, or nuclear energy. The two types of heat engines that are most commonly associated with renewable energy processes such as solar power and geothermal energy are the Rankine and Stirling cycle heat engines.

DID YOU KNOW?

Heat engines are often confused with the cycles they attempt to mimic. Typically, when describing the physical device, the term *engine* is used. When describing the model, the term *cycle* is used.

Rankine Cycle Heat Engine

Figure 2.2 illustrates the four processes as they occur in a closed-cycle *Rankine cycle heat engine*. The expansion can be accomplished through a cylinder with a piston, as in the locomotives one still sees in the old western movies, or through a turbine. In the Rankine cycle heat engine, as described for turbine use, the working fluid changes state and is also referred to as a *vapor gas* (i.e., hot air, not to be confused with fuel natural gas) *cycle*. Rankine cycle engines are all external-combustion devices (e.g., external-combustion gas turbine engines are widely used in units for electrical generating stations). As shown in Figure 2.2, the first phase pumps the working fluid from low to high pressure; because the fluid is a liquid at this stage, the pump requires little input energy. In the second phase, the high-pressure liquid enters a boiler, where it is heated at constant pressure by an external heat source to become a dry saturated vapor. In the third phase, the dry saturated vapor expands through the turbine, generating power. The temperature and pressure of the vapor decrease, and some condensation may occur. In the final phase, the wet vapor enters a condenser, where it is condensed at a constant pressure and temperature to become a saturated liquid. The pressure and temperature of the condenser are fixed by the temperature of the cooling coils as the fluid undergoes a phase change.

Stirling Cycle Heat Engine

The Stirling cycle heat engine processes include isothermal compression, isometric heat addition, isothermal expansion, and isometric heat rejection. It operates by cyclic compression and expansion of air or other gas (the working fluid) at different temperature levels such that there is a net conversion of heat energy to mechanical work. Like a steam engine, heat in the Stirling heat engine flows in and out through the engine wall. The Stirling heat engine is noted for its high efficiency (resulting from heat regeneration), quiet operation, and ease with which it can use almost any heat source. This compatibility with renewable energy sources has become increasingly significant as the price of conventional fuels rises, as well as in light of concerns such as climate change and peak oil. An example of this compatibility with renewable energy sources is placing a Stirling heat engine at the focus of a parabolic mirror, where it can convert solar energy to electricity with an efficiency better than that of photovoltaic cells.

PRINCIPLES OF BASIC ELECTRICITY*

People living and working in modern societies generally have little difficulty in recognizing electrical equipment; electrical equipment is everywhere and (if one pays attention to his or her surroundings) is easy to spot. Despite its great importance in our daily lives, however, few of us probably stop to think what life would be like without electricity. Like air and water, we tend to take electricity for granted, but we use electricity to do many jobs for us every day—from lighting, heating, and cooling our homes to powering our televisions and computers. Then there is the workplace—can we actually perform work without electricity? The typical industrial workplace is outfitted with electrical equipment that:

- Generates electricity (a generator or emergency generator)
- Stores electricity (batteries)
- Changes electricity from one level (voltage or current) to another (transformers)
- Transports or transmits and distributes electricity throughout the plant site (wiring distribution systems)
- Measures electricity (meters)
- Converts electricity into other forms of energy (mechanical energy, heat energy, light energy, chemical energy, radio energy)
- Protects other electrical equipment (fuses, circuit breakers, or relays)
- Operates and controls other electrical equipment (motor controllers)
- Converts some condition or occurrence into an electric signal (sensors)
- Converts some measured variable into a representative electrical signal (transducers or transmitters)

NATURE OF ELECTRICITY

The word "electricity" is derived from the Greek word *electron*, meaning "amber." Amber is a translucent (semitransparent), yellowish, fossilized mineral resin. The ancient Greeks used the term "electric force" to describe the mysterious forces of attraction and repulsion exhibited by amber when it was rubbed with a cloth. They did not understand the nature of this force, and they could not answer the question, "What is electricity?" This question still remains unanswered. Today, we often attempt to answer this question by describing the effect and not the force. That is, the standard answer given in physics is that electricity is the "force that moves electrons," which is about the same as defining a sail as "the force that moves a sailboat."

At the present time, we know little more about the fundamental nature of electricity than the ancient Greeks did, but we have made tremendous strides in harnessing and using it. As with many other unknown (or unexplainable) phenomena, elaborate theories concerning the nature and behavior of electricity have been advanced and have gained wide acceptance because of their apparent truth—and because they work.

Scientists have determined that electricity seems to behave in a constant and predictable manner in given situations or when subjected to given conditions. Faraday, Ohm, Lenz, and Kirchoff described the predictable characteristics of electricity and electric current in the form of certain

* Much of the information in this section is adapted from Spellman, F.R. and Drinan, J., *Electricity*, CRC Press, Boca Raton, FL, 2001.

rules. These rules are often referred to as *laws*. Thus, although electricity itself has never been clearly defined, its predictable nature and easily used nature have made it one of the most widely used power sources in modern times.

The bottom line: We can learn about electricity by becoming familiar with the rules, or laws, that apply to the behavior of electricity and by understanding the methods of producing, controlling, and using it. Learning about electricity, then, can be accomplished without ever having determined its fundamental identity.

You are probably scratching your head, puzzled. We understand the main question running through your brain at this exact moment: "This is a section in the text about the physics of electricity and the authors can't even explain what electricity is?" That is correct; we cannot. The point is, no one can definitively define electricity. Electricity is one of those subject areas where the old saying that "we don't know what we don't know about it" fits perfectly.

Only a few theories about electricity have so far stood the test of extensive analysis and much time (relatively speaking, of course). One of the oldest and the most generally accepted theories concerning electric current flow (or electricity) is known as the *electron theory*. The electron theory basically states that electricity or current flow is the result of the flow of free electrons in a conductor. Thus, electricity is the flow of free electrons or simply electron flow. In this text, this is how we define electricity: Electricity is the flow of free electrons. Electrons are extremely tiny particles of matter. To gain an understanding of electrons and exactly what is meant by electron flow, it is necessary to briefly review our earlier discussion about the structure of matter.

Structure of Matter

Matter is anything that has mass and occupies space. To study the fundamental structure or composition of any type of matter, it must be reduced to its fundamental components. All matter is made of *molecules*, or combinations of *atoms* (from the Greek word for "not able to be divided"), that are bound together to produce a given substance, such as salt, glass, or water. For example, if you keep dividing water into smaller and smaller drops, you would eventually arrive at the smallest particle that was still water. That particle is the molecule, which is defined as the *smallest bit of a substance that retains the characteristics of that substance.*

A molecule of water (H_2O) is composed of one atom of oxygen and two atoms of hydrogen. If the molecule of water was further subdivided, only unrelated atoms of oxygen and hydrogen would remain, and the water would no longer exist as such. Thus, the molecule is the smallest particle to which a substance can be reduced and still be called by the same name. This applies to all substances—solids, liquids, and gases.

Atoms are composed, in various combinations, of subatomic particles of *electrons*, *protons*, and *neutrons*. These particles differ in weight (a proton is much heavier than the electron) and charge. We are not concerned with the weights of particles in this text, but the *charge* is extremely important in electricity. The electron is the fundamental negative (–) charge of electricity. Electrons revolve about the nucleus or center of the atom in paths of concentric *orbits*, or shells. The proton is the fundamental positive (+) charge of electricity. Protons are found in the nucleus. The number of protons within the nucleus of any particular atom determines the atomic number of that atom; for example, the helium atom has two protons in its nucleus so the atomic number is 2. The neutron, which is the fundamental neutral charge of electricity, is also found in the nucleus.

Key Point: Molecules are made up of atoms, which are bound together to produce a given substance.

Most of the weight of the atom is in the protons and neutrons of the nucleus. Whirling around the nucleus are one or more negatively charged electrons. Normally, there is one proton for each electron in the entire atom so the net positive charge of the nucleus is balanced by the net negative charge of the electrons rotating around the nucleus (see Figure 2.3).

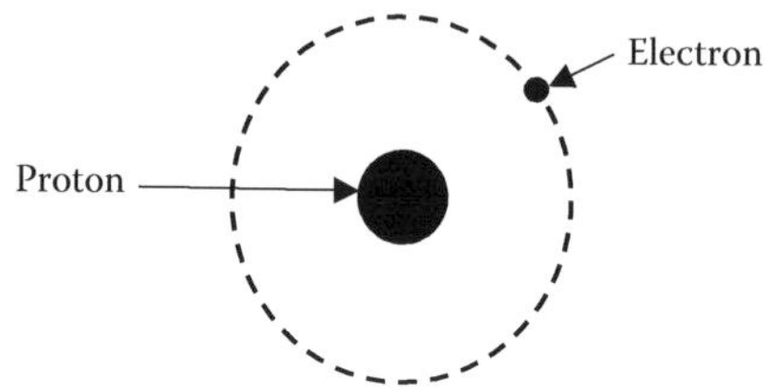

FIGURE 2.3 One proton + one electron = electrically neutral.

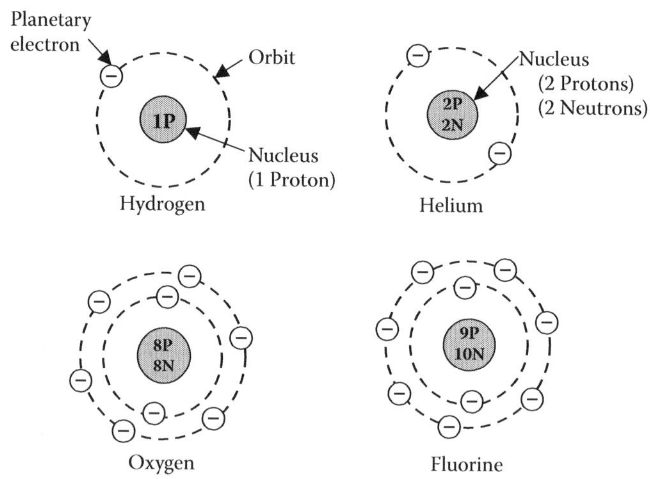

FIGURE 2.4 Atomic structure of elements.

We stated earlier that in an atom the number of protons is usually the same as the number of electrons. This is an important point because this relationship determines the kind of element in question (the atom is the smallest particle that makes up an element; an element retains its characteristics when subdivided into atoms). Figure 2.4 provides simplified drawings of atoms of various materials based on the concept of electrons orbiting about the nucleus. Hydrogen, for example, has a nucleus consisting of one proton, around which rotates one electron. The helium atom has a nucleus containing two protons and two neutrons with two electrons encircling the nucleus. Both of these elements are electrically neutral (or balanced) because each has an equal number of electrons and protons. Because the negative (–) charge of each electron is equal in magnitude to the positive (+) charge of each proton, the two opposite charges cancel.

Key Point: Most batteries are marked with the symbols + and – or even with the abbreviations POS (positive) and NEG (negative). The concept of a positive or negative polarity and its importance in electricity will become clear later; however, for the moment, you need to remember that an electron has a negative charge and that a proton has a positive charge.

A balanced (neutral or stable) atom has a certain amount of energy that is equal to the sum of the energies of its electrons. Electrons, in turn, have different energies called *energy levels*. The energy level of an electron is proportional to its distance from the nucleus; therefore, the energy levels of electrons in shells farther from the nucleus are higher than the energy levels of electrons in shells closer to the nucleus.

When an electric force is applied to a conducting medium, such as copper wire, electrons in the outer orbits of the copper atoms are forced out of orbit (they are liberated or freed) and travel along the wire. This electrical force, which forces electrons out of orbit, can be produced in a number of ways, such as by moving a conductor through a magnetic field; by friction, as when a glass rod is rubbed with cloth (silk); or by chemical action, as in a battery.

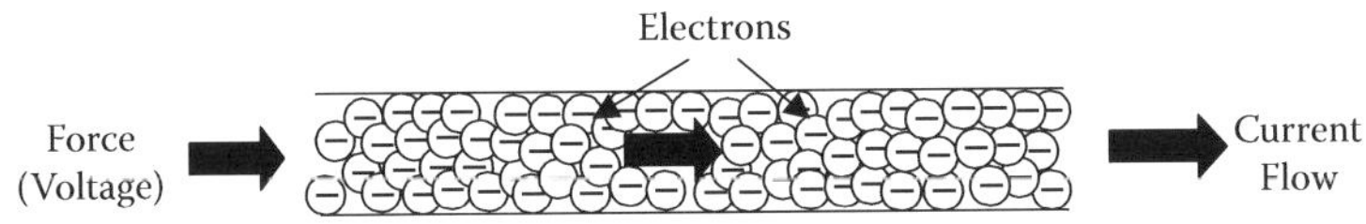

FIGURE 2.5 Electron flow in a copper wire.

When the electrons are forced from their orbits they are referred to as *free electrons*. Some of the electrons of certain metallic atoms are so loosely bound to the nucleus that they are relatively free to move from atom to atom. These free electrons constitute the flow of an electric current in electrical conductors. If the internal energy of an atom is raised above its normal state, the atom is said to be *excited*. Excitation may be produced by causing the atoms to collide with particles that are impelled by an electric force, as shown in Figure 2.5. In effect, energy is transferred from the electric source to the atom. The excess energy absorbed by the atom may be sufficient to cause loosely bound outer electrons (as shown in Figure 2.5) to leave the atom, resisting the force that acts to hold them within. An atom that has lost or gained one or more electrons is said to be *ionized*. If the atom loses electrons it becomes positively charged and is referred to as a *positive ion*. Conversely, if the atom gains electrons, it becomes negatively charged and is referred to as a *negative ion*.

Key Point: When an electric force is applied to a copper wire, free electrons are displaced from the copper atoms and move along the wire, producing an electric current, as shown in Figure 2.5.

Conductors, Semiconductors, and Insulators

Electric current moves easily through some materials but with greater difficulty through others. Substances that permit the free movement of a large number of electrons are *conductors*. The most widely used electrical conductor is copper because of its high conductivity and cost effectiveness. Electrical energy is transferred through a copper or other metal conductor by the movement of free electrons that migrate from atom to atom inside the conductor (see Figure 2.5). Each electron moves a very short distance to the neighboring atom, where it replaces one or more electrons by forcing them out of their orbits. The replaced electrons repeat the process in other nearby atoms until their movement is transmitted throughout the entire length of the conductor. A good conductor is said to have a low opposition, or *resistance*, to the electron (current) flow. Table 2.1 lists many of the metals commonly used as electric conductors. The best conductors appear at the top of the list, the poorer ones lower.

TABLE 2.1 Electrical Conductors

Silver
Copper
Gold
Aluminum
Zinc
Brass
Iron
Tin
Mercury

Note: If lots of electrons flow through a material with only a small force (voltage) applied, we refer to that material as a *conductor*. A convenient way in which to understand the purpose and function of an electrical conductor is to visualize a pipe or other conveyor of water or other liquid or gaseous substance. Just as a liquid or gas is conveyed or conducted through a pipe, so too are electrons (electricity) through a conductor. Force to push water or other liquid through a pipe is applied by a pump; in the case of electricity, voltage pushes the electrons. It is important to remember that in an electrical circuit voltage does not move; rather, voltage is the force that makes electrons move.

Note: The movement of each electron in copper wire, for example, takes a very small amount of time; it is almost instantaneous. This is an important point to keep in mind later in the book, when events in an electrical circuit seem to occur simultaneously.

Although electron motion is known to exist to some extent in all matter, some substances, such as rubber, glass, and dry wood, have very few free electrons. In these materials, large amounts of energy must be expended to break the electrons loose from the influence of the nucleus. Substances containing very few free electrons are called *insulators*. Insulators are important in electrical work because they prevent the current from being diverted from the wires, which in turn helps to prevent electrocutions and fires. Table 2.2 lists some materials that we often use as insulators in electrical circuits. These materials are listed in decreasing order of their ability to withstand high voltages without conducting. A material that is neither a good conductor nor a good insulator is called a *semiconductor*. Silicon and germanium are substances that fall into this category. Because of their peculiar crystalline structure, these materials may, under certain conditions, act as conductors and, under other conditions, as insulators. As the temperature is raised, however, a limited number of electrons become available for conduction.

Key Point: If the voltage is large enough, even the best insulators (rubber, plastic, wood) will break down and allow the electrons to flow.

TABLE 2.2
Common Insulators

Rubber
Mica
Wax or paraffin
Porcelain
Bakelite
Plastics
Glass
Fiberglass
Dry wood
Air

Static Electricity

Electricity at rest is often referred to as *static electricity*. More specifically, when two bodies of matter with unequal charges are near one another, an electric force is exerted between them because of their unequal charges; however, because they are not in contact, their charges cannot equalize. Static electricity (or electricity at rest) will flow when given the opportunity, because static electricity is an imbalance of negative and positive charges. An example of this phenomenon is when we walk across a dry carpet and then touch a doorknob—we usually feel a slight shock and might notice a spark at our fingertips. Another familiar example is static cling; for example, when we rub an air-filled balloon against the hair on our heads and then place the balloon against a wall, the balloon will stick to the wall, defying gravity, due to static cling. In the workplace, static electricity is prevented from building up by properly bonding or grounding equipment.

Charged Bodies

The fundamental law of charged bodies states: "Like charges repel each other and unlike charges attract each other." A positive charge and negative charge, being opposite or unlike, tend to move toward each other; that is, they are attracted to each other. In contrast, like bodies tend to repel each other. Electrons repel each other because of their like negative charges, and protons repel each other because of their like positive charges. Figure 2.6 demonstrates the law of charged bodies.

It is important to point out another significant aspect of the fundamental law of charged bodies: The force of attraction or repulsion existing between two magnetic poles decreases rapidly as the poles are separated from each other. More specifically, the force of attraction or repulsion varies directly as the product of the separate pole strengths and inversely as the square of the distance separating the magnetic poles, provided the poles are small enough to be considered as points. Let's look

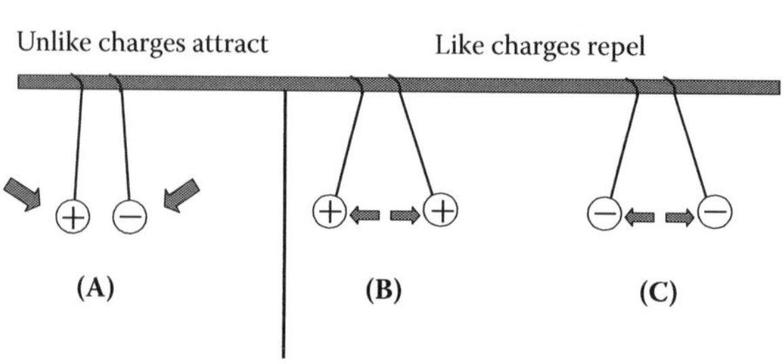

FIGURE 2.6 Reaction between two charged bodies. The opposite charges in (A) attract; the like charges in (B) and (C) repel each other.

at an example. If you increase the distance between two magnet north poles from 2 feet to 4 feet, the force of repulsion between them is decreased to one fourth of its original value. If either pole strength is doubled and the distance remains the same, the force between the poles will be doubled.

Coulomb's Law

Coulomb's law states that the amount of attracting or repelling force that acts between two electrically charged bodies in free space depends on two things:

1. Their charges
2. The distance between them

Specifically, Coulomb's law states that charged bodies attract or repel each other with a force that is directly proportional to the product of their charges and is inversely proportional to the square of the distance between them.

Note: The magnitude of electric charge a body possesses is determined by the number of electrons compared with the number of protons within the body. The symbol for the magnitude of electric charge is Q, expressed in units of *coulombs* (C). A charge of one positive coulomb means a body contains a charge of 6.25×10^{18}. A charge of one negative coulomb means a body contains a charge of 6.25×10^{18} more electrons than protons.

Electrostatic Fields

The fundamental characteristic of an electric charge is its ability to exert force. The space between and around charged bodies in which their influence is felt is the *electric field of force*. The electric field is always terminated on material objects and extends between positive and negative charges. This region of force can consist of air, glass, paper, or a vacuum and is referred to as an *electrostatic field*. When two objects of opposite polarity are brought near each other, the electrostatic field is concentrated in the area between them. The field is generally represented by lines referred to as *electrostatic lines of force*. These lines are imaginary and are used merely to represent the direction and strength of the field. To avoid confusion, the positive lines of force are always shown leaving, and the negative lines of force are shown entering. Figure 2.7 illustrates the use of lines to represent the field about charged bodies.

Note: A charged object will retain its charge temporarily if there is no immediate transfer of electrons to or from it. In this condition, the charge is said to be *at rest*. Remember, electricity at rest is called *static* electricity.

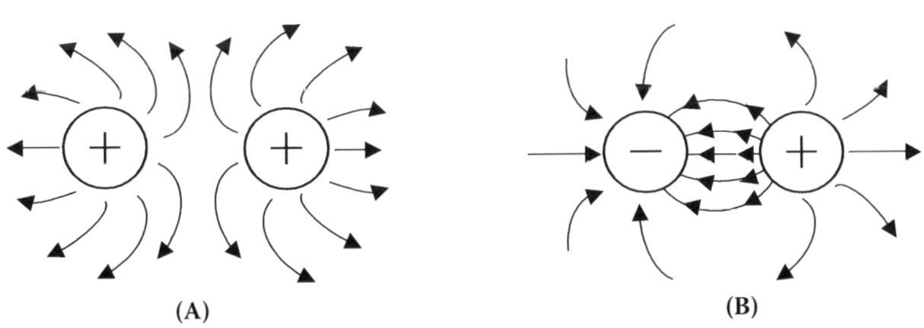

FIGURE 2.7 Electrostatic lines of force: (A) repulsion of like-charged bodies and their associated fields, and (B) attraction between unlike-charged bodies and their associated fields.

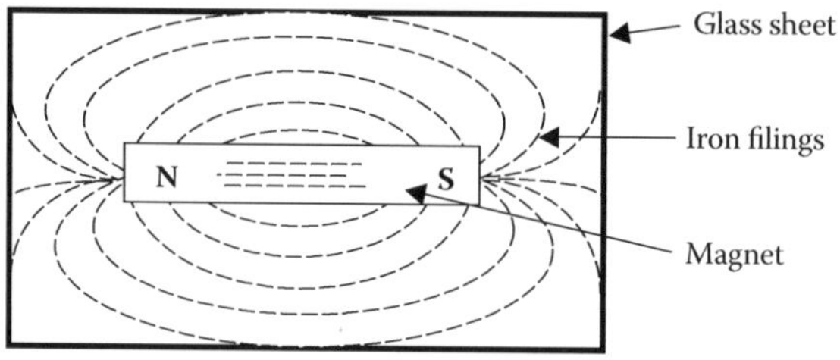

FIGURE 2.8 Magnetic field around a bar magnet. If the glass sheet is tapped gently, the filings will move into a definite pattern that describes the field of force around the magnet.

Magnetism

Most electrical equipment depends directly or indirectly on magnetism, a phenomenon associated with magnetic fields that has the power to attract such substances as iron, steel, nickel, or cobalt (metals known as magnetic materials). A substance is said to be a *magnet* if it has the property of magnetism; for example, when a piece of iron is magnetized it becomes a magnet. When magnetized, that piece of iron will have two points, opposite each other, that most readily attract other pieces of iron. (For this discussion, we will assume the piece of iron is a flat bar 6 inches long × 1 inch wide × 0.5 inch thick—in other words, a bar magnet; see Figure 2.8.) The points of maximum attraction (one on each end) are the *magnetic poles* of the magnet: the north (N) pole and the south (S) pole. Just as like electric charges repel each other and opposite charges attract each other, like magnetic poles repel each other and unlike poles attract each other. Although invisible to the naked eye, magnetic force can be shown to exist by sprinkling small iron filings on a glass covering a bar magnet, as shown in Figure 2.8. Figure 2.9 shows how the field looks without the iron filings. The group of magnetic field lines, which flow from the north pole of a magnet toward the south pole, is referred to as the *magnetic flux*; the symbol for magnetic flux is the Greek lowercase letter φ (phi).

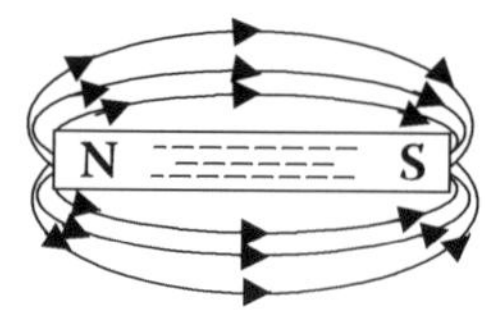

FIGURE 2.9 Magnetic field of force around a bar magnet, indicated by lines of force.

Note: A *magnetic circuit* is a complete path through which magnetic lines of force may be established under the influence of a magnetizing force. Most magnetic circuits are composed largely of magnetic materials to contain the magnetic flux. These circuits are similar to the *electric circuit* (an important point), which is a complete path through which current is caused to flow under the influence of an electromotive force.

The three types or groups of magnets are as follows:

1. *Natural magnets* are found in the natural state in the form of the mineral magnetite (an iron compound).
2. *Permanent magnets* (artificial magnets) are hardened steel or some alloy such as Alinco bars that have been permanently magnetized. The permanent magnet most people are familiar with is the horseshoe magnet; this red U-shaped magnet is the universal symbol of magnets, recognized throughout the world (see Figure 2.10).
3. *Electromagnets* (artificial magnets) are composed of soft iron cores around which are wound coils of insulated wire. When an electric current flows through the coil, the core becomes magnetized. When the current ceases to flow, the core loses most of the magnetism.

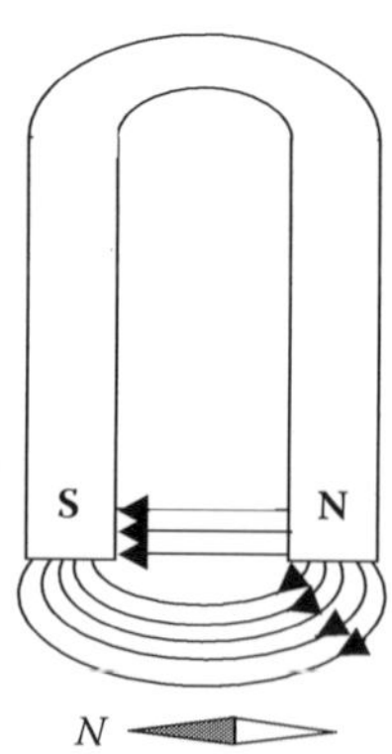

FIGURE 2.10 Horseshoe magnet.

Magnetic Materials

Natural magnets are no longer used in electrical circuitry because more powerful and more conveniently shaped permanent magnets can be produced artificially. Commercial magnets are made from such magnetic materials as special steels and alloys. Magnetic materials are those materials that are attracted or repelled by a magnet and that can be magnetized themselves. Iron, steel, and alloy bars are the most common magnetic materials. These materials can be magnetized by inserting the material (in bar form) into a coil of insulated wire and passing a heavy direct current through the coil. The same material may also be magnetized if it is stroked with a bar magnet. It will then have the same magnetic property that the magnet used to induce the magnetism has—namely, there will be two poles of attraction, one at either end. This process produces a permanent magnet by induction; that is, the magnetism is induced in the bar by the influence of the stroking magnet.

Once hard steel and other alloys are magnetized, they retain a large part of their magnetic strength and are considered to be permanent magnets. Conversely, materials that are relatively easy to magnetize—such as soft iron and annealed silicon steel—are said to have a *high permeability*. Such materials retain only a small part of their magnetism after the magnetizing force is removed and are called *temporary magnets*. The magnetism that remains in a temporary magnet after the magnetizing force is removed is called *residual magnetism*. Even though they are classified as permanent magnets, it is important to point out that hardened steel and certain alloys are relatively difficult to magnetize and are said to have a *low permeability* because the magnetic lines of force do not easily permeate or distribute themselves readily through the steel.

Key Point: *Permanent magnets* are made of hard magnetic materials (hard steel or alloys) that retain their magnetism when the magnetizing field is removed. A *temporary magnet* is one that has *no* ability to retain a magnetized state when the magnetizing field is removed.

Note: *Permeability* refers to the ability of a magnetic material to concentrate magnetic flux. Any material that is easily magnetized has high permeability. A measure of permeability for different materials in comparison with air or a vacuum is the *relative* permeability, symbolized by μ (mu).

Early magnetic studies classified magnetic materials as being magnetic or nonmagnetic, based on the strong magnetic properties of iron; however, because even weak magnetic materials can be important in some applications, current studies classify materials into one of three groups:

- *Paramagnetic materials* include aluminum, platinum, manganese, and chromium, materials that become only slightly magnetized even under the influence of a strong magnetic field. This slight magnetization is in the same direction as the magnetizing field. Relative permeability is slightly more than 1 (i.e., they are considered to be nonmagnetic materials).
- *Diamagnetic materials* include bismuth, antimony, copper, zinc, mercury, gold, and silver, materials that can also be slightly magnetized when under the influence of a very strong field. Relative permeability is less than 1 (i.e., they are considered to be nonmagnetic materials).
- *Ferromagnetic materials* include iron, steel, nickel, cobalt, and commercial alloys, materials that comprise the most important group in applications of electricity and electronics. Ferromagnetic materials are easy to magnetize and have high permeabilities, ranging from 50 to 3000.

Magnetic Earth

Earth is a huge magnet, and surrounding Earth is a magnetic field produced by Earth's magnetism. Most people would have no problem understanding or at least accepting this statement; however, they might question being told that Earth's north magnetic pole is actually its south magnetic pole and that the south magnetic pole is actually Earth's north magnetic pole. But, in terms of a magnet,

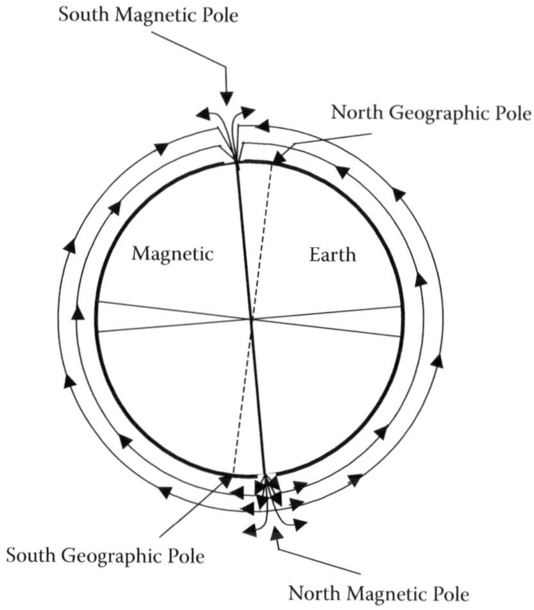

FIGURE 2.11 Earth's magnetic poles.

this is true. Figure 2.11 indicates the magnetic polarities of Earth. The geographic poles are also shown at each end of the axis of rotation of Earth. Clearly, the magnetic axis does not coincide with the geographic axis; therefore, the magnetic and geographic poles are not at the same place on the surface of Earth. Recall that magnetic lines of force are assumed to emanate from the north pole of a magnet and enter the south pole as closed loops. Because Earth is a magnet, lines of force emanate from its north magnetic pole and enter the south magnetic pole. A compass needle aligns itself in such a way that Earth's lines of force enter at its south pole and leave at its north pole. Because the north pole on a compass is defined as the point where the needle points in a northerly direction, it follows that the magnetic pole in the vicinity of the north geographic pole is in reality a south magnetic pole, and *vice versa*.

Difference in Potential

Because of the force of its electrostatic field, an electric charge has the ability to do the work of moving another charge by attraction or repulsion. The force that causes free electrons to move in a conductor as an electric current may be referred to as

- Electromotive force (emf)
- Voltage
- Difference in potential

When a difference in potential exists between two charged bodies that are connected by a wire (conductor), electrons (current) will flow along the conductor. This flow is from the negatively charged body to the positively charged body until the two charges are equalized and the potential difference no longer exists.

Note: The basic unit of potential difference is the *volt* (V). The symbol for potential difference is *V*, indicating the ability to do the work of forcing electrons (current flow) to move. Because the volt unit is used, potential difference is called *voltage*.

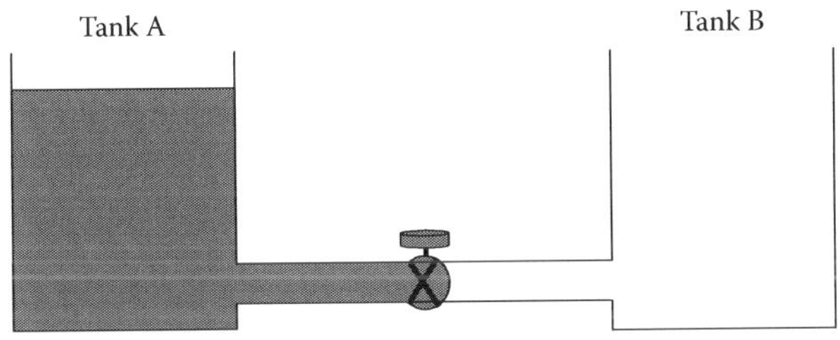

FIGURE 2.12 Water analogy for electric difference of potential.

Water Analogy

When training individuals in the concepts of basic electricity, especially with regard to potential difference (voltage), current, and resistance relationships in a simple electrical circuit, it has been common practice to use what is referred to as the *water analogy*. We use the water analogy later to explain (in simple, straightforward fashion) voltage, current, and resistance and their relationships in more detail, but for now we will use the analogy to explain the basic concept of electricity: the potential difference, or voltage. Because a difference in potential causes current flow (against resistance), it is important that this concept be understood first before exploring the concept of current flow and resistance. Consider the two water tanks connected by a pipe and valve in Figure 2.12. At first, the valve is closed and all the water is in Tank A; the water pressure across the valve is at maximum. When the valve is opened, the water flows through the pipe from A to B until the water level becomes the same in both tanks. The water then stops flowing in the pipe, because there is no longer a difference in water pressure (difference in potential) between the two tanks. Just as the flow of water through the pipe in Figure 2.12 is directly proportional to the difference in water level in the two tanks, current flow through an electric circuit is directly proportional to the difference in potential across the circuit.

Key Point: A fundamental law of current electricity is that the current is directly proportional to the applied voltage: If the voltage is increased, the current is increased. If the voltage is decreased, the current is decreased.

Principal Methods for Producing Voltage

There are many ways to produce electromotive force, or voltage. Some of these methods are much more widely used than others. The following is a list of the seven most common methods of producing electromotive force (USDOE, 1992):

1. *Friction* is the voltage produced by rubbing two materials together (static electricity or electrostatic force). Remember our discussion of static electricity? Let's refresh our memories. Have you ever walked across a carpet and received a shock when you touched a metal door knob? The soles of your shoes built up a charge when they rubbed on the carpet, and this charge was transferred to your body. Your body became positively charged, and when you touched the zero-charged door knob electrons were transferred to your body until both you and the door knob had equal charges.
2. *Pressure* (*piezoelectricity*) is the voltage produced by squeezing or applying pressure to crystals of certain substances (e.g., quartz, Rochelle salts, certain ceramics such as barium titanate). When pressure is applied to such substances, electrons can be driven out of orbit in the direction of the force. Electrons leave one side of the material and accumulate on the other side, building up positive and negative charges on opposite sides. When the pressure is released, the electrons return to their orbits. Some materials will react to bending pressure, while others will respond to twisting pressure. This generation of voltage is known as the *piezoelectric effect*. If external wires are connected while pressure and voltage are

present, electrons will flow and current will be produced. If the pressure is held constant, the current will flow until the potential difference is equalized. When the force is removed, the material is decompressed and immediately causes an electric force in the opposite direction. The power capacity of these materials is extremely small; however, these materials are very useful because of their extreme sensitivity to changes of mechanical force. One example is the crystal phonograph cartridge that contains a Rochelle salt crystal. A phonograph needle is attached to the crystal. As the needle moves in the grooves of a record, it swings from side to side, compressing and decompressing the crystal. The mechanical motion applied to the crystal generates a voltage signal that is used to reproduce sound.

3. *Heat* (*thermoelectricity*) is the voltage produced by heating the joint (junction) where two unlike metals are joined. Some materials readily give up their electrons and others readily accept electrons; for example, when two dissimilar metals such as copper and zinc are joined together, a transfer of electrons can take place. Electrons will leave the copper atoms and enter the zinc atoms. The zinc gets a surplus of electrons and becomes negatively charge. The copper loses electrons and takes on a positive charge. This creates a voltage potential across the junction of the two metals. The heat energy of normal room temperature is enough to make them release and gain electrons, causing a measurable voltage potential As more heat energy is applied to the junction, more electrons are released, and the voltage potential becomes greater. When heat is removed and the junction cools, the charges will dissipate and the voltage potential will decrease. This process is known as thermoelectricity. A device like this is generally referred to as a *thermocouple.* Thermocouple voltage is dependent on the heat energy applied to the junction of the two dissimilar metals. Thermocouples are widely used to measure temperature and as heat-sensing devices in automatic temperature-controlled equipment. Thermocouple power capacities are very small compared to some other sources, but they are somewhat greater than those of crystals. Generally speaking, a thermocouple can be subjected to higher temperatures than ordinary mercury or alcohol thermometers.
4. *Light* (*photoelectricity*) is the voltage produced by light (photons) striking photosensitive (light-sensitive) substances. When the photons in a light beam strike the surface of a material, they release their energy and transfer it to the atomic electrons of the material. This energy transfer may dislodge electrons from their orbits around the surface of the substance. Upon losing electrons, the photosensitive (light-sensitive) material becomes positively charge and an electric force is created. This phenomenon is called the *photoelectric effect* and has wide applications in electronics (e.g., photoelectric cells, photovoltaic cells, optical couplers, television camera tubes). Three uses of the photoelectric effect are described below.
   - *Photovoltaic*—The light energy in one of two plates that are joined together causes one plate to release electrons to the other; the plates build up opposite charges, as in a battery.
   - *Photoemission*—The photon energy from a beam of light can cause a surface to release electrons in a vacuum tube, and a plate then collects the electrons.
   - *Photoconduction*—The light energy applied to some materials that are normally poor conductors causes free electrons to be produced in the materials so they become better conductors.
5. *Chemical action* is the voltage produced by chemical reaction in a battery cell; an example is a voltaic chemical cell, in which a chemical reaction produces and maintains opposite charges on two dissimilar metals that serve as positive and negative terminals. The metals are in contact with an electrolyte solution. Connecting more than one of these cells will produce a battery.
6. *Magnetism* is the voltage produced in a conductor when the conductor moves through a magnetic field or a magnetic field moves through the conductor in such a manner as to cut

the magnetic lines of force of the field. A generator is a machine that converts mechanical energy into electrical energy by using the principle of *magnetic induction*. Magnetism is used to produce vast quantities of electric power.

7. *Thermionic emission* is produced by a thermionic energy converter, which consists of two electrodes placed near one another in a vacuum. One electrode is normally the cathode, or emitter, and the other is the anode, or plate. Ordinarily, electrons in the cathode are prevented from escaping from the surface by a potential-energy barrier. When an electron starts to move way from the surface, it induces a corresponding positive charge in the material, which tends to pull it back into the surface. To escape, the electron must somehow acquire enough energy to overcome this energy barrier. At ordinary temperatures, almost none of the electrons can acquire enough energy to escape; however, when the cathode is very hot, the electron energies are greatly increased by thermal motion. At sufficiently high temperatures, a considerable number of electrons are able to escape. The liberation of electrons from a hot surface is thermionic emission. The electrons that have escaped from the hot cathode form a cloud of negative charges near it called a *space charge*. If the plate is maintained positive with respect to the cathode by a battery, the electrons in the cloud are attracted to it. As long as the potential difference between the electrodes is maintained, there will be a steady current flow from the cathode to the plate. The simplest example of a thermionic device is a vacuum tube diode, in which the only electrodes are the cathode and plate, or anode. The diode can be used to convert alternating current (AC) flow to a pulsating direct current (DC) flow.

In the study of the basic electricity related to renewable energy production, we are most concerned with magnetism (e.g., generators powered by hydropower), light (photoelectricity produced by solar cells), and chemistry (chemical energy converted to electricity in batteries) as the means to produce voltage. Friction has little practical application, although we discussed it earlier with regard to static electricity. Pressure and heat do have useful applications, but we do not need to consider them in this text. Magnetism used in generators, electricity produced by solar light, and the chemistry involved in storing electricity in batteries, on the other hand, are the principal sources of voltage and are discussed at length in this text.

Electric Current

The movement or the flow of electrons is called *current*. To produce current, the electrons must be moved by a potential difference or pressure (voltage).

Note: The terms *current*, *current flow*, *electron flow*, and *electron current* all describe the same phenomenon.

For our purposes in this text, electron flow, or current, in an electric circuit is from a region of less negative potential to a region of more positive potential—from negative to positive.

Note: Current is represented by the letter *I*. The basic unit in which current is measured is the *ampere*, or *amp* (A). One ampere of current is defined as the movement of one coulomb past any point of a conductor during one second of time.

Recall that we used the water analogy to help us understand potential difference. We can also use the water analogy to help us understand current flow through a simple electric circuit. Consider Figure 2.13, which shows a water tank connected via a pipe to a pump with a discharge pipe. If the water tank contains an amount of water above the level of the pipe opening to the pump, the water exerts pressure (a difference in potential) against the pump. When sufficient water is available for pumping with the pump, water flows through the pipe against the resistance of the pump and pipe. The analogy should be clear—in an electric circuit, if a difference in potential exists, current will

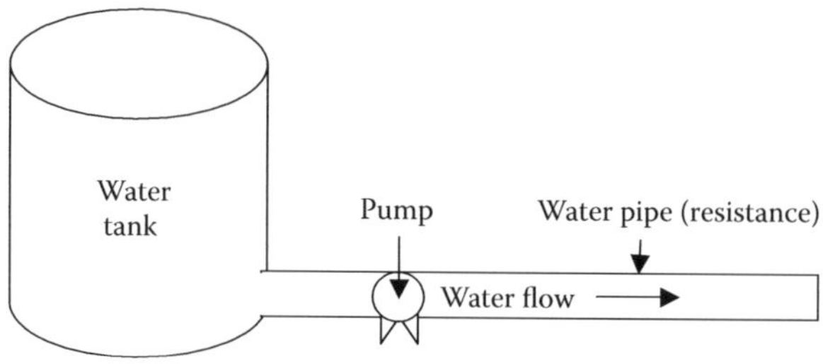

FIGURE 2.13 Water analogy: current flow.

flow in the circuit. Another simple way of looking at this analogy is to consider Figure 2.14, where the water tank has been replaced with a generator, the pipe with a conductor (wire), and water flow with the flow of electric current. Again, the key point illustrated by Figures 2.13 and 2.14 is that, to produce current, the electrons must be moved by a potential difference.

Electric current is generally classified into two general types:

- Direct current (DC)
- Alternating current (AC)

Direct current is current that moves through a conductor or circuit in one direction only. *Alternating current* periodically reverses direction.

Resistance

Earlier we pointed out that free electrons, or electric current, could move easily through a good conductor, such as copper, but that an insulator, such as glass, was an obstacle to current flow. In the water analogy shown in Figure 2.13 and the simple electric circuit shown in Figure 2.14, resistance is indicated by either the pipe or the conductor. Every material offers some resistance, or opposition, to the flow of electric current through it. Good conductors such as copper, silver, and aluminum offer very little resistance. Poor conductors, or insulators, such as glass, wood, and paper, offer a high resistance to current flow.

Note: The amount of current that flows in a given circuit depends on two factors: voltage and resistance.

Note: Resistance is represented by the letter *R*. The basic unit in which resistance is measured is the *ohm* (Ω). One ohm is the resistance of a circuit element, or circuit, that permits a steady current of 1 ampere (1 coulomb per second) to flow when a steady electromotive force (emf) of 1 volt is applied to the circuit. Manufactured circuit parts containing definite amounts of resistance are called *resistors*.

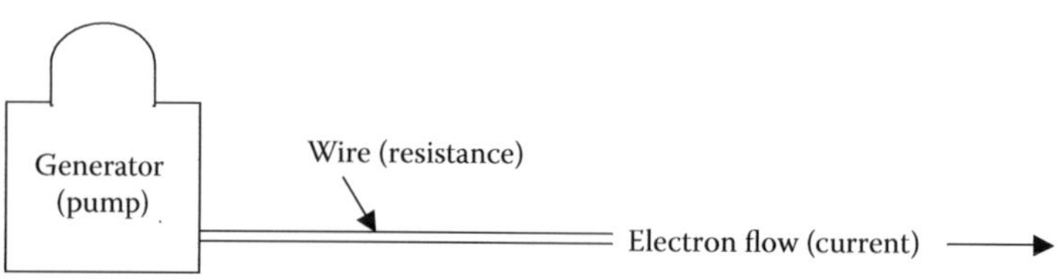

FIGURE 2.14 Simple electric circuit with current flow.

The size and type of material of the wires in an electric circuit are chosen so as to keep the electrical resistance as low as possible. In this way, current can flow easily through the conductors, just as water flows through the pipe between the tanks in Figure 2.12. If the water pressure remains constant, the flow of water in the pipe will depend on how far the valve is opened. The smaller the opening, the greater the opposition (resistance) to the flow and the smaller will be the rate of flow in gallons per second. In the electric circuit shown in Figure 2.14, the larger the diameter of the wire, the lower will be its electrical resistance (opposition) to the flow of current through it. In the water analogy, pipe friction opposes the flow of water between the tanks. This friction is similar to electrical resistance. The resistance of a pipe to the flow of water through it depends on: (1) the length of the pipe, (2) the diameter of the pipe, and (3) the nature of the inside walls (rough or smooth). Similarly, the electrical resistance of the conductors depends on: (1) the length of the wires, (2) the diameter of the wires, and (3) the material of the wires (e.g., copper, silver). It is important to note that temperature also affects the resistance of electrical conductors to some extent. In most conductors (e.g., copper, aluminum) the resistance increases with temperature. Carbon is an exception. In carbon, the resistance decreases as temperature increases.

Note: Electricity is a study that is frequently explained in terms of opposites. The term that is exactly the opposite of resistance is *conductance*. Conductance (*G*) is the ability of a material to pass electrons. The SI unit of conductance is the *siemens*. The commonly used unit of conductance is the *Mho*, which is ohm spelled backward. The relationship that exists between resistance and conductance is the *reciprocal*. A reciprocal of a number is obtained by dividing the number into one. If the resistance of a material is known, dividing its value into one will give its conductance. Similarly, if the conductance is known, dividing its value into one will give its resistance.

BATTERY-SUPPLIED ELECTRICITY

Battery-supplied direct current electricity has many applications and is widely used in household, commercial, and industrial operations. Applications include providing electrical energy for industrial vehicles, emergency diesel generators, material handling equipment (forklifts), portable electric/electronic equipment, backup emergency power for light packs, hazard warning signal lights, flashlights, and standby power supplies or uninterruptible power supplies (UPS) for computer systems. In some instances, they are used as the only source of power, whereas in others they are used as a secondary or standby power supply. In renewable energy applications, batteries are used to store electrical energy; for example, a battery pack can store electricity produced by a solar electric system. Today, they are commonly used in specialized applications such as plug-in hybrid electric vehicle energy storage systems. Batteries are often used in hybrid wind systems to store excess wind energy and then provide supplementary energy when the wind cannot generate sufficient power to meet the electric load. Batteries are also used in roadside solar-charging systems; that is, solar energy is used to charge a battery pack that supplies electrical power to some type of electrical signaling device, navigation buoys, or other remote application (e.g., emergency telephone) where it would be impractical and costly to run miles of electrical supply cable.

Battery Terminology

- A *voltaic cell* is a combination of materials used to convert chemical energy into electrical energy.
- A *battery* is a group of two or more connected voltaic cells.
- An *electrode* is a metallic compound, or metal, that has an abundance of electrons (negative electrode) or an abundance of positive charges (positive electrode).
- An *electrolyte* is a solution capable of conducting an electric current.

- *Specific gravity* is defined as the ratio comparing the weight of any liquid to the weight of an equal volume of water.
- An *ampere-hour* is defined as a current of 1 ampere flowing for 1 hour.

Voltaic Cell

The simplest cell (a device that transforms chemical energy into electrical energy) is the voltaic (or galvanic) cell (see Figure 2.15). It consists of a piece of carbon (C) and a piece of zinc (Zn) suspended in a jar that contains a solution of water (H_2O) and sulfuric acid (H_2SO_4).

Note: A simple cell consists of two strips, or *electrodes*, placed in a container that holds the *electrolyte*. A battery is formed when two or more cells are connected.

The electrodes are the conductors by which the current leaves or returns to the electrolyte. In the simple cell described above, they are carbon and zinc strips placed in the electrolyte. Zinc contains an abundance of negatively charged atoms, and carbon has an abundance of positively charged atoms. When the plates of these materials are immersed in an electrolyte, chemical action between the two begins. In the *dry cell* (Figure 2.16), the electrodes are the carbon rod in the center and the zinc container in which the cell is assembled. The electrolyte is the solution that acts upon the electrodes that are placed in it. The electrolyte may be a salt, an acid, or an alkaline solution. In the simple voltaic cell and in the automobile storage battery, the electrolyte is in a liquid form; in the dry cell, the electrolyte is a moist paste.

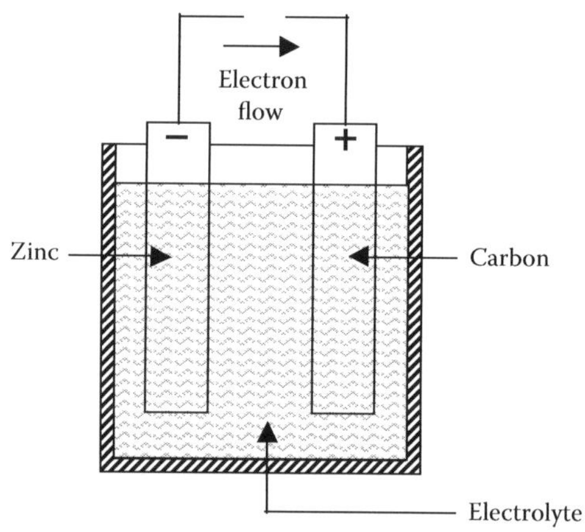

FIGURE 2.15 Simple voltaic cell.

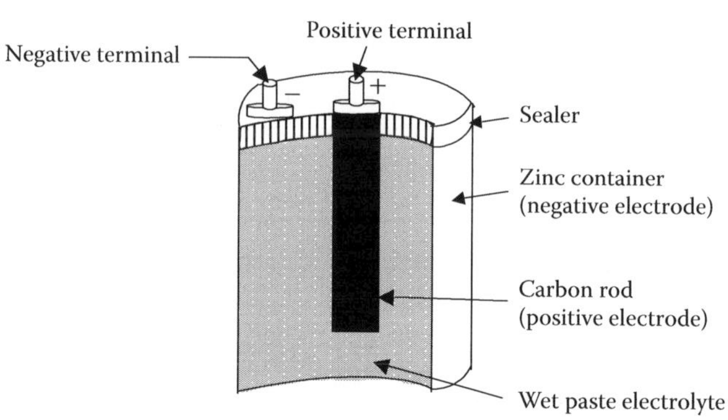

FIGURE 2.16 Dry cell (cross-sectional view).

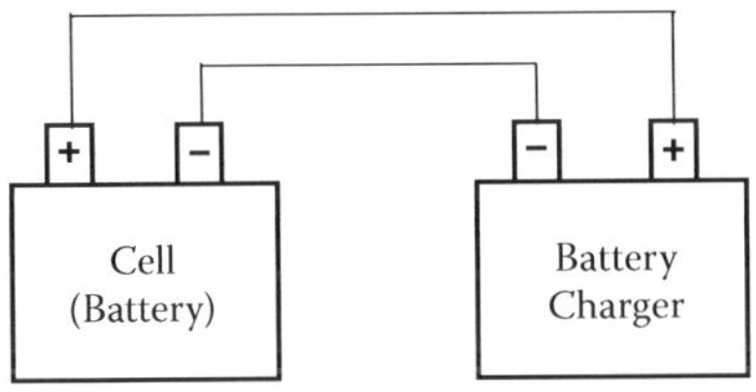

FIGURE 2.17 Hookup for charging a secondary cell with a battery charger.

Primary and Secondary Cells

Primary cells are normally those that cannot be recharged or returned to good condition after their voltage drops too low. Dry cells in flashlights and transistor radios are examples of primary cells. Some primary cells have been developed to the state where they can be recharged. A *secondary cell* is one in which the electrodes and the electrolyte are altered by the chemical action that takes place when the cell delivers current. These cells are rechargeable. During recharging, the chemicals that provide electric energy are restored to their original condition. Recharging is accomplished by forcing an electric current through them in the opposite direction of the discharge. Figure 2.17 shows how a secondary cell is recharged. Some battery chargers have a voltmeter and an ammeter that indicate the charging voltage and current. The automobile storage battery is the most common example of the secondary cell.

Battery

As stated previously, a cell is an electrochemical device capable of supplying the energy that results from an internal chemical reaction to an external electrical circuit. A battery consists of two or more cells placed in a common container. The cells are connected in series, in parallel, or in some combination of series and parallel, depending upon the amount of voltage and current required of the battery. The connection of cells in a battery is discussed in more detail later.

Battery Operation

The chemical reaction within a battery provides the voltage. This occurs when a conductor is connected externally to the electrodes of a cell which causes electrons to flow under the influence of a difference in potential across the electrodes from the zinc (negative) through the external conductor to the carbon (positive), returning within the solution to the zinc. After a short period of time, the zinc will begin to waste away because of the acid. The voltage across the electrodes depends on the materials from which the electrodes are made and the composition of the solution. The difference of potential between the carbon and zinc electrodes in a dilute solution of sulfuric acid and water is about 1.5 volts. The current that a primary cell may deliver depends on the resistance of the entire circuit, including that of the cell itself. The internal resistance of the primary cell depends on the size of the electrodes, the distance between them in the solution, and the resistance of the solution. The larger the electrodes and the closer together they are in solution (without touching), the lower the internal resistance of the primary cell and the more current it is capable of supplying to the load.

Note: When current flows through a cell, the zinc gradually dissolves in the solution and the acid is neutralized.

Combining Cells

In many operations, battery-powered devices may require more electrical energy than one cell can provide. Various devices may require either a higher voltage or more current, and in some cases both. Under such conditions it is necessary to combine, or interconnect, a sufficient number of cells

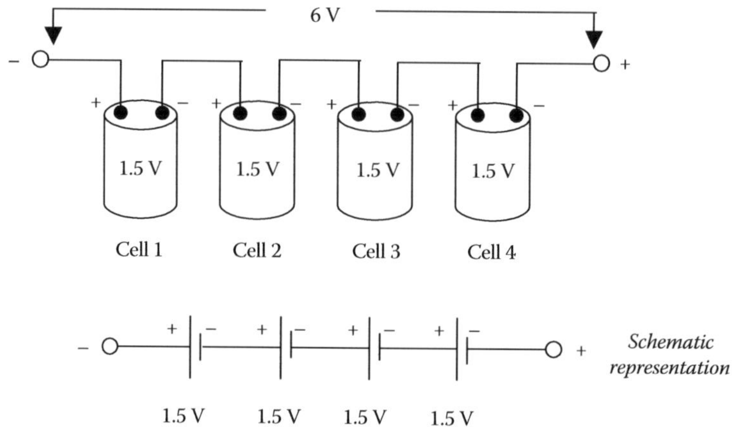

FIGURE 2.18 Cells in series.

to meet the higher requirements. Cells connected in series provide a higher voltage, and cells connected in parallel provide a higher current capacity. To provide adequate power when both voltage and current requirements are greater than the capacity of one cell, a series–parallel network of cells must be interconnected.

When cells are connected in *series* (Figure 2.18), the total voltage across the battery of cells is equal to the sum of the voltage of each of the individual cells. In Figure 2.18, the four 1.5-V cells in series provide a total battery voltage of 6 V. When cells are placed in series, the positive terminal of one cell is connected to the negative terminal of the other cell. The positive electrode of the first cell and negative electrode of the last cell then serve as the power takeoff terminals of the battery. The current flowing through such a battery of series cells is the same as from one cell because the same current flows through all of the series cells.

To obtain a greater current, cells can be connected in *parallel*, as shown in Figure 2.19. In this parallel connection, all of the positive electrodes are connected to one line, and all negative electrodes are connected to the other. Any point on the positive side can serve as the positive terminal of the battery and any point on the negative side can be the negative terminal. The total voltage output of a battery of three parallel cells is the same as that for a single cell (Figure 2.19), but the available

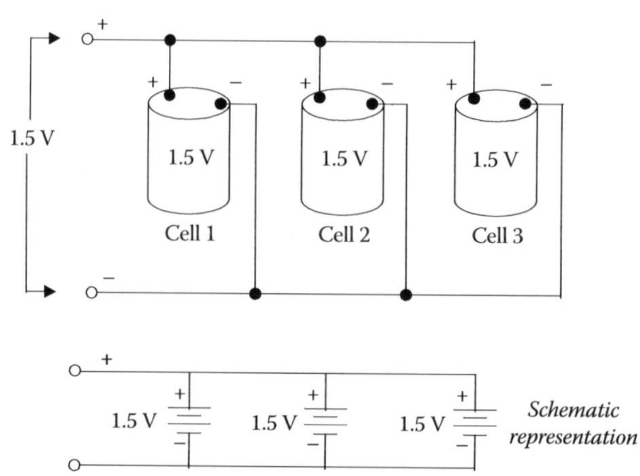

FIGURE 2.19 Cells in parallel.

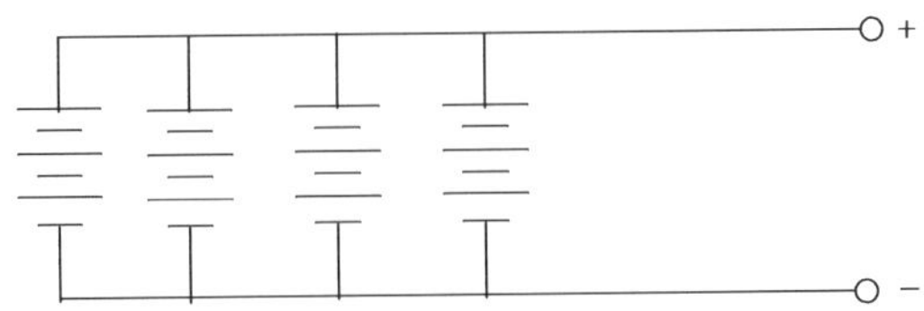

FIGURE 2.20 Series–parallel connected cells.

current is three times that of one cell; that is, the current capacity has been increased. Identical cells in parallel all supply equal parts of the current to the load; for example, of three different parallel cells producing a load current of 210 milliamperes (mA), each cell contributes 70 mA. Figure 2.20 is a schematic of a *series–parallel* battery network supplying power to a load requiring both a voltage and current greater than one cell can provide. To provide the required increased voltage, groups of three 1.5-V cells are connected in series. To provide the required increased amperage, four series groups are connected in parallel.

Types of Batteries

In the past 30 years, several different types of batteries have been developed. In this text, we briefly discuss the dry cell battery and those batteries that are currently used to store electrical energy in a reversible chemical reaction as applied to renewable energy production. The renewable energy source (solar, wind, or hydro) produces the energy, and the battery stores it for times of low or no renewable energy production. The types of batteries used in this application and discussed here include the lead–acid, alkaline, nickel–cadmium, mercury, nickel–metal hydride, lithium-ion, and lithium-ion polymer batteries. Keep in mind that a battery does not *create* energy; rather, it *stores* energy. For most renewable energy applications, the preferred battery type is the deep-cycle battery. A deep-cycle battery is designed to deliver a constant voltage as the battery discharges. A car-starting battery, in contrast, is designed to deliver sporadic current spikes. Battery-driven vehicles, such as forklifts, golf carts, and floor sweepers, commonly use deep-cycle batteries. Deep-cycle batteries can be charged with a lower current than regular batteries. Following are descriptions of the various types of batteries (National Renewable Energy Laboratory, 2009):

- *Dry cell*—In the dry cell, or carbon–zinc cell, the electrolyte is not in a liquid state but is a moist paste. A carbon rod placed in the center of the cell is the positive terminal. The case of the cell is made of zinc and is the negative terminal (see Figure 2.16). Between the carbon electrode and the zinc case is the electrolyte, a moist, chemical, paste-like mixture. The cell is sealed to prevent the liquid in the paste from evaporating. The voltage of a cell of this type is about 1.5 V.
- *Lead–acid battery*—The lead–acid battery is a secondary cell (e.g., storage battery or rechargeable battery) that stores chemical energy until it is released as electrical energy. The lead–acid battery differs from a primary cell battery mainly in that it may be recharged, whereas most primary cells cannot be recharged. As the name implies, the lead–acid battery consists of a number of lead–acid cells immersed in a dilute solution of sulfuric acid. Each cell has two groups of lead plates; one set is the positive terminal and the other is the negative terminal. Active materials within the battery (lead plates and sulfuric acid electrolyte) react chemically to produce a flow of direct current whenever current-consuming devices are connected to the battery terminal posts. This current is produced by a chemical reaction between the active material of the plates (electrodes) and the electrolyte (sulfuric acid). This type of cell produces slightly more than 2 V. Most automobile batteries contain six cells connected in series so the output voltage from the battery is slightly more than 12 V. In addition to being rechargeable, the main advantage of the lead–acid battery over

the dry cell battery is that it can supply current for a much longer time than the average dry cell. Lead–acid batteries can be designed to be high power and are inexpensive, safe, and reliable; also, a recycling infrastructure is in place for them. But, low specific energy, poor cold-temperature performance, and short calendar and cycle life are still impediments to their use. Advanced high-power, deep-cycle lead–acid batteries are being developed for hybrid electrical vehicle (HEV) applications; however, lead–acid batteries are used for residential solar electric systems because of their low maintenance requirements and cost.

Safety Note: Whenever a lead–acid storage battery is charging, the chemical action produces dangerous hydrogen gas; thus, the charging operation should take place only in a well-ventilated area.

- *Alkaline cell*—The alkaline cell is a secondary cell that gets its name from its alkaline electrolyte, potassium hydroxide. The alkaline battery has a negative electrode of zinc and a positive electrode of manganese dioxide. It generates 1.5 V.
- *Nickel–cadmium cell*—The nickel–cadmium cell, or NiCad cell, is the only dry cell that is a true storage battery with a reversible chemical reaction, allowing recharging many times. In the secondary nickel–cadmium dry cell, the electrolyte is potassium hydroxide, the negative electrode is nickel hydroxide, and the positive electrode is cadmium oxide. The operating voltage is 1.25 V. Because of its rugged characteristics (stands up well to shock, vibration, and temperature changes) and availability in a variety of shapes and sizes, it is ideally suited for powering portable communication equipment. NiCad batteries are very expensive. Moreover, although nickel–cadmium batteries are used in many electronic consumer products and have a higher specific energy and better life cycle than lead–acid batteries, they are low efficiency (65 to 80%). They do not deliver sufficient power and are not being considered for HEV applications. Cadmium is a heavy metal that is toxic and is very expensive to dispose of which reduces it desirability for use in hybrid applications.
- *Mercury cell*—The mercury cell was developed for space exploration activities, for which small transceivers and miniaturized equipment required a small power source. In addition to reduced size, the mercury cells have a good shelf life and are very rugged; they also produce a constant output voltage under different load conditions. Two types of mercury cells are available. One is a flat cell shaped like a button; the other is a cylindrical cell that looks like a standard flashlight cell. The advantage of the button-type cell is that several of them can be stacked inside one container to form a battery. A mercury cell produces 1.35 V.
- *Nickel–metal hydride battery*—Nickel–metal hydride batteries, used routinely in computer and medical equipment, offer reasonable specific energy and specific power capabilities. Their components are recyclable, but a recycling structure is not yet in place. Nickel–metal hydride batteries have a much longer life cycle than lead–acid batteries; they are safe and tolerate abuse. These batteries have been used successfully in production electric vehicles and recently in low-volume production of HEVs. The main challenges with nickel–metal hydride batteries are their high cost, high rate of self-discharge, very high gassing/waste consumption, heat generation at high temperatures, the need to control hydrogen loss, and their low cell efficiency (may be as low as 50%, typically 60 to 65%).
- *Lithium-ion battery*—Lithium-ion batteries are rapidly penetrating the laptop and cell phone markets because of their high specific energy. They also have high specific power, high energy efficiency, good high-temperature performance, and low self-discharge rates. Components of lithium-ion batteries can be recycled. These characteristics make lithium-ion batteries suitable for HEV applications; however, to make them commercially viable for HEVs, further development is necessary, including improvements in their calendar and cycle life, a greater degree of cell and battery safety, improved abuse tolerance, and acceptable costs.

- *Lithium-ion polymer battery*—Lithium-ion polymer batteries with high specific energy (i.e., high energy per unit mass), initially developed for cell phone applications, also have the potential to provide high specific power for HEV applications. Other key characteristics of the lithium–polymer battery are safety and a good cycle life. The battery could be commercially viable if the costs are lowered and even higher specific power batteries are developed.

Battery Characteristics

Batteries are generally classified by their various characteristics. Parameters such as internal resistance, specific gravity, capacity, shelf life, C rate, mid-point voltage (MPV), gravimetric energy density, volumetric energy density, constant-voltage charge, constant-current charge, and specific power are used to describe and classify batteries by operation and type.

When discussing *internal resistance*, it is important to keep in mind that a battery is a DC voltage generator. As such, the battery has internal resistance or equivalent series resistance (ESR). In a chemical cell, the resistance of the electrolyte between the electrodes is responsible for most of the internal resistance of the cell. Because any current in the battery must flow through the internal resistance, this resistance is in series with the generated voltage. With no current, the voltage drop across the resistance is zero, so fully generated voltage develops across the output terminals. This is the open-circuit voltage, or no-load voltage. If a load resistance is connected across the battery, the load resistance is in series with internal resistance. When current flows in this circuit, the internal voltage drop reduces the terminal voltage of the battery.

The ratio of the weight of a certain volume of liquid to the weight of the same volume of water is the *specific gravity* of the liquid. Pure sulfuric acid has a specific gravity of 1.835 because it weighs 1.835 times as much as water per unit volume. The specific gravity of a mixture of sulfuric acid and water varies with the strength of the solution from 1.000 to 1.830.

The specific gravity of the electrolyte solution in a lead–acid cell ranges from 1.210 to 1.300 for new, fully charged batteries. The higher the specific gravity, the less internal resistance of the cell and the higher the possible load current. As the cell discharges, the water formed dilutes the acid and the specific gravity gradually decreases to about 1.150, at which time the cell is considered to be fully discharged.

The specific gravity of the electrolyte is measured with a *hydrometer*, which has a compressible rubber bulb at the top, a glass barrel, and a rubber hose at the bottom of the barrel. When taking readings with a hydrometer, the decimal point is usually omitted; for example, a specific gravity of 1.260 is read simply as "twelve-sixty." A hydrometer reading of 1210 to 1300 indicates a full charge; about 1250, half-charge; and 1150 to 1200, complete discharge.

The *capacity* of a battery is measured in ampere-hours (Ah). The ampere-hour capacity is equal to the product of the current in amperes and the time in hours during which the battery is supplying this current; it is defined as the amount of current that a battery can deliver for 1 hour before the battery voltage reaches the end-of-life point. In other words, 1 ampere-hour is the equivalent of drawing 1 amp steadily for 1 hour or 2 amps steadily for 1/2 an hour. A typical 12-volt system may have 800 ampere-hours of battery capacity; that is, the battery can draw 100 amps for 8 hours if fully discharged and starting from a fully charged state. This is equivalent to 1200 watts for 8 hours (power in watts = amps × volts). The ampere-hour capacity varies inversely with the discharge current. The size of a cell is determined generally by its ampere-hour capacity. The capacity of a storage battery determines how long it will operate at a given discharge rate and depends on many factors; the most important of these are as follows:

- The area of the plates in contact with the electrolyte
- The quantity and specific gravity of the electrolyte
- The type of separators

- The general condition of the battery (e.g., degree of sulfating, buckled plates, warped separators, sediment in bottom of cells)
- The final limiting voltage

The *shelf life* of a cell is that period of time during which the cell can be stored without losing more than approximately 10% of its original capacity. The loss of capacity of a stored cell is due primarily to the electrolyte in a wet cell drying out and to chemical actions that alter the materials within the cell. The shelf life of a cell can be extended by keeping it in a cool, dry place.

The *C rate* of a current is numerically equal to the ampere-hour rating of the cell. Charge and discharge currents are typically expressed in fractions of multiples of the C rate.

The *mid-point voltage (MPV)* is the nominal voltage of the cell, the voltage that is measured when the battery has discharged 50% of its total energy.

The *gravimetric energy density* of a battery is a measure of how much energy a battery contains in comparison to its weight.

The *volumetric energy density* of a battery is a measure of how much energy a battery contains in comparison to its volume.

A *constant-voltage charger* is a circuit that recharges a battery by sourcing only enough current to force the battery voltage to a fixed value.

A *constant-current charger* is a circuit that charges a battery by sourcing a fixed current into the battery, regardless of battery voltage.

In batteries, *specific power* usually refers to the power-to-weight ratio, measured in kilowatts per kilogram (kW/kg).

DC CIRCUITS

An electric circuit includes an *energy source* (the source of the electromotive force or voltage, such as a battery or generator), a conductor (wire), a load, and a means of control (see Figure 2.21). The energy source could be a battery, as shown in Figure 2.21, or some other means of producing a voltage. The *load* that dissipates the energy could be a lamp, a resistor, or some other device that does useful work, such as an electric toaster, a power drill, a radio, a soldering iron, a laptop computer, or a battery pack that stores energy and serves as a backup to a renewable energy source such as solar or wind. Again, *conductors* are wires that offer low resistance to current; they connect all the loads in the circuit to the voltage source. No electrical device dissipates energy unless current flows through it. Because conductors, or wires, are not perfect conductors, they heat up (dissipate energy), so they are actually part of the load. For simplicity, however, we usually think of the connecting wiring as having no resistance, as it would be tedious to assign a very low resistance value to the wires every time we wanted to solve a problem. *Control devices* might be switches, variable resistors, circuit breakers, fuses, or relays.

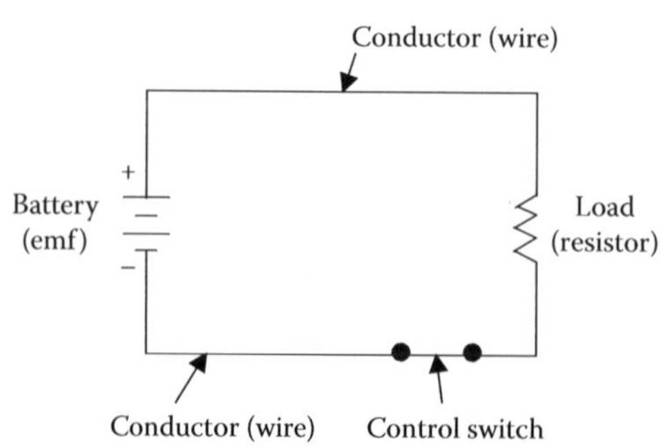

FIGURE 2.21 Simple closed circuit.

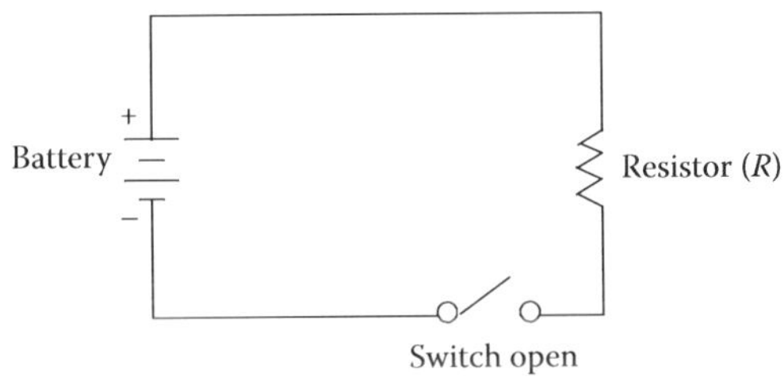

FIGURE 2.22 Open circuit.

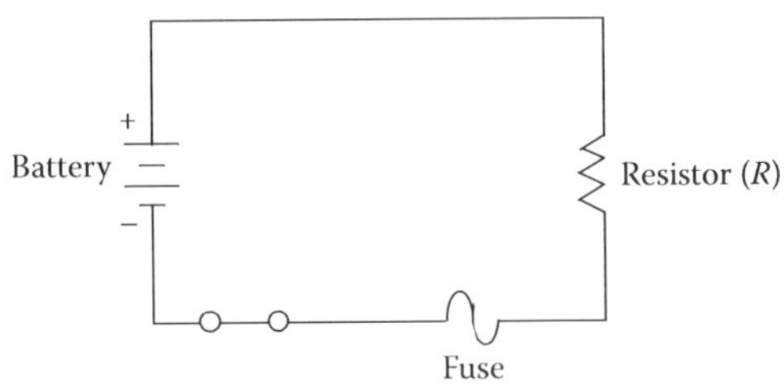

FIGURE 2.23 Simple fused circuit.

A complete pathway for current flow, or *closed circuit* (Figure 2.21), is an unbroken path for current from the electromotive force, through a load, and back to the source. An *open circuit* (see Figure 2.22) has a break in the circuit (e.g., open switch) that does not provide a complete path for current.

To protect a circuit, a fuse is placed directly into the circuit (Figure 2.23). A fuse will open the circuit whenever a dangerous large current begins to flow; a short-circuit condition occurs when an accidental connection between two points in a circuit offers very little resistance. A fuse will permit currents smaller than the fuse value to flow but will melt and therefore break or open the circuit if a larger current flows.

Key Point: Current flows from the negative (−) terminal of the battery (Figures 2.21 and 2.22), through the load to the positive (+) battery terminal, and continues through the battery from the positive (+) terminal back to the negative (−) terminal. As long as this pathway is unbroken, it is a closed circuit and current will flow; however, if the path is broken at *any* point, it is an open circuit and no current flows.

Schematic Representation

The simple circuits shown in Figures 2.21, 2.22, and 2.23 are displayed in *schematic* form. A schematic is a simplified drawing that represents the electrical, not the physical, situation in a circuit. The symbols used in schematic diagrams are the electrician's shorthand; they make the diagrams easier to draw and easier to understand. Consider the symbol shown in Figure 2.24 that is used to represent a battery power supply. The symbol is rather simple and straightforward, but it is very important. By convention, the shorter line in the symbol for a battery represents the *negative* terminal. It is important to remember this, because sometimes when you examine the schematic it is necessary to note the direction of current flow, which is from negative to positive. The battery symbol shown in Figure 2.24 represents a single cell, so only one short and one long line are used. The number of lines used to represent a battery vary (and they are not necessarily equivalent to the number of cells), but they are always in pairs, with long and short lines alternating. In the circuit shown in Figure 2.23, the current would flow in a *counterclockwise* direction. If the long and short lines of the battery symbol (see Figure 2.24) were reversed, the current in the circuit shown in Figure 2.23 would flow *clockwise*.

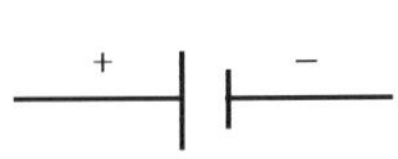

FIGURE 2.24 Schematic symbol for a battery.

Note: When studying electricity and electronics, many circuits are analyzed that consist mainly of specially designed resistive components. These components are *resistors*. Throughout the remaining analysis of the basic circuit, the resistive component will be a physical resistor; however, resistive components could be any of several electrical devices.

Keep in mind that, in the simple circuits shown in the figures to this point, we have only illustrated and discussed a few of the many symbols used in schematics to represent circuit components. (Other symbols will be introduced as we need them.) It is also important to keep in mind that a closed loop of wire (conductor) is not necessarily a circuit. A source of voltage must be included to make it an electric circuit. In any electric circuit where electrons move around a closed loop, current, voltage, and resistance are present. The physical pathway for current flow is actually the circuit. By knowing any two of the three quantities, such as voltage and current, the third (resistance) may be determined. This is done using Ohm's law, which is the foundation on which electrical theory is based.

OHM'S LAW

Simply put, *Ohm's law* defines the relationship between current, voltage, and resistance in electric circuits. Ohm's law can be expressed mathematically in three ways:

1. The *current* (I) in a circuit is equal to the voltage applied to the circuit divided by the resistance of the circuit. Stated another way, the current in a circuit is *directly* proportional to the applied voltage and *inversely* proportional to the circuit resistance. Ohm's law may be expressed by the equation:

$$I = E/R \tag{2.1}$$

where I = current in amps, E = voltage in volts, and R = resistance in ohms.

2. The *resistance* (R) of a circuit is equal to the voltage applied to the circuit divided by the current in the circuit:

$$R = E/I \tag{2.2}$$

3. The applied *voltage* (E) to a circuit is equal to the product of the current and the resistance of the circuit:

$$E = I \times R = IR \tag{2.3}$$

If any two of the quantities in Equations (2.1), (2.2), or (2.3) are known, the third may be easily found. Let's look at an example.

■ Example 2.1

Problem: Figure 2.25 shows a circuit containing a resistance of 6 ohms and a source voltage of 3 volts. How much current flows in the circuit?
Solution:

Given:

R = 6 ohms
E = 3 volts
I = ?

$$I = E/R = 3/6 = 0.5 \text{ amp}$$

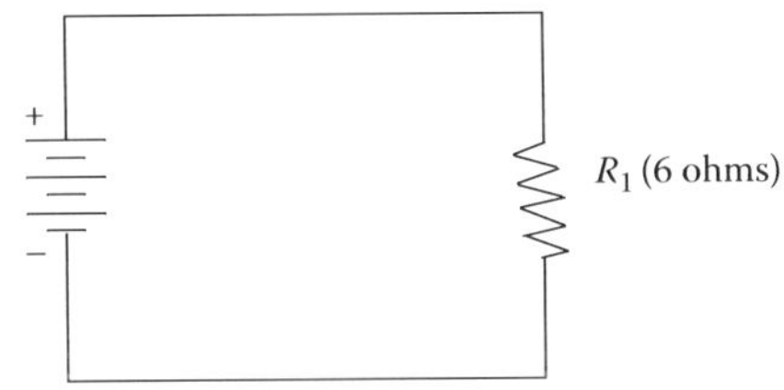

FIGURE 2.25 Determining current in a simple circuit.

To observe the effect of source voltage on circuit current, we can use the circuit shown in Figure 2.25 but double the voltage to 6 volts.

■ Example 2.2

Problem: How much current is flowing when $E = 6$ volts and $R = 6$ ohms?
Solution:

$$I = E/R = 6/6 = 1 \text{ amp}$$

Notice that as the source of voltage doubles, the circuit current also doubles. To verify that current is inversely proportional to resistance, assume that the resistor in Figure 2.25 has a value of 12 ohms.

Key Point: Circuit current is directly proportional to the applied voltage and will change by the same factor that the voltage changes.

■ Example 2.3

Problem: Given $E = 3$ volts and $R = 12$ ohms, what is I?
Solution:

$$I = E/R = 3/12 = 0.25 \text{ amp}$$

Comparing this current of 0.25 amp for the 12-ohm resistor to the 0.5-amp of current obtained with the 6-ohm resistor shows that doubling the resistance will reduce the current to one half the original value.

Key Point: Circuit current is inversely proportional to the circuit resistance.

Recall that if you know any two quantities of E, I, and R, you can calculate the third. In many circuit applications, current is known and either the voltage or the resistance will be the unknown quantity. To solve a problem in which current and resistance are known, the basic formula for Ohm's law must be transposed to solve for E, for I, or for R; however, the Ohm's law equations can be memorized and practiced effectively by using an Ohm's law circle (see Figure 2.26).

To find the equation for E, I, or R when two quantities are known, cover the unknown third quantity with your finger, as shown in Figure 2.27:

$$I = E/R \qquad R = E/I \qquad E = I \times R$$

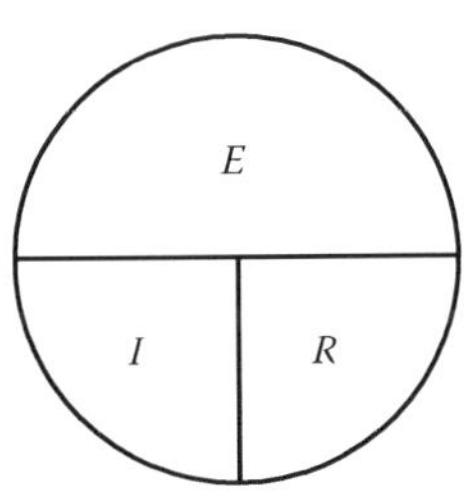

FIGURE 2.26 Ohm's law circle.

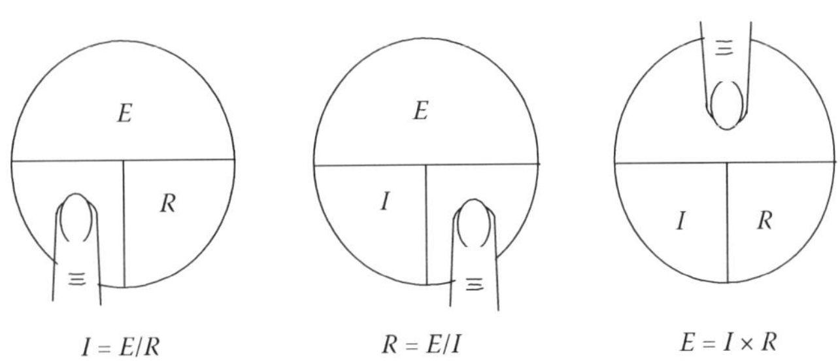

FIGURE 2.27 Putting the Ohm's law circle to work.

■ Example 2.4

Problem: Find I when $E = 120$ V and $R = 40$ ohms.

Solution: Place your finger on I, as shown in the figure. Use Equation (2.1) to find the unknown I:

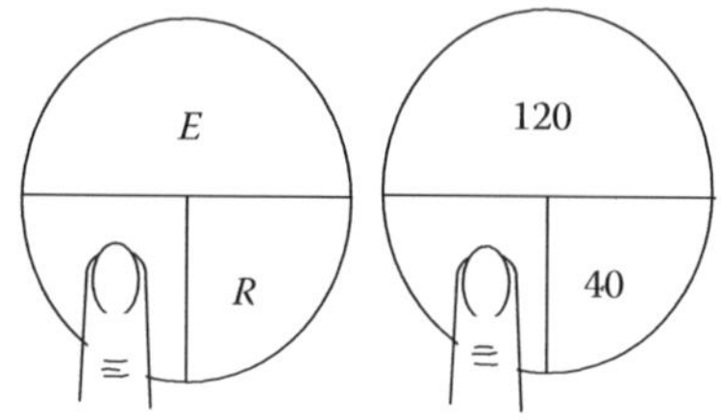

$$I = E/R = 120/40 = 3 \text{ A}$$

■ Example 2.5

Problem: Find R when $E = 220$ V and $I = 10$ A.

Solution: Place your finger on R, as shown in the figure:

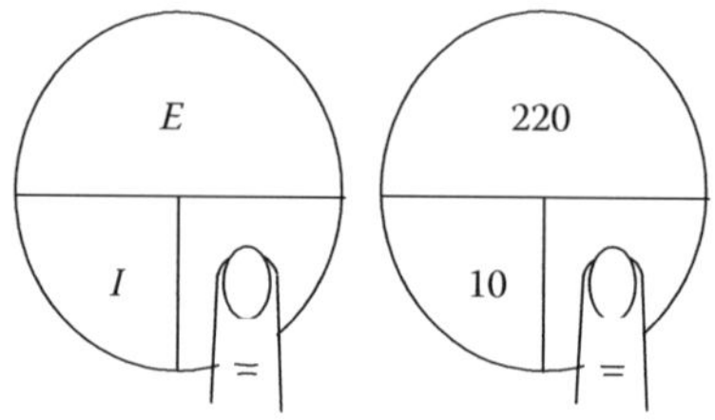

$$R = E/I = 220/10 = 22 \text{ ohms}$$

■ Example 2.6

Problem: Find E when $I = 2.5$ A and $R = 25$ ohms.

Solution: Place your finger on E, as shown in the figure.

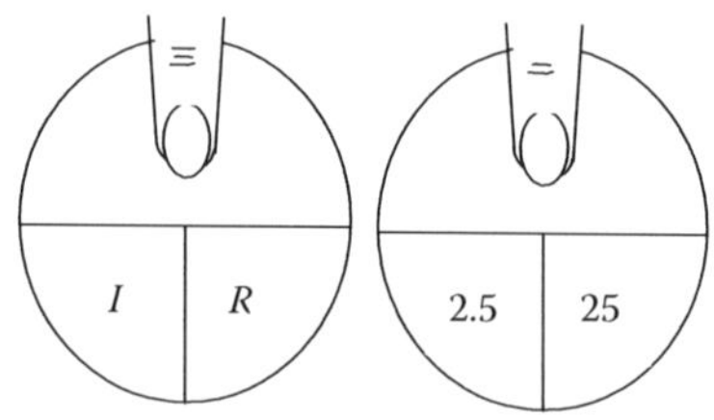

$$E = I \times R = 2.5 \times 25 = 62.5 \text{ V}$$

Note: In the previous examples, we have demonstrated how we can use the Ohm's law circle to help us solve simple voltage, current, and amperage problems. Beginning students, however, are cautioned not to rely entirely on the use of this circle when transposing simple formulas but rather to use it to supplement their knowledge of the algebraic method. Algebra is a basic tool in the solution of electrical problems, and the importance of knowing how to use it should not be underemphasized or bypassed after the operator has learned a shortcut method such as the one shown in this circle.

■ Example 2.7

Problem: An electric light bulb draws 0.5 A when operating on a 120-V DC circuit. What is the resistance of the bulb?

Solution: The first step in solving a circuit problem is to sketch a schematic diagram of the circuit itself, labeling each of the parts and showing the known values (see Figure 2.28). Because I and E are known, we use Equation (2.2) to solve for R:

$$R = E/I = 120/240 = 0.5 \text{ ohms}$$

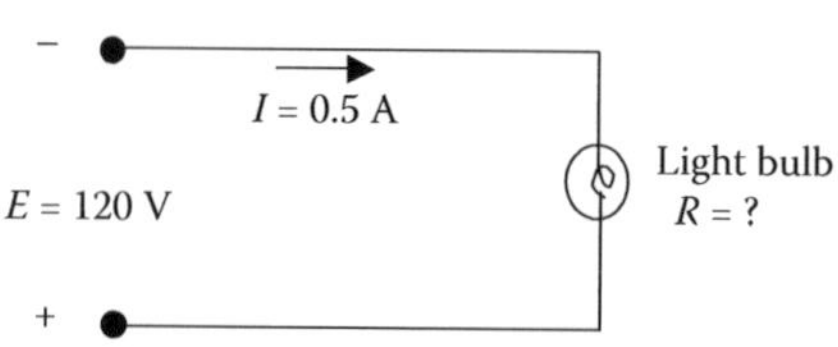

FIGURE 2.28 Circuit for Example 2.7.

ELECTRIC POWER

Power, whether electrical or mechanical, pertains to the rate at which work is being done, so the power consumption in your plant is related to current flow. A large electric motor or air dryer consumes more power (and draws more current) in a given length of time than, for example, an indicating light on a motor controller. *Work* is done whenever a force causes motion. If a mechanical force is used to lift or move a weight, work is done; however, force exerted *without* causing motion, such as the force of a compressed spring acting between two fixed objects, does not constitute work.

Key Point: Power is the rate at which work is done.

Electrical Power Calculations

The electric power (P) used in any part of a circuit is equal to the current (I) in that part multiplied by the voltage (E) across that part of the circuit. In equation form:

$$P = E \times I \tag{2.4}$$

where:

P = power (watts, W)
E = voltage (volts, V)
I = current (amps, A)

If we know the current (I) and the resistance (R) but not the voltage, we can find the power (P) by using Ohm's law for voltage, so by substituting Equation 2.3,

$$E = I \times R$$

into Equation 2.4, we obtain

$$P = (I \times R) \times I = I^2 \times R \tag{2.5}$$

In the same manner, if we know the voltage (V) and the resistance (R) but not the current (I), we can find P by using Ohm's law for current, so by substituting Equation 2.1:

$$I = E/R$$

Key Point: If we know any two quantities, we can calculate the third.

into Equation 2.4, we obtain

$$P = E \times \frac{E}{R} = \frac{E^2}{R} \tag{2.6}$$

■ **Example 2.8**

Problem: The current through a 200-ohm resistor to be used in a circuit is 0.25 A. Find the power rating of the resistor.
Solution: Because I and R are known, use Equation (2.5) to find P:

$$P = I^2 \times R = (0.25)^2 \times 200 = 0.0625 \times 200 = 12.5 \text{ W}$$

Key Point: The power rating of any resistor used in a circuit should be twice the wattage calculated by the power equation to prevent the resistor from burning out. Thus, the resistor used in Example 2.8 should have a power rating of 25 watts.

■ **Example 2.9**

Problem: How many kilowatts of power are delivered to a circuit by a 220-V generator that supplies 30 A to the circuit?
Solution:

$$P = E \times I = 220 \times 30 = 6600 \text{ W} = 6.6 \text{ kW}$$

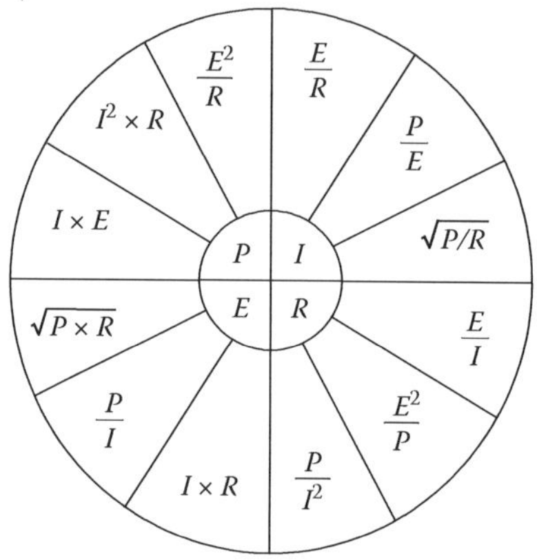

FIGURE 2.29 Ohm's law circle: summary of basic formulas.

■ Example 2.10

Problem: If the voltage across a 30,000-ohm resistor is 450 V, what is the power dissipated in the resistor?

Solution:

$$P = \frac{E^2}{R} = \frac{450^2}{30,000} = \frac{202,500}{30,000} = 6.75 \text{ W}$$

In this section, P was expressed in terms of various pairs of the other three basic quantities E, I, and R. In practice, you should be able to express any one of the three basic quantities, as well as P, in terms of any two of the others. Figure 2.29 is a summary of 12 basic formulas you should know. The four quantities E, I, R, and P, are at the center of the figure. Adjacent to each quantity are three segments. Note that in each segment, the basic quantity is expressed in terms of two other basic quantities, and no two segments are alike.

ELECTRIC ENERGY

Energy (the mechanical definition) is defined as the ability to do work (energy and time are essentially the same and are expressed in identical units). Energy is expended when work is done, because it takes energy to maintain a force when that force acts through a distance. The total energy expended to do a certain amount of work is equal to the working force multiplied by the distance through which the force moved to do the work. In electricity, total energy expended is equal to the *rate* at which work is done multiplied by the length of time the rate is measured. Essentially, energy (watts) is equal to the power (P) times time (t). The kilowatt-hour (kWh) is a unit commonly used for large amounts of electric energy or work. The amount of kilowatt-hours is calculated as the product of the power in kilowatts (kW) and the time in hours (h) during which the power is used:

$$\text{kWh} = \text{kW} \times \text{h} \tag{2.7}$$

■ Example 2.11

Problem: How much energy is delivered in 4 hours by a generator supplying 12 kW?

Solution:

$$\text{kWh} = \text{kW} \times \text{h} = 12 \times 4 = 48$$

The energy delivered is 48 kWh.

SERIES DC CIRCUIT CHARACTERISTICS

As previously mentioned, an electric circuit is made up of a voltage source, the necessary connecting conductors, and the effective load. If the circuit is arranged so the electrons have only *one* possible path, the circuit is a *series circuit*. A series circuit, then, is defined as a circuit that contains only one path for current flow. Figure 2.30 shows a series circuit having several loads (resistors).

Key Point: A *series circuit* is a circuit having only one path for current to flow along.

DC Resistance

To follow its electrical path, the current in a series circuit must flow through resistors inserted in the circuit (Figure 2.30). Thus, each additional resistor offers added resistance. In a series circuit, the total circuit resistance (R_T) is equal to the sum of the individual resistances:

$$R_T = R_1 + R_2 + R_3 \ldots R_n \tag{2.8}$$

where:

R_T = total resistance (ohms)
R_1, R_2, R_3 = resistance in series (ohms)
R_n = any number of additional resistors in equation

■ Example 2.12

Problem: Three resistors of 10 ohms, 12 ohms, and 25 ohms are connected in series across a battery whose emf is 110 volts (Figure 2.31). What is the total resistance?
Solution:

Given:
R_1 = 10 ohms
R_2 = 12 ohms
R_3 = 25 ohms
R_T = ?

$$R_T = R_1 + R_2 + R_3 = 10 + 12 + 25 = 47 \text{ ohms}$$

Transposition can be used in some circuit applications where the total resistance is known but the value of a circuit resistor has to be determined.

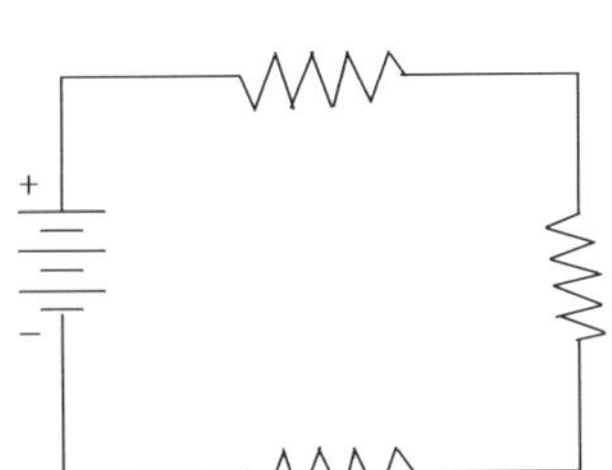

FIGURE 2.30 Series circuit.

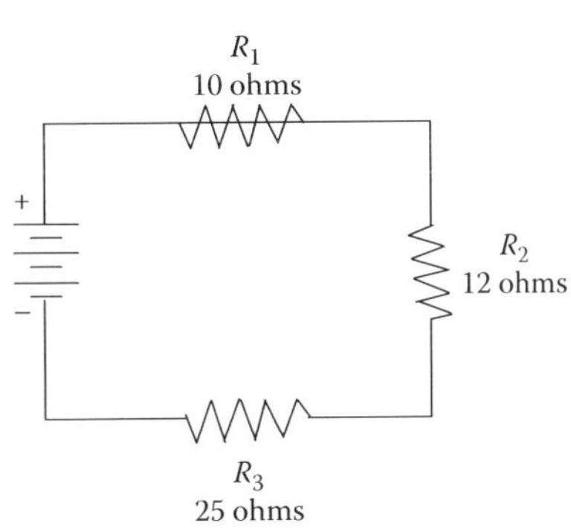

FIGURE 2.31 Solving for total resistance in a series circuit.

■ Example 2.13

Problem: The total resistance of a circuit containing three resistors is 50 ohms (see Figure 2.32). Two of the circuit resistors are 12 ohms each. Calculate the value of the third resistor.

Solution:

Given:

$R_T = 50$ ohms
$R_1 = 12$ ohms
$R_2 = 12$ ohms
$R_3 = ?$

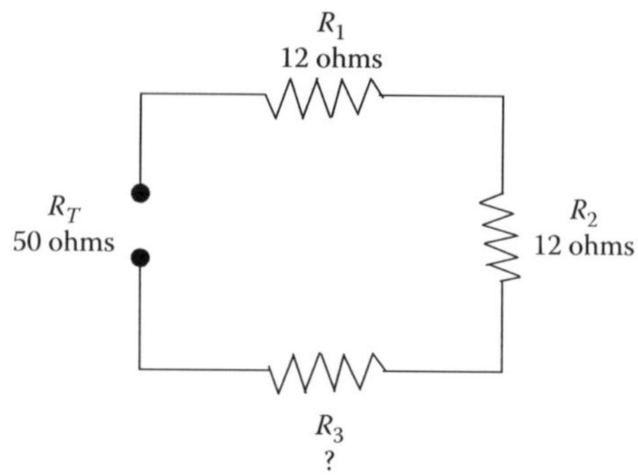

FIGURE 2.32 Calculating the value of one resistance in a series circuit.

$$R_T = R_1 + R_2 + R_3$$

Subtracting $(R_1 + R_2)$ from both sides of the equation, we obtain:

Key Point: When resistances are connected in series, the total resistance in the circuit is equal to the sum of the resistances of all parts of the circuit.

$$R_3 = R_T - R_1 - R_2$$

$$R_3 = 50 - 12 - 12$$

$$R_3 = 50 - 24 = 26 \text{ ohms}$$

DC Current

Because there is but one path for current in a series circuit, the same current (I) must flow through each part of the circuit. To determine the current throughout a series circuit, only the current through one of the parts must be known. The fact that the same current flows through each part of a series circuit can be verified by inserting ammeters into the circuit at various points as shown in Figure 2.33. As indicated in Figure 2.33, each meter indicates the same value of current.

Key Point: In a series circuit, the same current flows in every part of the circuit. *Do not* add the currents in each part of the circuit to obtain *I*.

Voltage

The voltage drop across the resistor in the basic DC circuit is the total voltage across the circuit and is equal to the applied voltage. The total voltage across a series circuit is also equal to the applied voltage but consists of the sum of two or more individual voltage drops. This statement can be proven by an examination of the circuit shown in Figure 2.34. In this circuit, a source potential (E_T)

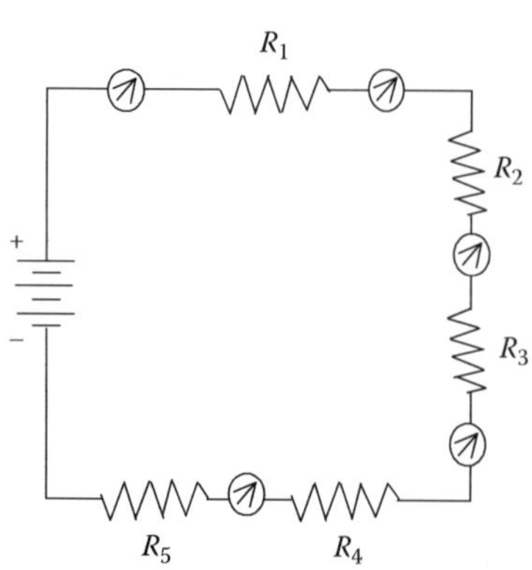

FIGURE 2.33 Current in a series circuit.

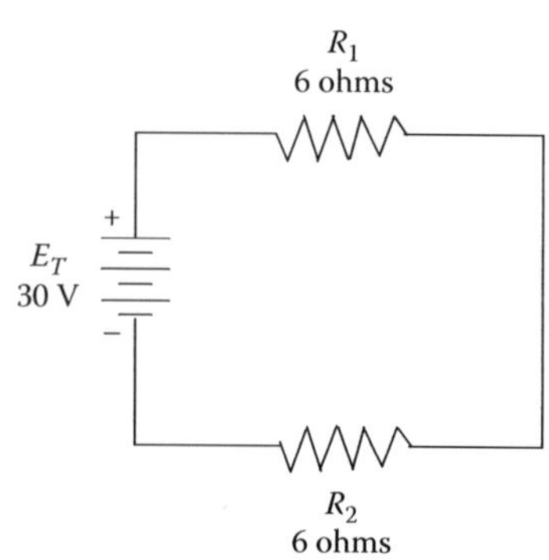

FIGURE 2.34 Calculating total resistance in a series circuit.

of 30 volts is impressed across a series circuit consisting of two 6-ohm resistors. The total resistance of the circuit is equal to the sum of the two individual resistances, or 12 ohms. Using Ohm's law, the circuit current may be calculated as follows:

$$I = \frac{E_T}{R_T} = \frac{30}{12} = 2.5 \text{ amps}$$

Because we know that the value of the resistors is 6 ohms each, and the DC current through the resistors is 2.5 amps, the voltage drops across the resistors can be calculated. The voltage (E_1) across R_1 is, therefore:

$$E_1 = I \times R_1$$

$$E_1 = 2.5 \text{ amps} \times 6 \text{ ohms} = 15 \text{ volts}$$

Because R_2 is the same ohmic value as R_1 and carries the same current, the voltage drop across R_2 is also equal to 15 volts. Adding these two 15-volt drops together gives a total drop of 30 volts, exactly equal to the applied voltage. For a series circuit, then:

$$E_T = E_1 + E_2 + E_3 + \ldots + E_n \tag{2.9}$$

where:

E_T = total voltage (V)
E_1 = voltage across resistance R_1 (V)
E_2 = voltage across resistance R_2 (V)
E_3 = voltage across resistance R_3 (V)

Key Point 1: The total voltage (E_T) across a DC series circuit is equal to the sum of the voltages across each resistance of the circuit.
Key Point 2: The voltage drops that occur in a series circuit are in direct proportions to the resistance across which they appear. This is the result of having the same current flow through each resistor. Thus, the larger the resistor, the larger will be the voltage drop across it.

■ Example 2.14

Problem: A series DC circuit consists of three resistors having values of 10, 20, and 40 ohms. Find the applied voltage if the current through the 20-ohm resistor is 2.5 amps.
Solution: To solve this problem, first draw a circuit diagram and label it as shown in Figure 2.35.

Given:
R_1 = 10 ohms
R_2 = 20 ohms
R_3 = 40 ohms
I = 2.5 amps

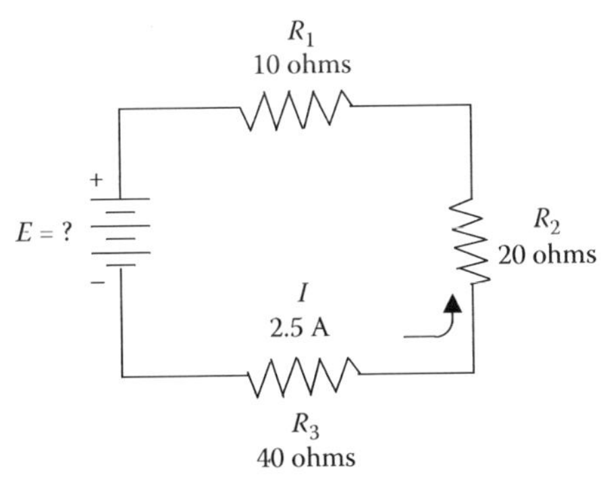

FIGURE 2.35 Solving for applied voltage in a series circuit.

Because the circuit involved is a DC series circuit, the same 2.5 amps of current flows through each resistor. Using Ohm's law, the voltage drops across each of the three resistors can be calculated:

$E_1 = 25$ V
$E_2 = 50$ V
$E_3 = 100$ V

Once the individual drops are known they can be added to find the total or applied voltage:

$$E_T = E_1 + E_2 + E_3$$

$$E_T = 25\text{ V} + 50\text{ V} + 100\text{ V} = 175\text{ V}$$

DC Power

Each resistor in a DC series circuit consumes power. This power is dissipated in the form of heat. Because this power must come from the source, the total power must be equal in amount to the power consumed by the circuit resistances. In a series circuit, the total power is equal to the sum of the powers dissipated by the individual resistors. Total power (P_T), then, is equal to

$$P_T = P_1 + P_2 + P_3 + \ldots + P_n \tag{2.10}$$

where:

P_T = total power (W)
P_1 = power used in the first part (W)
P_2 = power used in the second part (W)
P_3 = power used in the third part (W)
P_n = power used in the nth part (W)

■ Example 2.15

Problem: A DC series circuit consists of three resistors having values of 5, 15, and 20 ohms. Find the total power dissipation when 120 volts is applied to the circuit (see Figure 2.36).
Solution:

Given:

$R_1 = 5$ ohms
$R_2 = 15$ ohms
$R_3 = 20$ ohms
$E = 120$ volts

First find the total resistance:

$$R_T = R_1 + R_2 + R_3 = 5 + 15 + 20 = 40\text{ ohms}$$

Using total resistance and the applied voltage, calculate the circuit current:

$$I = \frac{E_T}{R_T} = \frac{120}{40} = 3\text{ amps}$$

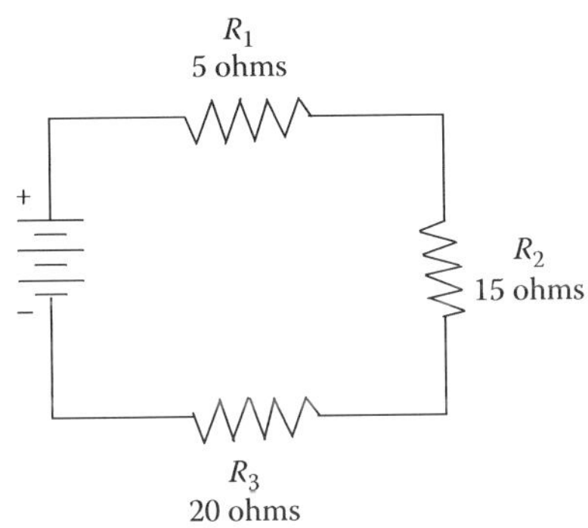

FIGURE 2.36 Solving for total power in a series circuit.

Using the power formula, calculate the individual power dissipations:

For resistor R_1:

$$P_1 = I^2 \times R_1 = (3)^2 \times 5 = 45 \text{ watts}$$

For resistor R_2:

$$P_2 = I^2 \times R_2 = (3)^2 \times 15 = 135 \text{ watts}$$

For resistor R_3:

$$P_3 = I^2 \times R_3 = (3)^2 \times 20 = 180 \text{ watts}$$

To obtain total power:

$$P_T = P_1 + P_2 + P_3 = 45 + 135 + 180 = 360 \text{ watts}$$

To check the answer, calculate the total power delivered by the source:

$$P = E \times I = 120 \text{ volts} \times 3 \text{ amps} = 360 \text{ watts}$$

Thus, the total power is equal to the sum of the individual power dissipations.

We found that Ohm's law can be used for total values in a DC series circuit as well as for individual parts of the circuit. Similarly, the formula for power may be used for total values:

$$P_T = E_T \times I \tag{2.11}$$

Summary of Rules for Series DC Circuits

To this point, we have covered many of the important factors governing the operation of basic DC series circuits. In essence, what we have really done is to lay a strong foundation to build upon in preparation for more advanced circuit theory that follows. Following is a summary of the important factors governing the operation of a DC series circuit:

- The same current flows through each part of a series circuit.
- Total resistance of a series circuit is equal to the sum of the individual resistances.
- Total voltage across a series circuit is equal to the sum of the individual voltage drops.
- Voltage drop across a resistor in a series circuit is proportional to the size of the resistor.
- Total power dissipated in a series circuit is equal to the sum of the individual dissipations.

General DC Series Circuit Analysis

Now that we have discussed the pieces required to put together the puzzle of DC series circuit analysis, we now move on to the next step in the process: series circuit analysis in total.

■ Example 2.16

Problem: Three resistors of 20, 20, and 30 ohms are connected across a battery supply rated at 100 volts terminal voltage. Completely solve the circuit shown in Figure 2.37.

Note: In solving the circuit, first find the total resistance and then calculate the circuit current. Once the current is known, the voltage drops and power dissipations can be calculated.

Solution: The total resistance is:

$$R_T = R_1 + R_2 + R_3 = 20 \text{ ohms} + 20 \text{ ohms} + 30 \text{ ohms} = 70 \text{ ohms}$$

By Ohm's law, the current is:

$$I = E/R_T = 100/70 = 1.43 \text{ amps (rounded)}$$

The voltage (E_1) across R_1 is:

$$E_1 = I \times R_1 = 1.43 \text{ amps} \times 20 \text{ ohms} = 28.6 \text{ volts}$$

The voltage (E_2) across R_2 is:

$$E_2 = I \times R_2 = 1.43 \text{ amps} \times 20 \text{ ohms} = 28.6 \text{ volts}$$

The voltage (E_3) across R_3 is:

$$E_3 = I \times R_3 = 1.43 \text{ amps} \times 30 \text{ ohms} = 42.9 \text{ volts}$$

The power dissipated by R_1 is:

$$P_1 = E_1 \times I = 28.6 \text{ volts} \times 1.43 \text{ amps} = 40.9 \text{ watts}$$

The power dissipated by R_2 is:

$$P_2 = E_2 \times I = 28.6 \text{ volts} \times 1.43 \text{ amps} = 40.9 \text{ watts}$$

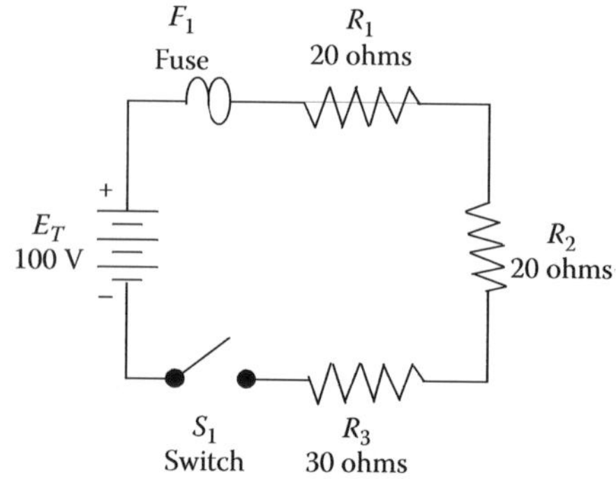

FIGURE 2.37 Solving for various values in a series circuit.

The power dissipated by R_3 is:

$$P_3 = E_3 \times I = 42.9 \text{ volts} \times 1.43 \text{ amps} = 61.3 \text{ watts}$$

The total power dissipated is:

$$P_T = E_T \times I = 100 \text{ volts} \times 1.43 \text{ amps} = 143 \text{ watts}$$

Note: When applying Ohm's law to a DC series circuit, consider whether the values used are component values or total values. When the information available enables the use of Ohm's law to find total resistance, total voltage, and total current, total values must be inserted into the formula.

To find total resistance:

$$R_T = \frac{E_T}{I_T}$$

To find total voltage:

$$E_T = I_T \times R_T$$

To find total current:

$$I_T = \frac{E_T}{R_T}$$

Kirchoff's Voltage Law

Kirchoff's voltage law states that the voltage applied to a closed circuit equals the sum of the voltage drops in that circuit. It should be obvious that this fact was used in the study of series circuits to this point. It was expressed as follows:

$$\begin{gathered}\text{Voltage applied} = \text{Sum of voltage drops} \\ E_A = E_1 + E_2 + E_3\end{gathered} \tag{2.12}$$

where E_A is the applied voltage and E_1, E_2, and E_3 are voltage drops.

Another way of stating Kirchoff's law is that the algebraic sum of the instantaneous emf values and voltage drops around any closed circuit is zero. Kirchoff's law can be used to solve circuit problems that would be difficult and often impossible to solve with only knowledge of Ohm's law. When Kirchoff's law is properly applied, an equation can be set up for a closed loop and the unknown circuit values may be calculated.

Polarity of Voltage Drops

When there is a voltage drop across a resistance, one end must be more positive or more negative than the other end. The polarity of the voltage drop is determined by the direction of current flow. In the circuit shown in Figure 2.38 the current is seen to be flowing in a counterclockwise direction due to the arrangement of the battery source (E). Notice that the end of resistor R_1 into which the current flows is marked negative (–). The end of R_1 at which the current leaves is marked positive (+). These polarity markings are used to show that the end of R_1 into which the current flows is at a higher negative potential than is the end of the resistor at which the current leaves. Point A is thus more negative than point B. Point C, which is at the same potential as point B, is labeled negative to

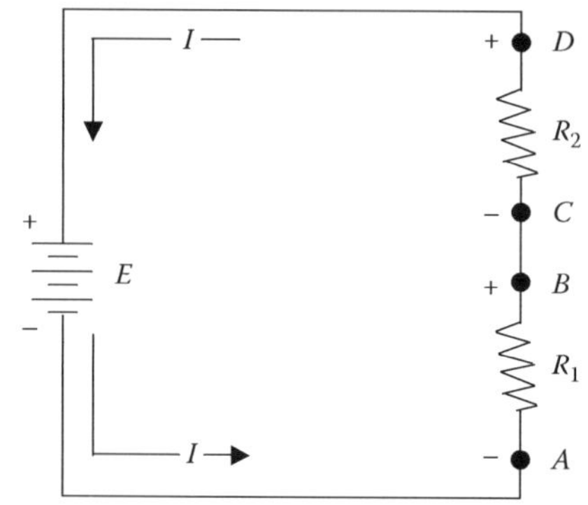

FIGURE 2.38 Polarity of voltage drops.

indicate that point C, though positive with respect to point A, is more negative than point D. To say a point is positive (or negative), without stating what it is positive *with respect to* has no meaning. Kirchoff's voltage law written as an equation is:

$$E_a + E_b + E_c + \ldots + E_n = 0 \tag{2.13}$$

where E_a, E_b, etc. are the voltage drops and emf values around any closed-circuit loop.

■ Example 2.17

Problem: Three resistors are connected across a 60-volt source. What is the voltage across the third resistor if the voltage drops across the first two resistors are 10 volts and 20 volts?
Solution: Draw a diagram like the one in Figure 2.39 and assume a direction of current as shown. Using this current, place the polarity markings at each end of each resistor and on the terminals of the source. Starting at point A, trace around the circuit in the direction of current flow, recording the voltage and polarity of each component. Starting at point A, these voltages would be as follows:

$$E_a + E_b + E_c + \ldots + E_n = 0$$

From the circuit:

$$(+E_?) + (+E_2) + (+E_3) - (E_A) = 0$$

Substituting values from circuit:

$$E_? + 10 + 20 - 60 = 0$$
$$E_? - 30 = 0$$
$$E_? = 30 \text{ volts}$$

Note: In much the same way, a problem can be solved in which the current is the unknown quantity.

Series Aiding and Opposing Sources

Sources of voltage that cause current to flow in the same direction are considered to be *series aiding* and their voltages add. Sources of voltage that would tend to force current in opposite directions are said to be *series opposing*, and the effective source voltage is the difference between the opposing voltages. When two opposing sources are inserted into a circuit, current flow would be in a direction determined by the larger source. Examples of series aiding and opposing sources are shown in Figure 2.40.

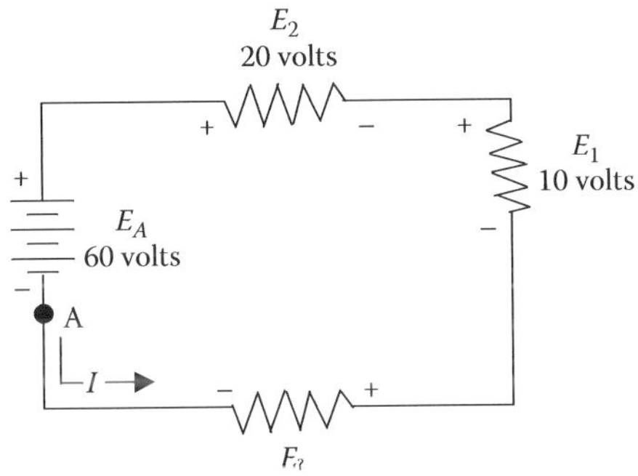

FIGURE 2.39 Determining unknown voltage in a series circuit.

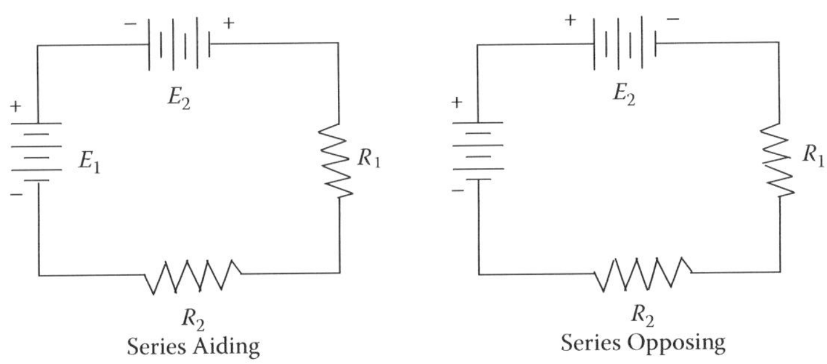

FIGURE 2.40 Aiding and opposing sources.

KIRCHOFF'S LAW AND MULTIPLE-SOURCE SOLUTIONS

Kirchoff's law can be used to solve multiple-source circuit problems. When applying this method, the exact same procedure is used for multiple-source circuits as for single-source circuits. This is demonstrated by the following example.

■ Example 2.18

Problem: Find the amount of current in the circuit shown in Figure 2.41.

Solution: Start at point A:

$$E_a + E_b + E_c + \ldots + E_n = 0$$

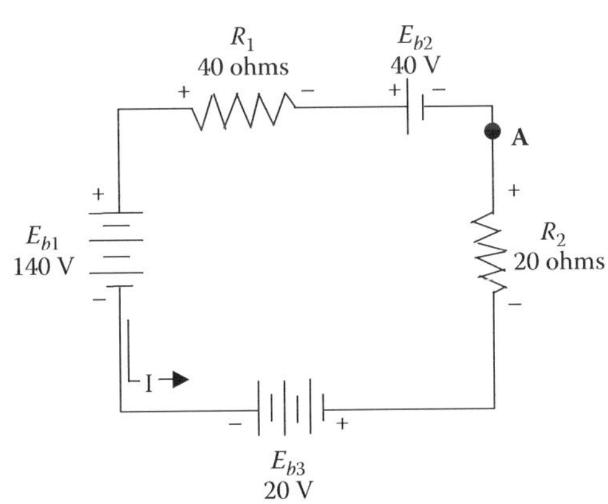

FIGURE 2.41 Solving for circuit current in a multiple source circuit.

From the circuit:

$$E_{b2} + E_1 - E_{b1} + E_{b3} + E_2 = 0$$

$$40 + 40I - 140 + 20 + 20I = 0$$

Combining like terms:

$$60I - 80 = 0$$

$$60I = 80$$

$$I = 1.33 \text{ amps}$$

Ground

The term *ground* is used to denote a common electrical point of zero potential. The reference point of a circuit is always considered to be at zero potential. The earth (ground) is said to be at zero potential. In Figure 2.42, point A is the zero reference or ground and is symbolized as such. Point C is 60 volts positive and point B is 20 volts positive with respect to ground. The common ground for much electrical/electronics equipment is the metal chassis. The value of ground is noted when considering its contribution to economy, simplification of schematics, and ease of measurement. When completing each electrical circuit, common points of a circuit at zero potential are connected directly to the metal chassis, thereby eliminating a large amount of connecting wire. An example of a grounded circuit is illustrated in Figure 2.43.

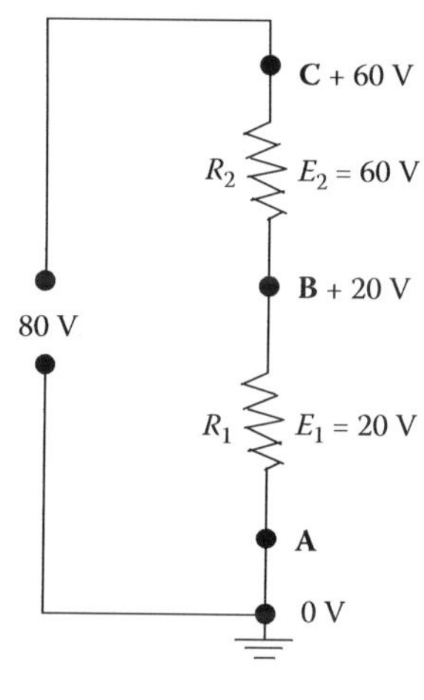

FIGURE 2.42 Use of ground symbols.

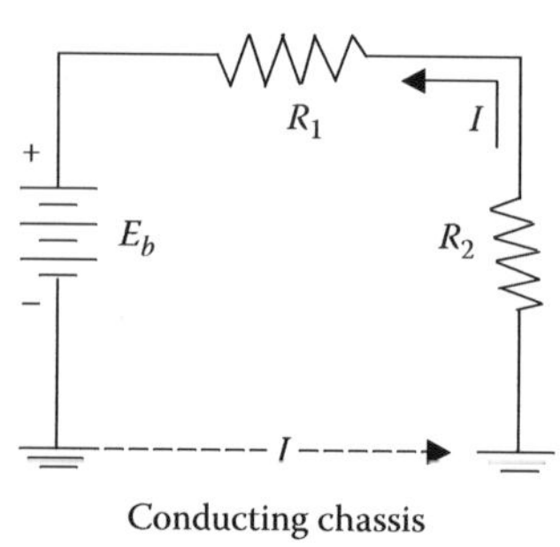

FIGURE 2.43 Ground used as a conductor.

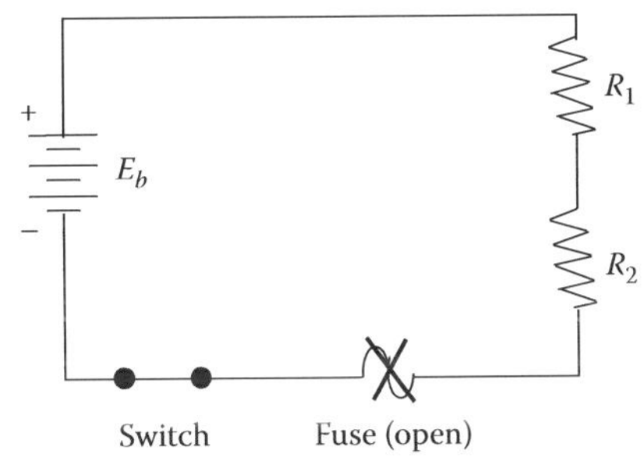

FIGURE 2.44 Open circuit with fuse blown.

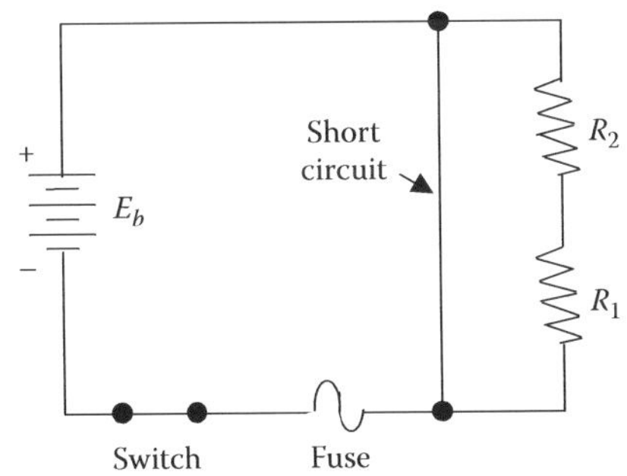

FIGURE 2.45 Short circuit.

Note: Most voltage measurements used to check proper circuit operation in electronic equipment are taken with respect to ground. One meter lead is attached to ground and the other meter lead is moved to various test points.

Open and Short Circuits

A circuit is *open* if a break in the circuit does not provide a complete path for DC or AC current. Figure 2.44 shows an open circuit, because the fuse is blown. To protect a circuit, a fuse is placed directly into the circuit (see Figure 2.45). A fuse will open the circuit whenever a dangerously large current begins to flow. A fuse will permit currents smaller than the fuse value to flow but will melt and therefore break or open the circuit if a larger current flows. A dangerously large current will flow when a *short circuit* occurs. A short circuit is usually caused by an accidental connection between two points in a circuit that offers very little resistance and passes an abnormal amount of current (see Figure 2.45). A short circuit often occurs as a result of improper wiring or broken insulation.

PARALLEL DC CIRCUITS

The principles we applied to solving simple DC series circuit calculations for determining the reactions of such quantities as voltage, current, and resistance also can be used in DC parallel and series–parallel circuits.

Parallel Circuit Characteristics

A *parallel circuit* is defined as one having two or more components connected across the same voltage source (see Figure 2.46). Recall that a series circuit has only one path for current flow. As additional loads (e.g., resistors) are added to the circuit, the total resistance increases and the

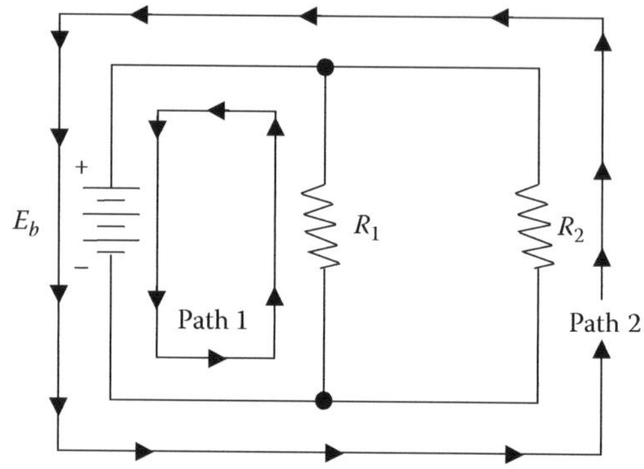

FIGURE 2.46 Basic parallel circuit.

total current decreases. This is *not* the case for a parallel circuit. In a parallel circuit, each load (or branch) is connected directly across the voltage source. In Figure 2.46, commencing at the voltage source (E_b) and tracing counterclockwise around the circuit, two complete and separate paths can be identified in which current can flow. One path is traced from the source through resistance R_1 and back to the source; the other, from the source through resistance R_2 and back to the source.

Voltage in Parallel Circuits

Recall that in a series circuit the source voltage divides proportionately across each resistor in the circuit. In a parallel circuit (see Figure 2.46), the same voltage is present across all the resistors of a parallel group. This voltage is equal to the applied voltage (E_b) and can be expressed in equation form as:

$$E_b = E_{R1} = E_{R2} = E_{Rn} \tag{2.14}$$

Key Point: In a parallel circuit, the voltage remains the same throughout the circuit.

We can verify Equation 2.14 by taking voltage measurements across the resistors of a parallel circuit, as illustrated in Figure 2.47. Notice that each voltmeter indicates the same amount of voltage; that is, the voltage across each resistor is the same as the applied voltage.

■ Example 2.19

Problem: Assume that the current through a resistor of a parallel circuit is known to be 4 milliamperes (mA) and the value of the resistor is 40,000 ohms. Determine the potential (voltage) across the resistor. The circuit is shown in Figure 2.48.

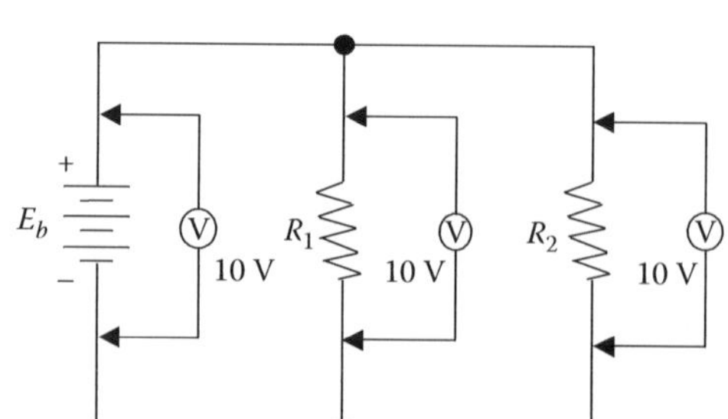

FIGURE 2.47 Voltage comparison in a parallel circuit.

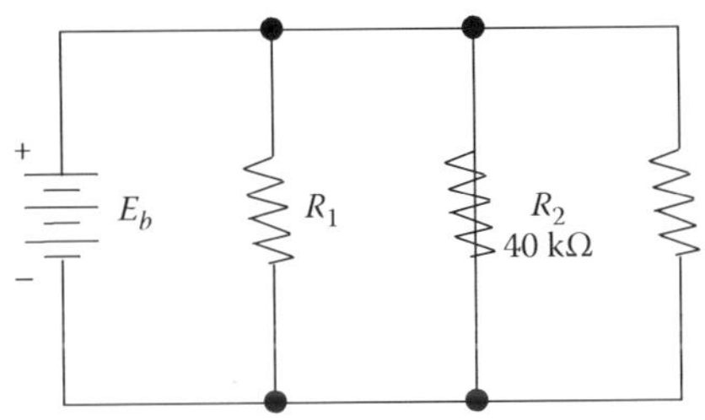

FIGURE 2.48 Example 2.19.

Solution: First find E_{R2} and then E_b.

Given:

$R_2 = 40\ \text{k}\Omega$
$I_{R2} = 4.0\ \text{ma}$

Select the proper equation:

$$E = I \times R$$

Substitute known values:

$$E_{R2} = I_{R2} \times R_2$$

$$E_{R2} = 4.0\ \text{mA} \times 40{,}000\ \text{ohms}$$

Use powers of ten:

$$E_{R2} = (4.0 \times 10^{-3}) \times (40 \times 10^{3})$$

$$E_{R2} = 4.0 \times 40 = 160\ \text{V}$$

Therefore:

$$E_b = 160\ \text{V}$$

Key Point: Ohm's law states that the current in a circuit is inversely proportional to the circuit resistance. This fact, important as a basic building block of electrical theory, obviously, is also important in the following explanation of current flow in parallel circuits.

Current in Parallel Circuits

In a series circuit, a single current flows. Its value is determined in part by the total resistance of the circuit; however, the source current in a parallel circuit divides among the available paths in relation to the value of the resistors in the circuit. Ohm's law remains unchanged. For a given voltage, current varies inversely with resistance. The behavior of current in a parallel circuit is best illustrated by example. In Figure 2.49, the resistors R_1, R_2, and R_3 are in parallel with each other and with the

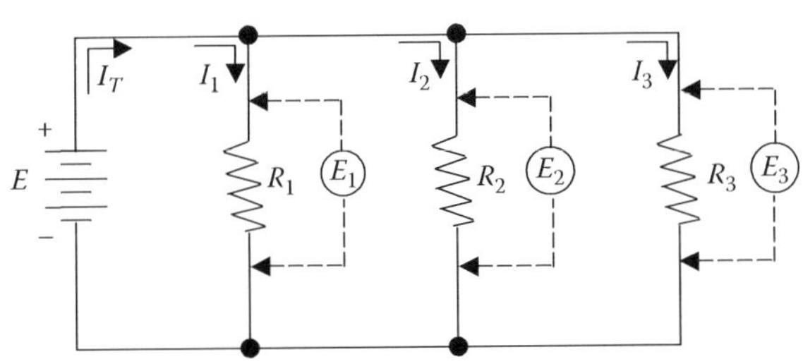

FIGURE 2.49 Parallel circuit.

battery. Each parallel path is then a branch with its own individual current. When total current I_T leaves voltage source E, part I_1 of current I_T will flow through R_1, part I_2 will flow through R_2, and I_3 will flow through R_3. Branch currents I_1, I_2, and I_3 can be different; however, if a voltmeter (used for measuring the voltage of a circuit) is connected across R_1, R_2, and R_3, the respective voltages E_1, E_2, and E_3 will be equal. Therefore,

$$E = E_1 = E_2 = E_3$$

Total current I_T is equal to the sum of all branch currents:

$$I_T = I_1 = I_2 = I_3 \tag{2.15}$$

This formula applies for any number of parallel branches, whether the resistances are equal or unequal.

By Ohm's law, each branch current equals the applied voltage divided by the resistance between the two points where the voltage is applied. Hence, for each branch we have the following equations (see Figure 2.49):

Branch 1

$$I_1 = \frac{E_1}{R_1} = \frac{V}{R_1}$$

Branch 2

$$I_2 = \frac{E_2}{R_2} = \frac{V}{R_2}$$

Branch 3

$$I_3 = \frac{E_3}{R_3} = \frac{V}{R_3}$$

With the same applied voltage, any branch that has less resistance allows more current through it than a branch with higher resistance.

■ Example 2.20

Problem: Two resistors, each drawing 2 amps, and a third resistor drawing 1 amp are connected in parallel across a 100-V line (see Figure 2.50). What is the total current?

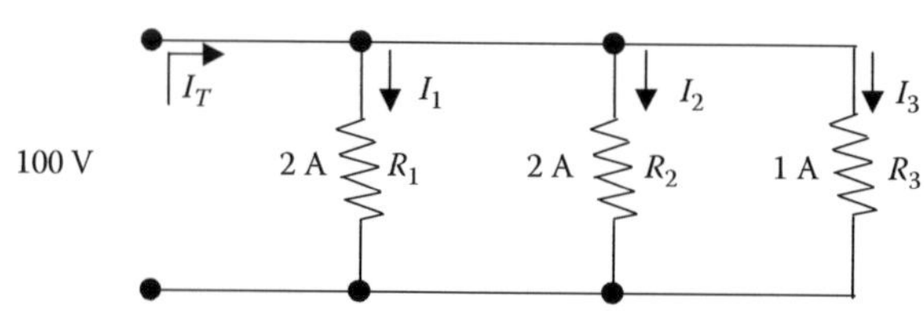

FIGURE 2.50 Example 2.20.

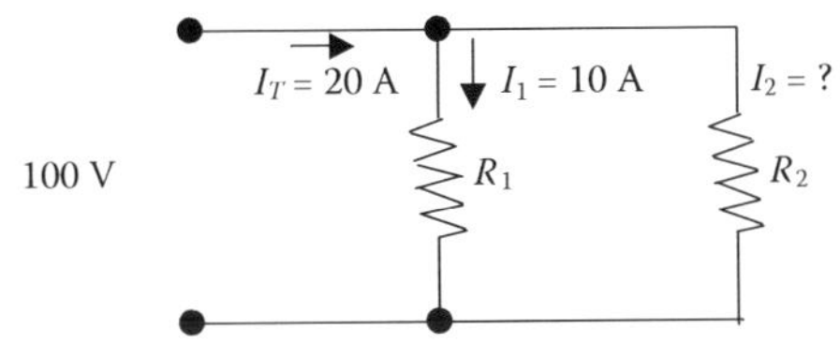

FIGURE 2.51 Example 2.21.

Solution: The formula for total current is:

$$I_T = I_1 + I_2 + I_3 = 2 + 2 + 1 = 5 \text{ amps}$$

The total current is 5 amps.

■ Example 2.21

Problem: Two branches, R_1 and R_2, across a 100-V power line draw a total line current of 20 A (Figure 2.51). Branch R_1 takes 10 A. What is current I_2 in branch R_2?

Solution: Starting with Equation (2.15), transpose to find I_2 and then substitute the given values:

$$I_T = I_1 + I_2$$
$$I_2 = I_T - I_1 = 20 - 10 = 10\ A$$

The current in branch R_2 is 10 A.

■ Example 2.22

Problem: A parallel circuit consists of two 15-ohm and one 12-ohm resistors across a 120-V line (see Figure 2.52). What current will flow in each branch of the circuit and what is the total current drawn by all the resistors?

Solution: There is a 120-V potential across each resistor, so:

$$I_1 = \frac{V}{R_1} = \frac{120}{15} = 8\text{ A}$$
$$I_2 = \frac{V}{R_1} = \frac{120}{15} = 8\text{ A}$$
$$I_3 = \frac{V}{R_3} = \frac{120}{12} = 10\text{ A}$$

Now find total current:

$$I_T = I_1 + I_2 + I_3 = 8 + 8 + 10 = 26\text{ A}$$

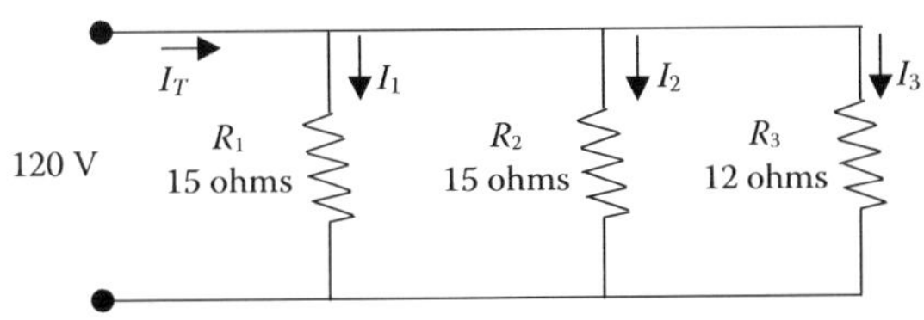

FIGURE 2.52 Example 2.22.

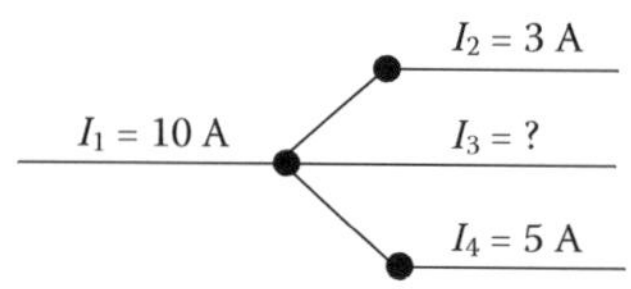

FIGURE 2.53 Example 2.23.

Parallel Circuits and Kirchoff's Current Law

The division of current in a parallel network follows a definite pattern. This pattern is described by *Kirchoff's current law*: The algebraic sum of the currents entering and leaving any junction of conductors is equal to zero. This can be stated mathematically as

$$I_a + I_b + \ldots + I_n = 0 \tag{2.16}$$

where I_a, I_b, etc. are the currents entering and leaving the junction. Currents entering the junction are assumed to be positive; currents leaving the junction, negative. When solving a problem using Equation (2.17), the currents must be placed into the equation with the proper polarity.

■ Example 2.23

Problem: Solve for the value of I_3 in Figure 2.53.
Solution: First, give the currents the proper signs:

$$I_1 = +10 \text{ amps} \qquad I_3 = ? \text{ amps}$$

$$I_2 = -3 \text{ amps} \qquad I_4 = -5 \text{ amps}$$

Then, place these currents into the equation with the proper signs, as follows:

$$I_a + I_b + \ldots + I_n = 0$$

$$I_1 + I_2 + I_3 + I_4 = 0$$

$$(+10) + (-3) + (I_3) + (-5) = 0$$

Combining like terms:

$$I_3 + 2 = 0$$

$$I_3 = -2 \text{ amps}$$

Thus, I_3 has a value of 2 amps, and the negative sign shows that it is a current leaving the junction.

Parallel Resistance

Unlike series circuits where total resistance (R_T) is the sum of the individual resistances, in a parallel circuit the total resistance is *not* the sum of the individual resistances. In a parallel circuit, we can use Ohm's law to find total resistance. We use the equation:

$$R = E/I \quad \text{or} \quad R_T = E_S/I_T$$

where R_T is the total resistance of all the parallel branches across the voltage source E_S, and I_T is the sum of all the branch currents.

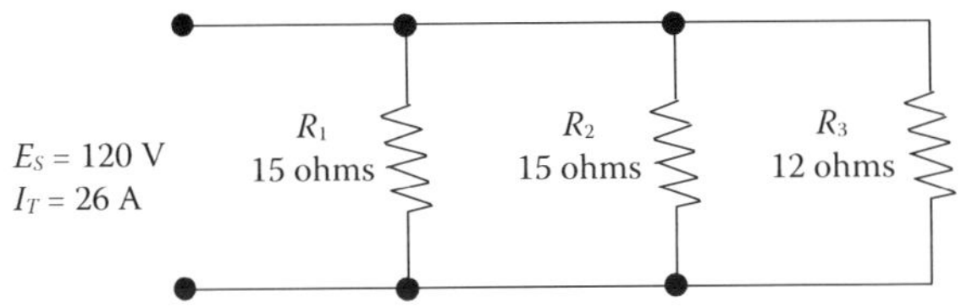

FIGURE 2.54 Example 2.24.

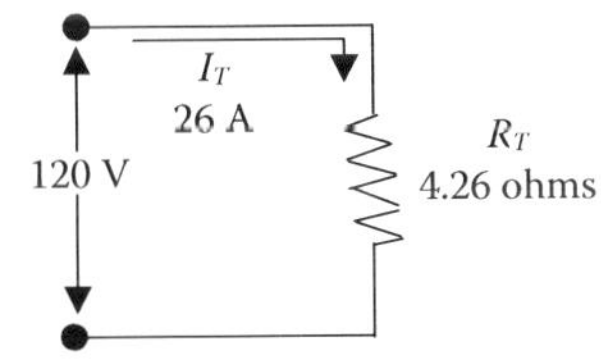

FIGURE 2.55 Circuit equivalent to that of Figure 2.54.

■ Example 2.24

Problem: What is the total resistance of the circuit shown in Figure 2.54?

Solution:

Given:

$E_5 = 120$ V

$I_T = 26$ A

$$R_T = \frac{E_5}{I_T} = \frac{120}{26} = 4.62 \text{ ohms}$$

Key Point: Notice that R_T is smaller than any of the three resistances in Figure 2.54. This fact may surprise you—it may seem strange that the total circuit resistance is *less* than that of the smallest resistor (R_3, 12 ohms). However, if we refer back to the water analogy we have used previously, it makes sense. Consider water pressure and water pipes, and assume it is possible to keep the water pressure constant. A small pipe offers more resistance to the flow of water than a larger pipe, but, if we add another pipe in parallel, one of even smaller diameter, the total resistance to water flow is *decreased.* In an electrical circuit, even a larger resistor in another parallel branch provides an additional path for current flow, so the *total* resistance is less. Remember, if we add one more branch to a parallel circuit, the total resistance decreases and the total current increases.

What we essentially demonstrated in working this particular problem is that the total load connected to the 120-V line is the same as the single equivalent resistance of 4.62 ohms connected across the line. It is probably more accurate to call this total resistance the "equivalent resistance," but by convention we use R_T, or total resistance, although the terms are often used interchangeably. We illustrate the equivalent resistance in the equivalent circuit shown in Figure 2.55.

Other methods can be used to determine the equivalent resistance of parallel circuits. The most appropriate method for a particular circuit depends on the number and value of the resistors; for example, consider the parallel circuit shown in Figure 2.56. For this circuit, the following simple equation is used:

$$R_{eq} = \frac{R}{N} \tag{2.17}$$

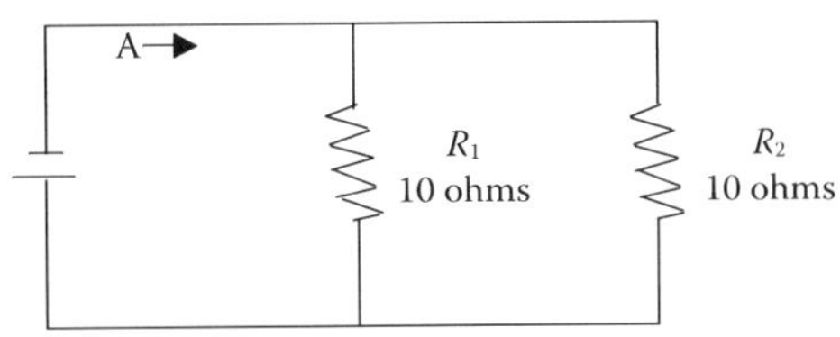

FIGURE 2.56 Two equal resistors connected in parallel.

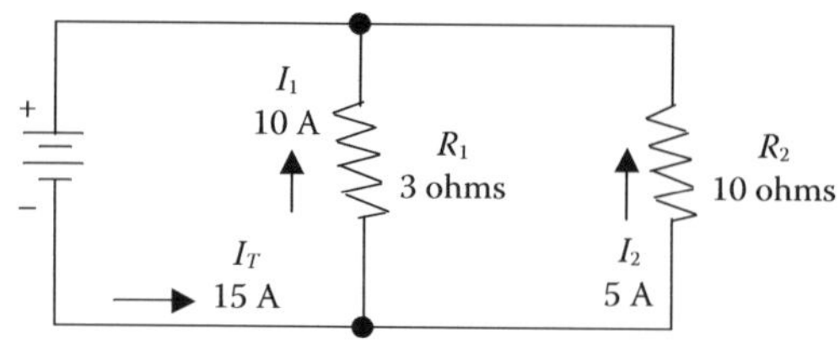

FIGURE 2.57 Example 2.26.

where:

R_{eq} = equivalent parallel resistance
R = ohmic value of one resistor
N = number of resistors

Note: Equation 2.17 is valid for any number of equal-value parallel resistors.

Thus,

$$R_{eq} = \frac{10 \text{ ohms}}{2} = 5 \text{ ohms}$$

When two equal value resistors are connected in parallel, they present a total resistance equivalent to a single resistor of one-half the value of either of the original resistors.

■ Example 2.25

Problem: When five 50-ohm resistors are connected in parallel, what is the equivalent circuit resistance?
Solution: Using Equation (2.17):

$$R_{eq} = \frac{R}{N} = \frac{50}{5} = 10 \text{ ohms}$$

What about parallel circuits containing resistances of unequal value? How is equivalent resistance determined? Example 2.26 demonstrates how this is accomplished.

■ Example 2.26

Problem: Refer to Figure 2.57 and determine R_{eq}.
Solution:

Given:

$R_1 = 3$ ohms
$R_2 = 6$ ohms
$E_a = 30$ volts

Known:

$I_1 = 10$ amps
$I_2 = 5$ amps
$I_T = 15$ amps

Solution:

$$R_{eq} = \frac{E_a}{I_T} = \frac{30}{15} = 2 \text{ ohms}$$

In Example 2.26, the equivalent resistance of 2 ohms is less than the value of either branch resistor. Remember, in parallel circuits the equivalent resistance will always be smaller than the resistance of any branch.

Reciprocal Method

When circuits are encountered in which resistors of unequal value are connected in parallel, the equivalent resistance may be computed by using the *reciprocal method.*

Note: A *reciprocal* is an inverted fraction; the reciprocal of the fraction 3/4, for example, is 4/3. We consider a whole number to be a fraction with 1 as the denominator, so the reciprocal of a whole number is that number divided into 1; for example, the reciprocal of R_T is $1/R_T$. The equivalent resistance in parallel is given by the formula:

$$\frac{1}{R_T} = \frac{1}{R_1} + \frac{1}{R_2} + \frac{1}{R_3} + \ldots + \frac{1}{R_n} \tag{2.18}$$

where R_T is the total resistance in parallel and R_1, R_2, R_3, and R_n are the branch resistances.

■ Example 2.27

Problem: Find the total resistance of 2-, 4-, and 8-ohm resistors in parallel (Figure 2.58).
Solution: Write the formula for three resistors in parallel:

$$\frac{1}{R_T} = \frac{1}{R_1} + \frac{1}{R_2} + \frac{1}{R_3}$$

Substitute the resistance values:

$$\frac{1}{R_T} = \frac{1}{2} + \frac{1}{4} + \frac{1}{8}$$

Add fractions:

$$\frac{1}{R_T} = \frac{4}{8} + \frac{2}{8} + \frac{1}{8} = \frac{7}{8}$$

Invert both sides of the equation:

$$R_T - \frac{8}{7} - 1.14$$

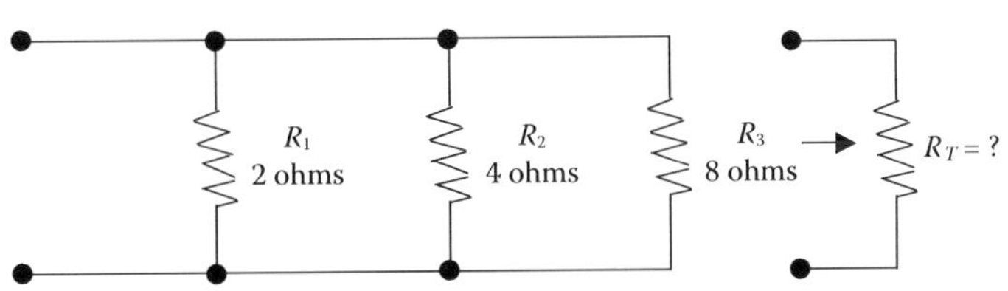

FIGURE 2.58 Example 2.27.

Note: When resistances are connected in parallel, the total resistance is always *less* than the smallest resistance of any single branch.

Product over Sum Method

When any two unequal resistors are in parallel, it is often easier to calculate the total resistance by multiplying the two resistances and then dividing the product by the sum of the resistances:

$$R_T = \frac{R_1 \times R_2}{R_1 + R_2} \tag{2.19}$$

where R_T is the total resistance in parallel, and R_1 and R_2 are the two resistors in parallel.

■ Example 2.28

Problem: What is the equivalent resistance of a 20-ohm and a 30-ohm resistor connected in parallel?
Solution:

Given:

$R_1 = 20$ ohms
$R_2 = 30$ ohms

$$R_T = \frac{R_1 \times R_2}{R_1 + R_2} = \frac{20 \times 30}{20 + 30} = \frac{600}{50} = 12 \text{ ohms}$$

Reduction to an Equivalent Circuit

In the study of basic electricity, it is often necessary to resolve a complex circuit into a simpler form. Any complex circuit consisting of resistances can be reduced to a basic equivalent circuit containing the source and total resistance. This process is called *reduction to an equivalent circuit.* An example of circuit reduction was demonstrated in Example 2.28 and is illustrated in Figure 2.59. The circuit shown in Figure 2.59A is reduced to the simple circuit shown in Figure 2.59B.

Power in Parallel Circuits

As in the series circuit, the total *power* consumed in a parallel circuit is equal to the sum of the power consumed in the individual resistors:

$$P_T = P_1 + P_2 + P_3 + \ldots + P_n \tag{2.20}$$

where P_T is the total power and P_1, P_2, P_3, and P_n are the branch powers.

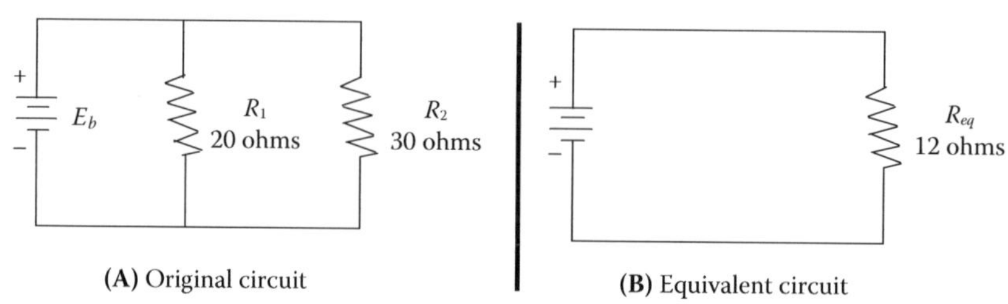

FIGURE 2.59 Parallel circuit with equivalent circuit.

Note: Because power dissipation in resistors consists of a heat loss, power dissipations are additive regardless of how the resistors are connected in the circuit.

Total power can also be calculated by the equation:

$$P_T = E \times I_T \tag{2.21}$$

where:

P_T is the total power.
E is the voltage source across all parallel branches.
I_T is the total current.

The power dissipated in each branch is equal to EI and to V^2/R.

Note: In both parallel and series arrangements, the sum of the individual values of power dissipated in the circuit equals the total power generated by the source. The circuit arrangements cannot change the fact that all power in the circuit comes from the source.

■ Example 2.29

Problem: Find the total power consumed by the circuit in Figure 2.60.
Solution:

$$P_{R1} = E_b \times I_{R1} = 50 \times 5 = 250 \text{ watts}$$

$$P_{R2} = E_b \times I_{R2} = 50 \times 2 = 100 \text{ watts}$$

$$P_{R3} = E_b \times I_{R3} = 50 \times 1 = 50 \text{ watts}$$

$$P_T = P_1 + P_2 + P_3 = 250 + 100 + 50 = 400 \text{ watts}$$

Note: The power dissipated in the branch circuits in Figure 2.60 is determined in the same manner as the power dissipated by individual resistors in a series circuit. The total power (P_t) is then obtained by summing up the powers dissipated in the branch resistors using Equation (2.21).

Because, in the example shown in Figure 2.60, the total current is known, we could determine the total power by the following method:

$$P_T = E_b \times I_T = 50 \text{ V} \times 8 \text{ A} = 400 \text{ W}$$

Rules for Solving Parallel DC Circuits

Problems involving the determination of resistance, voltage, current, and power in a parallel circuit are solved as simply as in a series circuit. The procedure is basically the same: (1) draw a circuit diagram, (2) state the values given and the values to be found, (3) state the applicable equations, and (4)

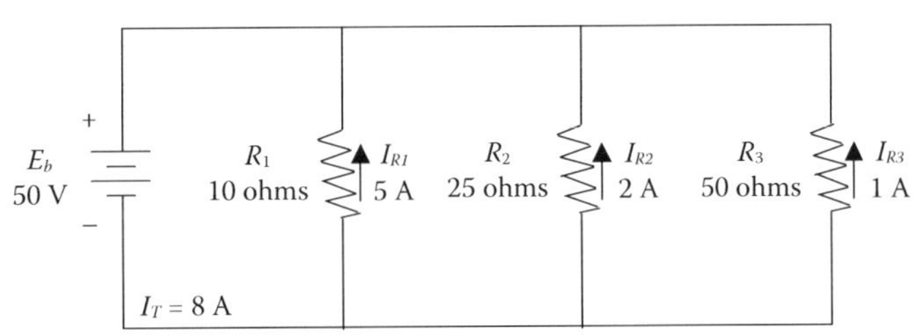

FIGURE 2.60 Example 2.29.

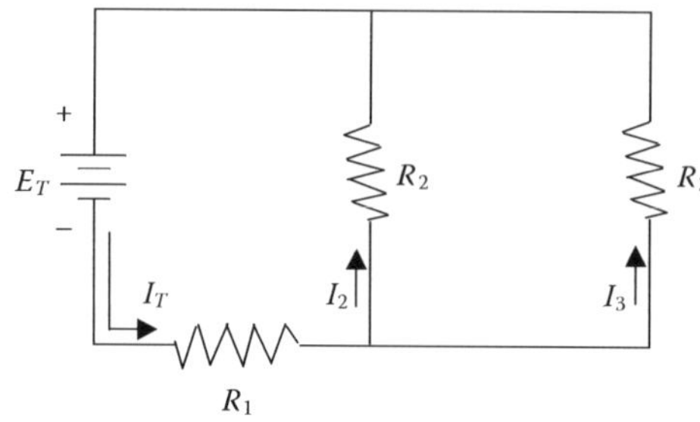

FIGURE 2.61 Series–parallel circuit.

substitute the given values and solve for the unknown. Along with following this problem-solving procedure, it is also important to remember to apply the rules for solving parallel DC circuits:

1. The same voltage exists across each branch of a parallel circuit and is equal to the source voltage.
2. The current through a branch of a parallel network is inversely proportional to the amount of resistance of the branch.
3. The total current of a parallel circuit is equal to the sum of the currents of the individual branches of the circuit.
4. The total resistance of a parallel circuit is equal to the reciprocal of the sum of the reciprocals of the individual resistances of the circuit.
5. The total power consumed in a parallel circuit is equal to the sum of the power consumption of the individual resistances.

SERIES–PARALLEL CIRCUITS

To this point we have discussed series and parallel DC circuits; however, the maintenance operator will seldom encounter a circuit that consists solely of either type of circuit. Most circuits consist of both series and parallel elements. A circuit of this type is referred to as a *series–parallel circuit*, or as a *combination circuit*. Solving a series–parallel (combination) circuit is simply a matter of applying the laws and rules discussed up to this point.

SOLVING A SERIES–PARALLEL CIRCUIT

At least three resistors are required to form a series–parallel circuit. An example of a series–parallel circuit is shown in Figure 2.61, where two parallel resistors, R_2 and R_3, are connected in series with resistor R_1 and voltage source E. In a circuit of this type, current I_T divides after it flows through R_1; part flows through R_2, and part flows through R_3. The current then joins at the junction of the two resistors and flows back to the positive terminal of the voltage source (E) and through the voltage source to the positive terminal.

When solving for values in a series–parallel circuit (current, voltage, and resistance), follow the rules that apply to a series circuit for the series part of the circuit and follow the rules that apply to a parallel circuit for the parallel part of the circuit. Solving series–parallel circuits is simplified if all parallel and series groups are first reduced to single equivalent resistances and the circuits are redrawn in simplified form. Recall, the redrawn circuit is called an *equivalent circuit*. The procedure for developing an equivalent circuit is shown in Figure 2.62.

Note: There are no general formulas for the solution of series–parallel circuits because there are so many different forms of these circuits.

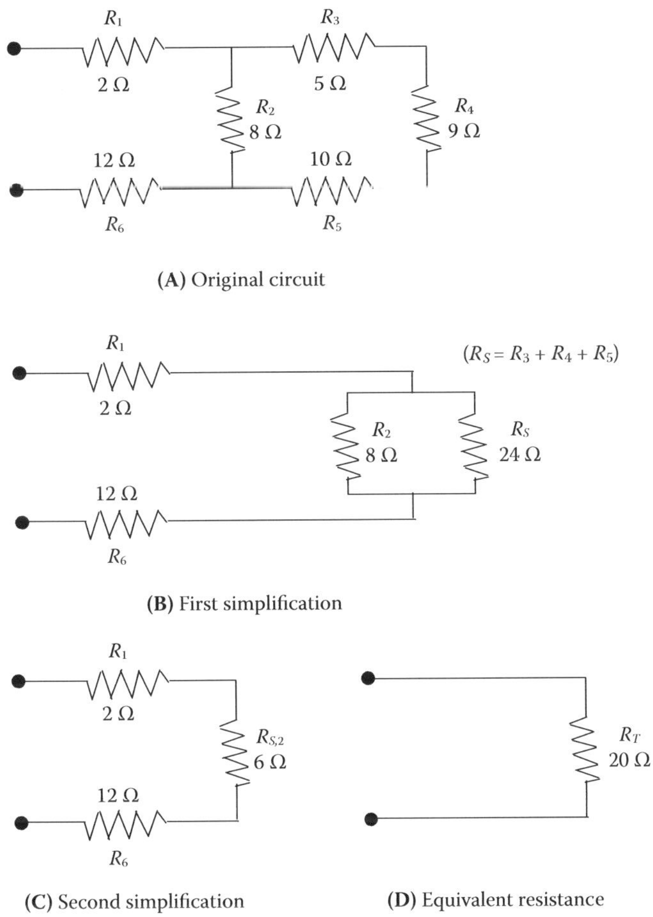

FIGURE 2.62 Developing an equivalent circuit.

■ Example 2.30

Problem: Find the total resistance (R_T), total circuit current (I_T), and branch currents of the circuit shown in Figure 2.63A.

Solution: Find the equivalent resistance of the parallel branch:

$$R_P = \frac{R_2 \times R_3}{R_2 + R_3} = \frac{15 \times 20}{15 + 20} = \frac{300}{35} = 8.6 \text{ ohms}$$

The equivalent circuit reduces to a series circuit (Figure 2.63B). Find the resistance of the equivalent series circuit:

$$R_T = R_1 + R_P = 10 + 8.6 = 18.6 \text{ ohms}$$

The equivalent circuit reduces to a single voltage source and a single resistance (Figure 2.63C). Find the actual current being supplied in the original series–parallel circuit (I_T):

$$I_T = \frac{V}{R_T} = \frac{60}{18.6} = 3.3 \text{ amps}$$

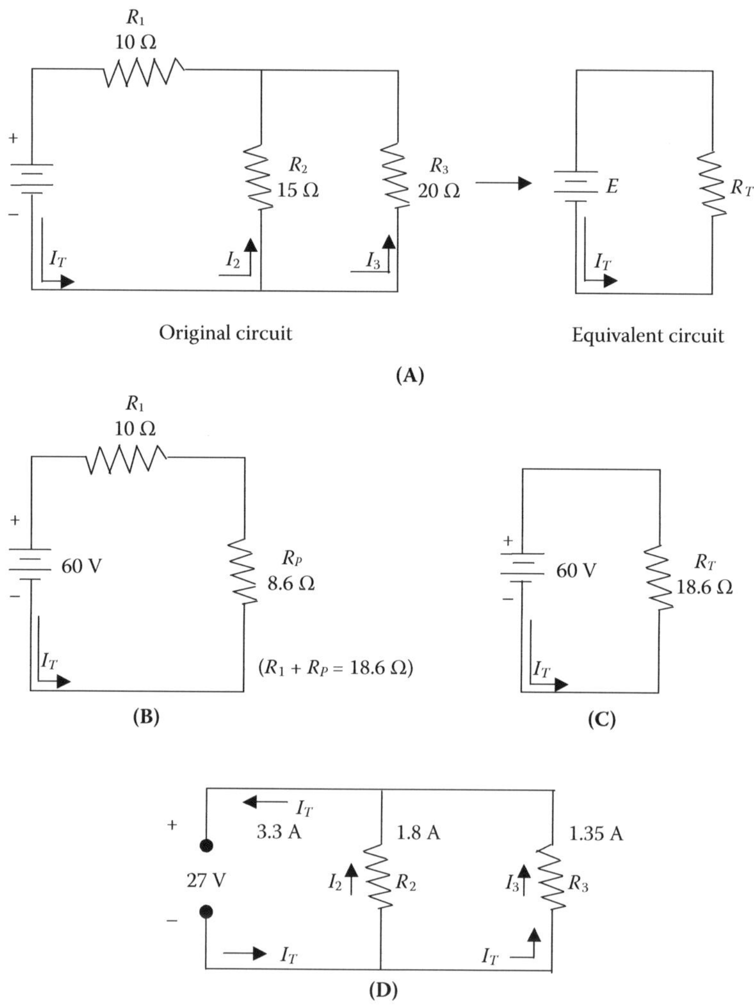

FIGURE 2.63 Example 2.30.

Find I_2 and I_3. The voltage across R_2 and R_3 is equal to the applied voltage (E) less the voltage drop across R_1 (Figure 2.63D):

$$V_2 = V_3 = V - (I_T \times R_1) = 60 - (3.3 \times 10) = 27 \text{ volts}$$

Then,

$$I_2 = \frac{V_2}{R_2} = \frac{27}{15} = 1.8 \text{ amps}$$

$$I_3 = \frac{V_3}{R_3} = \frac{27}{20} = 1.35 \text{ amps}$$

Note: The total current in the series–parallel circuit depends on the effective resistance of the parallel portion *and* on the other resistances.

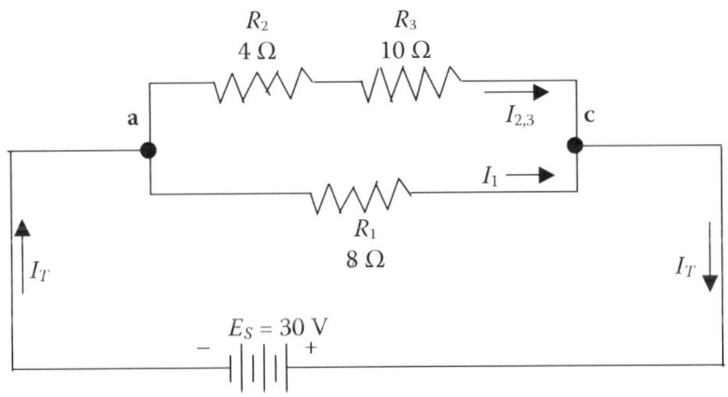

FIGURE 2.64 Example 2.31.

■ Example 2.31

Problem: Answer the following questions related to Figure 2.64:

1. The figure shows a circuit in which two resistors in series (R_2 and R_3) form one branch of a parallel circuit. The total current (I_T) flows into the parallel circuit, splitting into two branches at point a. At what point do the branch currents rejoin to form I_T?

   Answer: At point c.

2. Source voltage E_S (30 V) drops between points a and c. The largest voltage drop is across which resistor? Why?

   Answer: R_1, because the 30 V must be divided between R_2 and R_3.

3. What is the resistance of the top branch of the circuit in Figure 2.64?

   Answer: $R_2 + R_3 = 14$ ohms.

4. What is the value of R_T?

   Answer:

$$R_T = \frac{8 \times 14}{8 + 14} = \frac{112}{22} = 5.1 \text{ ohms}$$

5. What is the current through the top branch ($I_{2,3}$)?

   Answer:

$$I = \frac{E}{R} = \frac{30}{14} = 2.14 \text{ amps}$$

6. What is the value of P_T?

   Answer:

$$P_T = \frac{(E_S)^2}{R_T} = \frac{30^2}{5.1} = 176.5 \text{ watts}$$

CONDUCTORS

Recall that we pointed out earlier that electric current moves easily through some materials but with greater difficulty through others. Three good electrical conductors are copper, silver, and aluminum (generally, we can say that most metals are good conductors). At the present time, copper is the material of choice used in electrical conductors. Under special conditions, certain gases are also used as conductors; for example, neon gas, mercury vapor, and sodium vapor are used in various kinds of lamps. The function of the wire conductor is to connect a source of applied voltage to a load resistance with a minimum *IR* voltage drop in the conductor so most of the applied voltage can produce current in the load resistance. Ideally, a conductor must have a very low resistance; a typical value for a conductor such as copper is less than 1 ohm per 10 feet. Because all electrical circuits utilize conductors of one type or another, in this section we discuss the basic features and electrical characteristics of the most common types of conductors. Moreover, because conductor splices and connections (and insulation of such connections) are also an essential part of any electric circuit, they are also discussed.

Unit Size of Conductors

A standard (or unit size) of a conductor has been established to compare the resistance and size of one conductor with another. The unit of linear measurement used with regard to the diameter of a piece of wire is the *mil* (0.001 of an inch). A convenient unit of wire length used is the *foot.* Thus, the standard unit of size in most cases is the *mil-foot*; a wire will have unit size if it has a diameter of 1 mil and a length of 1 foot. The resistance in ohms of a unit conductor or a given substance is called the *resistivity* (or *specific resistance*) of the substance. As a further convenience, *gauge* numbers are also used to compare the diameter of wires. The Browne & Sharpe (B&S) gauge was used in the past; today, the *American wire gauge* (AWG) is more commonly used.

Square Mil

Figure 2.65 shows a square mil, which is a convenient unit of cross-sectional area for square or rectangular conductors. As shown in the figure, a square mil is the area of a square, the sides of which are 1 mil. To obtain the cross-sectional area in square mils of a square conductor, square one side measured in mils. To obtain the cross-sectional area in square mils of a rectangular conductor, multiply the length of one side by that of the other, each length being expressed in mils.

■ Example 2.32

Problem: Find the cross-sectional area of a large rectangular conductor 5/8 inch thick and 5 inches wide.
Solution: The thickness may be expressed in mils as $0.625 \times 1000 = 625$ mils and the width as $5 \times 1000 = 5000$ mils. The cross-sectional area is 625×5000, or 3,125,000 square mils.

Circular Mil

The *circular mil* is the standard unit of wire cross-sectional area used in most wire tables. To avoid the use of decimals (because most wires used to conduct electricity may be only a small fraction of an inch), it is convenient to express these diameters in mils. For example, the diameter of a wire is expressed as 25 mils instead of 0.025 inch. A circular mil is the area of a circle having a diameter of 1 mil, as shown in Figure 2.65B. The area in circular mils of a round conductor is obtained by squaring the diameter measured in mils; thus, a wire having a diameter of 25 mils has an area of 25^2, or 625, circular mils. By way of comparison, the basic formula for the area of a circle is

$$A = \pi \times R^2 \tag{2.22}$$

In this example, the area in square inches (in.^{2}) is

$$A = \pi \times R^2 = 3.14 \times (0.0125)^2 = 0.00049 \text{ in.}^2$$

If D is the diameter of a wire in mils, the area in square mils can be determined using:

$$A = \pi(D/2)^2 \tag{2.23}$$

which translates to:

$$A = 3.14/4D^2 = 0.785\ D^2 \text{ square mils}$$

Thus, a wire 1 mil in diameter has an area of:

$$A = 0.785 \times 1^2 = 0.785 \text{ square mils}$$

which is equivalent to 1 circular mil. The cross-sectional area of a wire in circular mils is therefore determined as:

$$A = \frac{0.785 \times D^2}{0.785} = D^2 \text{ circular mils}$$

where D is the diameter in mils; thus, the constant $\pi/4$ is eliminated from the calculation. Note that in comparing square and round conductors that the circular mil is a smaller unit of area than the square mil, so there are more circular mils than square mils in any given area. The comparison is shown in Figure 2.65C. The area of a circular mil is equal to 0.785 of a square mil.

Key Point: To determine the circular-mil area when the square-mil area is given, divide the area in square mils by 0.785. Conversely, to determine the square-mil area when the circular-mil area is given, multiply the area in circular mils by 0.785.

■ Example 2.33

Problem: A No. 12 wire has a diameter of 80.81 mils. (1) What is its area in circular mils? (2) What is its area in square mils?

Solution:

(1) Area = $D^2 = (80.81)^2 = 6530$ circular mils
(2) Area = $0.785 \times 6530 = 5126$ square mils

■ Example 2.34

Problem: A rectangular conductor is 1.5 inches wide and 0.25 inch thick. (1) What is its area in square mils? (2) What size of round conductor in circular mils is necessary to carry the same current as the rectangular bar?

Solution:

(1) 1.5 in. = $1.5 \times 1000 = 1500$ mils
0.25 in. = $0.25 \times 1000 = 250$ mils
Area = $1500 \times 250 = 375{,}000$ square mils

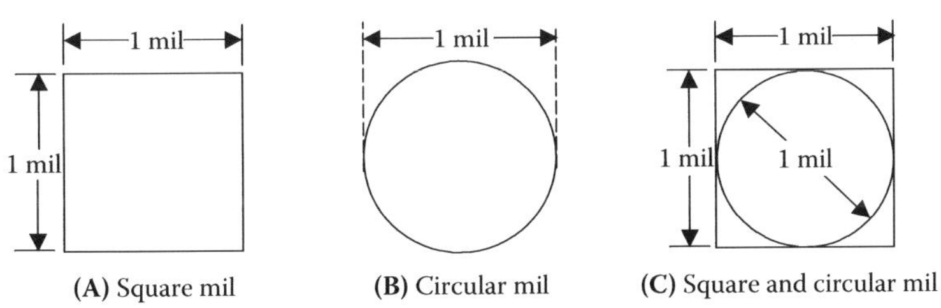

FIGURE 2.65 (A) Square mil; (B) circular mil; (C) comparison of circular and square mil.

(2) To carry the same current, the cross-sectional area of the rectangular bar and the cross-sectional area of the round conductor must be equal. There are more circular mils than square mils in this area; therefore,

$$A = \frac{375,000}{0.785} = 477,700 \text{ circular mils}$$

Note: Many electric cables are composed of stranded wires. The strands are usually single wires twisted together in sufficient numbers to make up the necessary cross-sectional area of the cable. The total area in circular mils is determined by multiplying the area of one strand in circular mils by the number of strands in the cable.

Circular-Mil-Foot

As shown in Figure 2.66, a *circular-mil-foot* is actually a unit of volume. More specifically, it is a unit conductor 1 foot in length and having a cross-sectional area of 1 circular mil. Because it is considered a unit conductor, the circular-mil-foot is useful in making comparisons between wires that are made of different metals. For example, a comparison of the *resistivity* of various substances can be made by determining the resistance of a circular-mil-foot of each of the substances.

Note: It is sometimes more convenient to employ a different unit of volume when working with certain substances. Accordingly, unit volume may also be expressed in cubic centimeters. Cubic inches may also be used. The unit of volume employed is given in tables of specific resistances.

Resistivity

All materials differ in their atomic structure and therefore in their ability to resist the flow of an electric current. As we have discussed, the measure of the ability of a specific material to resist the flow of electricity is its *resistivity*, or specific resistance—the resistance in ohms offered by the unit volume (the circular-mil-foot) of a substance to the flow of electric current. Resistivity is the reciprocal of conductivity (i.e., the ease by which current flows in a conductor). A substance that has a high resistivity will have a low conductivity, and *vice versa*. The resistance of a given length, for any conductor, depends on the resistivity of the material, the length of the wire, and the cross-sectional area of the wire according to the equation:

$$R = \rho \times \frac{L}{A} \tag{2.24}$$

where:

R = resistance of the conductor (ohms)
ρ = specific resistance or resistivity (cm × ohms/ft)
L = length of the wire (ft)
A = cross-sectional area of the wire (cm)

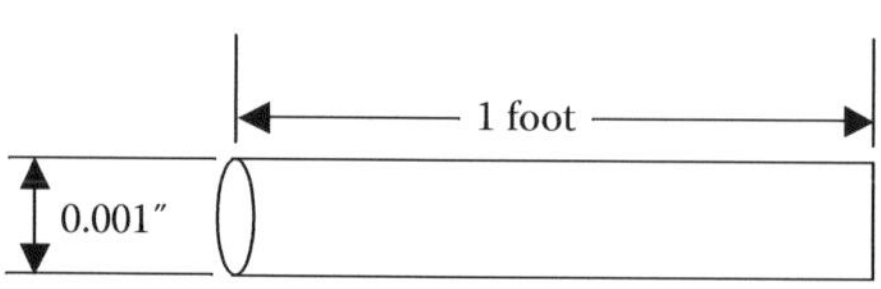

FIGURE 2.66 Circular-mil-foot.

TABLE 2.3
Resistivity (Specific Resistance)

Substance	Specific Resistance at 20°C (Ω·cmil/ft)
Silver	9.8
Copper (drawn)	10.37
Gold	14.7
Aluminum	17.02
Tungsten	33.2
Brass	42.1
Steel (soft)	95.8
Nichrome	660.0

The factor ρ (Greek letter *rho*) permits different materials to be compared for resistance according to their nature without regard to different lengths or areas. Higher values of ρ mean more resistance.

Many tables of resistivity are based on the resistance in ohms of a volume of the substance 1 foot long and 1 circular mil in cross-sectional area. The temperature at which the resistance measurement is made is also specified. If the kind of metal of which the conductor is made is known, the resistivity of the metal may be obtained from a table. The resistivity, or specific resistance, values of some common substances are given in Table 2.3.

Key Point: The resistivity of a substance is the resistance of a unit volume of the substance.

Note: Because silver, copper, gold, and aluminum have the lowest values of resistivity, they are the best conductors. Tungsten and iron have much higher resistivity values.

■ Example 2.35

Problem: What is the resistance of 1000 feet of copper wire having a cross-sectional area of 10,400 circular mils (No. 10 wire)? The wire temperature is 20°C.
Solution: Table 2.3 indicates that the resistivity (specific resistance) is 10.37. Substituting the known values in Equation (2.24), resistance R is determined as

$$R = \rho \times \frac{L}{A} = 10.37 \times \frac{1000}{10{,}400} = 1 \text{ ohm (approximately)}$$

Wire Measurement

Wires are manufactured in sizes numbered according to a table known as the American wire gauge (AWG). Table 2.4 lists the standard wire sizes that correspond to the AWG. The gauge numbers specify the size of round wire in terms of its diameter and cross-sectional area. Note the following:

- As the gauge numbers increase from 1 to 40, the diameter and circular area decrease. Higher gauge numbers indicate smaller wire sizes; thus, No. 12 wire is a smaller wire than No. 4 wire.
- The circular area doubles for every three gauge sizes; for example, No 12 wire has about twice the area of No. 15 wire.
- The higher the gauge number and the smaller the wire, the greater the resistance of the wire for any given length; therefore, 1000 ft of No. 12 wire has a resistance of 1.62 ohms while 1000 ft of No. 4 wire has a resistance of 0.253 ohm.

TABLE 2.4
Copper Wire Table

Gauge	Diameter	Circular mils	Ohms/1000 ft at 25°C
1	289.0	83,700.0	0.126
2	258.0	66,400.0	0.159
3	229.0	52,600.0	0.201
4	204.0	41,700.0	0.253
5	182.0	33,100.0	0.319
6	162.0	26,300.0	0.403
7	144.0	20,800.0	0.508
8	128.0	16,500.0	0.641
9	114.0	13,100.0	0.808
10	102.0	10,400.0	1.02
11	91.0	8230.0	1.28
12	81.0	6530.0	1.62
13	72.0	5180.0	2.04
14	64.0	4110.0	2.58
15	57.0	3260.0	3.25
16	51.0	2580.0	4.09
17	45.0	2050.0	5.16
18	40.0	1620.0	6.51
19	36.0	1290.0	8.21
20	32.0	1020.0	10.4
21	28.5	810.0	13.1
22	25.3	642.0	16.5
23	22.6	509.0	20.8
24	20.1	404.0	26.4
25	17.9	320.0	33.0
26	15.9	254.0	41.6
27	14.2	202.0	52.5
28	12.6	160.0	66.2
29	11.3	127.0	83.4
30	10.0	101.0	105.0
31	8.9	79.7	133.0
32	8.0	63.2	167.0
33	7.1	50.1	211.0
34	6.3	39.8	266.0
35	5.6	31.5	335.0
36	5.0	25.0	423.0
37	4.5	19.8	533.0
38	4.0	15.7	673.0
39	3.5	12.5	848.0
40	3.1	9.9	1070.0

Factors Governing Selection of Wire Size

Several factors must be considered when selecting the size of wire to be used for transmitting and distributing electric power. These factors include allowable power loss in the line, permissible voltage drop in the line, current-carrying capacity of the line, and ambient temperatures in which the wire is to be used:

- Allowable power loss (I^2R) in the line—This loss represents electrical energy converted into heat. The use of large conductors will reduce the resistance and therefore the I^2R loss; however, large conductors are heavier and require more substantial supports, so they are more expensive initially than small ones.
- Permissible voltage drop (*IR* drop) in the line—If the source maintains a constant voltage at the input to the line, any variation in the load on the line will cause a variation in line current and a consequent variation in the *IR* drop in the line. A wide variation in the *IR* drop in the line causes poor voltage regulation at the load.
- Current-carrying capacity of the line—When current is drawn through the line, heat is generated. The temperature of the line will rise until the heat radiated, or otherwise dissipated, is equal to the heat generated by the passage of current through the line. If the conductor is insulated, the heat generated in the conductor is not so readily removed as it would be if the conductor were not insulated.
- Conductors installed in relatively high ambient temperatures—When installed in such surroundings, the heat generated by external sources constitutes an appreciable part of the total conductor heating. Due allowance must be made for the influence of external heating on the allowable conductor current, and each case has its own specific limitations.

Copper vs. Other Metal Conductors

If it were not cost prohibitive, silver, the best conductor of electron flow (electricity), would be the conductor of choice in electrical systems. Instead, silver is used only in special circuits where a substance with high conductivity is required. The two most generally used conductors are copper and aluminum. Each has characteristics that make its use advantageous under certain circumstances. Likewise, each has certain disadvantages, or limitations. Copper has a higher conductivity and is more ductile (able to be drawn out into wire); it has relatively high tensile strength and can be easily soldered. It is more expensive and heavier than aluminum. Aluminum has only about 60% of the conductivity of copper, but its lightness makes possible long spans, and its relatively large diameter for a given conductivity reduces corona, which is the discharge of electricity from the wire when it has a high potential. The discharge is greater when smaller diameter wire is used than when larger diameter wire is used; however, aluminum conductors are not easily soldered, and aluminum's relatively large size for a given conductance does not permit the economical use of an insulation covering. A comparison of some of the characteristics of copper and aluminum is given in Table 2.5.

Note: Recent practice involves using copper wiring instead of aluminum wiring in homes and some industrial applications. Aluminum connections are not as easily made as they are with copper, and, over the years, many fires have been started because of improperly connected aluminum wiring; poor connections result in high resistance connections and excessive heat generation.

TABLE 2.5
Characteristics of Copper and Aluminum

Characteristic	Copper	Aluminum
Tensile strength (lb/in.^{2})	55,000	25,000
Tensile strength for same conductivity (lb)	55,000	40,000
Weight for same conductivity (lb)	100	48
Cross-section for same conductivity (cm)	100	160
Specific resistance (Ω/mil-ft)	10.6	17

TABLE 2.6
Properties of Conducting Materials (Approximate)

Material	Temperature Coefficient (α/°C)	Material	Temperature Coefficient (α/°C)
Aluminum	0.004	Iron	0.006
Carbon	–0.0003	Nichrome	0.0002
Constantan	0 (average)	Nickel	0.005
Copper	0.004	Silver	0.004
Gold	0.004	Tungsten	0.005

Wire Temperature Coefficients

The resistance of pure metals (e.g., silver, copper, aluminum) increases as the temperature increases. The *temperature coefficient* of resistance, α (*alpha*), indicates how much the resistance changes for a change in temperature. A positive value for α means R increases with temperature, a negative α means R decreases, and a zero α means R is constant, not varying with changes in temperature. Typical values of α are listed in Table 2.6. The amount of increase in the resistance of a 1-ohm sample of the copper conductor per degree rise in temperature (i.e., the temperature coefficient of resistance) is approximately 0.004. For pure metals, the temperature coefficient of resistance ranges between 0.004 and 0.006 ohm. Thus, a copper wire having a resistance of 50 ohms at an initial temperature of 0°C will have an increase in resistance of 50 × 0.004, or 0.2 ohms (approximate) for the entire length of wire for each degree of temperature rise above 0°C. At 20°C the increase in resistance is approximately 20 × 0.2, or 4 ohms. The total resistance at 20°C is 50 + 4, or 54 ohms.

Note: As shown in Table 2.6, carbon has a negative temperature coefficient. In general, α is negative for all semiconductors such as germanium and silicon. A negative value for α means less resistance at higher temperatures; therefore, the resistance of semiconductor diodes and transistors can be reduced considerably when they become hot with normal load current. Observe, also, that constantan has a value of zero for α. Thus, it can be used for precision wire-wound resistors that do not change resistance when the temperature increases.

Conductor Insulation

Electric current must be contained; it must be channeled from the power source to a useful load safely. To accomplish this, electric current must be forced to flow only where it is needed. Moreover, current-carrying conductors must not be allowed (generally) to come in contact with one another, their supporting hardware, or personnel working near them. To accomplish this, conductors are coated or wrapped with various materials. These materials have such a high resistance that they are, for all practical purposes, nonconductors. They are generally referred to as *insulators* or *insulating materials*. Numerous types of insulated conductors are available to meet the requirements of any job; however, only the necessary minimum of insulation is applied for any particular type of cable designed to do a specific job because insulation is expensive and has a stiffening effect. Also, it is required to withstand a great variety of physical and electrical conditions. Two fundamental but distinctly different properties of insulation materials (e.g., rubber, glass, asbestos, plastics) are insulation resistance and dielectric strength:

- *Insulation resistance* is the resistance to current leakage through and over the surface of insulation materials.
- *Dielectric strength* is the ability of the insulator to withstand potential difference and is usually expressed in terms of the voltage at which the insulation fails because of the electrostatic stress.

Various types of materials are used to provide insulation for electric conductors, including rubber, plastics, varnished cloth, paper, silk, cotton, and enamel.

Conductor Splices and Terminal Connections

When conductors join each other, or connect to a load, splices or terminals must be used. It is important that they be properly made, as any electric circuit is only as good as its weakest connection. The basic requirement of any splice or connection is that it be both mechanically and electrically as strong as the conductor or device with which it is used. High-quality workmanship and materials must be employed to ensure lasting electrical contact, physical strength, and insulation (if required).

Key Point: Conductor splices and connections are essential parts of any electric circuit.

Soldering Operations

Soldering operations are a vital part of electrical and electronics maintenance procedures. Soldering is a manual skill that must be learned by all personnel who work in the field of electricity. Obviously, practice is required to develop proficiency in the techniques of soldering. Both the solder and the material to be soldered (e.g., electric wire or terminal lugs) must be heated to a temperature that allows the solder to flow. If either is heated inadequately, cold solder joints result (i.e., high-resistance connections are created). Such joints do not provide either the physical strength or the electrical conductivity required. Moreover, it is necessary to select a solder that will flow at a temperature low enough to avoid damage to the part being soldered or to any other part or material in the immediate vicinity.

Solderless Connectors

Generally, terminal lugs and splicers that do not require solder are more widely used than those that do require solder because they are easier to mount correctly. Solderless connectors—made in a wide variety of sizes and shapes—are attached to their conductors by means of several different devices, but the principle of each is essentially the same. They are all crimped (squeezed) tightly onto their conductors. They offer adequate electrical contact, plus great mechanical strength.

Insulation Tape

The carpenter has his saw, the dentist his pliers, the plumber his wrench, and the electrician his insulation tape. Accordingly, one of the first things the rookie maintenance operator learns (a rookie who is also learning proper and safe techniques for performing electrical work) is the value of electrical insulation tape. Normally, the use of electrical insulating tape comes into play as the final step in completing a splice or joint, to place insulation over the bare wire at the connection point. Typically, the insulation tape used should be the same basic substance as the original insulation, usually a rubber-splicing compound. When using rubber (latex) tape as the splicing compound where the original insulation was rubber, it should be applied to the splice with a light tension so each layer presses tightly against the one underneath it. In addition to the rubber tape application (which restores the insulation to original form), restoring with friction tape is also often necessary. In recent years, plastic electrical tape has come into wide use. It has certain advantages over rubber and friction tape; for example, it will withstand higher voltages for a given thickness. Single, thin layers of certain commercially available plastic tape will tolerate several thousand volts without breaking down.

Key Point: Be advised that, although plastic electrical tape is widely used in industrial applications, it must be applied in more layers to ensure an extra margin of safety because it is thinner than rubber or friction tape.

ELECTROMAGNETISM

Earlier, we discussed the fundamental theories concerning simple magnets and magnetism. That discussion dealt mainly with forms of magnetism that were not related directly to electricity—permanent magnets, for instance. Further, only brief mention was made of those forms of magnetism having a direct relation to electricity (e.g., producing electricity with magnetism). In medicine, anatomy and physiology are so closely related that the medical student cannot study one at length without involving the other. A similar relationship holds for the electrical field; that is, magnetism and basic electricity are so closely related that one cannot be studied at length without involving the other. This close fundamental relationship is continually borne out in the study of generators, transformers, battery packs, and motors. To be proficient in electricity, we must become familiar with such general relationships that exist between magnetism and electricity:

- Electric current flow will always produce some form of magnetism.
- Magnetism is by far the most commonly used means for producing or using electricity.
- The occasional peculiar behavior of electricity is caused by magnetic influences.

Magnetic Field around a Single Conductor

In 1819, Hans Christian Oersted, a Danish scientist, discovered that a field of magnetic force exists around a single wire conductor carrying an electric current. In Figure 2.67, a wire is passed through a piece of cardboard and connected through a switch to a dry cell. When the switch is open (no current flowing), iron filings sprinkled on the cardboard will fall back haphazardly when tapped. If we close the switch, current will begin to flow in the wire. This time, when we tap the cardboard the magnetic effect of the current in the wire will cause the filings to fall back into a definite pattern of concentric circles, with the wire as the center of the circles. Every section of the wire has this field of force around it in a plane perpendicular to the wire, as shown in Figure 2.68

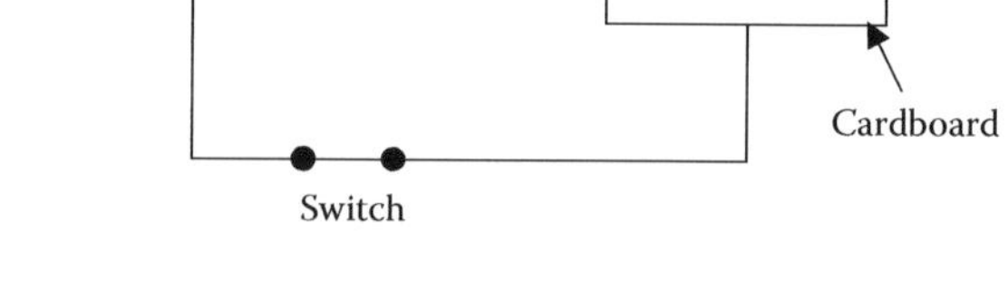

FIGURE 2.67 Circular pattern of magnetic force around a wire carrying an electric current.

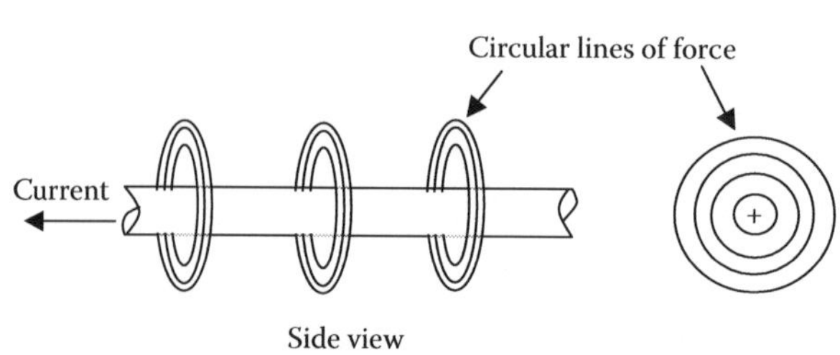

FIGURE 2.68 The circular fields of force around a wire carrying a current are perpendicular to the wire.

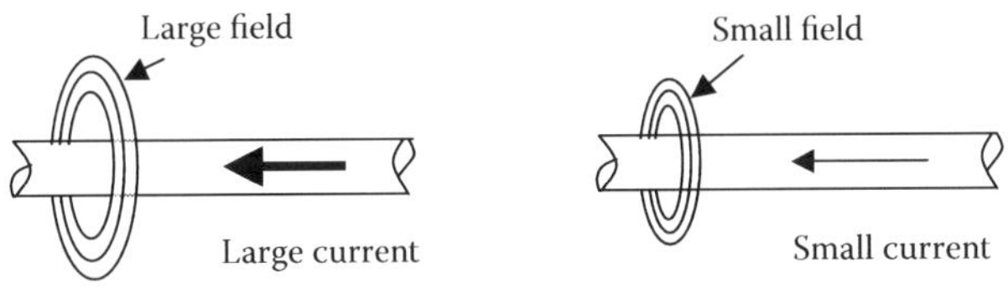

FIGURE 2.69 The strength of a magnetic field around a wire carrying a current depends on the amount of current.

The ability of the magnetic field to attract bits of iron (as demonstrated in Figure 2.47) depends on the number of lines of force present. The strength of the magnetic field around a wire carrying a current depends on the current, as it is the current that produces the field. The greater the current, the greater the strength of the field. A large current will produce many lines of force extending far from the wire, while a small current will produce only a few lines close to the wire, as shown in Figure 2.69.

Polarity of a Single Conductor

The relation between the direction of the magnetic lines of force around a conductor and the direction of current flow along the conductor may be determined by means of the *left-hand rule for a conductor.* If the conductor is grasped in the left hand with the thumb extended in the direction of electron flow (– to +), the fingers will point in the direction of the magnetic lines of force. This is the same direction that the north pole of a compass would point to if the compass were placed in the magnetic field.

Note: Arrows are generally used in electric diagrams to denote the direction of current flow along the length of wire. Where cross-sections of wire are shown, a special view of the arrow is used. A cross-sectional view of a conductor that is carrying current toward the observer is illustrated in Figure 2.70A. The direction of current is indicated by a dot, which represents the head of the arrow. A conductor that is carrying current away from the observer is illustrated in Figure 2.70B. The direction of current is indicated by a cross, which represents the tail of the arrow.

Field around Two Parallel Conductors

When two parallel conductors carry current in the same direction, the magnetic fields tend to encircle both conductors, drawing them together with a force of attraction, as shown in Figure 2.71A. Two parallel conductors carrying currents in opposite directions are shown in Figure 2.71B. The field

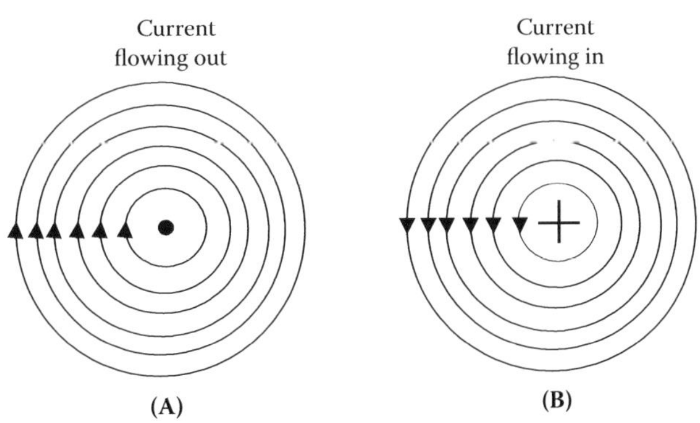

FIGURE 2.70 Magnetic field around a current-carrying conductor.

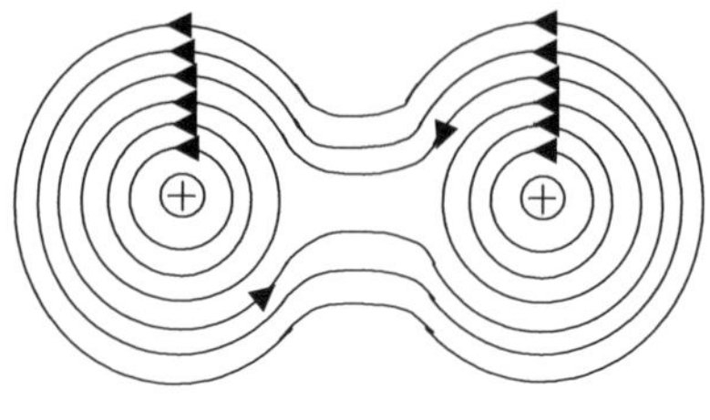

(A) Currents flowing in the same direction

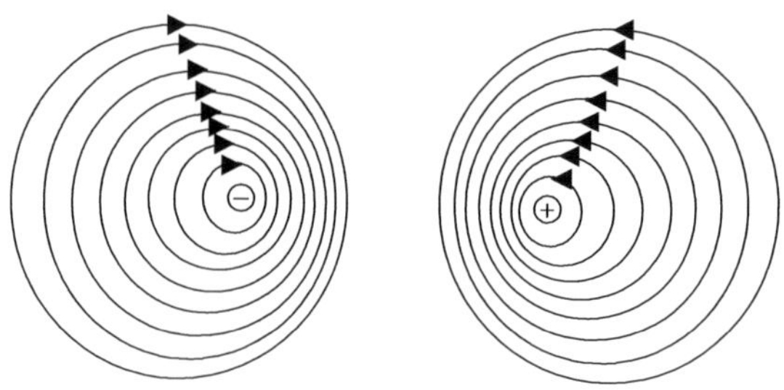

(B) Currents flowing in opposite directions

FIGURE 2.71 Magnetic field around two parallel conductors.

around one conductor is opposite in direction to the field around the other conductor. The resulting lines of force are crowded together in the space between the wires and tend to push the wires apart; two parallel adjacent conductors carrying currents in the same direction attract each other, and two parallel conductors carrying currents in opposite directions repel each other.

Magnetic Field of a Coil

The magnetic field around a current-carrying wire exists at all points along its length. Bending the current-carrying wire into the form of a single loop has two results. First, the magnetic field consists of more dense concentric circles in a plane perpendicular to the wire (see Figure 2.68), although the total number of lines is the same as for the straight conductor. Second, all the lines inside the loop are in the same direction. When this straight wire is wound around a core, as is shown in Figure 2.72, it becomes a coil and the magnetic field assumes a different shape. When current is passed through the coiled conductor, the magnetic field of each turn of wire links with the fields of adjacent turns. The combined influence of all the turns produces a two-pole field similar to that of a simple bar magnet. One end of the coil will be a north pole and the other end will be a south pole.

Polarity of an Electromagnetic Coil

In Figure 2.69, it was shown that the direction of the magnetic field around a straight conductor depends on the direction of current flow through that conductor; thus, a reversal of current flow through a conductor causes a reversal in the direction of the magnetic field that is produced. It follows that a reversal of the current flow through a coil also causes a reversal of its two-pole field. This occurs because that field is the product of the linkage between the individual turns of wire on the coil; therefore, if the field of each turn is reversed, it follows that the total field (coils' field) is also reversed. When the direction of electron flow through a coil is known, its polarity may be determined by use of the *left-hand rule for coils*. This rule is illustrated in Figure 2.72 and can be stated as follows: When the coil is grasped in the left hand, with the fingers wrapped around in the direction of electron flow, the thumb will point toward the north pole.

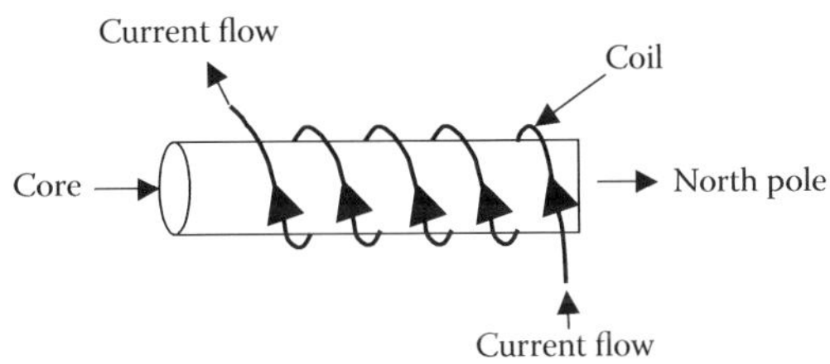

FIGURE 2.72 Current-carrying coil.

Strength of an Electromagnetic Field

The strength, or intensity, of the magnetic field of a coil depends on a number of factors:

- Number of turns of the conductor
- Amount of current flow through the coil
- Ratio of the coil length to its width
- Type of material in the core

Magnetic Units

The law of current flow in the electric circuit is similar to the law for establishing flux in the magnetic circuit. The *magnetic flux*, φ (*phi*), is similar to current in the Ohm's law formula and is comprised of the total number of lines of force existing in the magnetic circuit. The maxwell is the unit of flux; 1 line of force is equal to 1 maxwell.

Note: The maxwell is often referred to as simply a line of force, line of induction, or line.

The *strength* of a magnetic field in a coil of wire depends on how much current flows in the turns of the coil. The more current, the stronger the magnetic field. Also, the more turns, the more concentrated are the lines of force. The *force* that produces the flux in the magnetic circuit (comparable to electromotive force in Ohm's law) is known as *magnetomotive force*, or mmf. The practical unit of magnetomotive force is the ampere-turn (At). In equation form:

$$F \text{ (ampere-turns)} = N \times I \tag{2.25}$$

where:

F = magnetomotive force (At)
N = number of turns
I = current (A)

■ Example 2.36

Problem: Calculate the ampere-turns for a coil with 2000 turns and a 5-mA current.
Solution: Use Equation (2.25) and substitute $N = 2000$ and $I = 5 \times 10^{-3}$ A:

$$N \times I = 2000 \times (5 \times 10^{-3}) = 10 \text{ At}$$

The unit of *intensity* of magnetizing force per unit of length is designated as H and is sometimes expressed as Gilberts per centimeter of length. Expressed as an equation:

$$H = \frac{N \times I}{L} \tag{2.26}$$

where:

H = magnetic field intensity (ampere-turns per meter, At/m)
$N \times I$ = number of turns × current (ampere-turns, At)
L = length between poles of the coil (meters, m)

Note: Equation (2.26) is for a solenoid, and H is the intensity of an air core. With an iron core, H is the intensity through the entire core, and L is the length of or distance between poles of the iron core.

Properties of Magnetic Materials

In this section, we discuss two important properties of magnetic materials: permeability and hysteresis.

Permeability

When the core of an electromagnet is made of annealed sheet steel it produces a stronger magnet than if a cast iron core is used because annealed sheet steel is more readily acted upon by the magnetizing force of the coil than is the hard cast iron. Simply put, soft sheet steel is said to have greater permeability because of the greater ease with which magnetic lines are established in it. Recall that permeability is the relative ease with which a substance conducts magnetic lines of force. The permeability of air is arbitrarily set at 1. The permeability of other substances is the ratio of their ability to conduct magnetic lines compared to that of air. The permeability of nonmagnetic materials, such as aluminum, copper, wood, and brass, is essentially unity, or the same as for air.

Note: The permeability of magnetic materials varies with the degree of magnetization, being smaller for high values of flux density. *Reluctance*, which is analogous to resistance and is the opposition to the production of flux in a material—is inversely proportional to permeability. Iron has high permeability and, therefore, low reluctance. Air has low permeability and hence high reluctance.

Hysteresis

When the current in a coil of wire reverses thousands of times per second, a considerable loss of energy can occur. This loss of energy is caused by *hysteresis*. Hysteresis means "a lagging behind"; that is, the magnetic flux in an iron core lags behind the increases or decreases of the magnetizing force. The simplest method of illustrating the property of hysteresis is by graphical means, such as the hysteresis loop for a magnetic material shown in Figure 2.73. The range of field intensity is indicated by +H and –H. Similarly, the range of flux density is indicated by +B and –B. The current starts

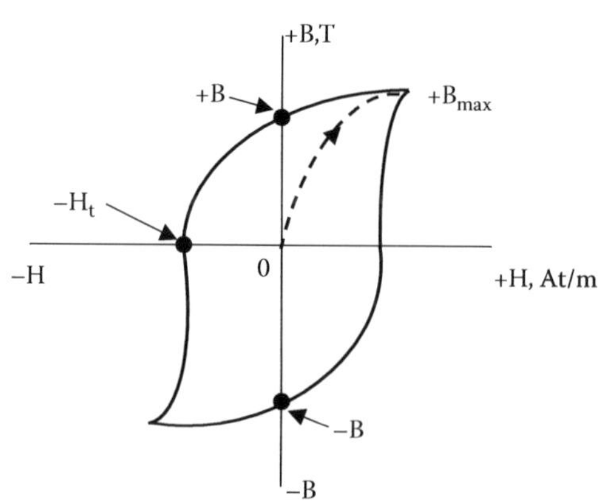

FIGURE 2.73 Hysteresis loop.

at the center 0 (zero) when the material is unmagnetized. Positive H values increase B to saturation at $+B_{max}$. Next H decreases to zero, but B drops only to the value of B_r because of hysteresis. The current that produced the original magnetization now is reversed so H becomes negative. B drops to zero and continues to $-B_{max}$. As the –H values decrease, B is reduced to $-B_r$ when H is zero. Now, with a positive swing of current, H becomes positive, producing saturation at $+B_{max}$ again. The hysteresis loop is now complete. The curve does not return to zero at the center because of hysteresis.

Electromagnets

An *electromagnet* is composed of a coil of wire wound around a core that is normally soft iron, because of its high permeability and low hysteresis. When direct current flows through the coil, the core will become magnetized with the same polarity that the coil would have without the core. If the current is reversed, the polarities of the coil and core are reversed. The electromagnet is of great importance in electricity simply because the magnetism can be turned on or turned off at will. The starter solenoid (an electromagnet) in automobiles and power boats is a good example. In an automobile or boat, an electromagnet is part of a relay that connects the battery to the induction coil and generates the very high voltage required to start the engine. The starter solenoid isolates this high voltage from the ignition switch. When no current flows in the coil, it is an air core, but when the coil is energized, a movable soft-iron core does two things. First, the magnetic flux is increased because the soft-iron core is more permeable than the air core. Second, the flux is more highly concentrated. All of this concentration of magnetic lines of force in the soft-iron core results in a very good magnet when current flows in the coil, but soft iron loses its magnetism quickly when the current is shut off. The effect of the soft iron is, of course, the same whether it is movable, as in some solenoids, or permanently installed in the coil. An electromagnet, then, consists basically of a coil and a core; it becomes a magnet when current flows through the coil. The ability to control the action of magnetic force makes an electromagnet very useful in many circuit applications. Many applications of electromagnets are discussed throughout this manual.

AC THEORY

Because voltage is induced in a conductor when lines of force are cut, the amount of the induced electromotive force (emf) depends on the number of lines cut in a unit time. To induce an emf of 1 volt, a conductor must cut 100,000,000 lines of force per second. To obtain this great number of cuttings, the conductor is formed into a loop and rotated on an axis at great speed (see Figure 2.74). The two sides of the loop become individual conductors in series, each side of the loop cutting lines of force and inducing twice the voltage than a single conductor would induce. In commercial generators, the number of cuttings and the resulting emf are increased by: (1) increasing the number of lines of force by using more magnets or stronger electromagnets, (2) using more conductors or loops, or (3) rotating the loops faster. (Both AC and DC generators are covered later.)

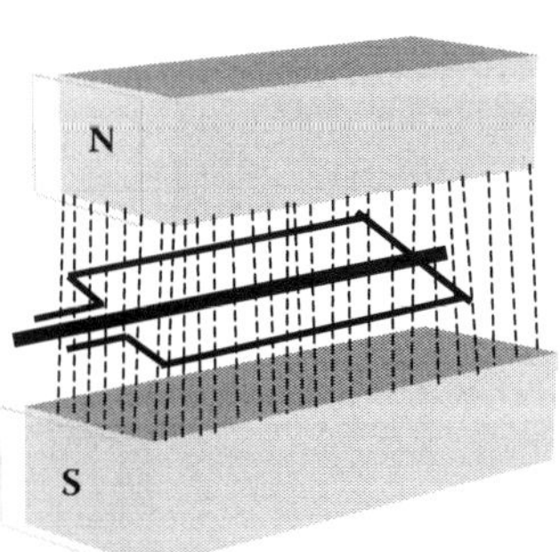

FIGURE 2.74 AC voltage produced by a loop rotating in a magnetic field.

How an AC generator operates to produce an AC voltage and current is a basic concept taught in elementary and middle-school science classes. Of course, today we accept technological advances as commonplace—we surf the Internet, text our friends, watch cable television, use our cell phones, and take outer space flight as a given. We consider the production of electricity, which makes all of these technologies possible, to be a right. These technologies are bottom shelf to us today; they are readily available to us so we use them—no big deal, right? Not worth thinking about. This point of view, though, surely was not typical of those who broke ground in developing technology and electricity.

During the groundbreaking years of electric technology development, the geniuses of the science of electricity (including George Simon Ohm) performed their technological breakthroughs in faltering steps. We tend to forget that those first faltering steps of scientific achievement in the field of electricity were performed with crude and, for the most part, homemade apparatus. (Does this sound something like the more contemporary garage and basement inventors who came up with the first basic user-friendly microcomputer and software packages and perhaps will come up with the renewable energy innovations of tomorrow?)

Indeed, the innovators of electricity had to fabricate nearly all of the laboratory equipment used in their experiments. At the time, the only convenient source of electrical energy available to these early scientists was the voltaic cell, invented some years earlier. Because of the fact that cells and batteries were the only sources of power available, some of the early electrical devices were designed to operate from *direct current* (DC); thus, direct current was used extensively at the time. When the use of electricity became widespread, certain disadvantages in the use of direct current became apparent. In a direct-current system, the supply voltage must be generated at the level required by the load. To operate a 240-volt lamp, for example, the generator must deliver 240 volts. A 120-volt lamp could not be operated from the same generator by any convenient means. A resistor could be placed in series with the 120-volt lamp to drop the extra 120 volts, but the resistor would waste an amount of power equal to that consumed by the lamp.

Another disadvantage of direct-current systems is the large amount of power lost due to the resistance of the transmission wires used to carry current from the generating station to the consumer. This loss could be greatly reduced by operating the transmission line at a very high voltage and low current. This is not a practical solution in a DC system, however, because the load would also have to operate at high voltage. As a result of the difficulties encountered with direct current, practically all modern power distribution systems use alternating current (AC).

Unlike DC voltage, AC voltage can be stepped up or down by a device called a *transformer*. Transformers allow the transmission lines to be operated at high voltage and low current for maximum efficiency. Then, at the consumer end, the voltage is stepped down to whatever value the load requires by using a transformer. Due to its inherent advantages and versatility, alternating current has replaced direct current in all but a few commercial power distribution systems.

Basic AC Generator

As shown in Figure 2.74, an AC voltage and current can be produced when a conductor loop rotates through a magnetic field and cuts lines of force to generate an induced AC voltage across its terminals. This describes the basic principle of operation of an alternating current generator, or *alternator*. An alternator converts mechanical energy into electrical energy. It does this by utilizing the principle of *electromagnetic induction*. The basic components of an alternator are an armature, around which many turns of conductor are wound and which rotates in a magnetic field, as well as some means of delivering the resulting alternating current to an external circuit. (We will cover generator construction in more detail later; in this section, we concentrate on the theory of operation.)

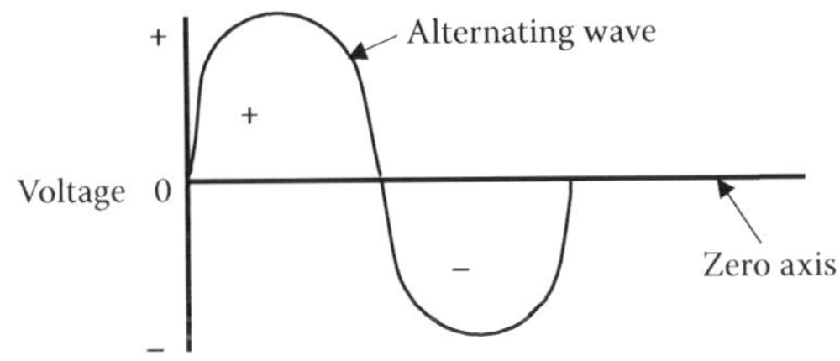

FIGURE 2.75 AC voltage waveform.

Cycle

An AC voltage is one that continually changes in magnitude and periodically reverses in polarity (see Figure 2.75). The zero axis is a horizontal line across the center. The vertical variations on the voltage wave show the changes in magnitude. The voltages above the horizontal axis have positive (+) polarity, while voltages below the horizontal axis have negative (–) polarity. Figure 2.76 shows a suspended loop of wire (conductor or armature) being rotated (moved) in a counterclockwise direction through the magnetic field between the poles of a permanent magnet. For ease of explanation, the loop has been divided into a thick and thin half. Notice that in Figure 2.76A, the thick half is moving along (parallel to) the lines of force; consequently, it is cutting none of these lines. The same is true of the thin half, moving in the opposite direction. Because the conductors are not cutting any lines of force, no emf is induced. As the loop rotates toward the position shown in Figure 2.76B, it cuts more and more lines of force per second because it is cutting more directly across the field (lines of force) as it approaches the position shown in Figure 2.76B. At the position shown in Figure 2.76B, the induced voltage is greatest because the conductor is cutting directly across the field.

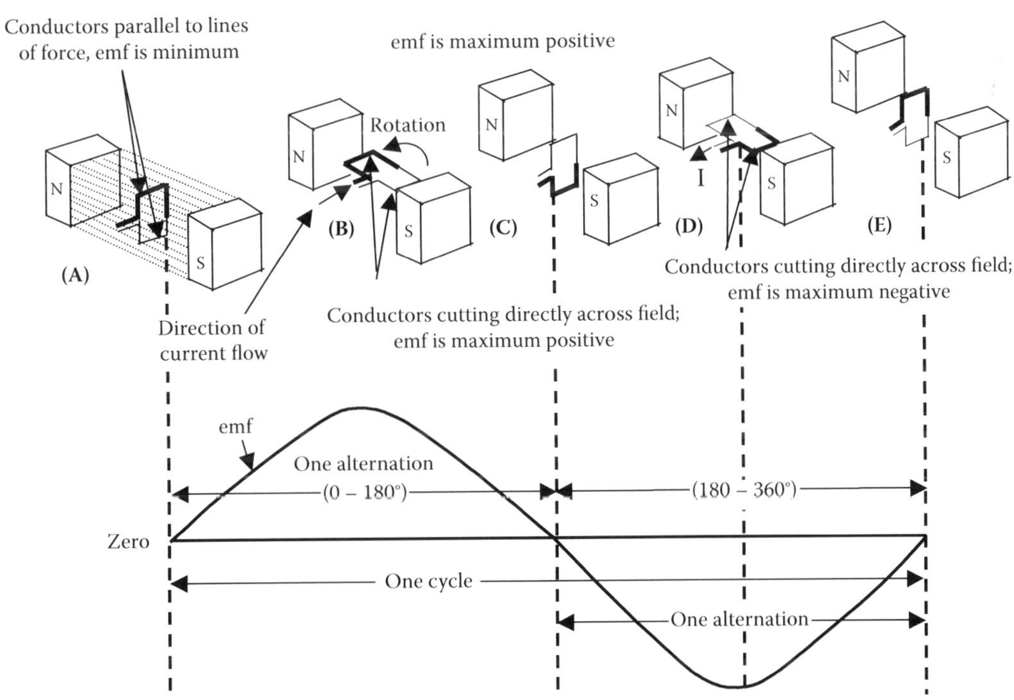

FIGURE 2.76 Basic alternating current generator.

As the loop continues to be rotated toward the position shown in Figure 2.76C, it cuts fewer and fewer lines of force per second. The induced voltage decreases from its peak value. Eventually, the loop is once again moving in a plane parallel to the magnetic field, and no voltage (zero voltage) is induced. The loop has now been rotated through half a circle (one alternation, or 180°). The sine curve shown in the lower part of Figure 2.76 shows the induced voltage at every instant of rotation of the loop. Notice that this curve contains 360°, or two alternations. In Figure 2.76, if the loop is rotated at a steady rate and if the strength of the magnetic field is uniform, the number of cycles per second (cps), or *hertz*, and the voltage will remain at fixed values. Continuous rotation will produce a series of sine-wave voltage cycles, or, in other words, an AC voltage. In this way, mechanical energy is converted into electrical energy.

Key Point: Two complete alternations in a period of time is called a *cycle*.

Frequency, Period, and Wavelength

The *frequency* of an alternating voltage or current is the number of complete cycles occurring in each second of time. It is indicated by the symbol f and is expressed in hertz (Hz). One cycle per second equals 1 hertz. Thus, 60 cycles per second (cps) equals 60 Hz. A frequency of 2 Hz (Figure 2.77B) is twice the frequency of 1 Hz (Figure 2.77A). The amount of time for the completion of 1 cycle is the *period*. It is indicated by the symbol T for time and is expressed in seconds. Frequency and period are reciprocals of each other:

$$f = \frac{1}{T} \tag{2.27}$$

Key Point: The higher the frequency, the shorter the period.

$$T = \frac{1}{f} \tag{2.28}$$

The angle of 360° represents the time for 1 cycle, or the period T. So, we can show the horizontal axis of the sine wave in units of either electrical degrees or seconds (see Figure 2.78). The *wavelength* is the length of one complete wave or cycle. It depends upon the frequency of the periodic variation and its velocity of transmission. It is indicated by the symbol λ (Greek lowercase lambda). Expressed as a formula:

$$\lambda = \frac{\text{Velocity}}{\text{Frequency}} \tag{2.29}$$

CHARACTERISTIC VALUES OF AC VOLTAGE AND CURRENT

Because an AC sine wave voltage or current has many instantaneous values throughout the cycle, it is convenient to specify magnitudes to compare one wave with another. The peak, average, or root-mean-square (RMS) value can be specified (see Figure 2.79). These values apply to current or voltage.

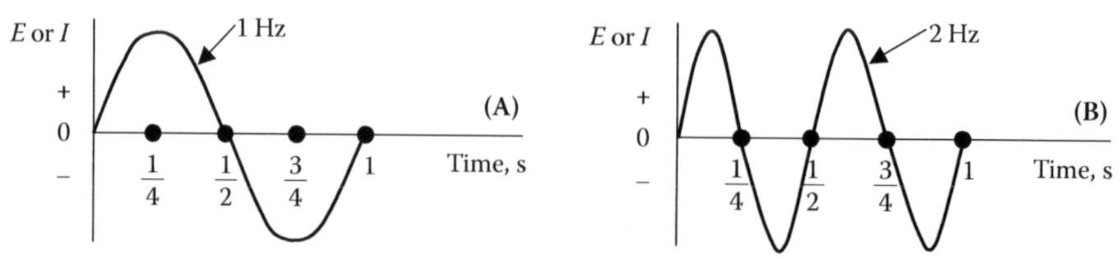

FIGURE 2.77 Comparison of frequencies.

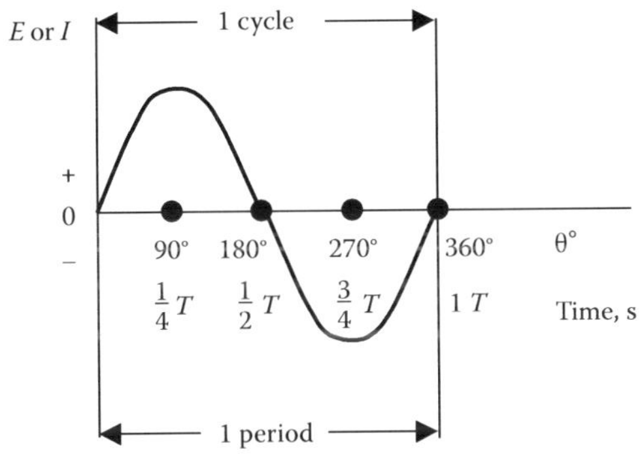

FIGURE 2.78 Relationship between electrical degrees and time.

Peak Amplitude

One of the most frequently measured characteristics of a sine wave is its amplitude. Unlike DC measurement, the amount of alternating current or voltage present in a circuit can be measured in various ways. In one method of measurement, the maximum amplitude of either the positive or the negative alternation is measured. The value of current or voltage obtained is called the *peak voltage* or the *peak current.* An oscilloscope is used to measure the peak value of current or voltage. The peak value is illustrated in Figure 2.79.

Peak-to-Peak Amplitude

A second method of indicating the amplitude of a sine wave consists of determining the total voltage or current between the positive and negative peaks. This value of current or voltage is the peak-to-peak value (see Figure 2.79). Because both alternations of a pure sine wave are identical, the peak-to-peak value is twice the peak value. Peak-to-peak voltage is usually measured with an oscilloscope, although some voltmeters have a special scale calibrated in peak-to-peak volts.

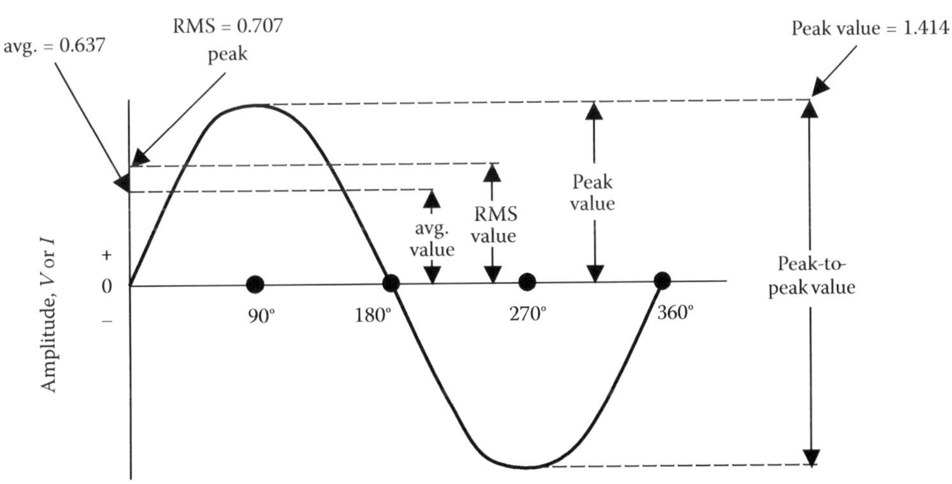

FIGURE 2.79 Amplitude values for AC sine wave.

Instantaneous Amplitude

The *instantaneous value* of a sine wave of voltage for any angle of rotation is expressed by the formula:

$$e = E_m \times \sin(\theta) \tag{2.30}$$

where:

e = instantaneous voltage
E_m = maximum or peak voltage
$\sin(\theta)$ = sine of the angle at which e is desired.

Similarly, the equation for the instantaneous value of a sine wave of current would be

$$i = I_m \times \sin(\theta) \tag{2.31}$$

where:

i = instantaneous current
I_m = maximum or peak current
$\sin(\theta)$ = sine of the angle at which i is desired

Note: The instantaneous value of voltage constantly changes as the armature of an alternator moves through a complete rotation. Because current varies directly with voltage, according to Ohm's law, the instantaneous changes in current also result in a sine wave for which the positive and negative peaks and intermediate values can be plotted exactly as we plotted the voltage sine wave. However, instantaneous values are not useful in solving most AC problems, so an *effective* value is used.

Effective or RMS Value

The effective value of an AC voltage or current of sine waveform is defined in terms of an equivalent heating effect of a direct current. Heating effect is independent of the direction of current flow. The alternating current of a sine waveform having a maximum value of 14.14 amps produces the same amount of heat in a circuit having a resistance of 1 ohm as a direct current of 10 amps. Knowing this, we can work out a constant value for converting any peak value to a corresponding effective value. This constant is represented by x in the simple equation below. Solve for x to three decimal places:

Key Point: Because all instantaneous values of induced voltage are somewhere between zero and E_m (maximum or peak voltage), the effective value of a sine wave voltage or current must be greater than zero and less than E_m.

$$14.14x = 10$$

$$x = 0.707$$

Key Point: Anytime an AC voltage or current is stated without any qualifications, it is assumed to be an effective value.

The effective value is also called the *root-mean-square* (RMS) value because it is the square root of the average of the squared values between zero and maximum. The effective value of an alternating current is stated in terms of an equivalent direct current. The phenomenon that is used as the standard comparison is the heating effect of the current. In many instances, it is necessary to convert from effective to peak or *vice versa* using a standard equation. Figure 2.79 shows that the peak value of a sine wave is 1.414 times the effective value; therefore, the equation we use is

$$E_m = E \times 1.414 \tag{2.32}$$

where:

E_m = maximum or peak voltage
E = effective or RMS voltage

and

$$I_m = I \times 1.414 \quad (2.33)$$

where:

I_m = maximum or peak current
I = effective or RMS current

Upon occasion it is necessary to convert a peak value of current or voltage to an effective value. This is accomplished by using the following equations:

$$E = E_m \times 0.707 \quad (2.34)$$

where:

E = effective voltage
E_m = maximum or peak voltage

and

$$I = I_m \times 0.707 \quad (2.35)$$

where:

I = effective current
I_m = maximum or peak current

Average Value

Because the positive alternation is identical to the negative alternation, the average value of a complete cycle of a sine wave is zero. In certain types of circuits, however, it is necessary to compute the average value of one alternation. Figure 2.79 shows that the average value of a sine wave is 0.637 × peak value; therefore,

$$\text{Average value} = 0.637 \times \text{peak value} \quad (2.36)$$

or

$$E_{avg} = E_m \times 0.637$$

where:

E_{avg} = average voltage of one alternation
E_m = maximum or peak voltage

Similarly,

$$I_{avg} = I_m \times 0.637 \quad (2.37)$$

where:

I_{avg} = average current in one alternation
I_m = maximum or peak current

Table 2.7 lists various sine wave amplitude values used to convert AC sine wave voltage and current.

TABLE 2.7
AC Sine Wave Conversion Table

Multiply the Value	by	To Get the Value
Peak	2	Peak-to-peak
Peak-to-peak	0.5	Peak
Peak	0.637	Average
Average	1.637	Peak
Peak	0.707	Root-mean-square (RMS, effective)
Root-mean-square (RMS, effective)	1.414	Peak
Average	1.110	Root-mean-square (RMS, effective)
Root-mean-square (RMS, effective)	0.901	Average

RESISTANCE IN AC CIRCUITS

If a sine wave of voltage is applied to a resistance, the resulting current will also be a sine wave. This follows Ohm's law, which states that the current is directly proportional to the applied voltage. Figure 2.80 shows a sine wave of voltage and the resulting sine wave of current superimposed on the same time axis. Notice that as the voltage increases in a positive direction the current increases along with it. When the voltage reverses direction, the current reverses direction. At all times the voltage and current pass through the same relative parts of their respective cycles at the same time. When two waves, such as those shown in Figure 2.80, are precisely in step with one another they are said to be *in phase*. To be in phase, the two waves must go through their maximum and minimum points at the same time and in the same direction. In some circuits, several sine waves can be in phase with each other. Thus, it is possible to have two or more voltage drops in phase with each other and in phase with the circuit current.

Note: It is important to remember that Ohm's law for DC circuits is applicable to AC circuits with resistance only.

Voltage waves are not always in phase; for example, Figure 2.81 shows voltage wave E_1, which starts at 0° (time 1). As voltage wave E_1 reaches its positive peak, a second voltage wave, E_2, begins to rise (time 2). Because these waves do not go through their maximum and minimum points at the same instant of time, a phase difference exists between the two waves, and the two waves are said to be *out of phase*. For the two waves in Figure 2.81, this phase difference is 90°.

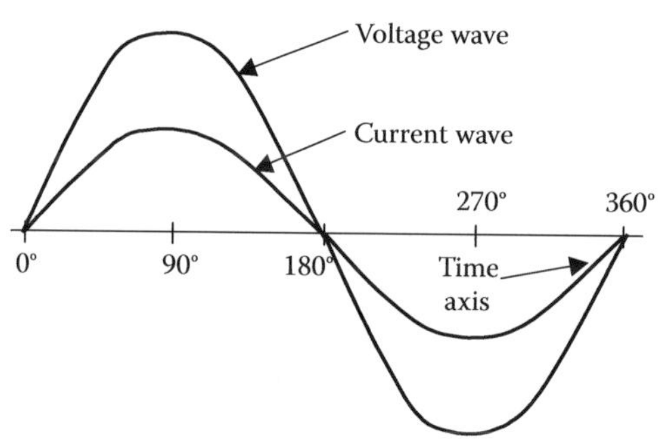

FIGURE 2.80 Voltage and current waves in phase.

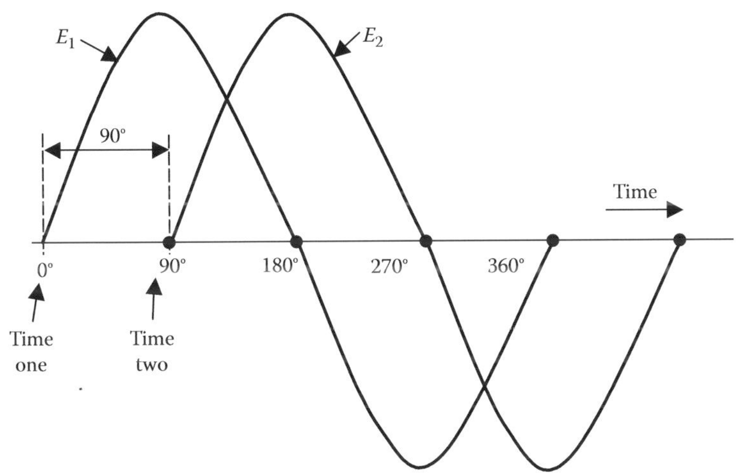

FIGURE 2.81 Voltage waves 90° out of phase.

Phase Relationships

Key Point: The amount by which one wave leads or lags another is measured in degrees.

In the preceding section, we discussed the important concepts of being in phase and phase difference. Another important phase concept is *phase angle*. The phase angle between two waveforms of the same frequency is the angular difference at a given instant of time. As an example, the phase angle between waves B and A (see Figure 2.82) is 90°. Take the instant of time at 90°. The horizontal axis is shown in angular units of time. Wave B starts at maximum value and reduces to zero value at 90°, while wave A starts at zero and increases to maximum value at 90°. Wave B reaches its maximum value 90° ahead of wave A, so wave B *leads* wave A by 90° (and wave A *lags* wave B by 90°). This 90° phase angle between waves B and A is maintained throughout the complete cycle and all successive cycles. At any instant of time, wave B has the value that wave A will have 90° later. Wave B is a cosine wave because it is displaced 90° from wave A, which is a sine wave.

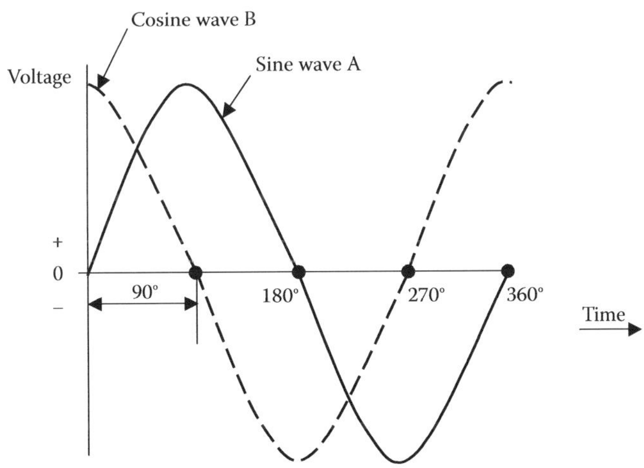

FIGURE 2.82 Wave B leads wave A by a phase angle of 90°.

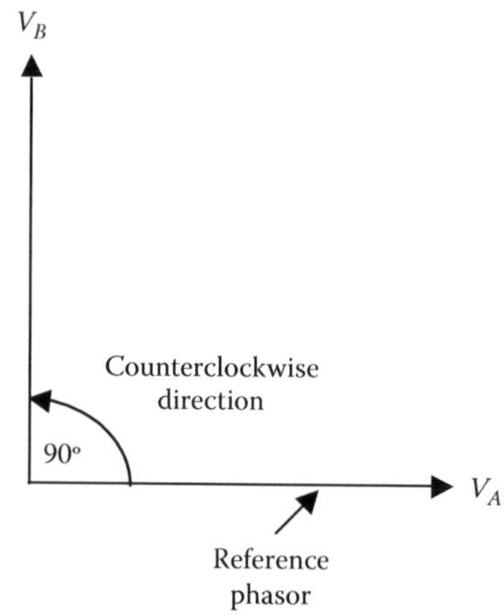

FIGURE 2.83 Phasor diagram.

To compare phase angles or phases of alternating voltages or currents, it is more convenient to use vector diagrams corresponding to the voltage and current waveforms. A *vector* is a straight line used to denote the magnitude and direction of a given quantity. Magnitude is denoted by the length of the line drawn to scale, and the direction is indicated by the arrow at one end of the line, together with the angle that the vector makes with a horizontal reference vector.

Note: In electricity, because different directions really represent *time* expressed as a phase relationship, an electrical vector is called a *phasor*. In an AC circuit containing only resistance, the voltage and current occur at the same time; that is, they are in phase. To indicate this condition by means of phasors all that is necessary is to draw the phasors for the voltage and current in the same direction. The value of each is indicated by the length of the phasor.

A vector, or phasor, diagram is shown in Figure 2.83, where vector V_B is vertical to show the phase angle of 90° with respect to vector V_A, which is the reference. Because lead angles are shown in the counterclockwise direction from the reference vector, V_B leads V_A by 90°.

INDUCTANCE

To this point we have learned the following key points about magnetic fields:

- A field of force exists around a wire carrying a current.
- This field has the form of concentric circles around the wire, in planes perpendicular to the wire and with the wire at the center of the circles.
- The strength of the field depends on the current. Large currents produce large fields; small currents produce small fields.
- When lines of force cut across a conductor, a voltage is induced in the conductor.

Moreover, to this point we have studied circuits that have been *resistive* (i.e., resistors presented the only opposition to current flow). Two other phenomena, inductance and capacitance, exist in DC circuits to some extent, but they are major players in AC circuits. Both inductance and capacitance present a kind of opposition to current flow that is called *reactance*, which we will cover later. Before we examine reactance, however, we must first study inductance and capacitance.

Inductance: What Is It?

Inductance is the characteristic of an electrical circuit that makes itself evident by opposing the starting, stopping, or changing of current flow. A simple analogy can be used to explain inductance. We are all familiar with how difficult it is to push a heavy load (a cart full of heavy items, for example). It takes more work to start the load moving than it does to keep it moving. This is because the load possesses the property of *inertia*. Inertia is the characteristic of mass that opposes a change in velocity; therefore, inertia can hinder us in some ways and help us in others. Inductance exhibits the same effect on current in an electric circuit as inertia does on velocity of a mechanical object. The effects of inductance are sometimes desirable, sometimes undesirable.

Key Point: Simply put, inductance is the characteristic of an electrical conductor that opposes a change in current flow.

Because inductance is the property of an electric circuit that opposes any change in the current through that circuit, if the current increases then a self-induced voltage opposes this change and delays the increase. On the other hand, if the current decreases then a self-induced voltage tends to aid (or prolong) the current flow, delaying the decrease. Thus, current can neither increase nor decrease as fast in an inductive circuit as it can in a purely resistive circuit. In AC circuits, this effect becomes very important because it affects the phase relationships between voltage and current. When inductance is a factor in a circuit, the voltage and current generated by the same armature are out of phase. We will examine these phase relationships later. Our objective now is to understand the nature and effects of inductance in an electric circuit.

Unit of Inductance

The unit for measuring inductance (L) is the *henry* (named for the American physicist Joseph Henry), which is abbreviated as h. Figure 2.84 shows the schematic symbol for an inductor. An inductor has an inductance of 1 henry if an emf of 1 volt is induced in the inductor when the current through the inductor is changing at the rate of 1 ampere per second. The relation between the induced voltage, inductance, and rate of change of current with respect to time can be stated mathematically as

$$E = L \times \frac{\Delta I}{\Delta t} \tag{2.38}$$

where:

E = induced emf (volts)
L = inductance (henrys)
ΔI = change in amperes occurring in Δt seconds

Note: The symbol Δ (Delta) means "change in."

The henry is a large unit of inductance that is used with relatively large inductors. The unit employed with small inductors is the millihenry (mh). For still smaller inductors, the unit of inductance is the microhenry (μh).

FIGURE 2.84 Schematic symbol for an inductor.

Self-Inductance

As previously explained, current flow in a conductor always produces a magnetic field surrounding, or linking with, the conductor. When the current changes, the magnetic field changes, and an emf is induced in the conductor. This emf is called a *self-induced emf* because it is induced in the conductor carrying the current.

Note: Even a perfectly straight length of conductor has some inductance.

The direction of the induced emf has a definite relation to the direction in which the field that induces the emf varies. When the current in a circuit is increasing, the flux linking with the circuit is increasing. This flux cuts across the conductor and induces an emf in the conductor in such a direction as to oppose the increase in current and flux. This emf is sometimes referred to as *counterelectromotive force* (cemf). The two terms are used synonymously throughout this manual. Likewise, when the current is decreasing, an emf is induced in the opposite direction and opposes the decrease in current.

Key Point: The effects described here are summarized by *Lenz's law*, which states that the induced emf in any circuit is always in a direction opposite that of the effect that produced it.

Shaping a conductor so the electromagnetic field around each portion of the conductor cuts across some other portion of the same conductor increases the inductance. This is shown in its simplest form in Figure 2.85A. A loop of conductor is looped so two portions of the conductor lie adjacent and parallel to one another. These portions are labeled Conductor 1 and Conductor 2. When the

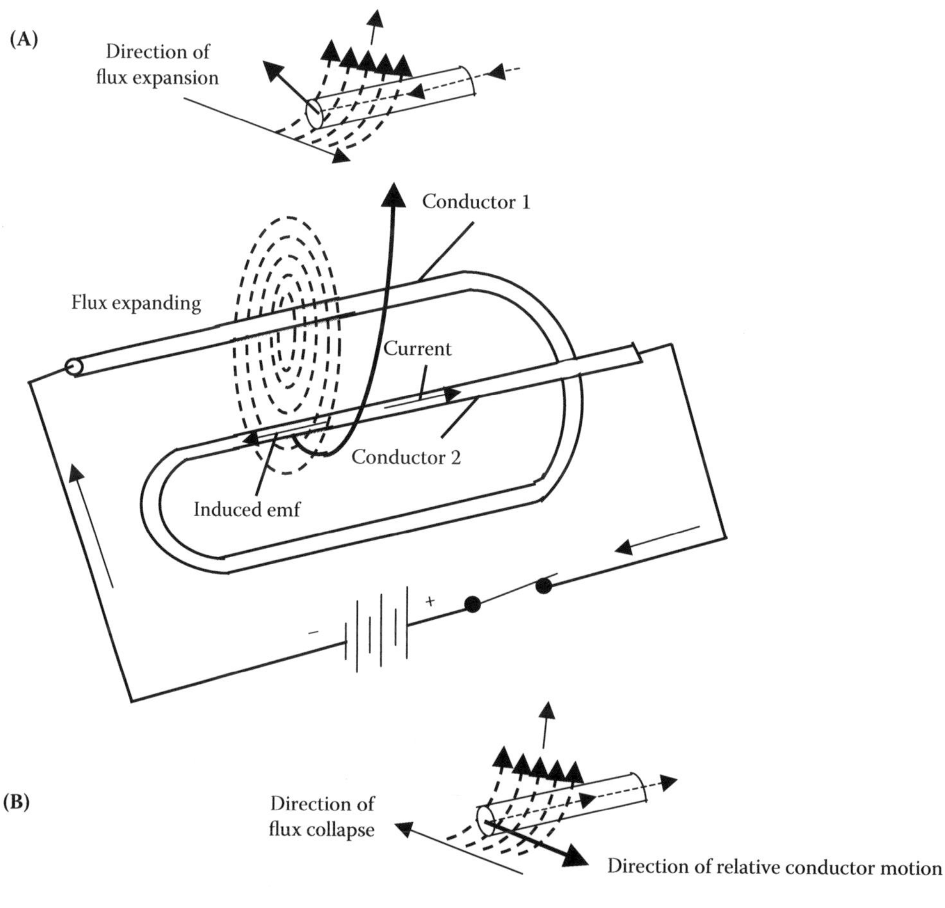

FIGURE 2.85 Self-inductance.

FIGURE 2.86 (A) Few turns, low inductance; (B) more turns, higher inductance.

FIGURE 2.87 (A) Wide spacing between turns, low inductance; (B) close spacing between turns, higher inductance.

FIGURE 2.88 (A) Small diameter, low inductance; (B) larger diameter, higher inductance.

switch is closed, electron flow through the conductor establishes a typical concentric field around all portions of the conductor. The field is shown in a single plane (for simplicity) that is perpendicular to both conductors. Although the field originates simultaneously in both conductors, it is considered as originating in Conductor 1, and its effect on Conductor 2 will be noted. With increasing current, the field expands outward, cutting across a portion of Conductor 2. The resultant induced emf in Conductor 2 is shown by the dashed arrow. Note that it is in opposition to the battery current and voltage, according to Lenz's law. In Figure 2.85B, the same section of Conductor 2 is shown, but with the switch open and the flux collapsing. Four major factors affect the self-inductance of a conductor, or circuit:

Key Point: From Figure 2.85, the important point to note is that the voltage of self-induction opposes both changes in current. It delays the initial buildup of current by opposing the battery voltage and delays the breakdown of current by exerting an induced voltage in the same direction in which the battery voltage acted.

1. *Number of turns*—Inductance depends on the number of wire turns. Wind more turns to increase inductance. Take turns off to decrease the inductance. Figure 2.86 compares the inductance of two coils made with different numbers of turns.
2. *Spacing between turns*—Inductance depends on the spacing between turns, or the length of the inductor. Figure 2.87 shows two inductors with the same number of turns. The turns of the first inductor have a wide spacing. The turns of the second inductor are closer together. The second coil, though shorter, has a larger inductance value because of its close spacing between turns.
3. *Coil diameter*—Coil diameter, or cross-sectional area, is highlighted in Figure 2.88. The larger diameter inductor has higher inductance. Both coils shown have the same number of turns, and the spacing between turns is the same, but the first inductor has a smaller diameter than the second one. The inductance of the second inductor is greater than that for the first inductor.
4. *Type of core material*—Permeability, as pointed out earlier, is a measure of how easily a magnetic field goes through a material. Permeability also tells us how much stronger the magnetic field will be with the material inside the coil. Figure 2.89 shows

Key Point: The inductance of a coil increases very rapidly as the number of turns is increased. It also increases as the coil is made shorter, the cross-sectional area is made larger, or the permeability of the core is increased.

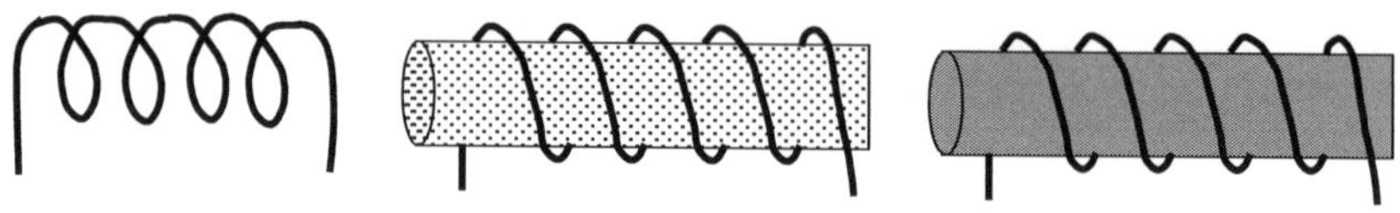

FIGURE 2.89 (A) Air core, low inductance; (B) powdered iron core, higher inductance; (C) soft iron core, highest inductance.

three identical coils. One has an air core, one has a powdered iron core in the center, and the other has a soft iron core. This figure illustrates the effects of core material on inductance. The inductance of a coil is affected by the magnitude of current when the core is a magnetic material. When the core is air, the inductance is independent of the current.

Growth and Decay of Current in an *RL* Series Circuit

If a battery is connected across a pure inductance, the current builds up to its final value at a rate that is determined by the battery voltage and the internal resistance of the battery. The current buildup is gradual because of the counter emf (cemf) generated by the self-inductance of the coil. When the current starts to flow, the magnetic lines of force move out, cut the turns of wire on the inductor, and build up a cemf that opposes the emf of the battery. This opposition causes a delay in the time it takes the current to build up to a steady value. When the battery is disconnected, the lines of force collapse, again cutting the turns of the inductor and building up an emf that tends to prolong the current flow.

Although the analogy is not exact, electrical inductance is somewhat like mechanical inertia. A boat begins to move on the surface of water at the instant a constant force is applied to it. At this instant, its rate of change of speed (acceleration) is greatest, and all the applied force is used to overcome the inertia of the boat. After a while, the speed of the boat increases but its acceleration decreases; the applied force is used up overcoming the friction of the water against the hull. As the speed levels off and acceleration drops to zero, the applied force equals the opposing friction force at this speed and the inertia effect disappears. In the case of inductance, it is electrical inertia that must be overcome.

Figure 2.90 shows a circuit that includes two switches, a battery, and a voltage divider containing a resistor (R) and an inductor (L). Switches S_1 and S_2 are mechanically interlocked ("ganged"), as indicated by the dashed line; when one closes, the other opens at exactly the same instant. Such an arrangement is an RL (resistive–inductive) series circuit. The source voltage of the battery is applied across the resistor and inductor when S_1 is closed. As S_1 is closed, as shown in Figure 2.90, a voltage (E) appears across the resistor and inductor. A current attempts to flow, but the inductor opposes this

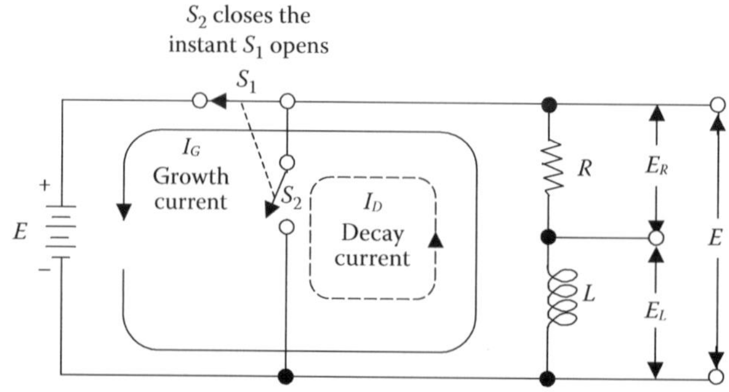

FIGURE 2.90 Growth and decay of current in an RL circuit.

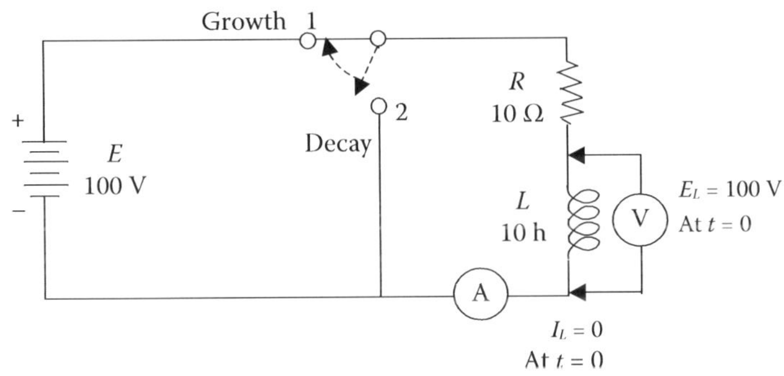

FIGURE 2.91 *L/R* time constant.

current by building up a cemf. At the first instant S_1 is closed, the cemf exactly equals the battery emf, and its polarity is opposite. Under this condition, no current will flow in resistor *R*. Because no current can flow when the cemf is exactly equal to the battery voltage, no voltage is dropped across *R*. As time goes on, more of the battery voltage appears across the resistor and less across the inductor. The rate of change of current is less and the induced emf is less. As the steady-state condition of the current flow is approached, the drop across the inductor approaches zero, and all the battery voltage is used to overcome the resistance of the circuit.

When S_2 is closed (source voltage *E* is removed from the circuit), the flux that has been established around *L* collapses through the windings and induces voltage E_L in *L* which has a polarity opposite that of *E* and is essentially equal to it in magnitude. The induced voltage, E_L, causes current I_d to flow through *R* in the same direction that it was flowing when S_1 was closed. Voltage E_R, which is initially equal to *E*, is developed across *R*. It rapidly falls to zero as voltage E_L across *L* falls to zero due to the collapsing flux.

Key Point: When a switch is first closed to complete a circuit, inductance opposes the buildup of current in the circuit.

Note: In many electronic circuits, the time required for the growth or decay of current is important, but these applications are beyond the scope of this manual; however, you need to learn the fundamentals of the *L/R* time constant, which are discussed in the following section.

L/R Time Constant

The time required for the current through an inductor to increase to 63.2% (63%) of the maximum current or to decrease to 36.7% (37%) is known as the *time constant* of the circuit. An *RL* circuit is shown in Figure 2.91. The value of the time constant in seconds is equal to the inductance in henrys divided by the circuit resistance in ohms. One set of values is given in Figure 2.92. *L/R* is the

Key Point: The time constant of an *RL* circuit is always expressed as a ratio between inductance (*L*) and resistance (*R*).

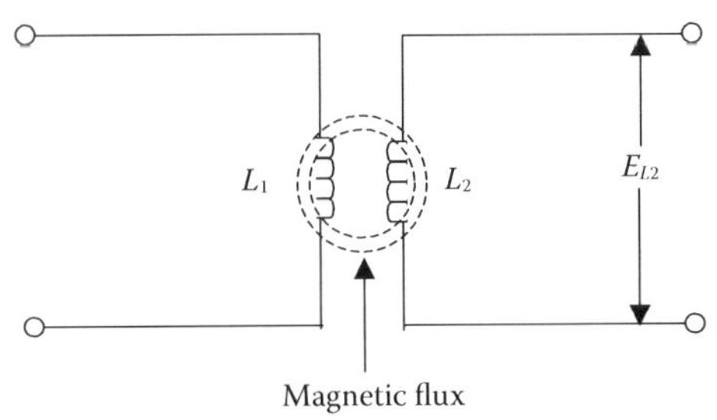

FIGURE 2.92 Mutual inductance between L_1 and L_2.

symbol used for this time constant. If L is in henrys and R is in ohms, t (time) is in seconds. If L is in microhenrys and R is in ohms, t is in microseconds. If L is in millihenrys and R is in ohms, t is in milliseconds. R in the L/R equation is always in ohms, and the time constant is on the same order of magnitude as L. Two useful relations used in calculating L/R time constants are as follows:

$$\frac{L \text{ (henrys)}}{R \text{ (ohms)}} = t \text{ (seconds)} \tag{2.39}$$

$$\frac{L \text{ (microhenrys)}}{R \text{ (ohms)}} = t \text{ (microseconds)} \tag{2.40}$$

Mutual Inductance

When the current in a conductor or coil changes, the varying flux can cut across any other conductor or coil located nearby, thus inducing voltages in both. A varying current in L_1, therefore, induces voltage across L_1 and across L_2 (Figure 2.92); see Figure 2.93 for the schematic symbol for two coils with mutual inductance. When the induced voltage (E_{L2}) produces current in L_2, its varying magnetic field induces voltage in L_1; hence, the two coils L_1 and L_2 have *mutual inductance* because current change in one coil can induce voltage in the other. The unit of mutual inductance is the henry, and the symbol is L_M. Two coils have a mutual inductance of 1 henry when a current change of 1 A/sec in one coil induces 1 volt in the other coil. The factors affecting the mutual inductance of two adjacent coils are dependent on

Key Point: The amount of mutual inductance depends on the relative position of the two coils. If the coils are separated a considerable distance, the amount of flux common to both coils is small and the mutual inductance is low. Conversely, if the coils are close together so that nearly all the flow of one coil links the turns of the other, then mutual inductance is high. The mutual inductance can be increased greatly by mounting the coils on a common iron core.

- Physical dimensions of the two coils
- Number of turns in each coil
- Distance between the two coils
- Relative positions of the axes of the two coils
- Permeability of the cores

Calculation of Total Inductance

In the study of advanced electrical theory, it is necessary to know the effect of mutual inductance in solving for total inductance in both series and parallel circuits; however, for our purposes, in this manual we do not attempt to make these calculations. Instead, we discuss the basic total inductance

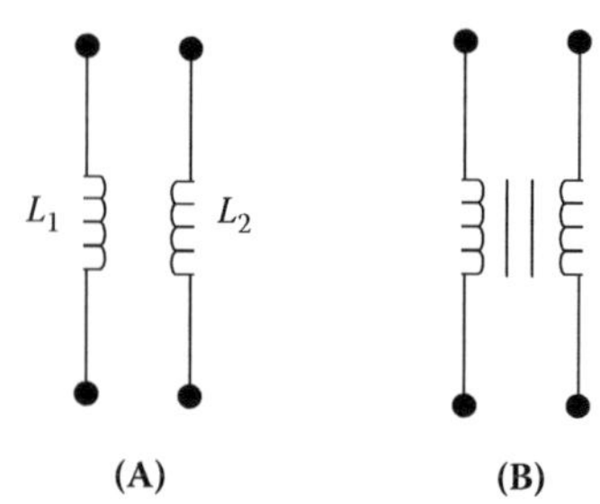

FIGURE 2.93 (A) Schematic symbol for two coils (air core) with mutual inductance; (B) two coils (iron core) with mutual inductance.

calculations that the maintenance operator should be familiar with. If inductors in series are located far enough apart, or well shielded to make the effects of mutual inductance negligible, the total inductance is calculated in the same manner as for resistances in series; we merely add them:

$$L_T = L_1 + L_2 + L_3 + \ldots + L_n \tag{2.41}$$

■ Example 2.37

Problem: If a series circuit contains three inductors with values of 40, 50, and 20 μh, what is the total inductance?
Solution:

$$L_T = 40 + 50 + 20 = 110 \text{ μh}$$

In a parallel circuit containing inductors (without mutual inductance), the total inductance is calculated in the same manner as for resistances in parallel:

$$\frac{1}{L_T} = \frac{1}{L_1} + \frac{1}{L_2} + \frac{1}{L_3} + \ldots \frac{1}{L_n} \tag{2.42}$$

■ Example 2.38

Problem: A circuit contains three totally shielded inductors in parallel. The values of the three inductances are 4, 5, and 10 mh. What is the total inductance?
Solution:

$$\frac{1}{L_T} = \frac{1}{4} + \frac{1}{5} + \frac{1}{10} = 0.25 + 0.2 + 0.1 = 0.55$$

$$L_T = \frac{1}{0.55} = 1.8 \text{ mh}$$

CAPACITANCE

No matter how complex the electrical circuit, it is composed of no more than three basic electrical properties: resistance, inductance, and capacitance. Accordingly, gaining a thorough understanding of these three basic properties is a necessary step toward the understanding of electrical equipment. We have covered resistance and inductance, and the last of the basic three, capacitance, is covered in this section. Earlier, we learned that inductance opposes any change in current. Capacitance is the property of an electric circuit that opposes any change of *voltage* in a circuit. If applied voltage is increased, capacitance opposes the change and delays the voltage increase across the circuit. If applied voltage is decreased, capacitance tends to maintain the higher original voltage across the circuit, thus delaying the decrease. Capacitance is also defined as that property of a circuit that enables energy to be stored in an electric field. Natural capacitance exists in many electric circuits; however, in this manual, we are concerned only with the capacitance that is designed into the circuit by means of devices called *capacitors*.

Key Point: The most noticeable effect of capacitance in a circuit is that voltage can neither increase nor decrease as rapidly in a capacitive circuit as it can in a circuit that does not include capacitance.

THE CAPACITOR

A capacitor, or condenser, is a manufactured electrical device that consists of two conducting plates of metal separated by an insulating material called a *dielectric* (see Figure 2.94). (The prefix *di-* means "through" or "across.") When a capacitor is connected to a voltage source, there is a short

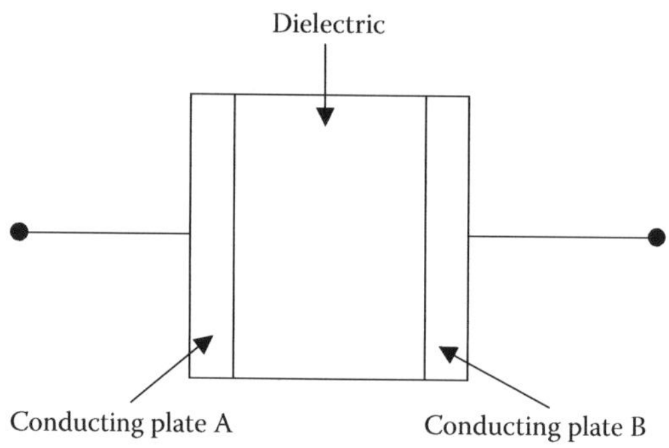

FIGURE 2.94 Capacitor.

FIGURE 2.95 (A) Schematic for a fixed capacitor; (B) variable capacitor.

current pulse. A capacitor stores this electric charge in the dielectric (it can be charged and discharged, as we shall see later). To form a capacitor of any appreciable value, however, the area of the metal pieces must be quite large and the thickness of the dielectric must be quite small. The symbol used to designate a capacitor is *C*; the unit of capacitance is the farad (F). The farad is that capacitance that will store 1 coulomb of charge in the dielectric when the voltage applied across the capacitor terminals is 1 volt. The schematic symbols for capacitors are shown in Figure 2.95.

Key Point: A capacitor is essentially a device that stores electrical energy.

The capacitor is used in a number of ways in electrical circuits. It may block DC portions of a circuit, as it is effectively a barrier to direct current (but not to alternating current). It may be part of a tuned circuit—one such application is in the tuning of a radio to a particular station. It may be used to filter AC out of a DC circuit. Most of these are advanced applications that are beyond the scope of this presentation; however, a basic understanding of capacitance is necessary to the fundamentals of AC theory.

Key Point: A capacitor does not conduct DC current. The insulation between the capacitor plates blocks the flow of electrons. We learned earlier that there is a short current pulse when we first connect the capacitor to a voltage source. The capacitor quickly charges to the supply voltage, and then the current stops.

The two plates of the capacitor shown in Figure 2.96 are electrically neutral because each plate has as many protons (positive charge) as electrons (negative charge); thus, the capacitor has no charge. Now connect a battery across the plates (Figure 2.97A). When the switch is closed (Figure 2.97B), the negative charge on plate A is attracted to the

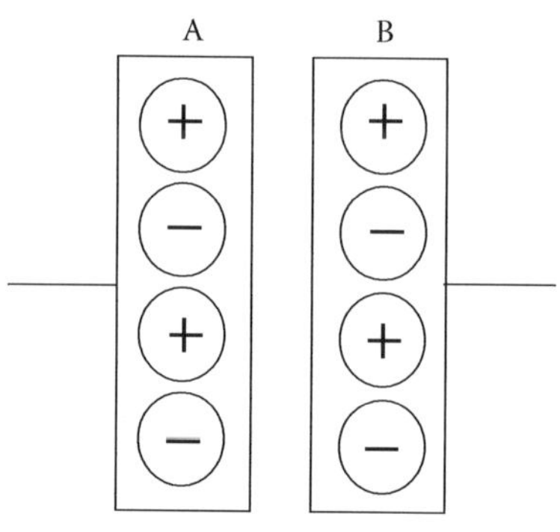

FIGURE 2.96 Two plates of a capacitor with a neutral charge.

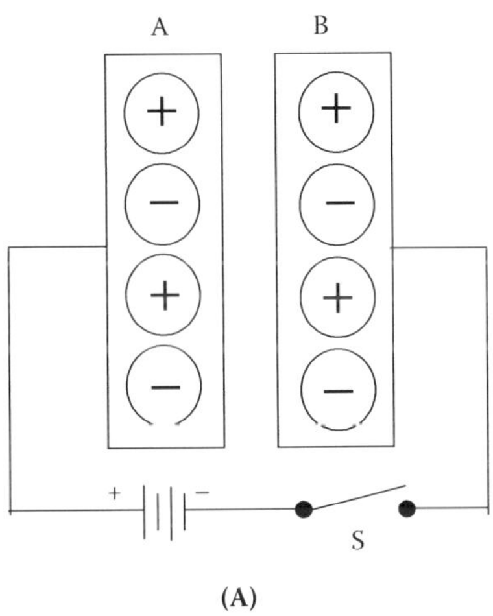

FIGURE 2.97 (A) Neutral capacitor; (B) charged capacitor.

positive terminal of the battery. This movement of charges will continue until the difference in charge between plates A and B is equal to the electromotive force (voltage) of the battery. The capacitor is now charged. Because almost none of the charge can cross the space between the plates, the capacitor will remain in this condition even if the battery is removed (Figure 2.98A); however, if a conductor is placed across the plates (Figure 2.99B), the electrons find a path back to plate A and the charges on each plate are again neutralized. The capacitor is now discharged.

Key Point: In a capacitor, electrons cannot flow through the dielectric, because it is an insulator. Because it takes a definite quantity of electrons to charge a capacitor, it is said to have *capacity*. This characteristic is referred to as *capacitance*.

Dielectric Materials

Somewhat similar to the phenomenon of permeability in magnetic circuits, various materials differ in their ability to support electric flux (lines of force) or to serve as dielectric material for capacitors. Materials are rated in their ability to support electric flux in terms of a number called a *dielectric*

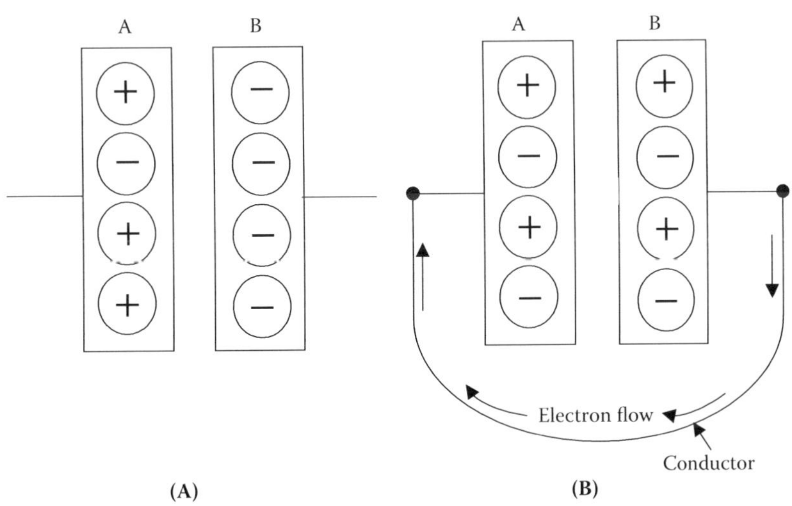

FIGURE 2.98 (A) Charged capacitor; (B) discharging a capacitor.

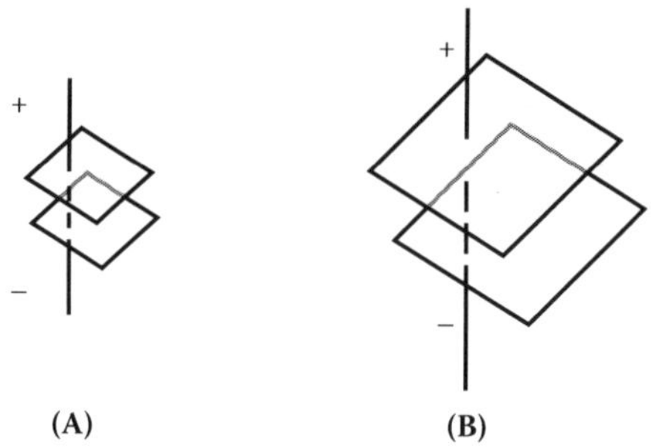

FIGURE 2.99 (A) Small plates, small capacitance; (B) larger plates, higher capacitance.

constant. Other factors being equal, the higher the value of the dielectric constant, the better the dielectric material. Dry air is the reference against which all other materials are rated. Dielectric constants for some common materials are given in Table 2.8.

Note: From Table 2.8 it is obvious that pure water is the best dielectric. Keep in mind that the key word here is "pure." Water capacitors are used today in some high-energy applications, in which differences in potential are measured in thousands of volts.

Unit of Capacitance

Capacitance is equal to the amount of charge that can be stored in a capacitor divided by the voltage applied across the plates:

$$C = \frac{Q}{E} \tag{2.43}$$

where:

C = capacitance (farads, F)
Q = amount of charge (coulombs, C)
E = voltage (volts, V)

■ Example 2.39

Problem: What is the capacitance of two metal plates separated by 1 centimeter of air if .001 coulomb of charge is stored when a potential of 200 volts is applied to the capacitor?
Solution: Given that Q = 0.001 coulomb, E = 200 volts, and

$$C = \frac{Q}{E}$$

TABLE 2.8
Dielectric Constants

Material	Constant	Material	Constant
Vacuum	1.0000	Rubber	2.5–35
Air	1.0006	Wood	2.5–8
Paraffin paper	3.5	Porcelain	5.1–5.9
Glass	5–10	Glycerin (15°C)	56
Quartz	3.8	Petroleum	2
Mica	3–6	Pure water	81

Convert to the power of ten:

$$C = \frac{10 \times 10^{-4}}{2 \times 10^{2}} = 5 \times 10^{-6} = 0.000005 \text{ farads}$$

Note: Although the capacitance value obtained in Example 2.39 appears small, many electronic circuits require much smaller capacitors. The farad is a cumbersome unit that is far too large for many applications. The microfarad (μf) is one millionth of a farad (1×10^{-6} farad) and is a more convenient unit.

Equation (2.43) can be rewritten as follows:

$$Q = C \times E \tag{2.44}$$

$$E = \frac{Q}{C} \tag{2.45}$$

Note: From Equation (2.44), do not get the mistaken idea that capacitance is dependent on charge and voltage. Capacitance is determined entirely by physical factors, which are covered later.

Factors Affecting Value of Capacitance

The capacitance of a capacitor depends on three main factors: plate surface area, distance between plates, and dielectric constant of the insulating material:

- *Plate surface area*—Capacitance varies directly with place surface area. We can double the capacitance value by doubling the plate surface area of the capacitor. Figure 2.99 shows a capacitor with a small surface area and another one with a large surface area. Adding more capacitor plates can increase the plate surface area. Figure 2.100 shows alternate plates connecting to opposite capacitor terminals.
- *Distance between plates*—Capacitance varies inversely with the distance between plate surfaces. The capacitance increases when the plates are closer together. Figure 2.101 shows capacitors with the same plate surface area but different spacing.
- *Dielectric constant of the insulating material*—An insulating material with a higher dielectric constant produces a higher capacitance rating. Figure 2.102 shows two capacitors. Both have the same plate surface area and spacing. Air is the dielectric in the first capacitor, and mica is the dielectric in the second one. The dielectric constant of mica is 5.4 times greater than the dielectric constant of air, so the mica capacitor has 5.4 times more capacitance than the other capacitor.

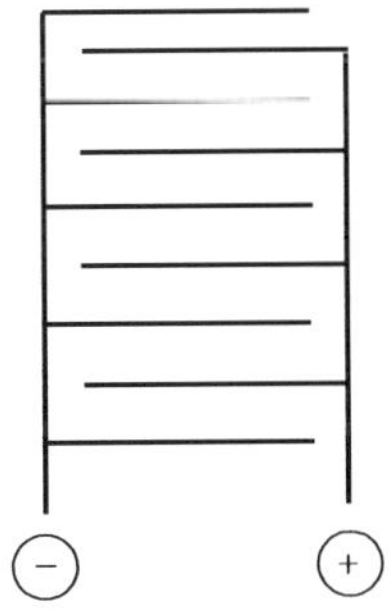

FIGURE 2.100 Sets of plates connected to produce a capacitor with more surface area.

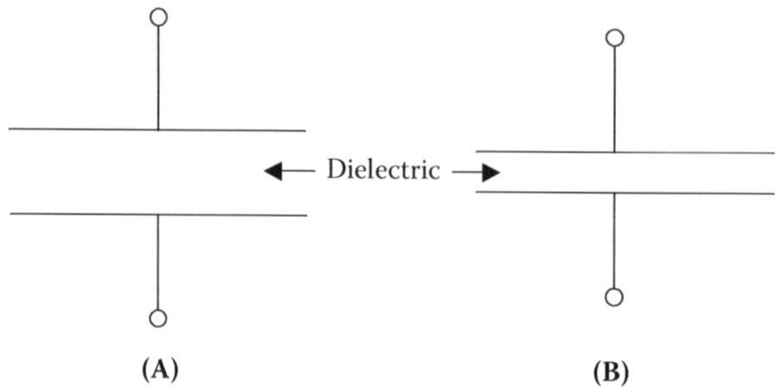

FIGURE 2.101 (A) Wide plate spacing, small capacitance; (B) narrow plate spacing, larger capacitance.

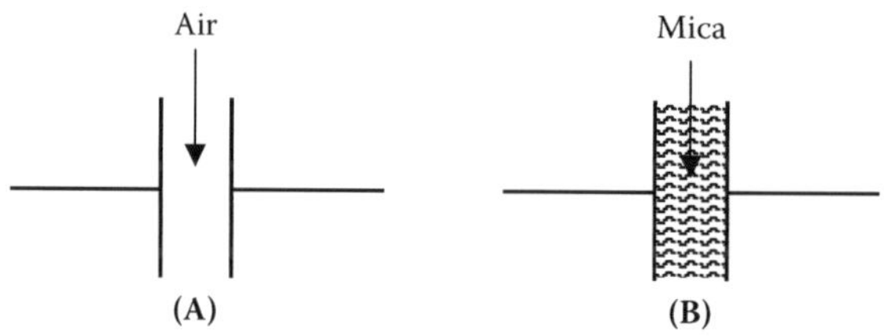

FIGURE 2.102 (A) Low capacitance; (B) higher capacitance.

Voltage Rating of Capacitors

There is a limit to the voltage that may be applied across any capacitor. If too large a voltage is applied, it will overcome the resistance of the dielectric and a current will be forced through it from one plate to the other, sometimes burning a hole in the dielectric. In this event, a short circuit exists and the capacitor must be discarded. The maximum voltage that may be applied to a capacitor is known as the *working voltage* and must never be exceeded. The working voltage of a capacitor depends on: (1) the type of material used as the dielectric, and (2) the thickness of the dielectric. As a margin of safety, the capacitor should be selected so its working voltage is at least 50% greater than the highest voltage to be applied to it; for example, if a capacitor is expected to have a maximum of 200 volts applied to it, its working voltage should be at least 300 volts.

Charge and Discharge of an *RC* Series Circuit

According to Ohm's law, the voltage across a resistance is equal to the current through it times the value of the resistance. This means that a voltage will be developed across a resistance *only when current flows through it.* As previously stated, a capacitor is capable of storing or holding a charge of electrons. When uncharged, both plates contain the same number of free electrons. When charged, one plate contains more free electrons than the other. The difference in the number of electrons is a measure of the charge on the capacitor. The accumulation of this charge builds up a voltage across the terminals of the capacitor, and the charge continues to increase until this voltage equals the applied voltage. The greater the voltage, the greater the charge on the capacitor. Unless a discharge path is provided, a capacitor keeps its charge indefinitely. Any practical capacitor, however, has some leakage through the dielectric so the voltage will gradually leak off. A voltage divider with resistance and capacitance may be connected in a circuit by means of a switch, as shown in Figure 2.103. Such a series arrangement is called an *RC* series circuit.

If S_1 is closed, electrons flow counterclockwise around the circuit containing the battery, capacitor, and resistor. This flow of electrons ceases when C is charged to the battery voltage. At the instant current begins to flow, there is no voltage on the capacitor and the drop across R is equal to

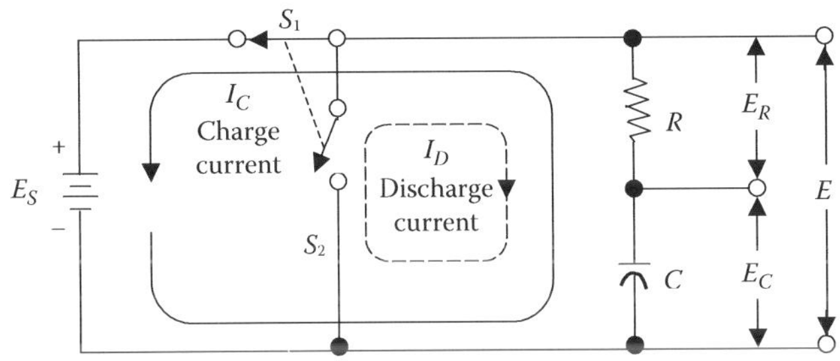

FIGURE 2.103 Charge and discharge of an *RC* series circuit.

the battery voltage. The initial charging current (I) is therefore equal to E_S/R. The current flowing in the circuit soon charges the capacitor. Because the voltage on the capacitor is proportional to its charge, a voltage (E_C) will appear across the capacitor. This voltage opposes the battery voltage—that is, these two voltages buck each other. As a result, voltage E_R across the resistor is equal to $E_S - E_C$, which is equal to the voltage drop (I_CR) across the resistor. Because E_S is fixed, I_C decreases as E_C increases. The charging process continues until the capacitor is fully charged and the voltage across it is equal to the battery voltage. At this instant, the voltage across R is zero and no current flows through it. In Figure 2.103, if S_2 is closed (S_1 opened), a discharge current (I_D) will discharge the capacitor. Because I_D is opposite in direction to I_C, the voltage across the resistor will have a polarity opposite to the polarity during the charging time; however, this voltage will have the same magnitude and will vary in the same manner. During discharge the voltage across the capacitor is equal and opposite to the drop across the resistor. The voltage drops rapidly from its initial value and then approaches zero slowly. The actual time it takes to charge or discharge is important in advanced electricity and electronics. Because the charge or discharge time depends on the values of resistance and capacitance, an *RC* circuit can be designed for the proper timing of certain electrical events. The *RC* time constant is covered in the next section.

RC Time Constant

The time required to charge a capacitor to 63% of maximum voltage or to discharge it to 37% of its final voltage is known as the *time constant* of the current. An *RC* circuit is shown in Figure 2.104. The time constant T for an *RC* circuit is

Key Point: An *RC* time constant expresses the charge and discharge times for a capacitor.

$$T = R \times C \tag{2.46}$$

The time constant of an *RC* circuit is usually very short because the capacitance of a circuit may be only a few microfarads or even picofarads.

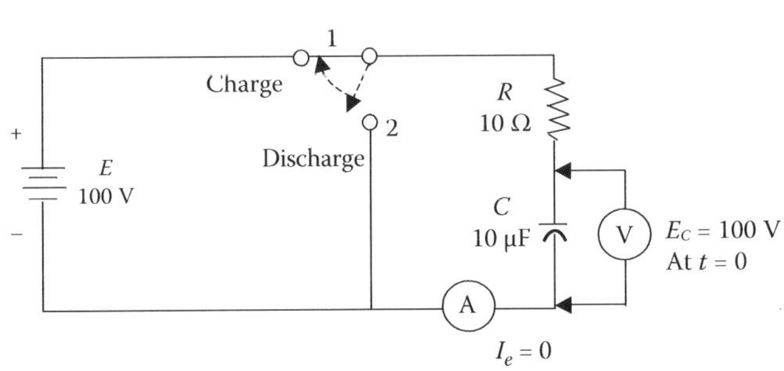

FIGURE 2.104 *RC* circuit.

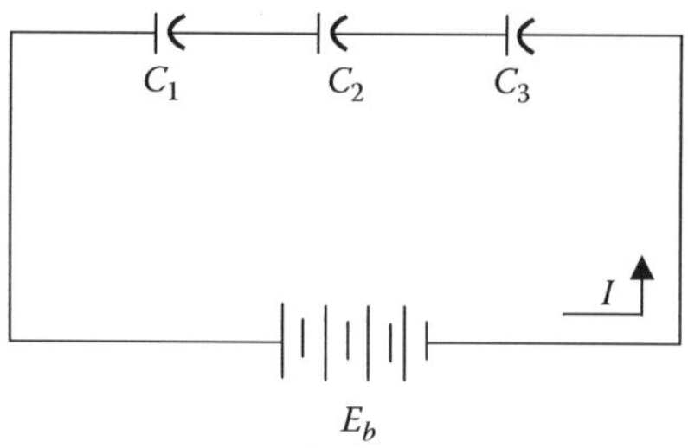

FIGURE 2.105 Series capacitive circuit.

Capacitors in Series and Parallel

Like resistors or inductors, capacitors may be connected in series, in parallel, or in a series–parallel combination. Unlike resistors or inductors, however, total capacitance in series, parallel, or series–parallel combinations is found in a different manner. Simply put, the rules are not the same for the calculation of total capacitance. This difference is explained as follows: Parallel capacitance is calculated like series resistance, and series capacitance is calculated like parallel resistance:

- When capacitors are connected in *series* (see Figure 2.105), the total capacitance (C_T) is

$$\frac{1}{C_T} = \frac{1}{C_1} + \frac{1}{C_2} + \frac{1}{C_3} + \ldots \frac{1}{C_n} \tag{2.47}$$

■ Example 2.40

Problem: Find the total capacitance of a 3-μF, a 5-μF, and a 15-μF capacitor in series.
Solution: Write Equation (2.47) for three capacitors in series:

$$\frac{1}{C_T} = \frac{1}{C_1} + \frac{1}{C_2} + \frac{1}{C_3} = \frac{1}{3} + \frac{1}{5} + \frac{1}{15} = \frac{9}{15} = \frac{3}{5}; \quad C_T = \frac{5}{3} = 1.7\ \mu\text{F}$$

- When capacitors are connected in *parallel* (see Figure 2.106), total capacitance C_T is the sum of the individual capacitances:

$$C_T = C_1 + C_2 + C_3 + \ldots + C_n \tag{2.48}$$

■ Example 2.41

Problem: Determine the total capacitance in a parallel capacitive circuit, given:

$C_1 = 2\ \mu\text{F}$
$C_2 = 3\ \mu\text{F}$
$C_3 = 0.25\ \mu\text{F}$

Solution: Write Equation (2.48) for three capacitors in parallel:

$$C_T = C_1 + C_2 + C_3 = 2 + 3 + 0.25 = 5.25\ \mu\text{F}$$

- If capacitors are connected in a combination of *series and parallel* (see Figure 2.107), the total capacitance is found by applying Equations (2.47) and (2.48) to the individual branches.

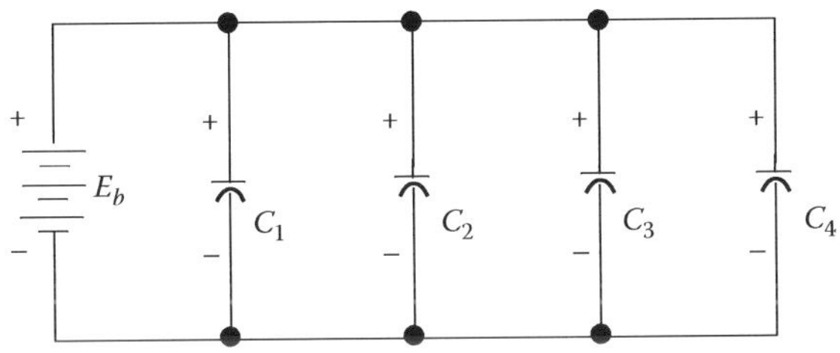

FIGURE 2.106 Parallel capacitive circuit.

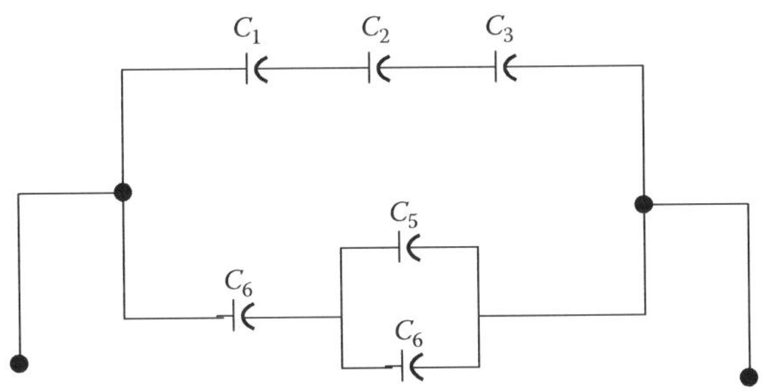

FIGURE 2.107 Series–parallel capacitance configuration.

Types of Capacitors

Capacitors used for commercial applications are divided into two major groups—fixed and variable—and are named according to their dielectric. Most common are air, mica, paper, and ceramic capacitors, plus the electrolytic type. These types are compared in Table 2.9. The fixed capacitor has a set value of capacitances that is determined by its construction. The construction of the variable capacitor allows a range of capacitances. Within this range, the desired value of capacitance is obtained by some mechanical means, such as by turning a shaft (as in turning a radio tuner knob, for example) or adjusting a screw to adjust the distance between the plates. The electrolytic capacitor consists of two metal plates separated by an electrolyte. The electrolyte, either paste or liquid, is in contact with the negative terminal, and this combination forms the negative electrode. The dielectric is a very thin film of oxide deposited on the positive electrode, which is aluminum sheet. Electrolytic capacitors are polarity sensitive (i.e., they must be connected in a circuit according to their polarity markings) used where a large amount of capacitance is required.

TABLE 2.9
Comparison of Capacitor Types

Dielectric	Construction	Capacitance Range
Air	Meshed plates	10–400 pF
Mica	Stacked plates	10–5000 pF
Paper	Rolled foil	0.001–1 μF
Ceramic	Tubular	0.5–1600 pF
	Disk	0.002–0.1 μF
Electrolytic	Aluminum	5–1000 μF
	Tantalum	0.01–300 μF

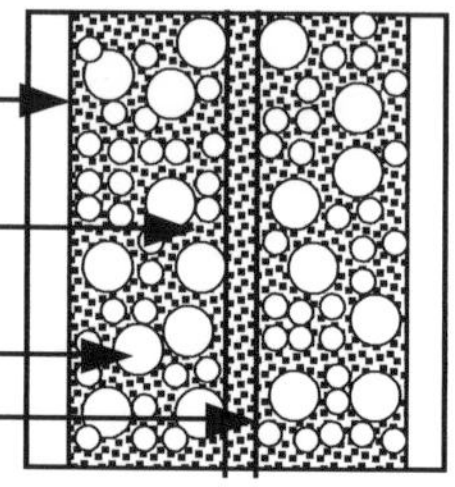

FIGURE 2.108 Ultracapacitor.

Ultracapacitors

Like batteries, ultracapacitors, also known as supercapacitors, pseudocapacitors, electric double-layer capacitors, or electrochemical double-layer capacitors (EDLCs), are energy storage devices (National Renewable Energy Laboratory, 2009). To meet the power, energy, and voltage requirements for a wide range of applications, they use electrolytes and various-sized cells configured into modules. As storage devices, however, they differ from batteries in that ultracapacitors (which are true capacitors in that energy is stored via charge separation at the electrode–electrolyte interface) store energy electrostatically, but batteries store energy chemically. Ultracapacitors provide quick bursts of energy and also offer an improvement of about two to three orders of magnitude in capacitance (as compared to an average capacitor), but with a lower working voltage. Moreover, they can withstand hundreds of thousands of charge/discharge cycles without degrading. As an alternative energy source, ultracapacitors have proven themselves as reliable energy storage components that can be used to power a variety of electronic and portable devices such as radios, flashlights, cell phones, and emergency kits. As ultracapacitor technology matures they are being developed to function as batteries; for example, the vehicle industry is deploying ultracapacitors as a replacement for chemical batteries.

Ultracapacitor Operation

An ultracapacitor polarizes an electrolytic solution to store energy electrostatically. Though it is an electrochemical device, no chemical reactions are involved in its energy storage mechanism. This mechanism is highly reversible and allows the ultracapacitor to be charged and discharged hundreds of thousands of times. As shown in Figure 2.108, an ultracapacitor is comprised of two nonreactive porous plates, or collectors, suspended within an electrolyte, with a voltage potential applied across the collectors. In an individual ultracapacitor cell, the applied potential on the positive electrode attracts the negative ions in the electrolyte, while the potential on the negative electrode attracts the positive ions. A dielectric separator between the two electrodes prevents the charge from moving between the two electrodes. Once the ultracapacitor is charged and energy stored, a load can use this energy. The amount of energy stored is very large compared to a standard capacitor because of the enormous surface area created by the porous carbon electrodes and the small separation (10 angstroms) created by the dielectric separate; however, it stores a much smaller amount of energy than does a battery. Because the rates of charge and discharge are determined solely by is physical properties, the ultracapacitor can release energy much faster (with more power) than a battery that relies on slow chemical reactions.

INDUCTIVE AND CAPACITIVE REACTANCE

Earlier, we learned that the inductance of a circuit acts to oppose any change of current flow in that circuit and that capacitance acts to oppose any change of voltage. In DC circuits these reactions are not important, because they are momentary and occur only when a circuit is first closed or opened.

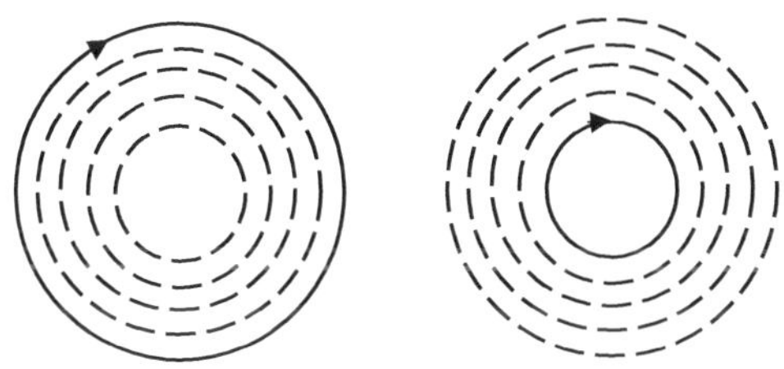

FIGURE 2.109 AC current producing a moving (expanding and collapsing) field; in a coil, this moving field cuts the wires of the coil.

In AC circuits, these effects become very important because the direction of current flow is reversed many times each second, and the opposition presented by inductance and capacitance is, for practical purposes, constant. In purely resistive circuits, either DC or AC, the term for opposition to current flow is *resistance*. When the effects of capacitance or inductance are present, as they often are in AC circuits, the opposition to current flow is called *reactance*. The total opposition to current flow in circuits that have both resistance and reactance is called *impedance*. In this section, we cover the calculation of inductive and capacitive reactance and impedance; the phase relationships of resistance, inductive, and capacitive circuits; and power in reactive circuits.

Inductive Reactance

Key Point: The *reactance* of a coil is the number of ohms of resistance that the coil seems to offer as a result of a counter emf induced in it. Its symbol is *X* to differentiate it from the DC resistance *R*.

In order to gain understanding of the reactance of a typical coil, we need to review exactly what occurs when AC voltage is impressed across the coil. The AC voltage produces an alternating current. When a current flows in a wire, lines of force are produced around the wire. Large currents produce many lines of force; small currents produce only a few lines of force. As the current changes, the number of lines of force will change. The field of force will seem to expand and contract as the current increases and decreases, as shown in Figure 2.109. As the field expands and contracts, the lines of force must cut across the wires that form the turns of the coil.

These cuttings induce an emf in the coil. This emf acts in the direction so as to oppose the original voltage and is called a *counter* (or back) *emf*. The effect of this counter emf is to reduce the original voltage impressed on the coil. The net effect will be to reduce the current below that which would flow if there were no cuttings or counter emf. In this sense, the counter emf is acting as a resistance in reducing the current. Although it would be more convenient to consider the current-reducing effect of a counter emf as a number of ohms of effective resistance, we don't do this. Instead, because a counter emf is not actually a resistance but merely *acts* as a resistance, we use the term *reactance* to describe this effect.

The inductive reactance of a coil depends primarily on: (1) the inductance of the coil, and (2) the frequency of the current flowing through the coil. The value of the reactance of a coil is therefore proportional to its inductance and the frequency of the AC circuit in which it is used. The formula for inductive reactance is

$$X_L = 2\pi fL \quad (2.49)$$

Because $2\pi = 2 \times 3.14 = 6.28$, Equation (2.49) becomes:

$$X_L = 6.28fL$$

where:

X_L = inductive reactance (ohms)
f = frequency (Hz)
L = inductance (h)

If any two quantities are known in Equation (2.49), the third can be found:

$$L = \frac{X_L}{6.28f} \tag{2.50}$$

$$f = \frac{X_L}{6.28L} \tag{2.51}$$

■ Example 2.42

Problem: The frequency of a circuit is 60 Hz and the inductance is 20 mh. What is X_L?
Solution:

$$X_L = 2\pi fL = 6.28 \times 60 \times 0.02 = 7.5 \text{ ohms}$$

■ Example 2.43

Problem: A 30-mh coil is in a circuit operating at a frequency of 1400 kHz. Find its inductive reactance.
Solution: Given that L = 30 mh and f = 1400 kHz, find X_L. First, change the units of measurement:

$$30 \text{ mh} = 30 \times 10^{-3} \text{ h}$$

$$1400 \text{ kHz} = 1400 \times 10^3$$

Now find the inductive reactance:

$$X_L = 6.28fL$$

$$X_L = 6.28 \times 1400 \times 10^3 \times 30 \times 10^{-3} = 263{,}760 \text{ ohms}$$

■ Example 2.44

Problem: Given L = 400 μh and f = 1500 Hz, find X_L.
Solution:

$$X_L = 2\pi fL = 6.28 \times 1500 \times 0.0004 = 3.78 \text{ ohms}$$

Note: If frequency or inductance varies, inductive reactance must also vary. The inductance of a coil does not vary appreciably after the coil is manufactured, unless it is designed as a variable inductor; thus, frequency is generally the only variable factor affecting the inductive reactance of a coil. The inductive reactance of the coil will vary directly with the applied frequency.

Capacitive Reactance

Previously, we learned that as a capacitor is charged electrons are drawn from one plate and deposited on the other. As more and more electrons accumulate on the second plate, they begin to act as an opposing voltage, which attempts to stop the flow of electrons just as a resistor would do. This opposing effect is called the *reactance* of the capacitor and is measured in ohms. As noted earlier, the basic symbol for reactance is X, and the subscript defines the type of reactance. Earlier, we used X_L to represent inductive reactance; the subscript L refers to inductance. Following the same pattern, we use X_C to represent capacitive reactance. The two factors affecting capacitive reactance (X_C) are

- Size of the capacitor
- Frequency

Key Point: Capacitive reactance, X_C, is the opposition to the flow of AC current due to capacitance in the circuit.

The larger the capacitor, the greater the number of electrons that may be accumulated on its plates. When the plate area is large, the electrons do not accumulate in one spot but spread out over the entire area of the plate and do not impede the flow of new electrons onto the plate; therefore, a large capacitor offers a small reactance. In a capacitor with a small plate area, the electrons cannot spread out, and they attempt to stop the flow of electrons coming onto the plate; therefore, a small capacitor offers a large reactance. The reactance, therefore, is *inversely* proportional to the capacitance.

If an AC voltage is impressed across the capacitor, electrons are accumulated first on one plate and then on the other. If the frequency of the changes in polarity is low, the time available to accumulate electrons is large. This means that a large number of electrons will be able to accumulate, which will result in a large opposing effect, or a large reactance. If the frequency is high, the time available to accumulate electrons will be small. This means that only a few electrons will accumulate on the plates, which will result in a small opposing effect, or a small reactance. The reactance, therefore, is *inversely* proportional to the frequency.

The formula for capacitive reactance is

$$X_C = \frac{1}{2\pi fC} \tag{2.52}$$

where:

X_C = capacitive reactance (ohms)
f = frequency (Hz)
C = capacitance (F)

Example 2.45

Problem: What is the capacitive reactance of a circuit operating at a frequency of 60 Hz, if the total capacitance is 130 μf?

Solution:

$$X_C = \frac{1}{2\pi fC} = \frac{1}{6.28 \times 60 \times 0.00013} = 20.4 \text{ ohms}$$

Phase Relationships for *R*, *L*, and *C* Circuits

Key Point: In an inductive circuit, voltage leads current by 90°; in a capacitive circuit, current leads voltage by 90°.

Unlike a purely resistive circuit, where current rises and falls with the voltage (i.e., it neither leads nor lags; current and voltage are in phase), current and voltage are not in phase in inductive and capacitive circuits. This is the case, of course, because events are not quite instantaneous in circuits that have either inductive or capacitive components. In the case of an inductor, voltage is first applied to the circuit, then the magnetic field begins to expand, and self-induction causes a countercurrent to flow in the circuit, opposing the original circuit current. In this case, *voltage leads current* by 90° (Figure 2.110). When a circuit includes a capacitor, a charge current begins to flow and then a difference in potential appears between the plates of the capacitor. In this case, *current leads voltage* by 90° (Figure 2.111).

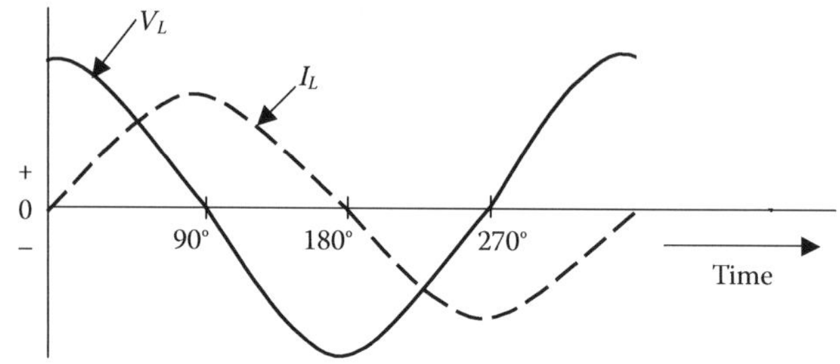

FIGURE 2.110 Inductive circuit: voltage leads current by 90°.

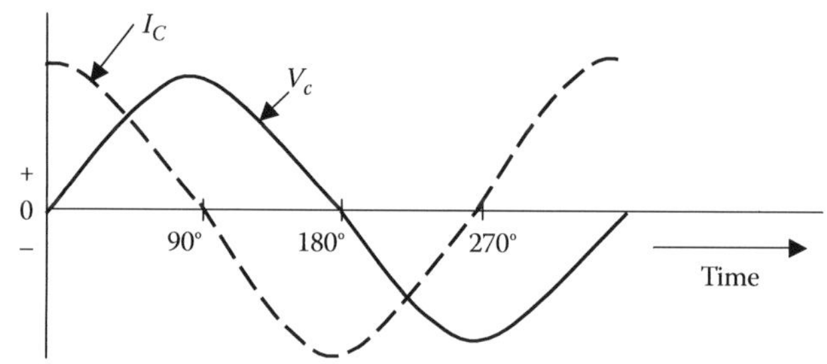

FIGURE 2.111 Capacitive circuit: current leads voltage by 90°.

Impedance

Impedance is the total opposition to the flow of alternating current in a circuit that contains resistance and reactance. In the case of pure inductance, inductive reactance (X_L) is the total opposition to the flow of current through it. In the case of pure resistance, R represents the total opposition. The combined opposition of R and X_L in series or in parallel to current flow is called *impedance*. The symbol for impedance is Z. The impedance of resistance in series with inductance is

$$Z = \sqrt{R^2 + X_L^2} \tag{2.53}$$

where:

Z = impedance (ohms)
R = resistance (ohms)
X_L = inductive reactance (ohms)

The impedance of resistance in series with capacitance is

$$Z = \sqrt{R^2 + X_C^2} \tag{2.54}$$

where:

Z = impedance (ohms)
R = resistance (ohms)
X_C = inductive capacitance (ohms)

When the impedance of a circuit includes R, X_L, and X_C, both resistance and net reactance must be taken into account. The equation for impedance that includes both X_L and X_C is

$$Z = \sqrt{R^2 + (X_L - X_C)^2} \tag{2.55}$$

Power in Reactive Circuits

The power in a DC circuit is equal to the product of volts and amps, but in an AC circuit this is true only when the load is resistive and has no reactance. In a circuit possessing inductance only, the true power is zero. The current lags the applied voltage by 90°. The *true power* in a capacitive circuit is also zero. True power is the average power actually consumed by the circuit, the average being taken over one complete cycle of alternating current. The *apparent power* is the product of the RMS volts and RMS amps. The ratio of true power to apparent power in an AC circuit is called the *power factor*. It may be expressed as a percent or as a decimal.

AC CIRCUIT THEORY

To this point, we have pointed out how combinations of inductance and resistance and then capacitance and resistance behave in an AC circuit, and we saw how *RL* and *RC* combinations affect the current, voltages, power, and power factor of a circuit:

1. The voltage drop across a resistor is *in phase* with the current through it.
2. The voltage drop across an inductor *leads* the current through it by 90°.
3. The voltage drop across a capacitor *lags* the current through it by 90°.
4. The voltage drops across inductors and capacitors are 180° *out of phase*.

Solving AC problems is complicated by the fact that current varies with time as the AC output of an alternator goes through a complete cycle. The various voltage drops in the circuit vary in phase—they are not at their maximum or minimum values at the same time. AC circuits frequently include all three circuit elements: resistance, inductance, and capacitance. In this section, all three of these fundamental circuit parameters are combined and their effect on circuit values studied.

Series *RLC* Circuit

Figure 2.112 shows both the sine waveforms and the vectors for purely resistive, inductive, and capacitive circuits. Only the vectors show the direction, because the magnitudes are dependent on the values chosen for a given circuit. (We are only interested in the *effective* root-mean-square values.) If the individual resistances and reactances are known, Ohm's law may be applied to find the voltage drops. For example, we know that $E_R = I \times R$ and $E_C = I \times X_C$. Then, according to Ohm's law, $E_L = I \times X_L$. In AC circuits, current varies with time, and the voltage drops across the various elements also vary with time; however, the same variation is not always present in each at the same time (except in purely resistive circuits) because current and voltage are not in phase. We are primarily concerned, in practical terms, with effective values of current and voltage; however, to understand basic AC theory, we need to know what occurs from instant to instant. In Figure 2.113, note first that current is the common reference for all three element voltages, because there is only one current in a series circuit, and it is common to all elements. The dashed line in Figure 2.113A represents the common series current. Voltage vectors for each element are provided to demonstrate their individual relationships to the common current. The total source voltage (E) is the vector sum of the individual voltages of IR, IX_L, and IX_C. The arrangement of these three voltages is summarized in Figure 2.113B. Because IX_L and IX_C are each 90° away from I, they are 180° from each other. Vectors in direct opposition (180° out of phase) may be subtracted directly. The total reactive voltage (E_X) is the difference of IX_L and IX_C; for example, $E_X = IX_L - IX_C = 45 - 15 = 30$ volts. The final relationship of line voltage and current, as seen from the source, is shown in Figure 2.113C. Had X_C been larger than X_L, the voltage would lag, rather than lead. When X_C and X_L are of equal value, the line voltage and current will be in phase.

Key Point: In a resistive circuit, the phase difference between voltage and current is zero.

Key Point: The voltage across a single reactive element in a series circuit can have a greater effective value than that of the applied voltage.

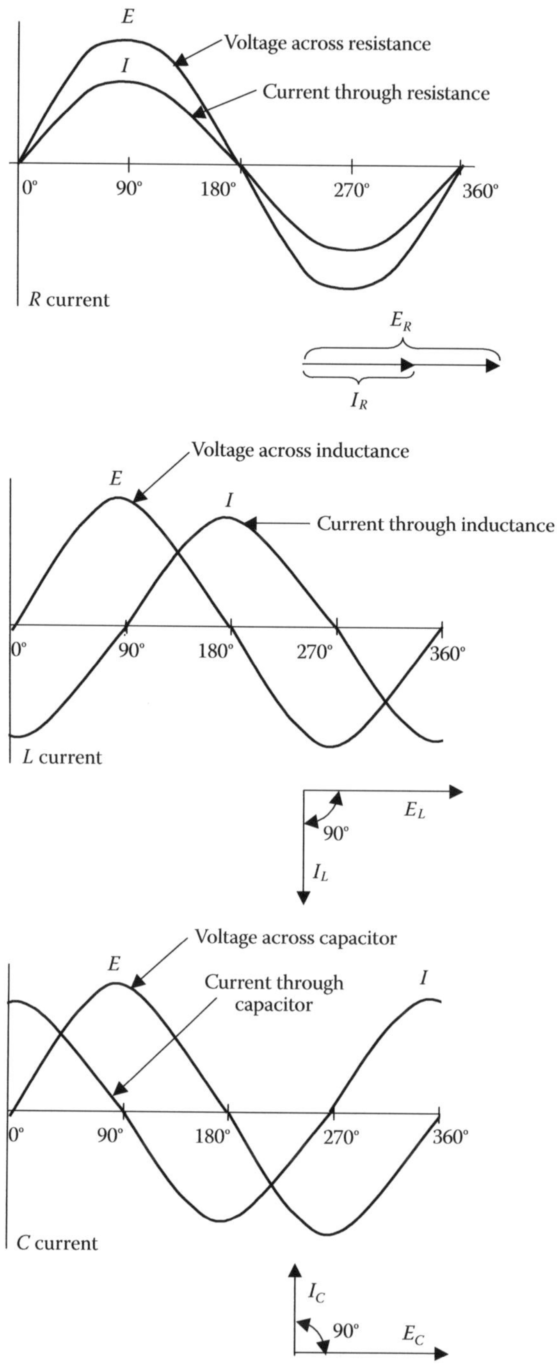

FIGURE 2.112 Sine waveforms and vector representation of *R*, *L*, and *C* circuits.

Note: One of the most important characteristics of an *RLC* circuit is that it can be made to respond most effectively to a single given frequency. When operated in this condition, the circuit is said to be in *resonance* with or *resonant* to the operating frequency. A circuit is at resonance when the inductive reactance (X_L) is equal to the capacitive reactance X_C. At resonance, impedance (*Z*) equals resistance (*R*).

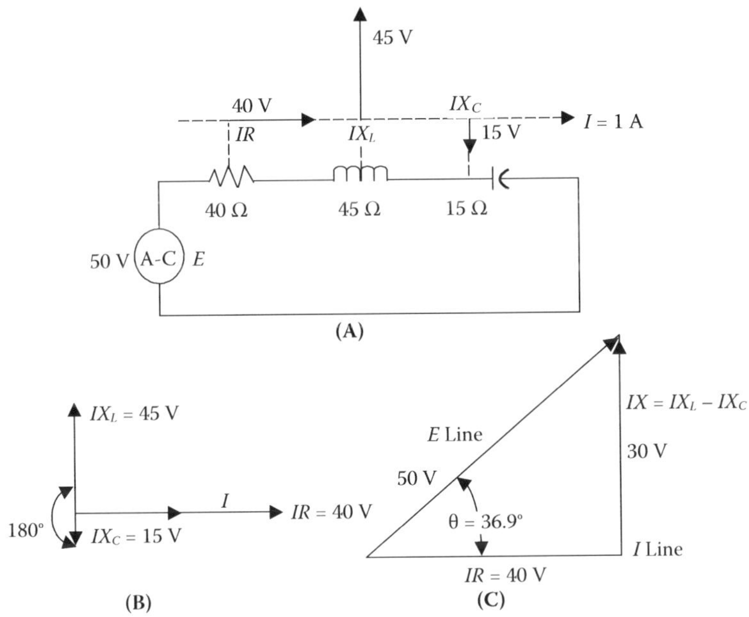

FIGURE 2.113 Resistance, inductance, and capacitance connected in a series.

In summary, the series *RLC* circuit illustrates three important points:

- The current in a series *RLC* circuit either leads or lags the applied voltage, depending on whether X_C is greater or less than X_L.
- A capacitive voltage drop in a series circuit always subtracts directly from an inductive voltage drop.
- The voltage across a single reactive element in a series circuit can have a greater effective value than that of the applied voltage.

Parallel *RLC* Circuits

The *true power* of a circuit is $P = E \times I \times \cos(\theta)$. For any given amount of power to be transmitted, the current (I) varies inversely with the power factor, $\cos(\theta)$. The addition of capacitance in parallel with inductance will, under the proper conditions, improve the power factor of the circuit and make possible the transmission of electric power with reduced line loss and improved voltage regulation. Figure 2.114A shows a three-branch, parallel AC circuit with a resistance in one branch, inductance in the second branch, and capacitance in the third branch. The voltage is the same across each parallel branch, so $V_T = V_R = V_L = V_C$. The applied voltage (V_T) is used as the reference line to measure phase angle θ. The total current (I_T) is the vector sum of I_R, I_L, and I_C. The current in the resistance (I_R) is in phase with the applied voltage (V_T) (Figure 2.114B). The current in the capacitor (I_C) leads total voltage V_T by 90°. I_L and I_C are exactly 180° out of phase and thus acting in opposite directions (Figure 2.114B). When $I_L > I_C$, I_T lags V_T (Figure 2.114C), so the parallel *RLC* circuit is considered inductive.

Power in AC Circuits

In circuits that have only resistance, but no reactance, the amount of power absorbed in the circuit is easily calculated by $P = I^2 \times R$. However, in dealing with circuits that include inductance and capacitance (or both), which is often the case in AC circuits, the calculation of power is a more

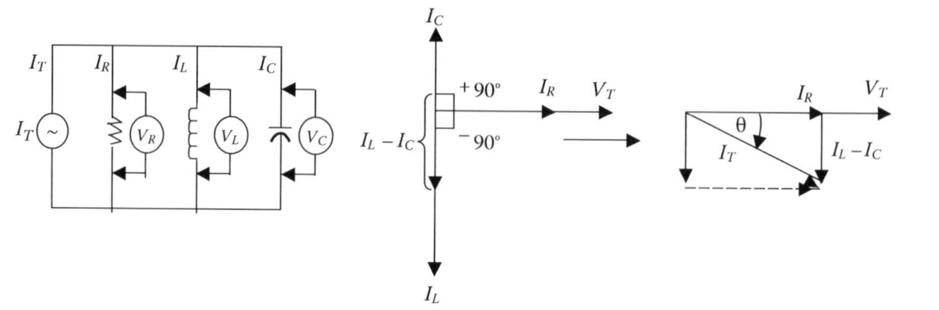

(A) Parallel *RLC* circuit diagram **(B)** Vector diagram, $I_L > I_C$ **(C)** Current-vector triangle, $I_L > I_C$

FIGURE 2.114 R, X_L, and X_C in parallel.

complicated process. Earlier, we explained that power is a measure of the rate at which work is done. The work of a resistor is to limit current flow to the correct, safe level. In accomplishing this, the resistor dissipates heat, and we say that power is consumed or absorbed by the resistor. Inductors and capacitors also oppose current flow, but they do so by producing current that opposes the line current. In either inductive or capacitive circuits, instantaneous values of power may be very large, but the power actually absorbed is essentially zero, since only resistance dissipates heat (absorbs power). Both inductance and capacitance return the power to the source.

Any component that has resistance, such as a resistor or the wiring of an inductor, consumes power. Such power is not returned to the source, because it is dissipated as heat. Previously, we stated that power consumed in the circuit is called *true power*, or *average power*. The two terms are interchangeable, but we generally use the term average power, because the overall value is more meaningful than the instantaneous values of power appearing in the circuit during a complete cycle. Not all apparent power is consumed by the circuit; however, because the alternator does deliver the power, it must be considered in the design. The average power consumption may be small, but instantaneous values of voltage and current are often very large. Apparent power is an important design consideration, especially in assessing the amount of insulation necessary. In an AC circuit that includes both reactance and resistance, some power is consumed by the load and some is returned to the source. How much of each depends on the phase angle, because current normally leads or lags voltage by some angle.

Key Point: In terms of the dissipation of power as heat in a circuit, *average power* is the power that is dissipated as heat, but *apparent power* includes both power that is dissipated as heat and power that is returned to the source.

Note: Recall that in a purely reactive circuit, current and voltage are 90° out of phase.

In an *RLC* circuit, the *R*/*Z* ratio is the cosine of the phase angle, θ; therefore, it is easy to calculate average power in an *RLC* circuit:

$$P = E \times I \times \cos(\theta) \tag{2.56}$$

where:

P = average power absorbed by the circuit
E = effective value of the voltage across the circuit
I = effective value of current in the circuit
θ = phase angle between voltage and current

Note: Recall that the equation for average power in a purely resistive circuit is $P = E \times I$. In a resistive circuit, $P = E \times I$, because the cos(θ) is 1 and need not be considered. In most cases, the phase angle will be neither 90° nor zero, but somewhere between those extremes.

■ Example 2.46

Problem: An *RLC* circuit has a source voltage of 500 volts, line current is 2 amps, and current leads voltage by 60°. What is the average power?
Solution: The cosine of 60° = 0.5, so

$$\text{Average power} = 500 \text{ V} \times 2 \text{ A} \times 0.5 = 500 \text{ W}$$

■ Example 2.47

Problem: An *RLC* circuit has a source voltage of 300 volts, line current is 2 amps, and current lags voltage by 31.8°. What is the average power?
Solution: The cosine of 31.8° = 0.8499, so

$$\text{Average power} = 300 \text{ V} \times 2 \text{ A} \times 0.8499 = 509.9 \text{ W}$$

■ Example 2.48

Problem: Given $E = 100$ volts, $I = 4$ amps, and $\theta = 58.4°$, what is the average power?
Solution: The cosine of 58.4° = 0.5240, so

$$\text{Average power} = 100 \text{ V} \times 4 \text{ A} \times 0.5240 = 209.6 \text{ W}$$

GENERATORS

DC Generators

Key Point: Mechanical power must be applied to the shaft constantly as long as the generator is supplying electrical energy to the external load circuit.

A DC generator is a rotating machine that converts mechanical energy into electrical energy. This conversion is accomplished by rotating an armature, which carries conductors, in a magnetic field, thus inducing an emf in the conductors. As stated previously, in order for an emf to be induced in the conductors, a relative motion must always exist between the conductors and the magnetic field in such a manner that conductors cut through the field. In most DC generators, the armature is the rotating member and the field is the stationary member. A mechanical force is applied to the shaft of the rotating member to cause the relative motion. Thus, when mechanical energy is put into the machine in the form of a mechanical force or twist on the shaft, causing the shaft to turn at a certain speed, electrical energy in the form of voltage and current is delivered to the external load circuit.

To gain a basic understanding of the operation of a DC generator, consider the following explanation. A simple DC generator consists of an armature coil with a single turn of wire (Figure 2.115A,B). (The armature coils used in large DC machines are usually wound in their final shape before being put on the armature, and the sides of the preformed coil are placed in the slots of the laminated armature core.) This armature coil cuts across the magnetic field to produce voltage. If a complete path is present, current will move through the circuit in the direction shown by the arrows in Figure 2.115A. In this position of the coil, commutator segment 1 is in contact with brush 1, while commutator segment 2 is in contact with brush 2. As the armature rotates a half turn in a clockwise direction, the contacts between the commutator segments and the brushes are reversed (Figure 2.115B). At this moment, segment 1 is in contact with brush 2 and segment 2 is in contact with brush 1. Because of this commutator action, that side of the armature coil that is in contact with either of the brushes is always cutting across the magnetic field in the same direction. Thus, brushes 1 and 2 have constant polarity, and a *pulsating DC current* is delivered to the external load circuit.

Note: In DC generators, voltage induced in individual conductors is AC. It is converted to DC (rectified) by the commutator, which rotates in contact with carbon brushes so current generated is in one direction (i.e., direct current).

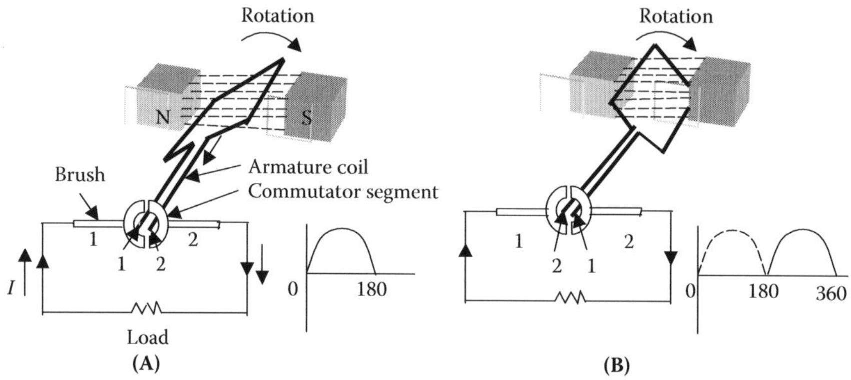

FIGURE 2.115 Basic operation of a DC generator.

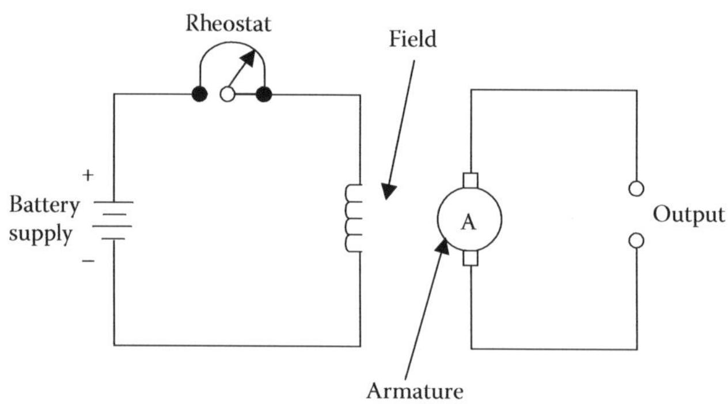

FIGURE 2.116 Separately excited DC generator.

The several different types of DC generators take their names from the type of field excitation used; that is, they are classified according to the manner in which the field windings are connected to the armature circuit. When the field of the generator is excited (or supplied) from a separate DC source (such as a battery) other than its own armature, it is called a *separately excited DC generator* (Figure 2.116). The field windings of a *shunt generator* (self-excited) are connected in series with a rheostat, across the armature in shunt with the load, as shown in Figure 2.117; the shunt generator is widely used in industry. The field windings of a *series generator* (self-excited) are connected in series with

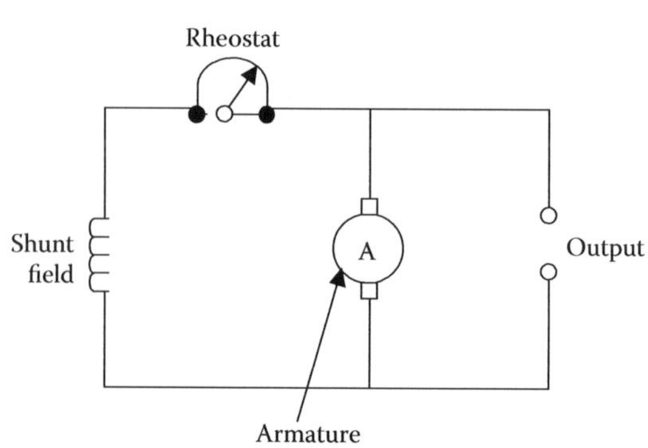

FIGURE 2.117 DC shunt generator.

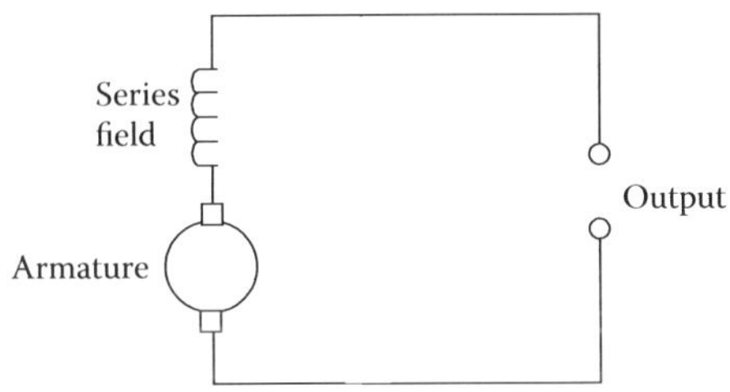

FIGURE 2.118 DC series generator.

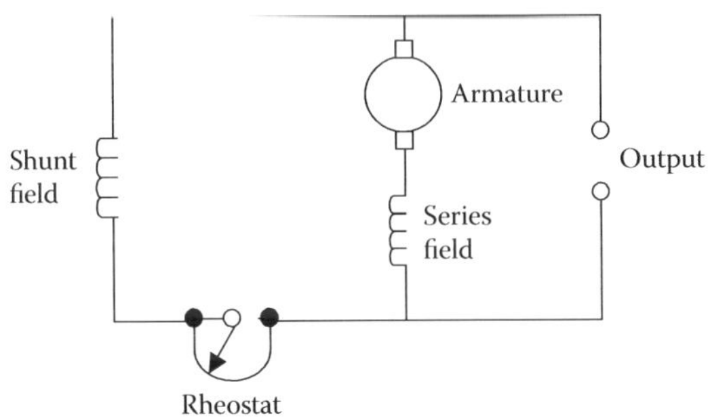

FIGURE 2.119 DC compound generator.

the armature and load, as shown in Figure 2.118. Series generators are seldom used. *Compound generators* (self-excited) contain both series and shunt field windings, as shown in Figure 2.119; compound generators are widely used in industry.

Note: As central generating stations increased in size along with number and power distribution distances, DC generating systems, because of the high power losses in long DC transmission lines, were replaced by AC generating systems to reduce power transmission costs.

AC Generators

Most electric power utilized today is generated by *alternating-current generators* (also called *alternators*). They are made in many different sizes, depending on their intended use. Regardless of size, however, all generators operate on the same basic principle—a magnetic field cutting through conductors or conductors passing through a magnetic field. They include: (1) a group of conductors in which the output voltage is generated, and (2) a second group of conductors through which direct current is passed to obtain an electromagnetic field of fixed polarity. The conductors in which the electromagnetic field originates are always referred to as the *field windings*.

In addition to the armature and field, there must also be motion between the two. To provide this, AC generators are built in two major assemblies, the *stator* and the *rotor*. The rotor rotates inside the stator. The revolving-field AC generator (see Figure 2.120) is the most widely used type. In this type of generator, direct current from a separate source is passed through windings on the rotor by means of slip rings and brushes. (Slip rings and brushes are adequate for the DC field supply because the power level in the field is much smaller than in the armature circuit.) This maintains a rotating electromagnetic field of fixed polarity. The rotating magnetic field, following the rotor, extends outward and cuts through the armature windings imbedded in the surrounding stator. As the rotor turns, AC voltages are induced in the

Key Point: In AC generators, frequency and electromagnetic wave cycles per second depend on how fast the rotor turns and the number of electromagnetic field poles. The voltage generated depends on the rotor speed, number of coils in the armature, and strength of the magnetic field.

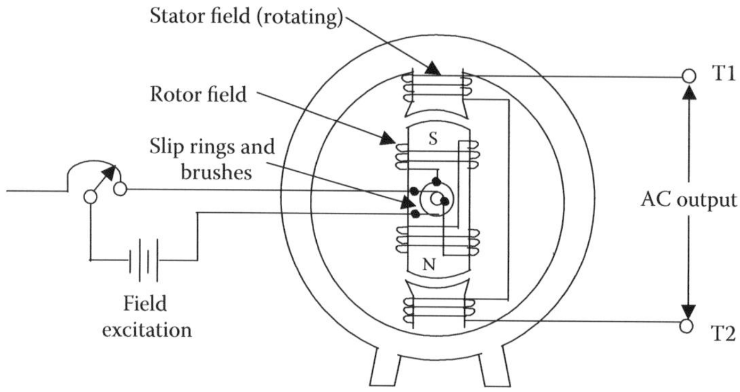

FIGURE 2.120 Essential parts of a rotating field AC generator.

windings because magnetic fields of first one polarity and then the other cut through them. The output power is taken from stationary windings and may be connected through fixed output terminals (T1 and T2 in Figure 2.120). This is advantageous, in that there are no sliding contacts and the whole output circuit is continuously insulated.

MOTORS

There is an almost endless variety of tasks that electric motors perform in the operation of industrial and domestic equipment. An *electric motor* is a machine used to change electrical energy to mechanical energy to do the work. (Recall that a generator does just the opposite; that is, a generator changes mechanical energy to electrical energy.) Previously, we pointed out that when a current passes through a wire a magnetic field is produced around the wire. If this magnetic field passes through a stationary magnetic field, the fields either repel or attract, depending on their relative polarity. If both are positive or negative, they repel. If they have opposite polarities, they attract. Applying this basic information to motor design, an electromagnetic coil, the armature, rotates on a shaft. The armature and shaft assembly are called the *rotor*. The rotor is assembled between the poles of a permanent magnet and each end of the rotor coil (armature) is connected to a commutator also mounted on the shaft. A commutator is composed of copper segments insulated from the shaft and from each other by an insulting material. As like poles of the electromagnet in the rotating armature pass the stationary permanent magnet poles, they are repelled, continuing the motion. As the opposite poles near each other, they attract, continuing the motion.

DC Motors

The construction of a DC motor is essentially the same as that of a DC generator; however, it is important to remember that the DC generator converts mechanical energy into the electrical energy back into mechanical energy. A DC generator may be made to function as a motor by applying a suitable source of DC voltage across the normal output electrical terminals. The various types of DC motors differ in the way their field coils are connected. Each has characteristics that are advantageous under given load conditions. *Shunt motors* (see Figure 2.121) have field coils connected in parallel with the armature circuit. This type of motor, with constant potential applied, develops variable torque at an essentially constant speed, even under changing load conditions. Such loads are found in machine-shop equipment, such as lathes, shapes, drills, milling machines, and so forth. *Series motors* (see Figure 2.122) have field coils connected in series with the armature circuit. This type of motor, with constant potential applied, develops variable torque but its speed varies widely under changing load conditions. The speed is low under heavy loads but becomes excessively high under light loads. Series motors are commonly used to drive electric hoists, winches, cranes, and certain types of vehicles (e.g., electric trucks). Also, series motors are used extensively to start internal combustion engines. *Compound motors* (see Figure 2.123) have one set of field coils in parallel with the armature circuit and another set of field coils in series with the armature circuit. This type of motor is a compromise between shunt and series motors. It develops an increased starting torque over that of the shunt motor and has less variation in speed than the series motor. The speed of a DC motor is variable. It is increased or decreased by a rheostat connected in series with the field or in parallel with the rotor. Interchanging either the rotor or field winding connections reverses direction.

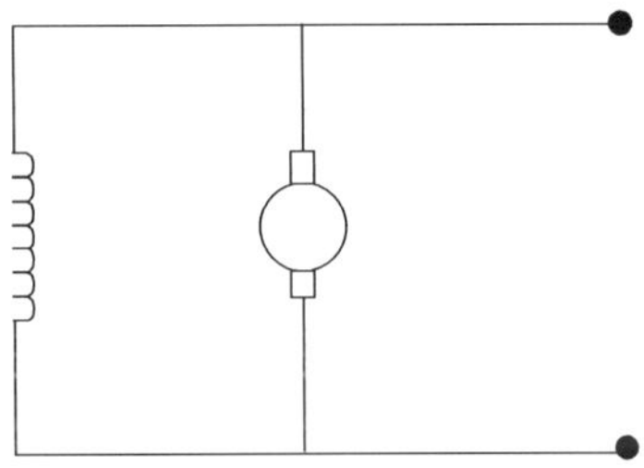

FIGURE 2.121 DC shunt motor.

FIGURE 2.122 DC series motor.

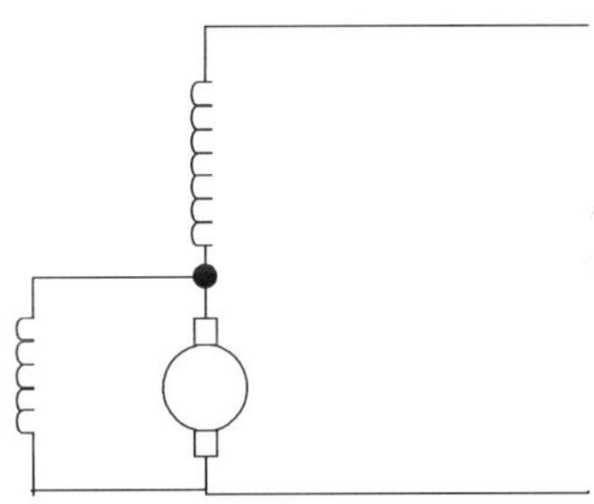

FIGURE 2.123 DC compound motor.

Brushless DC Motor

Brushless DC motors can be deployed in any area currently utilizing brushed DC motors. In renewable energy applications, the brushless DC motor can be found in such high-power applications as electric vehicles, hybrid vehicles, and some industrial machinery. The Segway® scooter and the Vectrix™ maxi-scooter use brushless DC motors. They are more commonly found in consumer devices, such as computer hard drives, CD/DVD players, and PC cooling fans. The brushless DC motor is a synchronous motor powered by DC current that has permanent magnets (usually rare-earth-type magnets) mounted on the spinning rotor; instead of a mechanical commutation system based on brushes, it has a commutation system that is electronically controlled by a solid-state controller. The commutator and brush system form a set of electrical switches, each opening and closing in sequence, such that electrical power always flows through the armature coil closest to the stationary stator.

AC Motors

Alternating current voltage can easily be transformed from low voltages to high voltages, or *vice versa*, and can be moved over a much greater distance without too much loss in efficiency. Most of the power-generating systems today, therefore, produce alternating current. Thus, it logically follows that a great majority of the electrical motors utilized today are designed to operate on alternating current. AC motors offer other advantages, though, in addition to the wide availability of AC power. In general, AC motors are less expensive than DC motors. Most types of AC motors do not employ brushes and commutators which eliminates many problems of maintenance and wear and eliminates dangerous sparking. AC motors are manufactured in many different sizes, shapes, and ratings for use on an even greater number of jobs. They are designed for use with either polyphase or single-phase power systems. This text cannot possibly cover all aspects of the subject of AC motors; consequently, it deals mainly with the operating principles of the two most common types—the induction motor and the synchronous motor.

Induction Motors (Polyphase)

The induction motor is the most common AC motor because of its simple, rugged construction and good operating characteristics. It consists of two parts: the *stator* (stationary part) and the *rotor* (rotating part). The most important type of polyphase induction motor is the three-phase motor.

Note: A three-phase system is a combination of three single-phase systems. In a three-phase, balanced system, the power comes from an AC generator that produces three separate but equal voltages, each of which is out of phase with the other voltages by 120°. Although single-phase circuits are widely used in electrical systems, most generation and distribution of AC current is three phase.

The driving torque of both DC and AC motors is derived from the reaction of current-carrying conductors in a magnetic field. In the DC motor, the magnetic field is stationary and the armature, with its current-carrying conductors, rotates. The current is supplied to the armature through a commutator and brushes. In induction motors, the rotor currents are supplied by electromagnet induction. The stator windings, connected to the AC supply, contain two or more out-of-time-phase currents, which produce corresponding magnetomotive force (mmf), establishing a rotating magnetic field across the air gap. This magnetic field rotates continuously at constant speed regardless of the load on the motor. The stator winding corresponds to the armature winding of a DC motor or to the primary winding of a transformer. The rotor is not connected electrically to the power supply. The induction motor derives its name from the fact that mutual induction (or transformer action) takes place between the stator and the rotor under operating conditions. The magnetic revolving field produced by the stator cuts across the rotor conductors, inducing a voltage in the conductors. This induced voltage causes rotor current to flow. Hence, motor torque is developed by the interaction of the rotor current and the magnetic revolving field.

Synchronous Motors

Like induction motors, synchronous motors have stator windings that produce a rotating magnetic field, but, unlike the induction motor, the synchronous motor requires a separate source of direct current from the field. It also requires special starting components, including a salient-pole field with starting grid winding. The rotor of the conventional type synchronous motor is essentially the same as that of the salient-pole AC generator. The stator windings of induction and synchronous motors are essentially the same. In operation, the synchronous motor rotor locks into step with the rotating magnetic field and rotates at the same speed. If the rotor is pulled out of step with the rotating stator field, no torque is developed and the motor stops. Because a synchronous motor develops torque only when running at synchronous speed, it is not self-starting and hence needs some device to bring the rotor to synchronous speed. A synchronous motor may be started by a DC motor on a common shaft. After the motor is brought to synchronous speed, AC current is applied to the stator windings. The DC starting motor now acts as a DC generator, which supplies DC field excitation for the rotor. The load then can be coupled to the motor.

Single-Phase Motors

Single-phase motors are so called because their field windings are connected directly to a single-phase source. These motors are used extensively in fractional horsepower sizes in commercial and domestic applications. The advantages of using single-phase motors in small sizes are that they are less expensive to manufacture than other types, and they eliminate the need for three-phase AC lines. Single-phase motors are used in fans, refrigerators, portable drills, grinders, and so forth. A single-phase induction motor with only one stator winding and a cage rotor is like a three-phase induction motor with a cage rotor except that the single-phase motor has no magnetic revolving field at start and hence no starting torque. However, if the rotor is brought up to speed by external means, the induced currents in the rotor will cooperate with the stator currents to produce a revolving field, which causes the rotor to continue to run in the direction in which it was started. Several methods

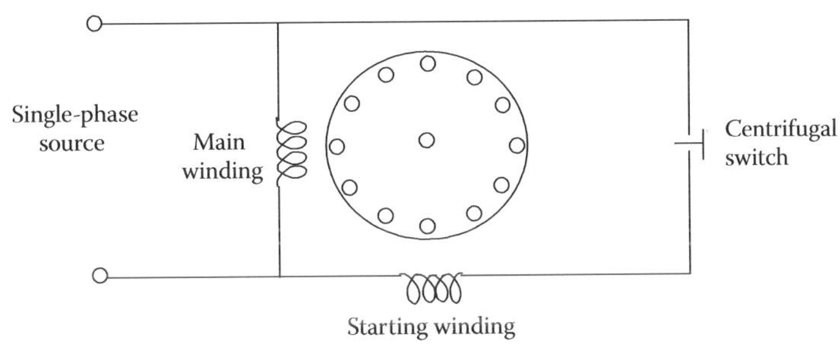

FIGURE 2.124 Split-phase motor.

are used to provide the single-phase induction motor with starting torque. These methods categorize the motor as split-phase, capacitor, shaded-pole, or repulsion-start. Another class of single-phase motors is the AC series (universal) type.

Split-Phase Motor

The split-phase motor (Figure 2.124), has a stator composed of slotted lamination that contains a starting winding and a running winding. The starting winding has fewer turns and smaller wire than the running winding, hence higher resistance and less reactance. The main winding occupies the lower half of the slots and the starting winding occupies the upper half. When the same voltage is applied to both windings, the current in the main winding lags behind the current in the starting winding. The angle (θ) between the main winding and the starting winding is enough phase difference to provide a weak rotating magnetic field that produces a starting torque. When the motor reaches a predetermined speed, usually 75% of synchronous speed, a centrifugal switch mounted on the motor shaft opens, thereby disconnecting the starting winding. Because of their low starting torque, fractional-horsepower, split-phase motors are used in a wide variety of equipment, such as washers, oil burners, ventilating fans, and woodworking machines. The direction of rotation of the split-phase motor can be reversed by interchanging the starting winding leads.

Note: If two stator windings of unequal impedance are spaced 90° apart but connected in parallel to a single-phase source, the field produced will appear to rotate. This is the principle of *phase splitting.*

Capacitor Motor

The capacitor motor is a modified form of split-phase motor. Its capacitor is in series with the starting winding (see Figure 2.125). The capacitor motor operates with an auxiliary winding and series capacitor permanently connected to the line (see Figure 2.125). The capacitance in series may be of one value for starting and another value for running. As the motor approaches synchronous speed, the centrifugal switch disconnects one section of the capacitor. If the starting winding is cut out after the motor has increased in speed, the motor is a *capacitor-start motor.* If the starting winding and capacitor are designed to be left in the circuit continuously, the motor is a *capacitor-run motor.*

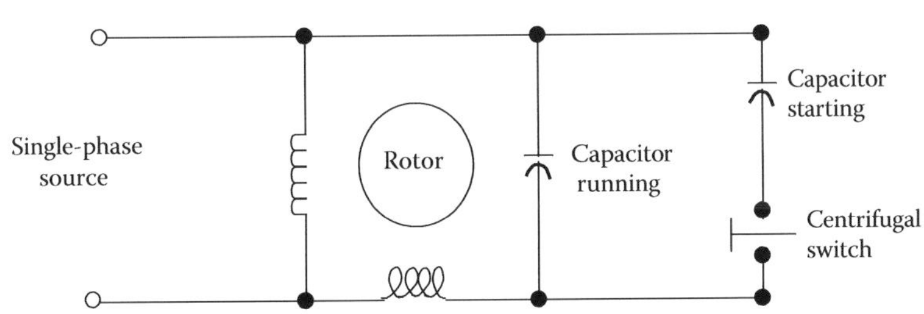

FIGURE 2.125 Capacitor motor.

Capacitor motors are used to drive grinders, drill presses, refrigerator compressors, and other loads that require relatively high starting torque. The direction of rotation of the capacitor motor can be reversed by interchanging the starting winding leads.

Shaded-Pole Motor

A shaded-pole motor employs a salient-pole stator and a cage rotor. The projecting poles on the stator resemble those of DC machines except that the entire magnetic circuit is laminated and a portion of each pole is split to accommodate a short-circuited coil called a *shading coil* (Figure 2.126). The coil is usually a single band or strap of copper. The effect of the coil is to produce a small sweeping motion of the field flux from one side of the pole piece to the other as the field pulsates. This slight shift in the magnetic field produces a small starting torque. Thus, shaded-pole motors are self-starting. This motor is generally manufactured in very small sizes, up to 1/20 horsepower, for driving small fans, small appliances, and clocks.

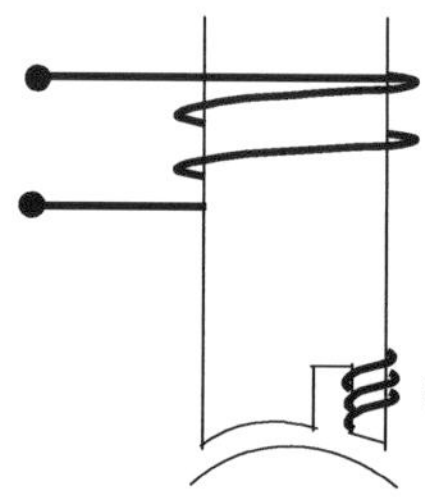

FIGURE 2.126 Shaded pole.

In operation, during that part of the cycle when the main pole flux is increasing, the shading coil is cut by the flux, and the resulting induced emf and current in the shading coil tend to prevent the flux from rising readily through it. Thus, the greater portion of the flux rises in that portion of the pole that is not in the vicinity of the shading coil. When the flux reaches its maximum value, the rate of change of flux is zero, and the voltage and current in the shading coil are also zero. At this time, the flux is distributed more uniformly over the entire pole face. Then, as the main flux decreases toward zero, the induced voltage and current in the shading coil reverse their polarity, and the resulting mmf tends to prevent the flux from collapsing through the iron in the region of the shading coil. The result is that the main flux first rises in the unshaded portion of the pole and later in the shaded portion. This action is equivalent to a sweeping movement of the field across the pole face in the direction of the shaded pole. The rotor conductors are cut by this moving field, and the force exerted on them causes the rotor to turn in the direction of the sweeping field. The shaded-pole method of starting is used in very small motors, up to about 1/25 hp, for driving small fans, small appliances, and clocks.

Repulsion-Start Motor

Like a DC motor, the repulsion-start motor has a form-wound rotor with commutator and brushes. The stator is laminated and contains a distributed single-phase winding. In its simplest form, the stator resembles that of the single-phase motor. In addition, the motor has a centrifugal device that removes the brushes from the commutator and places a short-circuiting ring around the commutator. This action occurs at about 75% of synchronous speed. Thereafter, the motor operates with the characteristics of the single-phase induction motor. This type of motor is made in sizes ranging from 1/2 to 15 hp and is used in applications requiring a high starting torque.

AC Series Motor

The AC series motor operates on either AC or DC circuits. When an ordinary DC series motor is connected to an AC supply, the current drawn by the motor is low due to the high series-field impedance. The result is low running torque. To reduce the field reactance to a minimum, AC series motors are built with as few turns as possible. Armature reaction is overcome by using *compensating windings* (see Figure 2.127) in the pole pieces. As in DC series motors, the speed in AC series motors increases to a high value with a decrease in load. The torque is high for high armature currents, so the motor has a good starting torque. AC series motors operate more efficiently at low frequencies. Fractional horsepower AC series motors are called *universal motors*. They do not have compensating windings. They are used extensively to operate fans and portable tools, such as drills, grinders, and saws.

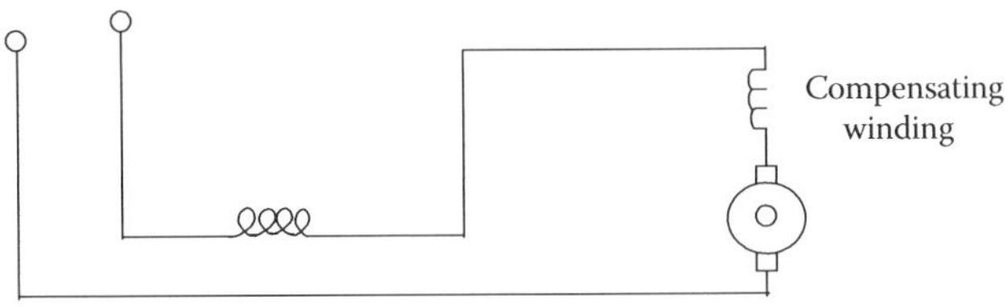

FIGURE 2.127 AC series motor.

TRANSFORMERS

A transformer is an electric control device (with no moving parts) that raises or lowers voltage or current in an electric distribution system. The basic transformer consists of two coils electrically insulated from each other and wound on a common core (Figure 2.128). Magnetic coupling is used to transfer electric energy from one coil to another. The coil that receives energy from AC sources is the primary coil. The coil that delivers energy to AC loads is the secondary coil. The core of transformers used at low frequencies is generally made of magnetic material, usually laminated sheet steel. Cores of transformers used at higher frequencies are made of powdered iron and ceramics, or nonmagnetic materials. Some coils are simply wound on nonmagnetic hollow forms such as cardboard or plastic so the core material is actually air.

In operation, an alternating current will flow when an AC voltage is applied to the primary coil of a transformer. This current produces a field of force that changes as the current changes. The changing magnetic field is carried by the magnetic core to the secondary coil, where it cuts across the turns of that coil. In this way, an AC voltage in one coil is transferred to another coil, even though there is no electrical connection between them. The number of lines of force available in the primary is determined by the primary voltage and the number of turns on the primary, with each turn producing a given number of lines. If there are many turns on the secondary, each line of force will cut many turns of wire and induce a high voltage. If the secondary contains only a few turns, there will be few cuttings and low induced voltage. The secondary voltage, then, depends on the number of secondary turns as compared with the number of primary turns. If the secondary has twice as many turns as the primary, the secondary voltage will be twice as large as the primary voltage. If the secondary has half as many turns as the primary, the secondary voltage will be one-half as large as the primary voltage. A voltage ratio of 1:4 means that for each volt on the primary there are 4 volts on the secondary. This is called a *step-up* transformer. A step-up transformer receives a low voltage on the primary coil and delivers a high voltage from the secondary coil. In contrast, a voltage ratio of 4:1 means that for 4 volts on the primary coil there is only 1 volt on the secondary. This is called a *step-down transformer.* A step-down transformer receives a high voltage on the primary coil and delivers a low voltage from the secondary.

Key Point: The voltage on the coils of a transformer is directly proportional to the number of turns on the coils.

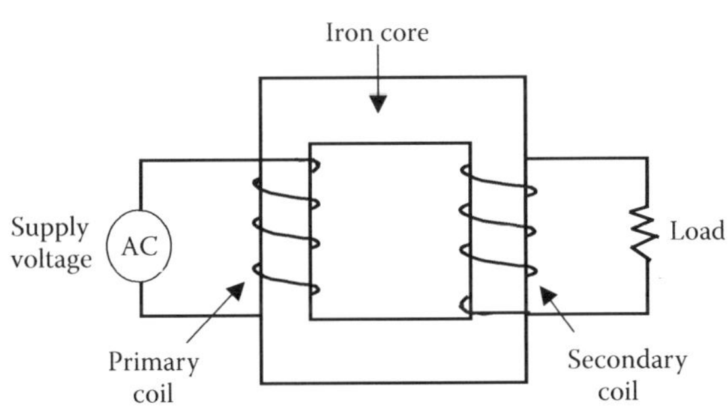

FIGURE 2.128 Basic transformer.

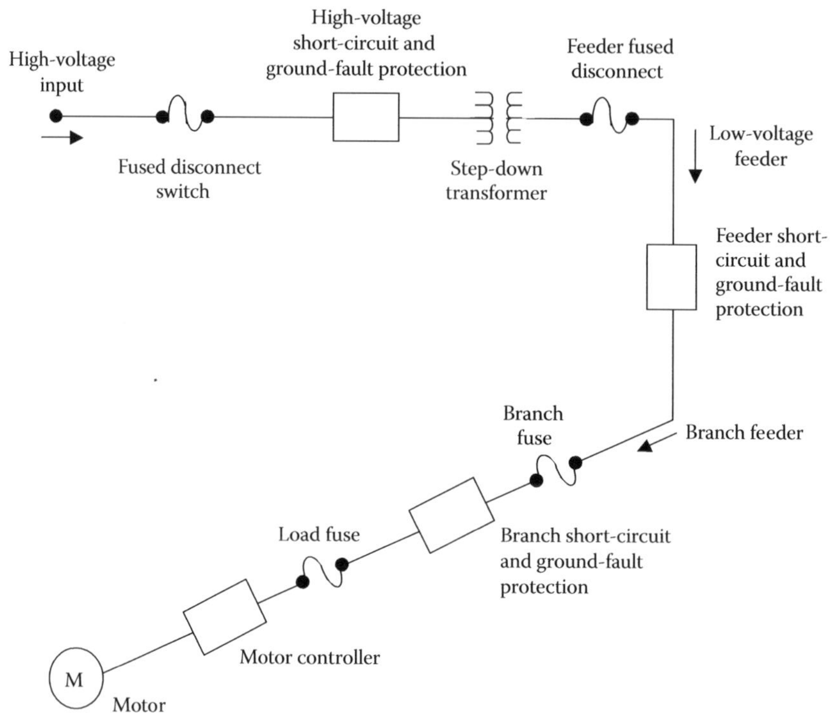

FIGURE 2.129 Motor power distribution system.

POWER DISTRIBUTION SYSTEM PROTECTION

Interruptions are very rare in a power distribution system that has been properly designed. Still, protective devices are necessary because of the load diversity. Most installations are quite complex. In addition, externally caused variations might overload them or endanger personnel. Figure 2.129 shows the general relationship between protective devices and different components of a complete system. Each part of the circuit has its own protective device or devices that protect not only the load but also the wiring and control devices themselves. These disconnect and protective devices are described in the following sections.

FUSES

The passage of an electric current produces heat. The larger the current, the more heat is produced. To prevent large currents from accidentally flowing through expensive apparatus and burning it up, a fuse is placed directly into the circuit, as in Figure 2.129, so as to form a part of the circuit through which all the current must flow. The fuse will permit currents smaller than the fuse value to flow but will melt and therefore break the circuit if a larger, dangerous current ever appears; for example, a dangerously large current will flow when a short circuit occurs. A short circuit is usually caused by an accidental connection between two points in a circuit which offer very little resistance to the flow of electrons. If the resistance is small, there will be nothing to stop the flow of the current, and the current will increase enormously. The resulting heat generated might cause a fire. If the circuit is protected by a fuse, the heat caused by the short-circuit current will melt the fuse wire, thus breaking the circuit and reducing the current to zero.

Key Point: A fuse is a thin strip of easily melted material. It protects a circuit from large currents by melting quickly, thereby breaking the circuit.

Fuses are rated by the number of amps of current that can flow through them before they melt and break the circuit; thus, a fuse can be, for example, 10, 15, 20, or 30 amps. We must be sure that any fuse inserted in a circuit is rated low enough to melt, or "blow," before the apparatus is damaged. In a plant building wired to carry a current of 10 A, it is best to use a fuse no larger than 10 A so a current larger than 10 A could never flow. Some equipment, such as an electric motor, requires more current during starting than for normal running; fast-time or medium-time fuses that provide running protection might blow during the initial period when high starting current is required. *Delayed-action* fuses are used to handle these situations.

Circuit Breakers

Circuit breakers are protective devices that open automatically at a preset ampere rating to interrupt an overload or short circuit. Unlike fuses, they do not require replacement when they are activated. They are simply reset to restore power after the overload has been cleared. Circuit breakers are made in both plug-in and bolt-on designs. Plug-in breakers are used in load centers. Bolt-on breakers are used in panelboards and exclusively for high interrupting current applications. Circuit breakers are rated according to current and voltage, as well as short-circuit-interrupting current. A single handle opens or closes contacts between two or more conductors. Breakers are single pole, but single-pole units can be ganged to form double- or triple-pole devices opened with a single handle.

Key Point: A circuit breaker is designed to break the circuit and stop the current flow when the current exceeds a predetermined value.

Several types of circuit breakers are commonly used. They may be thermal or magnetic, or a combination of the two. Thermal breakers are tripped when the temperature rises because of heat created by the overcurrent condition. Bimetallic strips provide the time delay for overload protection. Magnetic breakers operate on the principle that a sudden current rise will create enough magnetic field to turn an armature, tripping the breaker and opening the circuit. Magnetic breakers provide the instantaneous action needed for short-circuit protection. Magnetic breakers are also used where ambient temperatures might adversely affect the action of a thermal breaker. Thermal–magnetic breakers combine features of both types. An important feature of the circuit breaker is its arc chutes, which allow the breaker to extinguish very hot arcs harmlessly. Some circuit breakers must be reset by hand, while others reset themselves automatically. If the overload condition still exists when the circuit breaker is reset, the circuit breaker will trip again to prevent damage to the circuit.

Control Devices

Control devices are electrical accessories (switches and relays) that govern the power delivered to any electrical load. In its simplest form, the control applies voltage to, or removes it from, a single load. In more complex control systems, the initial switch may set into action other control devices (relays) that govern motor speeds, servomechanisms, temperatures, and numerous other pieces of equipment. In fact, all electrical systems and equipment are controlled in some manner by one or more controls. A controller is a device or group of devices that serves to govern, in some predetermined manner, the device to which it is connected. In large electrical systems, it is necessary to have a variety of controls for operation of the equipment. These controls range from simple push buttons to heavy-duty contactors that are designed to control the operation of large motors. The push button is manually operated; a contactor is electrically operated.

ELECTRICAL DRAWINGS

Working drawings for the fabrication and troubleshooting of electrical machinery, switching devices, and chassis for electronic equipment, cabinets, housings, and other mechanical elements associated with electrical equipment are based on the same principles as given earlier. Those qualified to

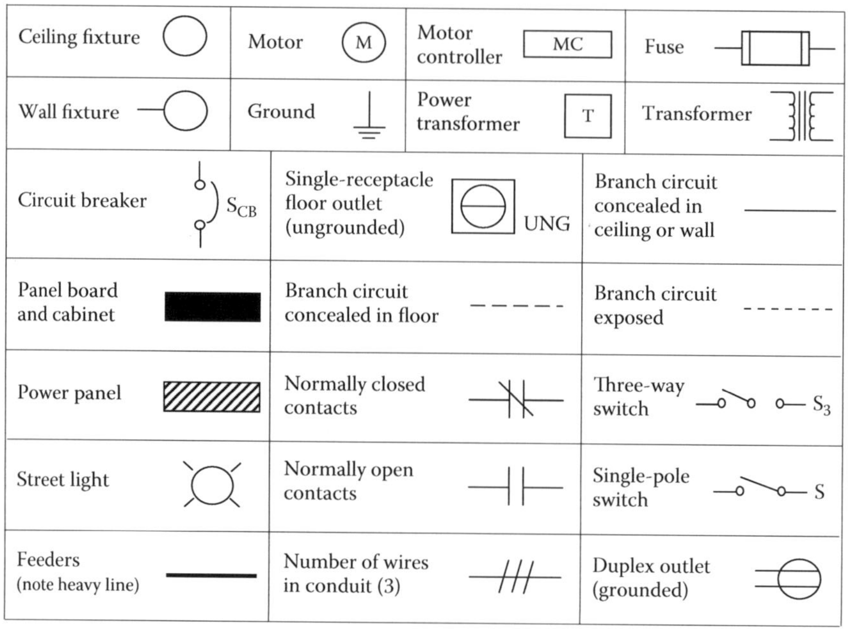

FIGURE 2.130 Common electrical symbols.

operate, maintain, and repair electrical equipment must understand electrical systems. The electrician or student of electricity must be able to read electrical drawings to understand the system and to determine what is wrong when electrical equipment fails to run properly. This section introduces electrical drawings, and the functions of important electrical components and how they are shown on drawings are explained.

Electrical Symbols

Figure 2.130 shows some of the most common symbols used on electrical drawings. It is not necessary to memorize these symbols, but the maintenance operator should be familiar with them as an aid to reading electrical drawings.

Types of Electrical Drawings

Key Point: A circuit drawing does not indicate the physical location of any load or circuit.

The two kinds of electrical drawings used for troubleshooting operations are *architectural drawings* and *circuit drawings*. An architectural drawing shows the physical locations of the electric lines in a plant building or between buildings. A circuit drawing shows the electrical loads served by each circuit.

Architectural Drawings

Figure 2.131 shows three types of architectural drawings. The *plot plan* shows the electric distribution to all the plant buildings. The *floor plan* shows where branch circuits are located in one building or pumping station, where equipment is located, and where outside and inside tie-ins to water, heat, and electric power are located. The *riser diagram* shows how the wiring goes to each floor of the building.

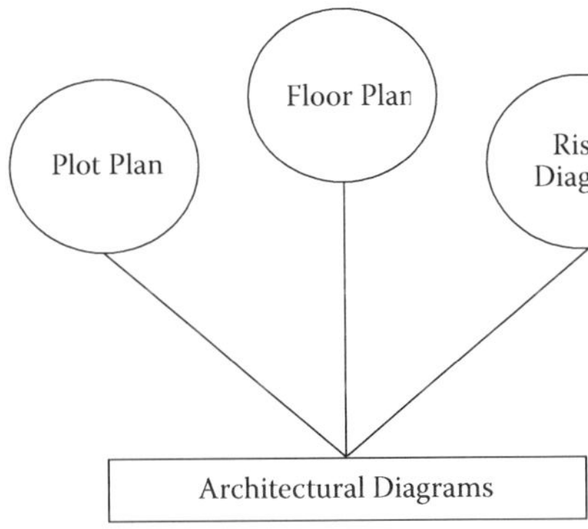

FIGURE 2.131 Types of architectural diagrams.

Circuit Drawings

A circuit drawing shows how a single circuit distributes electricity to various loads (e.g., pump motors, grinders, bar screens, mixers). Unlike an architectural drawing, a circuit drawing does not show the location of these loads. Figure 2.132 depicts a typical single-line circuit drawing and shows the power distribution to 11 loads. The number in each circle indicates the power rating of the loads in horsepower.

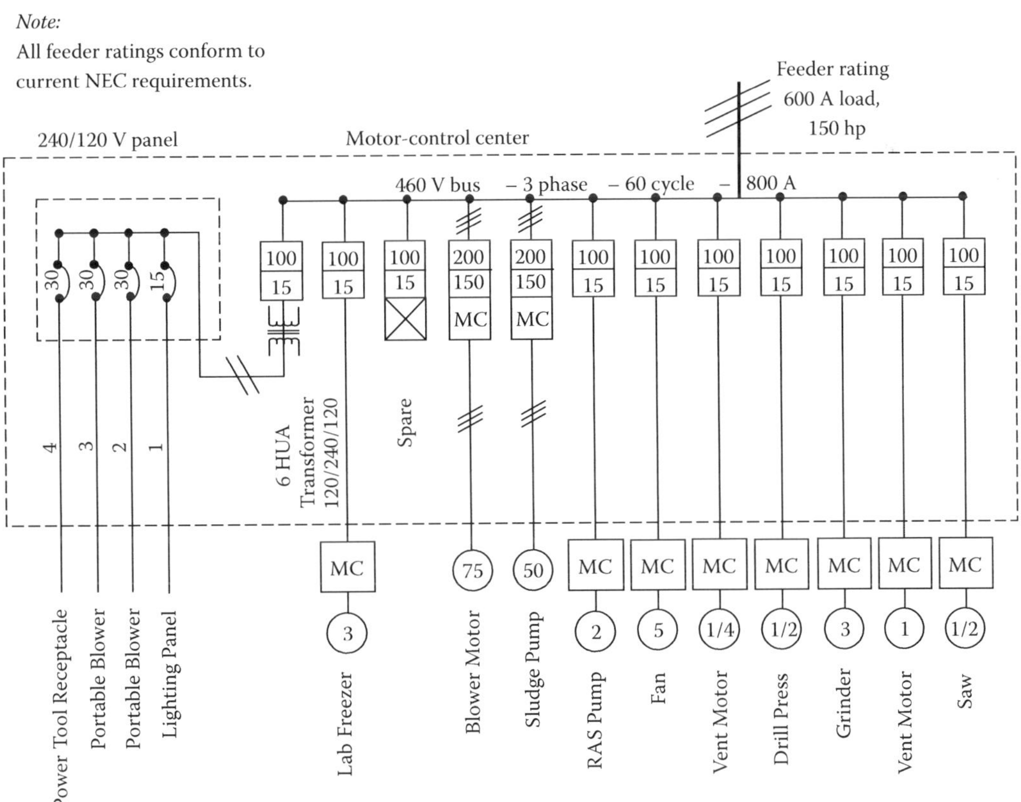

FIGURE 2.132 Single-line circuit diagram.

Note: Electrical loads in all plants can be divided into two categories: critical and noncritical. Critical loads are essential to the operation of the plant and cannot be turned off (e.g., critical unit processes). Noncritical loads include pieces of equipment that would not disrupt the operation of the plant or pumping station or compromise safety if they were turned off for a short period of time (e.g., air conditioners, fan systems, electric water heaters, certain lighting systems).

The numbers in the rectangles show the current ratings of circuit breakers. The upper number is the current in amps that the circuit breaker will allow as a momentary surge.

Note: Circuit breakers are typically equipped with surge protection for three-phase motors and other devices. When a three-phase motor is started, current demand is six to ten times normal value. After start, current flow decreases to its normal rated value. Surge protection is also provided to allow slight increases in current flow when the load varies or increases slightly. The lower number is the maximum current the circuit breaker will allow to flow continuously. Most circuit breakers are also equipped with an instantaneous trip value for protection against short circuits.

Ladder Drawing

A ladder drawing is a type of schematic diagram that shows a control circuit. The parts of the control circuit lie on horizontal lines, like the rungs of a ladder. Figure 2.133 is an example of a ladder diagram. The purpose of a ladder drawing, such as the one shown in Figure 2.133, is to cut maintenance and troubleshooting time. This is accomplished when the designer follows certain guidelines when making electrical drawings and layouts. Let's take a closer look at the ladder drawing for a control circuit shown in Figure 2.133. Note the numbering of elementary circuit lines. Normally, closed contacts are indicated by a bar under the line number. Moreover, note that the line numbers are enclosed in a geometric figure to prevent mistaking the line numbers for circuit numbers. All contacts and the conductors connected to them are properly numbered. Typically, numbering is carried throughout the entire electrical system. This may involve going through one or more terminal blocks. The incoming and outgoing conductors as well as the terminal blocks carry the proper electrical circuit numbers. When possible, connections to all electrical components are taken back to one common checkpoint. All electrical elements on a machine should be correctly identified with the same markings as shown on the ladder drawing in Figure 2.133; for example, if a given solenoid is marked "SOL A2" on the drawing, the actual solenoid on the machine should carry the marking "SOL A2."

Key Point: The size of electrical drawings is important. This can be understood first in the problem of storing drawings. If every size and shape were allowed, the task of systematic and protective filing of drawings could be tremendous. Pages that are 8-1/2 × 11 or 9 × 12 inches and multiples thereof are generally accepted. The drawing size can also be a problem for the troubleshooter or maintenance operator. If the drawing is too large, it is unwieldy to handle at the machine. If it is too small, it is difficult to read the schematic.

DISCUSSION AND REVIEW QUESTIONS

1. Explain the difference between a series and a parallel circuit.
2. What is the sum of all voltages in a series circuit is equal to?
3. For any total voltage rise in a circuit, there must be an equal total ______________.

Refer to the figure for questions 4 through 8.

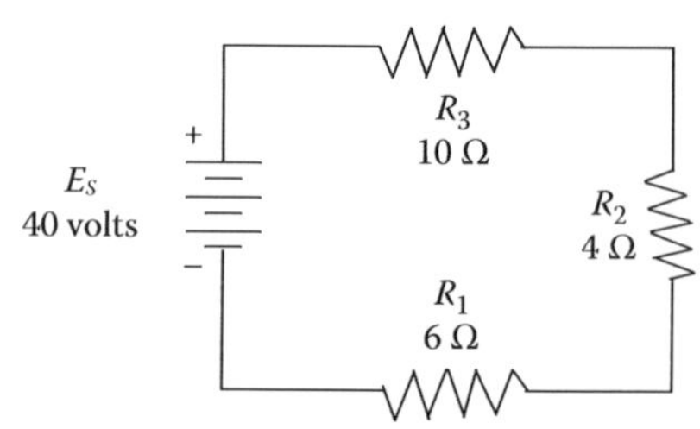

4. Is the direction of current flow clockwise or counterclockwise?
5. $I_T = ?$
6. E dropped across $R_1 = ?$
7. What is the power absorbed by R_2?
8. $P_T = ?$

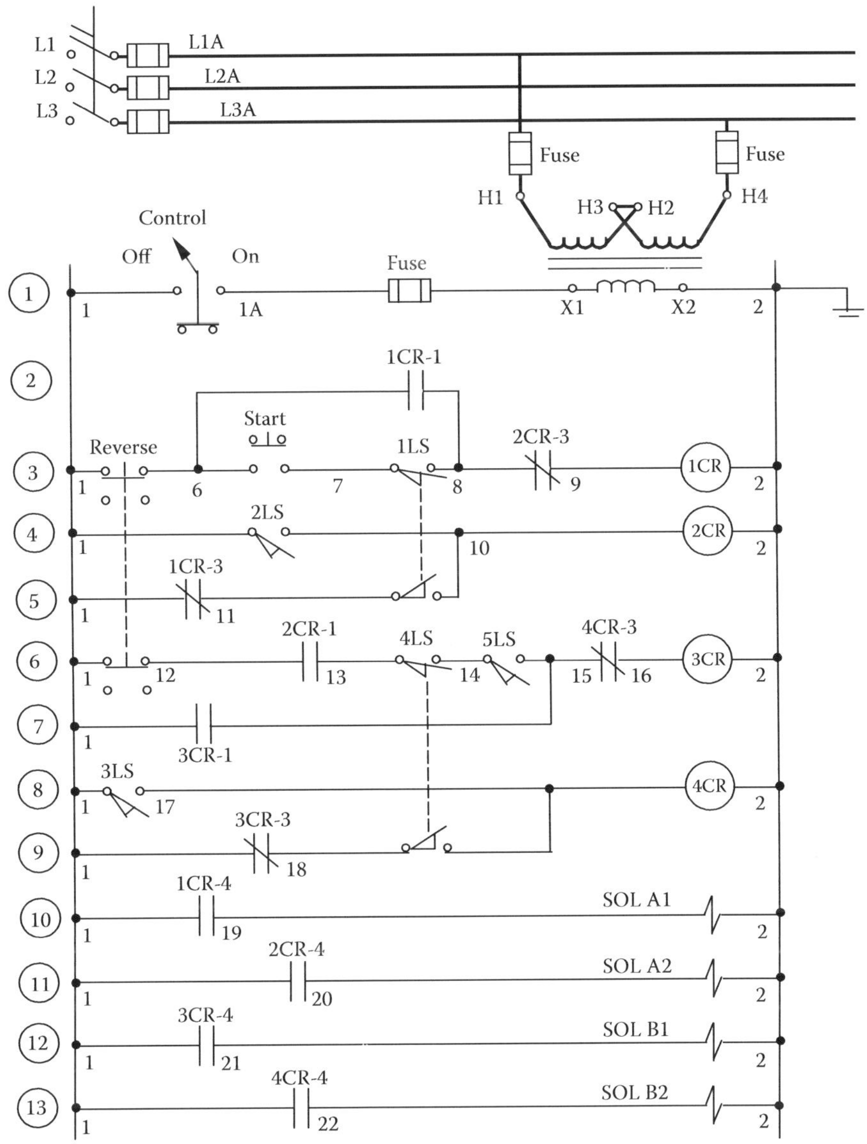

FIGURE 2.133 Typical ladder drawing.

9. The equivalent resistance (R_T) of parallel branches is __________ than the smallest branch resistance because all of the branches must take __________ current from the source than any one branch.
10. A short circuit has __________ resistance, resulting in __________ current.
11. What is the total resistance of four 30-ohm resistors connected in series?
12. There is only __________ voltage across all components in parallel.
13. The sum of the __________ values of power dissipated in parallel resistances equals the __________ power produced by the source.

Refer to the figure for questions 14 through 16.

14. $R_T = ?$
15. $I_T = ?$
16. $P_T = ?$

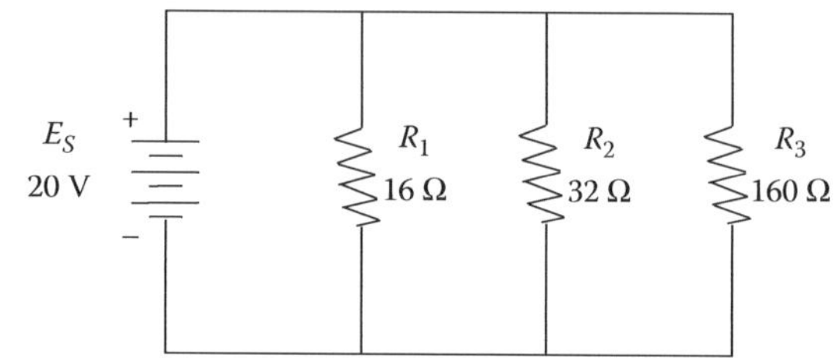

Refer to the figure for questions 17 through 20.

17. What is the total resistance (R_T) of the circuit?
18. What is the total current (I_T) in the circuit?
19. What is the voltage across R_3?
20. What is the total power consumed in the circuit?

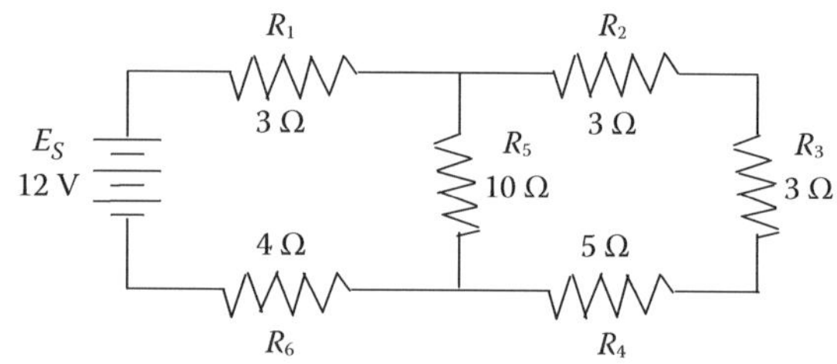

REFERENCES AND RECOMMENDED READING

Callen, H.B. (1985). *Thermodynamics and an Introduction to Thermostatistics*, 2nd ed. New York: John Wiley & Sons.

Feynman, R.P., Leighton, R.B., and Sands, M. (1963). *The Feynman Lectures on Physics*. New York: Addison-Wesley.

Halliday, D. and Resnick, R. (1978). *Physics*, 3rd ed. New York: John Wiley & Sons.

Hinrichs, R.A. and Kleinbach, M. (2006). *Energy: Its Use and the Environment*, 4th ed. Belmont, CA: Brooks & Cole.

Kroemer, H. and Kittel, C. (1980). *Thermal Physics*, 2nd ed. New York: W.H. Freeman.

National Renewable Energy Laboratory. (2009). *Energy Storage: Batteries*. Washington, D.C.: U.S. Department of Energy (http://www.nrel.gov/vehiclesandfuels/energystorage/batteries.html).

Organ, A.J. (1992). *Thermodynamics and Gas Dynamics of the Stirling Cycle Machine*. London: Cambridge University Press.

Organ, A.J. (1997). *The Regenerator and the Stirling Engine*. New York: John Wiley & Sons.

Spellman, F.R. and Drinan, J.E. (2001). *Electricity*. Boca Raton, FL: CRC Press.

USDOE. (1992). *DOE Fundamentals Handbook: Electrical Science*, Vol. 1. Washington, D.C.: U.S. Department of Energy.

U.S. Energy Information Administration. (2008). *Use of Energy in the United States Explained*. Washington, D.C.: U.S. Department of Energy (http://tonto.eia.doe.gov/energyexplained/print.cfm?page=us_energy_use).

U.S. Energy Information Administration. (2009). *What Is Energy? Explained*. Washington, D.C.: U.S. Department of Energy (http://tonto.eia.doe.gov/energyexplained/print.cfm?page=about_sources_of_energy).

3 Solar Energy

One generation passes away, and another generation cometh: but the Earth abides forever. The sun also rises.

—Ecclesiastes 1:4–5

Busy old fool, unruly Sun,
Why dost thou thus,
Through windows and through curtains, call on us?

—John Dunne ("The Sun Rising," 1896)

The Sun, with all the planets revolving around it, and depending on it, can still ripen a bunch of grapes as though it had nothing else in the universe to do.

—Galileo Galilei

Solar panel, Zion National Park, Utah. (Photo by Frank R. Spellman.)

Solar One Nevada, Boulder City, Nevada. (Photo by Frank R. Spellman.)

INTRODUCTION

It is fitting to begin our discussion of the various kinds of renewable energy with the sun—the medium-sized star that symbolizes life, power, strength, force, clarity, and, yes, energy. The sun nourishes our planet. Solar energy is a renewable resource because it is continuously supplied to the Earth by the sun. When we consider the sun and solar energy, we quickly realize that there is nothing new about renewable energy. The sun was the first energy source; it has been around for 4.5 billion years, as long as anything else we are familiar with. On Earth, without the sun there would be nothing—absolutely nothing, zilch. The old 5-5-5 rule sums up the essence of life on Earth: We can survive 5 minutes without oxygen, 5 days without water, and 5 weeks without food. Without the sun, though, the 5-5-5-rule has no validity. Without the sun, nothing related to life has validity. Without the sun, there is nothing: no air, no water, no food…no life. The sun, the ultimate energy source, provided light and heat to the first humans. During daylight, people searched for food. They hunted and gathered and probably stayed together for their safety. When nightfall arrived, we can only imagine that they huddled together for warmth and reassurance under the light of the stars and moon, waiting for the sun and its live-giving and life-sustaining light to return.

Solar energy (a term used interchangeably with solar power) uses various technologies to take advantage of the power of the sun to produce energy. Solar energy is one of the best renewable energy sources available because it is one the cleanest sources of energy. Direct solar radiation absorbed in solar collectors can provide space heating and hot water. Passive solar can be used to enhance the solar energy used in buildings for space heating and lighting requirements. Solar energy can also be used to produce electricity, and this is the type of renewable energy that we focus on in this section. Radiant energy from the sun, in the form of photons, strikes the surface of the Earth with the average equivalent of about 168 kWh of energy (equal to 575,000 Btu thermal energy) per square foot per year; it varies, of course, with location, cloud cover, and orientation with the surface (Hanson 2004). The question, then, becomes one of how much of this energy can be used by consumers in the United States. Table 3.1 shows the energy consumption by energy source for 2008; in that year, solar energy accounted for 0.097 quadrillion Btu, a figure that is expected to steadily increase, which should be reflected in later figures when they are released. Not only do solar technologies diversify the energy supply but they also reduce the country's dependence on imported fuel. Moreover, solar energy technologies have a beneficial environmental impact in that they help to improve air quality and offset greenhouse gas emissions. An additional (and significant) benefit of a growing solar technology industry is that it stimulates our economy by creating jobs in solar manufacturing and installation. The two solar electric technologies with the greatest potential (based on cost-effectiveness) are concentrating solar power (CSP) and photovoltaics (PV) (USDOE, 2010b).

CONCENTRATING SOLAR POWER*

Concentrating solar power (CSP) offers a utility-scale, reliable, firm, dispatchable, renewable energy option that can help meet a nation's demand for electricity. Nine trough power plants producing more than 400 megawatts (MW) of electricity have been operating reliably in the California Mojave Desert since the 1980s (USDOE, 2009c). CSP plants produce power by first using mirrors to concentrate and focus sunlight onto a thermal receiver, similar to a boiler tube. The receiver absorbs and converts sunlight into heat. Ultimately, this high-temperature fluid is used to spin a turbine or power an engine that drives a generator that produces electricity. Concentrating solar power systems can be classified by how they collect solar energy: linear concentrators, dish/engine systems, or power tower systems (NREL, 2010a). All of these are typically engineered with tracking devices for following the sun, both seasonally and throughout the day, to maximize the electrical output of the system (Chiras, 2002).

* Much of the information in this section is from USDOE, *Concentrating Solar Power: Energy from Mirrors*, U.S. Department of Energy, Washington, D.C., 2001.

TABLE 3.1
U.S. Energy Consumption by Energy Source, 2008

Energy Source	Energy Consumption (quadrillion Btu)
Total	99.438
Fossil fuels (coal, natural gas, petroleum)	83.532
Electricity net imports	0.113
Nuclear electric power	8.427
Renewable	7.367
Biomass (total)	3.852
Biofuels	1.372
Waste	0.436
Wood and wood-derived fuels	2.044
Geothermal energy	0.360
Hydroelectric conventional	2.512
Solar thermal/PV energy	**0.097**
Wind energy	0.546

Source: EIA, *U.S. Energy Consumption by Energy Source*, Energy Information Administration, Washington, D.C., 2007 (www.eia.doe.gov/cneaf/solar.renewables/page/trends/table1.html).

Linear Concentrators

Linear CSP collectors capture or collect the sun's energy using long, rectangular, curved (U-shaped) mirrors. The mirrors, tilted toward the sun, focus sunlight on tubes (or receivers) that run the length of the mirrors. The reflected sunlight heats a fluid flowing through the tubes. The hot fluid then is used to boil water to create superheated steam that spins a conventional steam-turbine generator to produce electricity. Alternatively, steam can be generated directly in the solar field, eliminating the need for costly heat exchangers. Linear concentrating collector fields consist of a large number of collectors in parallel rows that are typically aligned in a north–south orientation to maximize both annual and summertime energy collection. A single-axis sun-tracking system allows the mirrors to track the sun from east to west during the day, ensuring that the sun reflects continuously onto the receiver tubes (USDOE, 2009a). The two major types of linear concentrator systems are *parabolic trough systems*, where receiver tubes are positioned along the focal line of each parabolic mirror, and *linear Fresnel reflector systems*, where one receiver tube is positioned above several mirrors to allow the mirrors greater mobility in tracking the sun.

Parabolic Trough Systems

Parabolic trough systems have been developed in the United States, Spain, and Japan (Chiras, 2002); in the United States today, the predominant CSP systems in operation are linear concentrators using parabolic trough collectors (USDOE, 2009a). Parabolic trough systems have long, curved arrays of mirrors (usually coated silver or polished aluminum). All rays of light that enter parallel to the axis of a parabolic-shaped mirror will be reflected to one point, the focus; hence, it is possible to concentrate virtually all of the radiation incident upon the mirror in a relatively small area (Perez-Blanco, 2009). The receiver tube (Dewar tube) runs the length of the mirror and is positioned along the focal line of each parabola-shaped reflector (Duffie and Beckman, 1991; Patel, 1999). The tube is fixed to the mirror structure, and the heated fluid—either a heat-transfer fluid (usually oil) or water or steam—flows through and out of the field of solar mirrors to where it will be used to create steam; in the case of a water/steam receiver, the flow is sent directly to the turbine (i.e., the turbine is the prime mover of the electrical generator that produces electrical current flow in an accompanying electrical distribution system). Temperatures in these systems range from 150 to 750°F (80 to 400°C). Currently, the largest individual trough systems can generate 80 MW of electricity; however, individual systems being developed will generate 250 megawatts. In addition, individual systems can be collocated in power parks. The potential capacity of such power parks would be constrained only by transmission capacity and the availability of contiguous land area. Trough designs can incorporate thermal energy storage (discussed in more detail later). In a thermal storage system, the collector field is oversized to heat a storage system during the day that can be used in the evening or during cloudy weather to generate additional steam to produce electricity.

Parabolic trough plants can be designed as hybrids, meaning that they use fossil fuel to supplement the solar output during periods of low solar radiation. In such a design, a natural-gas-fired heater or gas steam boiler/reheater is used. In the future, troughs may be integrated with existing or new combined-cycle, natural gas and coal-fired plants (USDOE, 2009a).

Fresnel Reflector Systems

Another type of linear concentrator technology is the Fresnel reflector system. Flat or slightly curved mirrors mounted on trackers on the ground are configured to reflect sunlight onto a water-filled receiver tube fixed in space above these mirrors. A small parabolic mirror is sometimes added atop the receiver to further focus the sunlight. The key advantage of Fresnel reflectors systems is their simplicity as compared to other systems.

Dish/Engine Systems

Dish/engine systems use a mirrored, dish-shaped, parabolic mirror similar to a very large satellite dish. The dish-shaped surface directs and concentrates sunlight onto a thermal receiver (the interface between the dish and the engine or generator), which absorbs and collects the heat and transfers it to the engine generator. Solar dish/engine systems convert the energy from the sun into electricity at a very high efficiency. The power conversion unit includes the thermal receiver and the engine or generator. It absorbs the concentrated beams of solar energy, converts them to heat, and transfers the heat to the engine or generator. A thermal receiver can be a bank of tubes with a cooling fluid—usually hydrogen or helium—that typically is the heat-transfer medium and also the working fluid for an engine. Alternative thermal receivers are heat pipes, where the boiling and condensing of an intermediate fluid transfer the heat to the engine. The most common type of heat engine used today in dish/engine systems is the Stirling engine (conceived in 1816). This system uses the fluid heated by the receiver to move pistons and create mechanical power. The mechanical power is then used to run a generator or alternator to produce electricity (USDOE, 1998). Disk/engine systems use dual-axis collectors to track the sun. As mentioned, the ideal concentrator is parabolic in shape, with either a single reflective surface or multiple reflectors, or facets. Many options exist for receiver and engine types, including the Stirling cycle, microturbine, and high concentration photovoltaic modules. Because of the high concentration ratios achievable with parabolic dishes and the small size of the receiver, solar dishes are efficient (as high as 30%) at collecting solar energy and produce very high temperatures (900 to 2700°F, or 500 to 1500°C). Solar dish/engine systems have environmental, operational, and potential economical advantages over more conventional power generation options because (USDOE, 1998):

- They produce zero emissions when operating on solar energy.
- They operate more quietly than diesel or gasoline engines.
- They are easier to operate and maintain than conventional engines.
- They start up and shut down automatically.
- They operate for long periods with minimal maintenance.

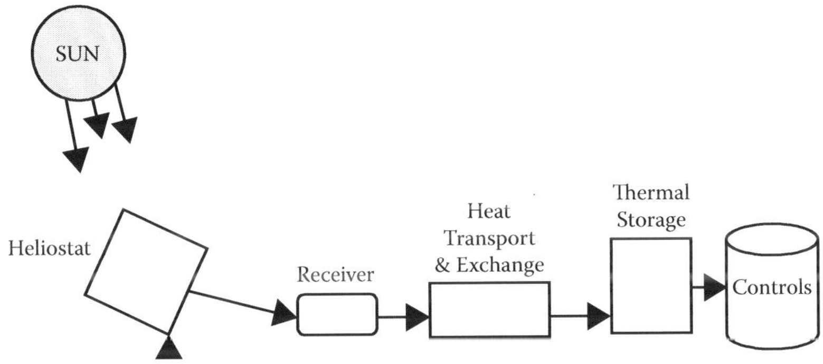

FIGURE 3.1 Solar power tower main components.

Solar dish/engine systems are well suited and often used for nontraditional power generation because of their size and durability. Individual units range in size from 10 to 25 kW. They can operate independently of power grids in sunny locations to produce power for pumping water, for example, or for people living in isolated locations.

The high temperatures produced by dish/engine systems can be used to produce steam, which can be used either for electricity production or various high-temperature industrial processes (Chiras, 2002). Dish/engine systems can also be linked together to provide utility-scale power to a transmission grid. Such systems could be located near consumers, substantially reducing the need for building or upgrading transmission capacity. Largely because of their high efficiency, the costs of these systems are expected to be lower than those of other solar systems for these applications (USDOE, 1998).

Power Tower Systems

The power tower or the central receiver system consists of a large field of flat, sun-tracking mirrors known as *heliostats* (silver-laminated acrylic membranes) focused to concentrate sunlight onto a receiver on the top of a tower. A heat-transfer fluid heated in the receiver is used to generate steam, which, in turn, is used in a conventional turbine generator to produce electricity. Some power towers use water or steam as the heat-transfer fluid. Other advanced designs are experimenting with molten (liquefied) nitrate salt because of its superior heat-transfer and energy-storage (heat retention) capabilities (NREL, 2010b). This energy storage (thermal storage) capability allows the system to continue to dispatch electricity during cloudy weather or at night. A power tower system is comprised of five main components: heliostat, receiver, heat transport and exchange unit, thermal storage, and controls (Figure 3.1). Individual commercial plants can be sized to produce up to 200 MW of electricity.

Thermal Energy Storage

Thermal energy storage (TES) has become a critical aspect of any concentrating solar power system deployed today (USDOE, 2009b). One challenge facing the widespread use of solar energy is reduced or curtailed energy production when the sun sets or is blocked by clouds. Thermal energy storage provides a workable solution to this problem. In a CSP system, the sun's rays are reflected onto a receiver, creating heat that is then used to generate electricity. If the receiver contains oil or molten salt as the heat-transfer medium, then the thermal energy can be stored for later use. This allows CSP systems to be a cost-competitive option for providing clean, renewable energy. Current steam-based receivers cannot store thermal energy for later use.

TABLE 3.2
Thermal Energy Storage Materials

	Specific Heat	Density		Heat Capacity	
Material	**(Btu/lb °F)**	**(lb/ft³)**	**(kg/m³)**	**(Btu/ft³ °F)**	**(kJ/m³ °C)**
Water	1.00	62	1000	62	4186
Iron	0.12	490	7860	59	3521
Copper	0.09	555	8920	50	3420
Aluminum	0.22	170	2700	37	2430
Concrete	0.23	140	2250	32	2160
Stone	0.21	170	2700	36	2270
White pine	0.67	2	435	18	1220
Sand	0.19	95	1530	18	1540
Air	0.24	0.075	1.29	0.02	1.3

An important criterion in selecting a storage medium, such as the oil or molten salt referred to above, is the *specific heat* of the substance. Specific heat (c) is a measure of the heat energy required to increase the temperature of a unit quantity of a substance by a unit of temperature. For example, at a temperature of 15°C, the heat energy required to raise the temperature of water 1 Kelvin (equal to 1°C) is 4186 joules per kilogram (J/kg). Different materials absorb different amounts of heat when undergoing the same temperature increase. The relationship between a temperature change (ΔT) and the amount of heat (Q) added (or subtracted is given by

$$Q = mc\Delta T \tag{3.1}$$

where c is the specific heat and m is the mass of the substance. Table 3.2 lists some specific heat values for various materials per pound and per cubic foot. Another important parameter related to specific heat and thermal energy storage materials is *heat capacity*, which is defined as the ratio of the heat energy absorbed by a substance to the increase in temperature of the substance. It is given by

$$\text{Heat Capacity} = \text{Specific Heat} \times \text{Density} \tag{3.2}$$

Other media for thermal storage are *phase change materials*, which are classified as latent heat storage units. Specifically, a phase change material is a substance with a high heat of fusion, which is the amount of thermal energy that must be absorbed or evolved for 1 mole of a substance to change states from a solid to a liquid or *vice versa* without a temperature change. A phase change material melts and solidifies at a certain temperature and is capable of storing and releasing large amounts of energy. Heat is absorbed or released when the material changes from solid to liquid and *vice versa*. One common group of substances being used as phase change materials for solar applications (and elsewhere) are eutectic salts, such as Glauber's salt. This salt melts at 91°F with the addition of 108 Btu/lb. Conversely, when the temperature drops below 91°F, 108 Btu/lb of heat energy is released as the salt solidifies (Hinrichs and Kleinbach, 2006).

Several TES technologies have been tested and implemented since 1985. These include the two-tank direct system, two-tank indirect system, and single-tank thermocline system.

Two-Tank Direct System

Solar thermal energy in this system is stored in the same fluid used to collect it. The fluid is stored in two tanks—one at high temperature and the other at low temperature. Fluid from the low-temperature tank flows through the solar collector or receiver, where solar energy heats it to a high temperature; it then flows back to the high-temperature tank for storage. Fluid from the high-temperature tank flows through a heat exchanger, where it generates steam for electricity production. The fluid exits the heat exchanger at the low temperature and returns to the low-temperature tank. Two-tank direct storage was used in early parabolic trough power plants and at the Solar Two power tower in California. The trough plants used mineral oil as the heat-transfer and storage fluid, and the Solar Two power tower used molten salt.

Two-Tank Indirect System

The two-tank indirect system functions in the same way as the two-tank direct system, except that different fluids are used for the heat-transfer fluid and for the storage fluid. This system is used in plants where the heat-transfer fluid is too expensive or not suited for use as the storage fluid. The storage fluid from the low-temperature tank flows through an extra heat exchanger, where it is heated by the high-temperature heat-transfer fluid. The high-temperature storage fluid then flows back to the high-temperature storage tank. The fluid exits this heat exchanger at a low temperature and returns to the solar collector or receiver, where it is heated back to the high temperature. Storage fluid from the high-temperature tank is used to generate steam in the same manner as the two-tank direct system. The indirect system requires an extra heat exchanger, which adds cost to the system. This system will be used in many of the parabolic power plants in Spain and has also been proposed for several U.S. parabolic plants. The plants will use organic oil as the heat-transfer fluid and molten salt as the storage fluid.

Single-Tank Thermocline System

This system stores thermal energy in a solid medium—most commonly silica sand—located in a single tank. At any time during operation, a portion of the medium is at high temperature and a portion is at lower temperature. The hot- and cold-temperature regions are separated by a temperature gradient or thermocline (see Figure 3.2). High-temperature heat-transfer fluid flows into the top of the thermocline and exits the bottom at low temperature. This process moves the thermocline downward and adds thermal energy to the system for stage. Reversing the flow moves the thermocline upward and removes thermal energy from the system to generate steam and electricity. Buoyancy effects create thermal stratification of the fluid within the tank, which helps to stabilize and maintain the thermocline.

Note: Using a solid storage medium and only requiring one tank reduces the cost of the single-tank thermocline system relative to the two-tank systems.

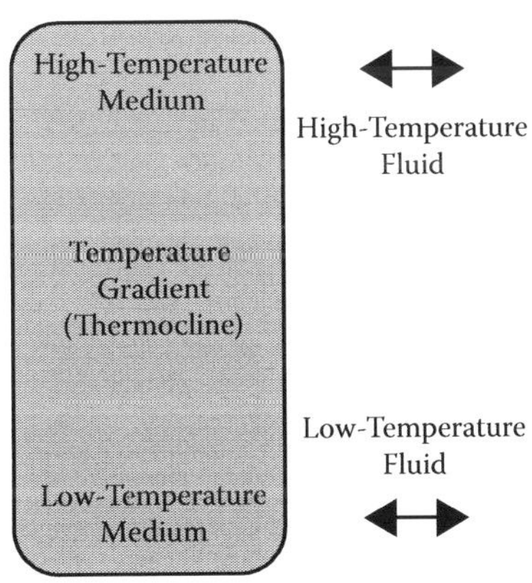

FIGURE 3.2 Single-tank thermocline thermal energy storage system.

PHOTOVOLTAICS

Photovoltaics (PV) has been and will continue to be one of the more glamorous technologies in the energy field (Hinrichs and Kleinbach, 2006). Photovoltaic (*photo* from the Greek word for "light" and *volt* for electricity pioneer Alessandro Volta) technology makes use of the abundant energy in the sun, and it has little impact on our environment. Photovoltaics is the direct conversion of light (photons) into electricity (voltage) at the atomic level. Some materials exhibit a property known as the *photoelectric effect* (discovered and described by Becquerel in 1839) that causes them to absorb photons of light and release electrons. When these free electrons are captured, an electric current (flow of free electrons) results which can be used as electricity. The first photovoltaic module (billed as a solar battery) was built by Bell laboratories in 1954. In the 1960s, the space program began to make the first serious use of the technology to provide power aboard spacecraft. Use of this technology in the space program produced giant advancements in its reliability and helped to lower costs; however, it was the oil embargo of the 1970s (the so-called energy crisis) that focused attention on using photovoltaic technology for applications other than the space program. Photovoltaics can be used in a wide range of products, from small consumer items to large commercial solar electric systems.

Figure 3.3 illustrates the photoelectric effect when light shines on a negative plate; electrons are emitted with an amount of kinetic energy inversely proportional to the wavelength of the incident light. Figure 3.4 illustrates the operation of a basic *photovoltaic cell*, also called a *solar cell*. Solar cells are made of silicon and other semiconductor materials such as germanium, gallium arsenide, and silicon carbide and are used in the microelectronics industry. For solar cells, a thin semiconductor wafer is specially treated to form an electric field, positive on one side and negative on the other. When light energy strikes the solar cell, electrons are jarred loose from the atoms in the semiconductor material (see Figure 3.3 and Figure 3.4). If electrical conductors are attached to the positive and negative sides, forming an electrical circuit, the electrons can be captured in the form of an electrical current (recall that electron flow is electricity). This electricity can then be used to power a load, such as for a light, tool, or toaster, for example.

A *photovoltaic module* is comprised of a number of solar cells electrically connected to each other and mounted on a support panel or frame (see Figure 3.5). Solar panels used to power homes and businesses are typically made from solar cells combined into modules that hold about 40 cells. Modules are designed to supply electricity at a certain voltage (e.g., 12 volts). The current produced is directly dependent on how much light strikes the module. Multiple modules can be wired together to form an array. In general, the larger the area of a module or array, the more electricity will be produced. Photovoltaic modules and arrays produce direct-current (DC) electricity. They can be connected in both series and parallel electrical arrangements to produce any required voltage and current combination.

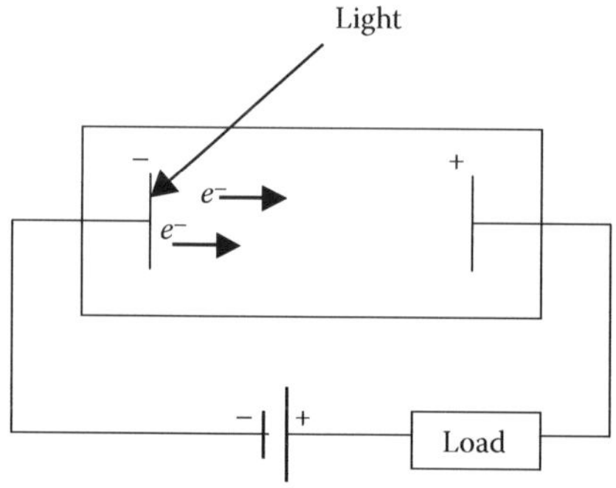

FIGURE 3.3 Photoelectric effect.

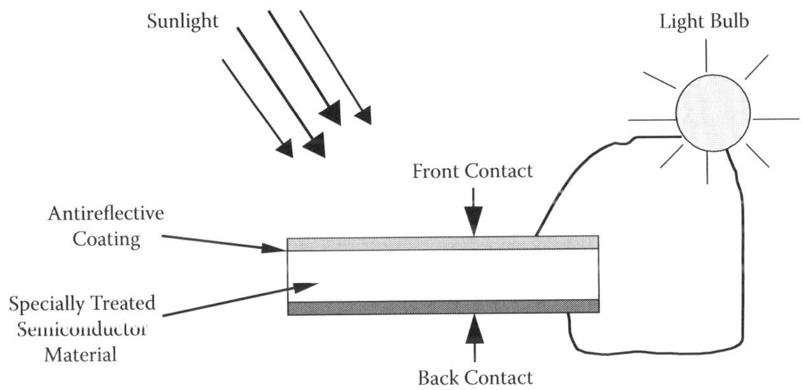

FIGURE 3.4 Operation of basic photovoltaic cell.

DOMESTIC AND INDUSTRIAL USE OF SOLAR ENERGY

When thinking about a potential renewable energy source, we first want to know if it is reliable, clean, and affordable. This is especially the case for consumers and business owners contemplating solar applications for both home and business use. Until recently, solar energy has been used in both homes and industrial locations, but only on a somewhat limited basis. This trend is changing, especially in those locations where sunlight is prevalent throughout the year. With the soaring costs of fossil fuel supplies and their pending decreased availability, solar power is beginning to receive more attention. Solar technologies for domestic use include photovoltaics, passive heating, daylighting, and water heating. Industrial and commercial facilities may use the same solar technologies, in addition to solar energy technologies that would be impractical for a home, such as ventilation air preheating, solar process heating, and solar cooling.

SOLAR HOT WATER

Just as the sun heats the surface layers of exposed bodies of water—ponds, lakes, streams, and oceans—it can also heat water used in buildings and swimming pools. Harnessing the ability of the sun to heat water requires a solar collector and a storage tank. The most common collector is the

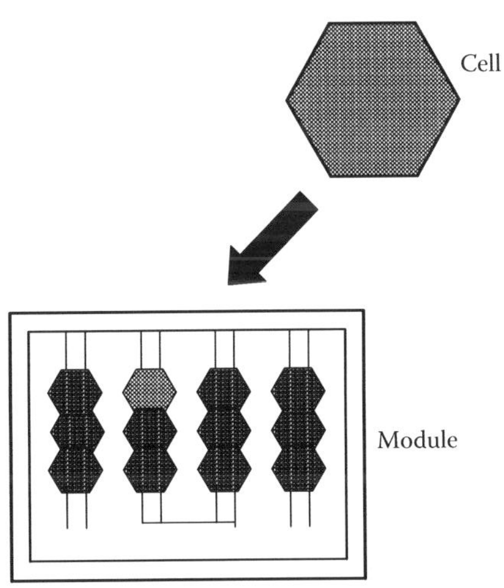

FIGURE 3.5 Single solar cell and solar cell module.

flat-plate collector (Figure 3.5). The flat-plate collector system is the most commonly used collector and is ideal for applications that require water temperatures under 140°F (60°C). Mounted on the roof, the system consists of a thin, flat, rectangular, insulated box with a transparent glass cover that faces the sun. Small copper tubes run through the box and carry the fluid (water or another fluid such as an antifreeze solution) to be heated. The tubes, typically arranged in series (i.e., water flows in one end and out the other end), are attached to an absorber plate, which, along with the tubes, is painted black to absorb the heat. As heat builds up in the collector, it heats the fluid (water or propylene glycol) passing through the tubes. The motive force for passing fluid through the tubes can be accomplished either by active or passive means. In an active system, the most common type of system, water heaters rely on an electric pump and controller to circulate water or other heat-transfer fluid through the collectors. Passive solar water-heater systems rely on gravity and the tendency for water to naturally circulate as it is heated. Because passive systems have no moving parts, they are reliable and easier to maintain and often have a longer life than active systems. The storage tank that holds the hot liquid (Figure 3.6) is usually a large well-insulated tank, but modified water heaters can also be used to store the hot liquid. When the fluid used is other than water, the water is heated by passing it through a coil of tubing in the tank, which is full of hot fluid. Active solar systems usually include a storage tank along with a conventional water heater. In two-tank systems, the solar water heater preheats water before it enters the conventional water heater.

Space Heating

In order to maintain indoor air quality (IAQ), many large buildings require ventilated air. Although fresh air can be great for indoor air quality, heating this air in cold climates can use large amounts of energy and be very expensive. An elegantly simple solar ventilation system can preheat the air, though, saving both energy and money. This type of system typically uses a transpired collector. Transpired air collector systems essentially consist of a dark-colored, perforated facade installed on the south-facing wall of a building. A fan or the existing ventilation system draws ventilation air into the building through the perforated absorber plate on the facade and up the plenum (the air space between the absorber and the south wall). Solar energy absorbed by the dark absorber and transferred to the air flowing through it can preheat the intake air by as much as 40°F (22°C). Reduced heating costs will pay for the systems in 3 to 12 years (USDOE, 2006).

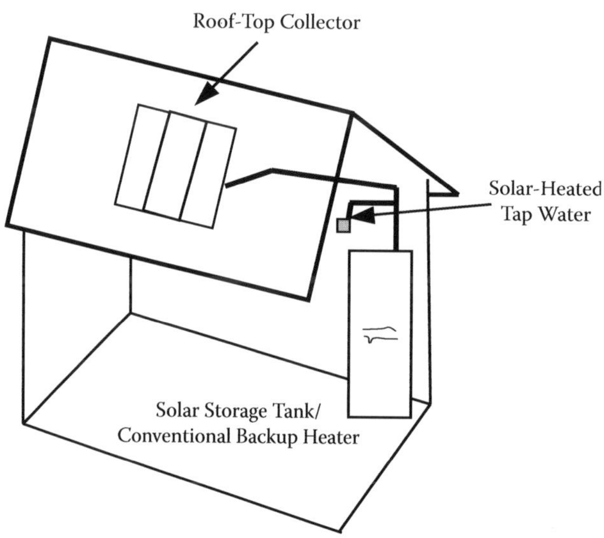

FIGURE 3.6 Simplified representation of a home solar water-heating system.

In addition to meeting a portion of the heating load of a building with clean, free solar energy, the transpired collector helps save energy and money in other ways. It recaptures heat loss through the south-facing wall of a building, as heat that escapes through the south wall is captured in the air space between the structural wall and the transpired collector and returned to the interior. Also, by introducing make-up air though ceiling-mounted ducts, the system eliminates the wasteful air stratification that often plagues high-ceiling buildings (USDOE, 2006).

To illustrate an active, real-world example of transpired collectors used as solar preheaters for outdoor ventilation air, it is probably best to use an actual example of a working system in an industrial application. The following case study is such an example; it has been monitored extensively by experts.

General Motors Battery Plant*

The General Motors battery plant in Oshawa, Canada, is a 100,000-ft^2 facility in which automotive batteries are manufactured. The plant was built in the 1970s and consists of an open shop floor and a 28-foot-high ceiling. GM operates two full-time production shifts within the plant and conducts maintenance activities at night and on weekends, so the building is continuously occupied.

Until the early 1990s, GM relied solely on a team-operated fan coil system for space heating, but the system was not capable of providing the necessary quantities of heated outdoor air. As a result, the plant was not being adequately ventilated. In 1991, plant management installed transpired collector to correct the ventilation problems. Over the next two years, the transpired collector system was modified slightly to improve airflow; the original fans and motors were replaced with vane axial fans and high-efficiency motors, and the original ducting was replaced with upgraded fabric ducting.

The GM plant collector is comprised of 4520 ft^2 of absorber sheeting. The lower 21 feet of the transpired collector is black, perforated, aluminum wall cladding with 1.6-mm holes totaling 2% porosity. The average depth of the plenum between the transpired collector and the plant's structural wall is 6 inches. The canopy at the top of the wall acts as both a manifold for the air flow and a solar heat collection device. The canopy face is made of perforated plate with 1% porosity. The transpired collector covers about 50% of the total area of the plant's south-facing wall; the remainder of the south facade has shipping doors and other obstructions that make it unsuitable for mounting collector cladding.

The GM transpired collector has two fan/distribution systems, each consisting of a constant-speed fan, a recirculation damper system, and a fabric distribution duct. The total airflow delivered by the system's fans is 40,000 cfm. Both recirculated air and air drawn through the solar collectors make up this flow; the percentages of each depend on the temperature of the air coming from the collector. The GM battery plant's transpired collector has been monitored extensively since it was installed. The data in this case study reflect the performance of the system during the 1993–1994 heating season. An in-depth report on the monitoring program is available (Enermodal Engineering, Ltd., 1995). The data show that the annual energy savings for the 4520-ft^2 collector was 940 million Btu/year: 678 MBtu resulted from the thermal energy gained directly from the outside air as it passed through the absorber, and 262 MBtu resulted from heat loss recaptured by the wall from inside the building. Other possible energy-saving mechanisms—such as destratification and heat recapture—likely contributed to improved system performance; however, these effects are highly structure specific and have not been incorporated into the savings reported here.

The cost of the transpired collector system at the GM plant was \$66,530, or \$14.72/ft^2 of installed collector. The cost per square foot is higher than typical installations for two reasons: (1) this system was installed soon after the technology was introduced, before design and installation procedures had been streamlined, and (2) the cost includes the fan and ducting modifications that were implemented during the first 2 years of operation.

* Much of the information in this section is from USDOE, *Transpired Collectors (Solar Preheaters for Outdoor Ventilation Air)*, Federal Technology Alert, U.S. Department of Energy, Washington, D.C., 1998.

It is important to point out that the GM transpired collector had a number of operational problems. After the system was initially installed, employees complained about fan noise and cold drafts, and they occasionally disabled the system. The fan and duct upgrades described previously eliminated the problems on one of the fan systems; the other fan continues to generate noise, and employees still disable it when working in the immediate vicinity. The manufacturer has addressed these complaints by specifying smaller, but more numerous, fans in subsequent installations.

Also, both bypass dampers and a recirculation damper required additional maintenance. The recirculation damper became stuck in the full recirculation mode, and a new modulating motor was installed to fix the problem. The bypass dampers occasionally became bound, which led to unacceptably high leakage rates. These dampers were kept closed manually throughout the 1993–1994 heating season.

Water Heating

Solar water-heating systems are designed to provide large quantities of hot water for nonresidential buildings. A typical system includes solar collectors that work along with a pump, heat exchanger, and one or more large storage tanks. The two main types of solar collectors used for nonresidential buildings, evacuated-tube collector and linear concentrator, can operate at high temperatures with high efficiency. An evacuated-tube collector is a set of many double-walled glass tubes and reflectors that heat the fluid inside the tubes. A vacuum between the two walls insulates the inner tube, retaining the heat. As described earlier, linear concentrators use long, rectangular, curved (U-shaped) mirrors tilted to focus sunlight on tubes that run along the length of the mirrors. The concentrated sunlight heats the fluid within the tubes.

Space Cooling

Space cooling can be accomplished using thermally activated cooling systems (TACS) driven by solar energy. Because of a high initial cost, TACS are not widespread. The two systems currently in operation are solar absorption systems and solar desiccant systems. Solar absorption system use thermal energy to evaporate a refrigerant fluid to cool the air. In contrast, solar desiccant systems use thermal energy to regenerate desiccants that dry the air, thereby cooling the air. These systems also work well with evaporative coolers in more humid climates.

LET THERE BE NATURAL LIGHT*

Lighting is an essential element of any building. Proper lighting improves the aesthetics of indoor spaces and provides illumination for tasks and activities. An efficient lighting strategy, including natural daylighting, can provide proper levels of illumination and reduce energy costs. *Daylighting* is a passive solar design used to illuminate a living space or an industrial or commercial working space. In the Northern Hemisphere, southern exposures receive the greatest amount of sunlight, while the opposite is true in the Southern Hemisphere. In other words, installing large windows in a south-facing wall or roof allows natural light to penetrate and reduces the use of expensive electrical lighting. Because the proper use of daylighting requires the integration of natural and artificial lighting sources early in the building design process, it is important to have an understanding of the fundamentals of lighting and lighting technology. Accordingly, in this section, we first present a discussion of lighting technology before discussing the principles and practices associated with daylighting.

* Much of the information in this section is from USDOE, *Commercial Building Initiative*, U.S. Department of Energy, Washington, D.C., 2010 (www1.eere.energy.gov/buildings/commercial/printable_versions/lighting.html).

Lighting Technologies

Lighting technologies include:

- *Lamps*—Lighting sources, such as fluorescent and incandescent light bulbs, as well as solid-state lighting
- *Solid-state lighting (SSL)*—LED (light-emitting diode) lighting, which has the potential to save energy, enhance the quality of building environments, and contribute to energy and climate change solutions
- *Ballasts*—Components of electric-discharge lamps such as fluorescent lamps that transform and control electrical power to the light
- *Luminaires* (fixtures)—Complete lighting units that contain the bulbs and, if necessary, the ballasts
- *Lighting controls*—Devices such as timers and sensors that can save energy by turning lights off when not needed
- *Daylighting*—Use of natural light in a building

Lamps

Commonly called light bulbs, lamps produce light. When comparing lamps, it is important to understand the following performance characteristics:

- *Color Rendering Index (CRI)* is a measurement of the accuracy of a light source in rendering colors compared to a reference light source with the same correlated color temperature. The highest attainable CRI is 100. Lamps with CRI values above 70 are typically used in office and living environments.
- *Correlated color temperature (CCT)* is a measurement on the Kelvin (K) scale that indicates the warmth or coolness of the color of a lamp. The higher the color temperature, the cooler the color. Typically, a CCT rating below 3200 K is considered warm, whereas a rating above 4000 K is considered cool.
- *Efficacy* is the ratio of the light output to the power, measured in lumens per watt (lm/W). The higher the efficacy, the more efficient the lamp.

Incandescent Lamps

A standard incandescent lamp consists of a fairly large, thin, frosted glass envelope. Inside the glass is an inert gas such as argon or nitrogen. At the center of the lamp is a tungsten filament. Electricity heats the filament. The heated tungsten emits visible light through a process called *incandescence.* Most standard light bulbs are incandescent lamps. They have a CRI value of 100 and CCT values between 2600 and 3000, making them attractive lighting sources for many applications; however, these bulbs are typically inefficient (10 to 15 lm/W). They convert only about 10% of the energy into light while transforming the rest into heat.

Another type of incandescent lamp is the halogen lamp. Halogen lamps also have a CRI value of 100, but they are slightly more energy efficient, and they maintain their light output over time. A halogen lamp also uses a tungsten filament, but the filament is encased inside a much smaller quartz envelope. The gas inside the envelope is from the halogen group. If the temperature is high enough, the halogen gas will combine with tungsten atoms as they evaporate and redeposit them on the filament. This recycling process means the filament lasts much longer. In addition, it is now possible to run the filament hotter, which means that more light per unit of energy can be obtained. Because the quartz envelope is so close to the filament, it becomes about four times hotter than a standard incandescent lamp. As a result of this wasted heat energy, halogen lamps, which are popular in torchieres (i.e., portable electric lamps with a reflector bowl that directs light upward to given indirect illumination), are not very energy efficient. The exposed heat from halogen torchieres can also pose a serious fire risk, especially near flammable objects. Because of this inefficiency and risk, manufacturers have developed torchieres that can use other types of lamps, such as compact fluorescent lamps.

Fluorescent Lamps

A fluorescent lamp (70 to 100 lm/W) is a sealed glass tube containing a small amount of mercury and an inert gas, such as argon, kept under very low pressure. In these electric-discharge lamps, a fluorescing coating (phosphor powder) on the glass transforms some of the ultraviolet energy generated into light. Fluorescent lamps require a ballast to start and maintain their operation. Early fluorescent lamps were sometimes criticized for not producing warm enough colors, making them appear too white or not complimentary to skin tones; a cool white fluorescent lamp had a CRI value of 62. Today, some fluorescent lamps have CRI values of 80 and above, and they simulate natural daylighting and incandescent light. They also are available in a variety of CCT values, ranging from 2900 to 7000. The "T" designation for fluorescent lamps stands for tubular, the shape of the lamp. The number after the "T" gives the diameter of the lamp in eighths of an inch. The T8 lamp (available straight or U-shaped) has become the standard for new construction. It is also commonly serves as a retrofit replacement for 40-watt T12 lamps, improving efficacy, CRI, and efficiency. In some cases, T10 lamps offer advantages over both T12 and T8 lamps, including higher efficiency, higher CRI values, a wider selection of CCT values, and compatibility with several ballast types. Another lamp type is the T5FT fluorescent lamp. These lamps produce maximum light output at higher ambient temperatures than those that are linear or U-shaped. Linear fluorescent lamps often are less expensive than compact fluorescent lamps. They can also produce more light, are easier to dim, and last longer.

Compact Fluorescent Lamps

Compact fluorescent lamps (CFLs) are small-diameter fluorescent lamps folded for compactness, with an efficacy of 50 to 75 lm/W for 27 to 40 watts. There are several styles of CFLs: two-, four-, and six-tube lamps, as well as circular lamps. Some CFLs have permanently connected tubes and ballasts; others have separate tubes and ballasts. Some CFLs feature a round adaptor, allowing them to screw into common electrical sockets and making them ideal replacements for incandescent lamps. They last up to 10 times longer than incandescent lamps, and they use about one-fourth the energy, producing 90% less heat. Typical 60- to 100-watt incandescent lamps are no more than 5.3 inches long, but standard CFLs are longer than 6 inches. For this reason, sub-CFLs have been developed that are no more than 4.5 inches long and fit into most incandescent fixtures.

High-Intensity Discharge Lamps

Compared to fluorescent and incandescent lamps, high-intensity discharge (HID) lamps produce a large quantity of light in a small package. HID lamps produce light by striking an electrical arc across tungsten electrodes housed inside a specially designed inner glass tube. This tube is filled with both gas and metals. The gas aids in the starting of the lamps; the metals then produce the light when they are heated to a point of evaporation. Like fluorescent lamps, HID lamps require a ballast to start and maintain their operation. Types of HID lamps include mercury vapor (CRI range, 15 to 55), metal halide (CRI range, 65 to 80), and high-pressure sodium (CRI range, 22 to 75). Mercury vapor lamps (25 to 45 lm/W), which originally produced a bluish-green light, were the first commercially available HID lamps. Today, they are also available as a color-corrected, whiter light, but they are still often being replaced by the newer, more efficient high-pressure sodium and metal halide lamps. Standard high-pressure sodium lamps have the highest efficacy of all HID lamps, but they produce a yellowish light. High-pressure sodium lamps that produce a whiter light are now available, but efficiency is somewhat sacrificed. Metal halide lamps are less efficient but produce an even whiter, more natural light. Colored metal halide lamps are also available.

Low-Pressure Sodium Lamps

Low-pressure sodium lamps have the highest efficacy of all commercially available lighting sources, producing up to 180 lm/W. Even though they emit a yellow light, a low-pressure sodium lamp should not be confused with a standard high-pressure sodium lamp—a high-intensity discharge lamp. Low-pressure sodium lamps operate much like a fluorescent lamp and require a ballast. The lamps are also physically large (about 4 feet long for the 180-watt size), so light distribution from fixtures is less controllable. The lamp requires a brief warm-up period to reach full brightness. With a CRI value of 0, low-pressure sodium lamps are used where color rendition is not important but energy efficiency is. They are commonly used for outdoor, roadway, parking lot, and pathway lighting. Low-pressure sodium lamps are preferred around astronomical observatories because the yellow light can be filtered out of the random light surrounding the telescope.

Solid-State Lighting

Compared to incandescent and fluorescent lamps, solid-state lighting (SSL) creates light with less directed heat. A semiconducting material converts electricity directly into light, which makes the light very energy efficient. Solid-state lighting includes a variety of light-producing semiconductor devices, such as light-emitting diodes (LEDs) and organic light-emitting diodes (OLEDs). Warm white LEDs have an efficacy of 50 lm/W, while cool white LEDs can achieve efficacies up to 100 lm/W. Until recently, LEDs—basically tiny light bulbs that fit easily into an electrical circuit—were used as simple indicator lamps in electronics and toys, but recent research has achieved efficiencies equal to those of fluorescent lamps. Also, the costs of semiconductor materials, which used to be quite expensive, are lower today, making LEDs a more cost-effective lighting option.

Ongoing research shows that LEDs have great potential as energy-efficient lighting for residential and commercial building use. New uses for LEDs include small-area lighting (e.g., task lighting and under-shelf fixtures), decorative lighting, and pathway and step marking. As LEDs become more powerful and effective, they will be used in more general illumination applications, perhaps with entire walls and ceilings becoming the lighting system. They are already being used successfully in many general illumination applications, such as traffic signals and exit signs. OLEDs currently are used in very thin, flat display screens, such as those in portable televisions; in some vehicle dashboard readouts; and in postage-stamp-sized data screens built into pilots' helmet visors. Because OLEDs emit their own light and can be incorporated into arrays on very thin, flexible materials, they also could be used to fashion large, extremely thin panels for light sources in buildings.

DID YOU KNOW?

One way to create white light with LEDs is to mix the three primary colors of light (red, green, and blue, or RBG). Phosphor-conversion white light can be produced by blue, violet, or near-ultraviolet LEDs coated with yellow or multichromatic phosphors; the combined light emission appears white.

Ballasts

Ballasts consume, transform, and control electrical power for electric-discharge lamps, providing the necessary circuit conditions for starting and operating them. Electric-discharge lamps include fluorescent, high-intensity discharge, and low-pressure sodium. When comparing ballasts, it is important to understand the following performance characteristics:

- *Ballast factor* (BF)—The ratio of the light output of a lamp or lamps operated by a specific ballast to the light output of the same lamps operated by a reference ballast. It can be used to calculate the actual light output of that specific lamp–ballast combination. The BF is typically different for each lamp type. Ballasts with extremely high BF values could reduce lamp life and accelerate lumen deficiency because of high lamp current. Extremely low BF values also could reduce lamp life because they reduce lamp current.
- *Ballast efficacy factor* (BEF)—The ratio of ballast factor (as a percentage) to power (in watts). BEF comparisons should be made only among ballasts operating the same type and number of lamps.
- *System efficacy*—The ratio of the light output to the power. Efficacy is measured in lumens per watt (lm/W) for a particular lamp ballast system.

The three basic types of ballasts are magnetic, hybrid, and electronic:

- *Magnetic ballasts* contain a magnetic core of several laminated steel plates wrapped with copper windings. These ballasts operate lamps at line frequency (60 Hz in North America). Of all ballasts, magnetic ones are the least expensive and also the least efficient. They have greater power losses than electronic ballasts, but magnetic ballasts manufactured today are 10% more efficient than the older, high-loss magnetic ballasts that used aluminum windings. Magnetic ballasts are available with dimming capability; however, they cannot be dimmed below 20% and still use more electricity than electronic ballasts.
- *Hybrid ballasts* (cathode-disconnect ballasts) use a magnetic core-and-coil transformer and an electronic switch for the electrode heating circuit. Like magnetic ballasts, they operate lamps at line frequency (60 Hz in North America). After they start the lamp, these ballasts disconnect the electrode heating circuit. Hybrid ballasts cost more than magnetic ballasts, but they are more energy efficient.
- *Electronic ballasts* appeared in the early 1980s, when manufacturers began to replace the core-and-coil transformer with solid-state, electronic compounds that could operate lamps at 20 to 50 kHz. These electronic ballasts experience half the power loss of magnetic ballasts. Also, lamp efficacy is increased by approximately 10 to 15% compared to 60-Hz operation. Electronic ballasts are the most expensive, but they are also the most efficient. Operating lamps with electronic ballasts reduces electricity use by 10 to 15% over magnetic ballasts of the same light output. They are quieter and lighter, and they virtually eliminate lamp flicker. Electronic ballasts are also available as dimming ballasts. These ballasts allow the light level to be controlled between 1 and 100%. Several types of electronic ballasts are available for use with fluorescent lamps. Electronic ballasts have been successfully used with lower watt high-intensity discharge (HID) lamps (primarily 35 to 100 W). These ballasts provide an energy savings over magnetic ballasts of 8 to 20%. Their lighter weight also helps in some HID applications, such as track lighting.

Luminaires (Lighting Fixtures)

A lamp or lamp–ballast combination may produce light very efficiently, but if it is installed in an inefficient luminaire the overall system efficiency may still be poor. The best luminaire manufacturers will design their fixtures around specific lamps to optimize the amount of light delivered to the work area; for example, a luminaire designed specifically for a compact fluorescent lamp can deliver almost 10 times as much illumination as an incandescent fixture fitted with the same compact fluorescent lamp. Luminaire components include reflectors, diffusers (which absorb some of the light from a lamp), and polarizing panels. Reflectors can be used to direct more of the light produced by the lamp out of the luminaire onto the work area. Polarizing panels can sometimes increase the contrast of a visual task.

When comparing luminaires, it is important to understand the following luminaire performance characteristics:

- *Illuminance*—The amount of light that reaches a surface; it is measured in footcandles (lumens per square foot) or lux (lumens per square meter)
- *Luminaire efficacy/efficiency rating* (LER)—The light output (lumens) per watt (W) of electricity use

Before selecting luminaires or lighting fixtures for an office building, factory, warehouse, or even a parking lot, it is a good idea to consult a certified lighting designer. A lighting designer will not only help find the most energy-efficient luminaires but also provide lighting that makes for a comfortable and more productive work environment. Today, energy-efficient commercial lighting design includes more than just the ambient or general lighting of a workspace, such as the use of ceiling luminaires. When designing or retrofitting the lighting, the general illuminance can be reduced if task lighting is implemented properly into the overall design. Task lighting can result in significant energy savings and improved visibility for workers.

DID YOU KNOW?

The most efficient light source technology for exit signs is light-emitting diodes (LEDs). The most popular parking lot luminaires use energy-efficient, high-intensity discharge lamps or low-pressure sodium lamps. The most efficient light source technology for outdoor use, though, is outdoor photovoltaic lighting.

Lighting Controls

Lighting controls help conserve energy and make a lighting system more flexible. The most common light control is the on/off switch. Other types of light control technologies include manual dimming, photosensors, occupancy sensors, clock switches or timers, and centralized controls:

- *Manual dimming*—These controls allow occupants of a space to adjust the light output or illuminance. This can result in energy savings through reductions in input power, as well as reductions in peak power demand and enhanced lighting flexibility. Slider switches allow the occupant to change the lighting over the complete output range. Preset scene controls change the dimming setting for various lights all at once with the press of a button; it is possible to have different settings for the morning, afternoon, and evening. Remote control dimming is also available; this type of technology is well suited for retrofit projects, where it is useful to minimize rewiring.
- *Photosensors*—These devices automatically adjust the light output of a lighting system based on detected illuminance. The technology behind photosensors is the photocell. A photocell is a light-responding silicon chip that converts incident radiant energy into electrical current. Whereas some photosensors just turn lights off and on, others can also

DID YOU KNOW?

Fluorescent lighting fixtures require special dimming ballasts and compatible control devices. Some dimming systems for high-intensity discharge lamps also require special dimming ballasts.

DID YOU KNOW?

Nearly all photosensors are used to decrease the electric power for lighting. In addition to lowering the electric power demand, dimming the lights also reduces the heat load on the cooling system of a building. Any solar heat gain that occurs in a building during the day must be taken into account for a whole building energy usage analysis.

dim lights. Automatic dimming can help with lumen maintenance. Lumen maintenance involves dimming luminaires when they are new, which minimizes the wasteful effects of over-design. The power supplied to them is gradually increased to compensate for light loss over the life of the lamp.

- *Occupancy sensors*—These sensors turn lights on and off based on their detection of motion within a space. Some sensors can also be used in conjunction with dimming controls to keep the lights from turning completely off when a space is unoccupied. This control scheme may be appropriate when occupancy sensors control separate zones in a large space, such as in a laboratory or in an open office area. In these situations, the lights can be dimmed to a predetermined level when the space is unoccupied. Sensors can also be used to enhance the efficiency of centralized controls by switching off lights in unoccupied areas during normal working hours as well as after hours. The three basic types of occupancy sensors are listed below.
  - *Passive infrared (PIR) sensors*—These sensors react to the movement of a heat-emitting body through their field of view. Wall-box-type PIR occupancy sensors are best suited for small, enclosed spaces such as private offices, where the sensor replaces the light switch on the wall and no extra wiring is required. They should not be used where walls, partitions, or other objects might block the ability of the sensor to detect motion.
  - *Ultrasonic sensors*—These sensors emit an inaudible sound pattern and read the reflection. They react to changes in this reflected sound pattern. These sensors detect very minor motion better than most infrared sensors; therefore, they are good to use in spaces such as restrooms with stalls, which can block the field of view, because the hard surfaces will reflect the sound pattern.
  - *Dual-technology (hybrid) sensors*—These occupancy sensors use both passive infrared and ultrasonic technologies to minimize the risk of false triggering (lights coming on when the space is unoccupied). They also tend to be more inexpensive.
- *Clock switches or timers*—These control lighting for a preset period of time. They come equipped with an internal mechanical or digital clock, which will automatically adjust for the time of year. The user determines when the light should be turned on and when they should be turned off. Clock switches can be used in conjunction with photosensors.
- *Centralized controls*—These are building controls or building automation systems that can be used to automatically turn on, turn off, or dim electrical lights around a building. In the morning, the centralized control system can be used to turn on the lights before employees arrive. During the day, a central control system can be used to dim the lights during periods of high power demand. And, at the end of the day, the lights can be turned off automatically. A centralized lighting control system can significantly reduce energy use in buildings where lights are left on when not needed.

Daylighting

As mentioned earlier, daylighting is the practice of placing windows or other openings and reflective surfaces so that during the day natural light provides effective internal lighting. When properly designed and effectively integrated with the electric lighting system, daylighting can offer significant energy savings by offsetting a portion of the electric lighting load. A related benefit is the reduction in cooling capacity and use due to lowering a significant component of internal gains. In addition to energy savings, daylighting generally improves occupant satisfaction and comfort. Windows also provide visual relief, contact with nature, time orientation, the possibility of ventilation, and emergency egress.

Daylight Zone

High daylight potential is particularly found in those spaces that are predominantly occupied in the daytime. Site solar analyses should assess the access to daylight by considering what can be viewed from the various potential window orientations. What proportion of the sky is seen from typical task locations in the room? What are the exterior obstructions and glare sources? Is the building design going to shade a neighboring building or landscape feature that is dependent on daylight or solar access? It is important to establish which spaces will most benefit from daylight and which spaces have little or no need for daylight. Within the spaces that can use daylight, place the most critical visual tasks in positions near the window. Try to group tasks by similar lighting requirements and occupancy patterns. Avoid placing the windows in the direct line of sight of the occupant as this can cause extreme contrast and glare. It is best to orient the occupant at 90° from the window. Where privacy is not a major concern, consider interior glazing (known as *relights* or *borrow lights*), which allows light from one space to be shared with another. This can be achieved with transom lights, vision glass, or translucent panels if privacy is required. The floor plan configuration should maximize the perimeter daylight zone. This may result in a building with a higher skin-to-volume ratio than a typical compact building design. A standard widow can produce useful illumination to a depth of about 1.5 times the height of the window. With light shelves or other reflector systems this can be increased to 2.0 times or more. As a general rule of thumb, the higher the window is placed on the wall, the deeper the daylight penetration.

Window Design Considerations

The daylight that arrives at a work surface comes from three sources:

1. *Exterior reflected component*—This includes ground surfaces, pavement, adjacent buildings, wide windowsills, and objects. Remember that excessive rough reflectance will result in glare.
2. *Direct sun/sky component*—Typically, the direct sun component is blocked from occupied spaces because of heat gain, glare, and ultraviolet degradation issues. The sky dome then becomes an important contribution to daylighting the place.
3. *Internal reflected component*—Once the daylight enters the room, the surrounding wall, ceiling, and floor surfaces are important light reflectors. Using high-reflectance surfaces will better bounce the daylight around the room, and it will reduce extreme brightness contrast. Window-frame materials should be light colored to reduce contrast with the view and should have a nonspecular finish to eliminate glare spots. The window jambs and sills can be beneficial light reflectors. Deep jambs should be splayed (angled toward the interior) to reduce the contrast around the perimeter of the window.

DID YOU KNOW?

The most important interior light-reflecting surface is the ceiling. High-reflectance paints and ceiling tiles are now available with 90% or higher reflectance values. Tilting the ceiling plane toward the daylight source increases the daylight that is reflected from this surface. The rear wall of a small room should also have a highly reflective finish. Side walls, followed by the floor, have less impact on the reflected daylight in the space.

Major room furnishings such as office cubicles or partitions can have a significant impact on reflected light so select light-colored materials. Suggested reflectance levels for various room surfaces are

- Ceilings, >80%
- Walls, 50 to 70%
- Floors, 20 to 40%
- Furnishings, 25 to 45%

Because light essentially has no scale for architectural purposes, the proportions of a room are more important than the dimensions. A room that has a higher ceiling compared to the room depth will have deeper penetration of daylight, whether from side lighting (windows) or top lighting (skylights and clerestories). Raising the window head height will result in deeper penetration and more even illumination in the room. Punched window openings, such as small, square windows separated by wall area, result in uneven illumination and harsh contrast between the window and adjacent wall surfaces. A more even distribution is achieved with horizontal strip windows.

DID YOU KNOW?

There is no direct sunlight on the polar-side wall of a structure from the autumnal equinox to the spring equinox in parts of the globe north of the Tropic of Cancer and in parts of the globe south of the Tropic of Capricorn.

Effective Aperture (EA)

One method of assessing the relationship between visible light and the size of the window is the effective aperture method. The *effective aperture* (EA) is defined as the product of the *visible transmittance* (VT) and the *window-to-wall ratio* (WWR). The window-to-wall ratio is the proportion of window area compared to the total wall area where the window is located; for example, if a window covers 25 square feet in a 100-square-foot wall, then the WWR is 25/100, or 0.25. A good starting target for the EA is in the range of 0.20 to 0.30. For a given EA number, a higher WWR (larger window) results in a lower visible transmittance. For example, a WWR = 0.5 (half the wall is glazing) has a VT = 0.6 and EA = 0.3. For WWR = 0.75, VT = 0.4 for the same EA of 0.3.

DID YOU KNOW?

Typically, lowering the visible transmittances will also lower the shading coefficient, but you must verify this with glazing manufacturer data because this is not always the case.

Light Shelves

A light shelf is an effective horizontal light-reflecting overhang placed above eye level with a transom window placed above it. Light shelves enhance the lighting from windows of the equator-facing side of a building. Exterior shelves are more effective shading devices than interior shelves. A combination of exterior and interior will work best in providing an even illumination gradient.

Toplighting Strategies

Large single-level floor areas and the top floors of multiple-story buildings can benefit from toplighting. Types of toplighting include skylights, clerestories, monitors, and sawtooth roofs. Horizontal skylights can be an energy problem because they tend to receive maximum solar gain at the peak of the day; the daylight contribution peaks at midday and falls off severely in the morning and afternoon. Some high-performance skylight designs incorporate reflectors or prismatic lenses that reduce peak daylight and heat gain while increasing early and late afternoon daylight contributions. Another option is light pipes, high-reflectance ducts that channel light from a skylight down to a diffusing lens in the room. These may be advantageous in deep roof constructions.

Clerestory windows are vertical glazings located high on an interior wall. South-facing clerestories can be effectively shaded from direct sunlight by a properly designed horizontal overhang. In this design, the interior north wall can be sloped to better reflect the light down into the room. Use light-colored overhangs and adjacent roof surfaces to improve the reflected component. If exterior shading is not possible, consider interior vertical baffles to better diffuse the light. A south-facing clerestory will produce higher daylight illumination than a northern-facing clerestory. East- and west-facing clerestories present the same problems as east and west windows: difficult shading and potentially high heat gains.

A roof monitor consists of a flat roof section raised above the adjacent roof with vertical glazing on all sides. This design often results in an excessive glazing area, which results in higher heat losses and gains than a clerestory design. The multiple orientations of the glazing can also create shading problems. The sawtooth roof is an old design often seen in industrial buildings. Typically, one sloped surface is opaque and the other is glazed. A contemporary sawtooth roof may have solar collectors or photovoltaic cells on the south-facing slope and daylight glazing on the north-facing slope. Unprotected glazing on the south-facing sawtooth surface may result in high heat gains. In these applications, an insulated diffusing panel may be a good choice.

DID YOU KNOW?

A building designed for daylighting but without an integrated electrical lighting system will be a net energy loser because of the increased thermal loads. Only when the electric lighting load is reduced will we realize more than offsetting savings in electrical and cooling loads. The benefits from daylighting are maximized when both occupancy and lighting sensors are used to control the electric lighting system.

THE BOTTOM LINE ON SOLAR ENERGY

Solar energy has some obvious advantages in that the source is free; however, the initial investment in operating equipment is not free or inexpensive. Solar energy is also environmentally friendly, requires almost no maintenance, and reduces our dependence on foreign energy supplies. Probably the greatest downside of solar energy use is that in areas without direct sunlight during certain times of the year solar panels cannot capture enough energy to provide heat for homes or offices. Geographically speaking, the higher latitudes do not receive as much direct sunlight as tropical areas. Because of the position of the sun in the sky, solar panels must be placed in sun-friendly locations, such as the U.S. desert southwest and the Sahara region of northern Africa. Another downside of solar energy is the efficiency of these systems, which may be seriously affected by how well the system has been installed. The real bottom line on solar energy is obvious: Its advantages outnumber its disadvantages.

DISCUSSION AND REVIEW QUESTIONS

1. The sun has been described in many ways. How would you describe the sun?
2. Is there a difference between solar energy and solar power?
3. What is the 5-5-5 rule?
4. How are concentrating solar power systems (CSPs) classified?

5. What are the two types of linear concentrator systems?
6. What are the advantages of using solar dish engines?
7. What is specific heat?
8. Explain daylighting.
9. Explain ballast factor.
10. Do you think solar energy will ever replace fossil fuels completely? Why? Why not?

REFERENCES AND RECOMMENDED READING

Associated Press. (2008). UN says half the world's population will live in urban areas by end of 2008. *International Herald Tribune*, February 26.

Baker, M.S. (1990). Modeling complex daylighting with DOE 2.1-C. *DOE-2 User News*, 11(1), 1–15.

Baylon, D. and Storm, P. (2008). Comparison of commercial LEED buildings and non-LEED buildings within the 2002–2004 Pacific Northwest commercial building stock, in *Proceedings for ACEEE Summer Study on Energy Efficiency in Buildings*. Stockholm, Sweden: European Council for an Energy Efficient Economy.

Bierman, A. (2007). Photosensors: dimming and switching systems for daylight harvesting. *National Lighting Product Information Program*, 11(1) (http://www.lighignreserach.org/programs/NLPIP/PDF/View/SR_Photosensors.pdf).

Birt, B. and Newsham, G.R. (2009). *Post-Occupancy Evaluation of Energy and Indoor Environment Quality in Green Building: A Review*, NRCC-51211. Ottawa, Ontario, Canada: Institute for Research in Construction, National Research Council Canada.

Brookfield, H.C. (1989). *Sensitivity to Global Change: A New Task for Old/New Geographers*, 19th Norma Wilkinson Memorial Lecture. Reading, U.K.: University of Reading.

Brundtland, G.H. (1987). *Our Common Future*. New York: Oxford University Press.

Brunger, A. and Hollands, K. (1996). Back-of-Plate Heat Transfer in Unglazed Perforated Collectors Operated under Non-Uniform Air Flow Conditions, paper presented at the 22nd Annual Conference of the Solar Energy Society of Canada, Inc., Orillia, Ontario, Canada.

Calder, W.A. (1996). *Size, Function, and Life History*. Mineola, NY: Dover Publications.

Chiras, D.D. (2002). *The Solar House: Passive Heating and Cooling*. White River Junction, VT: Chelsea Green Publishing, p. 33.

Christensen, C. (1982). As quoted in *Making It Happen: A Positive Guide to the Future*, Richardson, J., Ed. Warwick, NY: Roundtable Press.

Christensen, C., Hancock, E., Barker, G., and Kutscher, C. (1990). Cost and performance predictions for advanced active solar concepts, in *Proceedings of the American Solar Energy Society Annual Meeting*, March 19–22, Austin, TX.

Debres, K. (2005). Burgers for Britain: a cultural geography of McDonald's UK. *Journal of Cultural Geography*, 22(2).

Dengler, J. and Wittwear, V. (1994). Glazings with granular aerogels. *SPIE*, 255, 718–727.

Duffie, J. and Beckman, W. (1991). *Solar Engineering of Thermal Processes*. New York: John Wiley & Sons.

Dumortier, D. (1997). Evaluation of luminous efficacy models according to sky types and atmospheric conditions, in *Proceedings Lux Europa Conference*, May 11–14, Amsterdam, pp. 1068–1080.

Enermodal Engineering, Ltd. (1995). *Performance of the Perforated-Plate/Canopy Solarwall at GM Canada*. Ottawa, Ontario, Canada: Government of Canada Publications.

Fenchel, T. (1974). Intrinsic rate of natural increase: the relationship with body size. *Oecologia*, 14, 317–326.

Fontoynont, M., Place, W., and Bauman, P. (1984). Impact of electric lighting efficiency on the energy saving potential of daylighting from roof monitors, *Energy and Buildings*, 6(4), 375–386.

Galasiu, A.D., Newsham, G.R., Suvagau, C., and Sander, D.M. (2007). Energy saving lighting control system for open-plan offices: a field study. *Leukos*, 4(1), 7–29.

Galasiu, A.D., Atif, M.R., and MacDonald, R.A. (2004). Impact of window blinds on daylight-linked dimming and automatic on/off lighting controls. *Solar Energy*, 76(5), 523–544.

Gunnewick, L. (1994). An Investigation of the Flow Distribution Through Unglazed Transpired-Plate Solar Air Heaters, Master's thesis. Waterloo, Canada: Department of Mechanical Engineering, University of Waterloo.

Hanson, B.J. (2004). *Energy Power Shift: Benefiting from Today's New Technologies*. Maple, WI: Lakota Scientific Press.

Hardin, G. (1986). Cultural carrying capacity: a biological approach to human problems. *BioScience*, 36, 599–606.
Heschong, L. and McHugh, J. (2006). Skylights: calculating illumination levels and energy impacts. *Journal of the Illuminating Engineering Society*, Winter, pp. 90–100.
Hickman, C.P., Roberts, L.S., and Hickman, F.M. (1990). *Biology of Animals*. St. Louis, MO: Times Mirror/Mosby.
Hinrichs, R.A. and Kleinback, M.H. (2006). *Energy: Its Use and the Environment*, 4th ed. New York: Thomson Learning.
HMG. (2006). *Sidelighting Photocontrols Field Study: A Field Study Sponsored by Southern California Edison Company, Pacific Gas & Electric Company, and Northwest Energy Efficiency Alliance*. Sacramento, CA: Heschong Mahone Group (http://www.h-m-g.com/Projects/Photocontrols/sidelighting_photocontrols_field.htm).
Holdren, J.P. (1991). Population and the energy problem. *Population and Environment*, 12, 231–255.
Hubbell, S.P. and Johnson, L.K. (1977). Competition and next spacing in a tropical stingless bee community. *Ecology*, 58, 949–963.
James J. Hirsch & Associates. (2009). *eQUEST Introductory Tutorial*. Camarillo, CA: James J. Hirsch & Associates.
James J. Hirsch & Associates and E.O. Lawrence Berkeley National Laboratory. (2004). *DOE-2.2: Building Energy Use and Cost Analysis Program*. Vol. 3. *Topics*. University of California, Berkeley: Simulation Research Group, E.O. Lawrence Berkeley National Laboratory.
Janak, M. (1997). Coupling building energy and lighting simulation, in *Proceedings of the Fifth International IPBSA Conference*, September 8–10, Prague, pp. 313–319.
Janak, M. and Macdonald, I. (1999). Current state-of-the-art of integrated thermal and lighting simulation and future issues, in *Proceedings of Building Simulation '99, an IBPSA Conference*, September 13–15, Kyoto.
Krebs, R.E. (2001). *Scientific Laws, Principles, and Theories*. Westport, CT: Greenwood Press.
Kuhlken, R. (2002). Intensive agricultural landscapes of Oceania. *Journal of Cultural Geography*, 19(2), 161–195.
Kutscher, C. (1992). An Investigation of Heat Transfer for Air Flow Through Low Porosity Perforated Plates, PhD thesis. Boulder, CO: University of Colorado.
Lauoadi, A. and Aresnault, C. (2004). *Validation of Skyvision*. IRC-RR-167. National Research Council.
Lauoadi, A. and Atif, M.R. (1999). Predicting optical and thermal characteristics of transparent single-glazed domed skylights. *ASHRAE Transactions*, 105(2), 325–333.
Kutscher, C., Christensen, C., and Barker, G. (1993). Unglazed transpired solar collectors: heat loss theory. *Journal of Solar Energy Engineering*, 115(3), 182–188.
Library of Congress. (2010). *The World's First Hydroelectric Power Plant Began Operation September 30, 1882*, America's Story from America's Library Series. Washington, D.C.: The Library of Congress (http://www.americaslibrary.gov/jb/gilded/jb_gilded_hydro_1.html).
Lund, J.W. (2007). Characteristics, development and utilization of geothermal resources. *Institute of Technology*, 28(2), 1–9.
Moore, F. (1991). *Concepts and Practice of Architectural Daylighting*. New York: Van Nostrand Reinhold, pp. 98–99.
Newsham, G.R. (2004). *American National Standard Practice for Office Lighting*, RP-1. Washington, D.C.: American National Standards Institute–Illuminating Engineering Society of North America.
NREL. (2010a). *Concentrating Solar Power*. Washington, D.C.: National Renewable Energy Laboratory (http://www.nrel.gov/learning/re_csp.html).
NREL. (2010b). *Concentrating Solar Power*. Washington, D.C.: National Renewable Energy Laboratory (http://www.nrel.gov/csp/pdfs/capability_factsheet_overview_of_csp_capabilities.pdf).
O'Connor, J., Lee, E., Rubinstein, L.E., and Selkowitz, S. (1997). *Tips for Daylighting with Windows*. Berkeley, CA: Lawrence Berkeley National Laboratory.
Patel, M. (1999). *Wind and Solar Power Systems*. Boca Raton, FL: CRC Press.
Perez-Blanco, H. (2009). *The Dynamics of Energy: Supply, Conversion, and Utilization*. Boca Raton, FL: CRC Press.
Place, W. et al. (1984). The predicted impact of roof aperture design on the energy performance of office buildings. *Energy and Building*, 6(4), 361–373.
Reid, W. et al. (1988). *Bankrolling Successes*. Washington, D.C.: Environmental Policy Institute and National Wildlife Federation.

Rubinstein, F., Ward, G., and Verderber, R. (1989). Improving the performance of photo-electrically controlled lighting systems. *Journal of the Illuminating Engineering Society*, 18(1), 70–94.

Searchinger, T. et al. (2008). Use of U.S. croplands for biofuels increases greenhouse gases through emissions from land-use change. *Science*, 319(5867), 1238–1240.

Simon, J.L. (1980). Resources, population, environment: an oversupply of false bad news. *Science*, 208, 1431–1437.

UCS. (2008). *Land Use Changes and Biofuels*. Cambridge, MA: Union of Concerned Scientists (http://www.ucsusa.org/clean_vehicles/technologies_and_fuels/biofuels/Land-Use-Changes-and-Biofuels.html).

USDOE. (1998). *Solar Buildings: Transpired Air Collectors*. Washington, D.C.: U.S. Department of Energy (http://www1.eere.energy.gov/femp/pdfs/FTA_trans_coll.pdf).

USDOE. (2006). *Geothermal Technologies Program: A History of Geothermal Energy in the United States*. Washington, D.C.: U.S. Department of Energy (http://www1.eere.energy.gov/geothermal/history.html).

USDOE. (2008a). *History of Hydropower*. Washington, D.C.: U.S. Department of Energy (http://www1.eere.energy.gov/windandhydro/hydro_history.html).

USDOE. (2008b). *Concentration Solar Power Commercial Application Study*. Washington, D.C.: U.S. Department of Energy (http://www1.eere.energy.gov/solar/pdfs/csp_water_study.pdf).

USDOE. (2009a). *Linear Concentrators Research and Development*. Washington, D.C.: U.S. Department of Energy (http://www1.eere.energy.gov/solar/linear_concentrator_rnd.html).

USDOE. (2009b). *Thermal Storage Research and Development*. Washington, D.C.: U.S. Department of Energy (http://www1.eere.energy.gov/solar/thermal_storage_rnd.html).

USDOE. (2010a). *Dish/Engine Systems for Concentrating Solar Power*. Washington, D.C.: U.S. Department of Energy (http://www.eere.energy.gov/basics/renewable_energy/dish_engine.html).

USDOE. (2010b). *Solar Energy Technologies Program*. Washington, D.C.: U.S. Department of Energy (http://www1.eere.energy.gov/solar/).

USDOE. (2010c). *Fossil Fuels*. Washington, D.C.: U.S. Department of Energy (http://www.energy.gov/energysources/fossilfuels.htm).

USEPA. (2009). *The Science of Ozone Layer Depletion*. Washington, D.C.: U.S. Environmental Protection Agency (http://www.epa.gov/ozone/science/index.html).

USGS. (2010). *Hawaiian Volcano Observatory*. Washington, D.C.: U.S. Geological Survey (http://hvo.wr.usgs.gov/).

Vitousek, P.M., Ehrlich, A.H., and Matson, P.A. (1986). Human appropriation of the products of photosynthesis. *BioScience*, 36, 368–373.

Volterra, V. (1926). Variazioni e fluttuazioni del numero d'indivudui in specie di animali conviventi. *Memorie Accademia dei Lincei*, 2, 31–113.

World Commission on Environment and Development. (1987). *Our Common Future*. New York: Oxford University Press.

Wilson, E.O. (2001). *The Future of Life*. New York: Knopf.

Winkelman, F. and Selkowitz, S. (1985). Daylighting simulation in the DOE-2 building energy analysis program. *Energy and Buildings*, 8, 271–286.

4 Wind Energy

The wind goeth toward the south, and turneth about unto the north; it whirleth about continually, and the wind returneth again.

—Ecclesiastes 1:6

The Good, Bad, and Ugly of Wind Energy

Good: As long as Earth exists, the wind will always exist. The energy in the winds that blow across the United States each year could produce more than 16 billion GJ of electricity—more than one and one-half times the electricity consumed in the United States in 2000.

Bad: Turbines are expensive. Wind doesn't blow all the time, so they have to be part of a larger plan. Turbines make noise. Turbine blades kill birds.

Ugly: Some look upon giant wind turbine blades cutting through the air as grotesque scars on the landscape, as visible polluters.

The bottom line: Do not expect Don Quixote, mounted in armor on his old nag, Rocinate, with or without Sancho Panza, to lead the charge to build those windmills. Instead, expect—you can count on it, bet on it, and rely on it—that the charge will be made by the rest of us to satisfy our growing, inexorable need for renewable energy. What other choice do we have?

INTRODUCTION*

Obviously, wind energy or power is all about wind. In simple terms, wind is the response of the atmosphere to uneven heating conditions. Earth's atmosphere is constantly in motion. Anyone observing the constant cloud movement and weather changes around them is well aware of this phenomenon. Although its physical manifestations are obvious, the importance of the dynamic state of our atmosphere is much less obvious. The constant motion of Earth's atmosphere is both horizontal (*wind*) and vertical (*air currents*). This air movement is the result of thermal energy produced from heating of the surface of the Earth and the air molecules above. Because of differential heating of the surface of the planet, energy flows from the equator poleward. The energy resources contained in the wind in the United States are well known and mapped in detail (Hanson, 2004). It is clear that air movement plays a critical role in transporting the energy of the lower atmosphere, bringing the warming influences of spring and summer and the cold chill of winter, and wind and air currents are fundamental to how nature functions. Still, though, the effects of air movements on our environment are often overlooked. All life on Earth has evolved or has been sustained with mechanisms dependent on air movement; for example, pollen is carried by the winds for plant reproduction, animals sniff the wind for essential information, and wind power was the motive force during the earliest stages of the Industrial Revolution. We can also see other effects of winds. Wind causes weathering (erosion) of the Earth's surface, wind influences ocean currents, and the wind carries air pollutants and contaminants such as radioactive particles that impact our environment.

* Parts of this section are adapted from Spellman, F.R. and Whiting, N.E., *Environmental Science and Technology*, Government Institutes Press, Lanham, MD, 2006.

DID YOU KNOW?

Wind speed is generally measured in meters per second (m/s), but Americans usually think in terms of miles per hour (mph). To convert m/s to mph, a good rule of thumb is to double the value in m/s and add 10%.

AIR IN MOTION

With respect to winds and air currents, as in all dynamic situations, forces are necessary to produce motion and changes in motion. The atmosphere, which is made up of various gases, is subject to two primary forces: gravity and pressure differences from temperature variations. *Gravity* (gravitational forces) holds the atmosphere close to the Earth's surface. Newton's law of universal gravitation states that every body in the universe attracts another body with a force equal to

$$F = G\left(\frac{m_1 m_2}{r^2}\right)$$

where F is the magnitude of the gravitational force between the two bodies, G is the gravitational constant ≈ 6.67×10^{-11} N (m^2/kg^2), m_1 and m_2 are the masses of the two bodies, and r is the distance between the two bodies.

The force of gravity decreases as an inverse square of the distance between the two bodies. Thermal conditions affect density, which in turn affects vertical air motion and planetary air circulation (and how air pollution is naturally removed from the atmosphere). Although forces acting in other directions can overrule gravitational force, gravity constantly acts vertically downward, on every gas molecule, which accounts for the greater density of air near the Earth.

Atmospheric air is a mixture of gases, so the gas laws and other physical principles govern its behavior. Pressure is equal to the force per unit area (pressure = force/area). Because the pressure of a gas is directly proportional to its temperature, temperature variations in the air give rise to differences in pressure or force. These differences in pressure cause air movement, both large and small scale, from a high-pressure region to a low-pressure region.

Horizontal air movements (*advective winds*) result from temperature gradients, which give rise to density gradients and, subsequently, pressure gradients. The force associated with these pressure variations (*pressure gradient force*) is directed perpendicular to lines of equal pressure (called *isobars*) and is directed from high to low pressure. In Figure 4.1, the pressures over a region are mapped by taking barometric readings at different locations. Lines drawn through the points (locations) of equal pressure are the isobars. All points on an isobar are of equal pressure, indicating no air movement along the isobar. The wind direction is at right angles to the isobar in the direction of the lower pressure. In Figure 4.1, notice that air moves down a pressure gradient toward a lower isobar like a ball rolls down a hill. If the isobars are close together, the pressure gradient force is large, and such areas are characterized by high wind speeds. If isobars are widely spaced, the winds are light because the pressure gradient is small.

Localized air circulation gives rise to *thermal circulation* (a result of the relationship based on a law of physics whereby the pressure and volume of a gas are directly related to its temperature). A change in temperature causes a change in the pressure and volume of a gas. With a change in volume comes a change in density (density = mass/volume), so regions of the atmosphere with different temperatures may have different air pressures and densities. As a result, localized heating sets up air motion and gives rise to *thermal circulation.* To gain an understanding of this phenomenon, consider Figure 4.2.

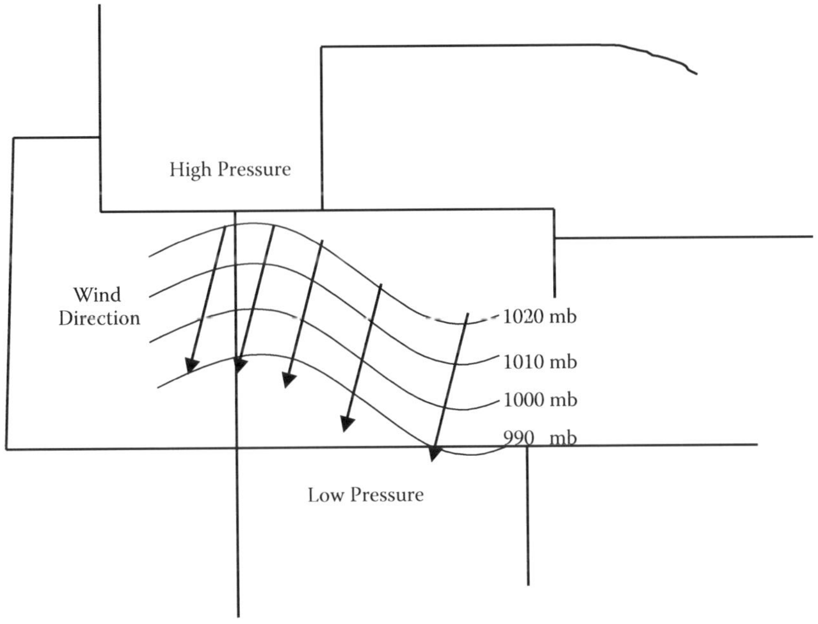

FIGURE 4.1 Isobars drawn through locations having equal atmospheric pressures. The air motion, or wind direction, is at right angles to the isobars and moves from a region of high pressure to a region of low pressure.

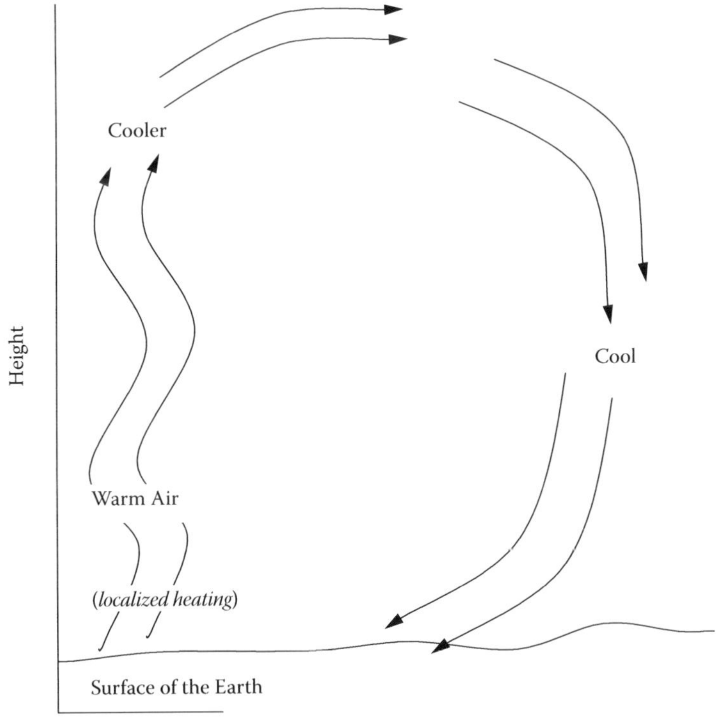

FIGURE 4.2 Thermal circulation of air. Localized heating, which causes air in the region to rise, initiates the circulation. As the warm air rises and cools, cool air near the surface moves horizontally into the region vacated by the rising air. The upper, still cooler, air then descends to occupy the region vacated by the cool air.

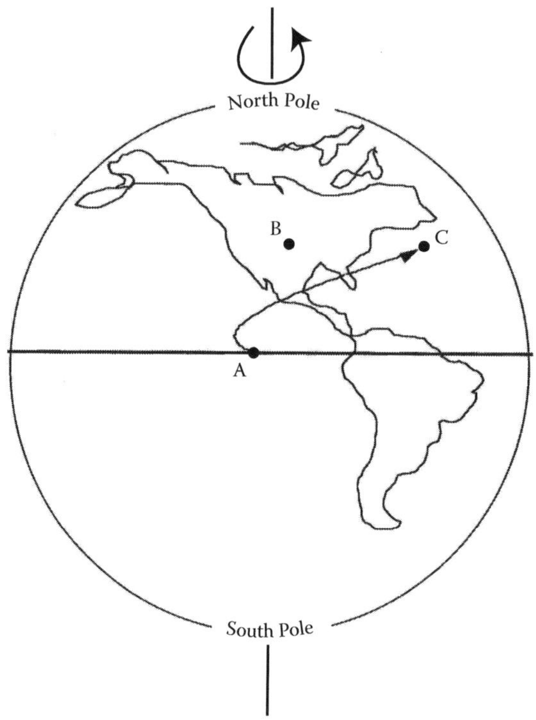

FIGURE 4.3 The effect of the Earth's rotation on the trajectory of a propelled particle.

Once the air has been set into motion, secondary (velocity-dependent) forces come into play. These secondary forces are caused by Earth's rotation and contact of the air with the rotating Earth. The *Coriolis force*, named after its discoverer, French mathematician Gaspard Coriolis (1772–1843), is an effect on the atmosphere and on all objects on the Earth's surface caused by this rotation. The large-scale north or south movement of air appears to be deflected from the expected path. Air moving poleward in the Northern Hemisphere appears to be deflected toward the east; air moving southward appears to be deflected toward the west. The reverse is true for the Southern Hemisphere. Figure 4.3 illustrates the Coriolis effect on a propelled particle (analogous to the apparent affect of an air mass flowing from Point A toward Point B): The Earth's rotation from east to west acts on the air particle as it travels north over the Earth's surface. The particle will actually reach Point C because even though it is moving in a straight line (deflected) the earth rotates east to west beneath it.

Friction (drag or resistance) can also deflect or decelerate air movement and can be both internal and external. Internal friction is caused by air molecules running into each other; external friction is due to contact of the air with terrestrial surfaces.

WIND ENERGY*

Wind energy is power produced by the movement of air. Since early recorded history, people have been harnessing the energy of the wind to, for example, mill grain and pump water. Wind energy propelled boats along the Nile River as early as 5000 B.C. By 200 B.C., simple windmills in China were pumping water, and windmills with woven reed sails were grinding grain in Persia and the Middle East. The use of wind energy spread around the world, and by the 11th century people in the

* Much of the information in this section is from EERE, *History of Wind Energy*, Energy Efficiency and Renewable Energy, U.S. Department of Energy, Washington, D.C., 2005 (http://www1.eere.energy.gov/windandhydro/wind_history.html).

TABLE 4.1
U.S. Energy Consumption by Energy Source, 2008

Energy Source	**Energy Consumption (quadrillion Btu)**
Total	99.438
Fossil fuels (coal, natural gas, petroleum)	83.532
Electricity net imports	0.113
Nuclear electric power	8.427
Renewable	7.367
Biomass (total)	3.852
Biofuels	1.372
Waste	0.436
Wood and wood-derived fuels	2.044
Geothermal energy	0.360
Hydroelectric conventional	2.512
Solar thermal/PV energy	0.097
Wind energy	**0.546**

Source: EIA, *U.S. Energy Consumption by Energy Source*, Energy Information Administration, Washington, D.C., 2007 (www.eia.doe.gov/cneaf/solar.renewables/page/trends/table1.html).

Middle East were using windmills extensively for food production; returning merchants and crusaders carried this idea back to Europe. The Dutch refined the windmill and adapted it for draining lakes and marshes in the Rhine River delta. When settlers took this technology to the New World in the later 19th century, they used windmills to pump water for farms and ranches and, later, to generate electricity for homes and industry.

The first known wind turbine designed to produce electricity was built in 1888 by Charles F. Brush, in Cleveland, Ohio; it was a 12-kW unit that charged batteries in the cellar of a mansion. The first wind turbine used to generate electricity outside of the United States was built in Denmark in 1891 by Poul la Cour, who used electricity from his wind turbines to electrolyze water to make hydrogen for the gas lights at the local schoolhouse. By the 1930s and 1940s, hundreds of thousands of wind turbines were used in rural areas of the United States that were not yet served by the grid. The oil crisis in the 1970s created a renewed interest in wind, until the U.S. government stopped giving tax credits.

Today, several hundred thousand windmills are in operation around the world, many of which are used for pumping water. The use of wind energy as a pollution-free means of generating electricity on a significant scale is attracting the most interest in the subject today. As a matter of fact, due to current and pending shortages and high costs of fossil fuels to generate electricity, as well as the green movement toward the use of cleaner fuels, wind energy is the world's fastest-growing energy source and could power industry, businesses, and homes with clean, renewable electricity for many years to come. In the United States, wind-based electricity generating capacity has increased markedly since the 1970s. Today, though, it still represents only a small fraction of total electric capacity and consumption (see Table 4.1), despite the advent of $4/gal gasoline, increases in the cost of electricity, high heating and cooling costs, and worldwide political unrest or uncertainty in oil-supplying countries. Traveling the wind corridors of the United States (primarily Arizona, New Mexico, Texas, Missouri, and north through the Great Plains to the Pembina Escarpment and Turtle Mountains of North Dakota) gives some indication of the considerable activity and seemingly exponential increase in wind energy development and wind turbine installations.

DID YOU KNOW?

We can classify wind energy as a form of solar energy. As mentioned, winds are caused by uneven heating of the atmosphere by the sun, irregularities of the Earth's surface, and rotation of the Earth. As a result, winds are strongly influenced and modified by local terrain, bodies of water, weather patterns, and vegetative cover, among other factors. The energy of winds harvested by wind turbines can be used to generate electricity.

WIND POWER BASICS

The terms *wind energy* and *wind power* reflect the process by which the wind is used to generate mechanical power or electricity. Wind turbines convert the kinetic energy in the wind into mechanical power. This mechanical power can be used for specific tasks (such as grinding grain or pumping water) or a generator can convert this mechanical power into electricity (EERE, 2006a). We have been harnessing the energy of the wind for hundreds of years. From old Holland to farms in the United States, windmills have been used for pumping water or grinding grain; today, the modern equivalent of a windmill—a wind turbine—can use the energy of the wind to generate electricity. The blades of a wind turbine spin like aircraft propeller blades. Wind turns the blades, which in turn, spin a shaft connected to a generator and produce electricity (Wind Energy Programmatic EIS, 2009). Unlike fans, which use electricity to make wind, wind turbines use wind to make electricity.

WIND TURBINE TYPES

Whether referred to as *wind-driven generators*, *wind generators*, *wind turbines*, *wind-turbine generators*, or *wind energy conversion systems*, modern wind turbines fall into two basic groups: the horizontal-axis wind turbine (HAWT), like the traditional farm windmills used for water pumping (see Figure 4.4), and the vertical-axis wind turbine (VAWT), like the eggbeater-style, Darrieus

FIGURE 4.4 Typical farm windmill used for pumping water.

rotor model, named after its French inventor, the only vertical-axis machine with any commercial success. Wind hitting the vertical blades, called *aerofoils*, generates lift to create rotation. No yaw (rotation about vertical axis) control is needed to keep them facing into the wind. The heavy machinery in the nacelle (cowling) is located on the ground. Blades are closer to the ground, where wind speeds are lower. Most large modern wind turbines are horizontal-axis turbines; therefore, they are highlighted and described in detail in this text.

> **DID YOU KNOW?**
> Whenever wind energy is considered as a possible source of renewable energy, it is important to consider the amount of land area required, accessibility to generators, and aesthetics.

Horizontal-Axis Wind Turbines

Wind turbines are available in a variety of sizes and power ratings. Utility-scale turbines range in size from 100 kW to as large as several megawatts. Horizontal-axis wind turbines typically have two or three blades. Downwind horizontal-axis wind turbines have a turbine with the blades behind (downwind from) the tower. No yaw control is needed because they naturally orient themselves in line with the wind; however, these downwind HAWTs experience a shadowing effect, in that when a blade swings behind the tower the wind it encounters is briefly reduced and the blade flexes. Upwind HAWTs usually have three-blades in front (upwind) of the tower. These upwind wind turbines require a somewhat complex yaw control to keep them facing into the wind. They operate more smoothly and deliver more power and thus are the most commonly used modern wind turbines. The largest machine has blades that span more than the length of a football field, stands 20 building stories high, and produces enough electricity to power 1400 homes.

> **DID YOU KNOW?**
> Groups of wind turbines are referred to as *wind farms* or *wind parks.*

Inside the HAWT

Basically, a horizontal-axis wind turbine consists of three main parts: a turbine, a nacelle, and a tower. Several other important parts are contained within the tower and nacelle, including anemometer, blades, brake, controller, bear box, generator, high-speed shaft, low-speed shaft, pitch, rotor, tower, wind direction, wind vane, yaw drive, and yaw motor (see Figure 4.5).

TURBINE COMPONENTS

Anemometer—An anemometer is a device that measures the wind speed and transmits wind speed data to the controller.

Blades—Wind blowing over the two or three blades causes the blades to lift and rotate; they capture the kinetic energy of the wind and help the turbine rotate.

Brake—A disc brake can be applied mechanically, electrically, or hydraulically to stop the rotor in emergencies.

Controller—The controller starts up the machine at wind speeds of about 8 to 16 mph and shuts off the machine at about 55 mph. Turbines do not operate at wind speeds above about 55 mph because they might be damaged by the high winds.

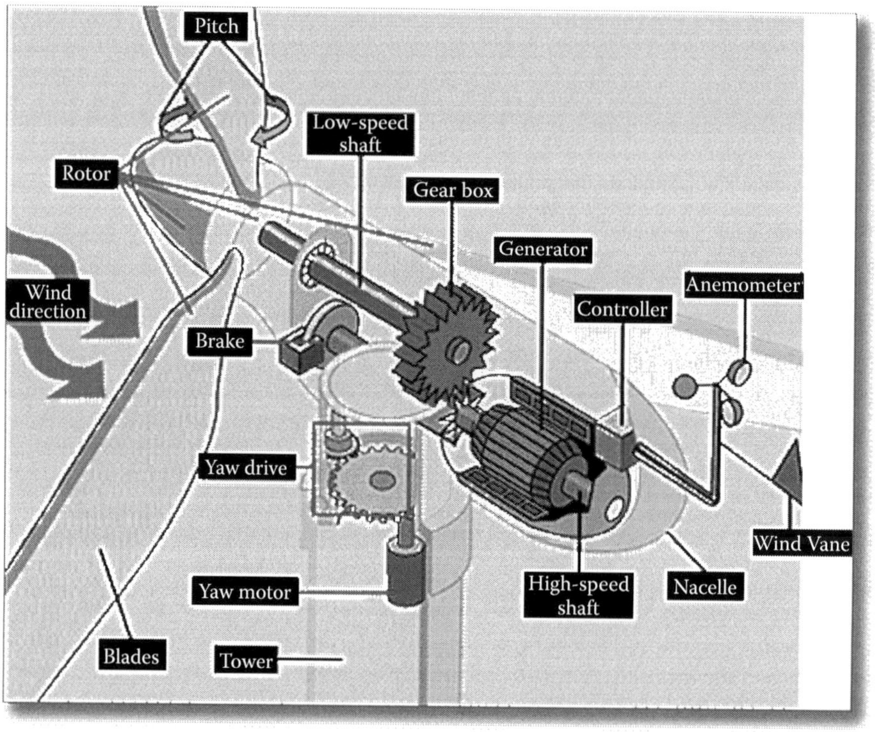

FIGURE 4.5 Horizontal-axis wind turbine components. (From EERE, *How Wind Turbines Work*, Energy Efficiency & Renewable Energy, U.S. Department of Energy, Washington, D.C.; http://www1.eere.energy.gov/Windandhydro/wind_how.html.)

Gear box—Gears connect the low-speed shaft to the high-speed shaft and increase the rotational speeds from about 30 to 60 rotations per minute (rpm) to about 1000 to 1800 rpm, the rotational speed required by most generators to produce electricity. The gear box is a costly (and heavy) part of the wind turbine, and engineers are exploring direct-drive generators that operate at lower rotational speeds and do not require gear boxes.

Generator—An off-the-shelf induction generator produces 60-hertz AC electricity.

High-speed shaft—This shaft drives the generator.

Low-speed shaft—This shaft turns at about 30 to 60 rotations per minute.

Nacelle—The nacelle unit sits atop the tower and contains the gear box, generator, low- and high-speed shafts, controller, and brake.

Pitch—Bales are turned, or pitched, out of the wind to control the rotor speed and keep the rotor turning in winds that are too high or too low to produce electricity.

Rotor—The blades and the hub together are called the rotor.

Tower—The tower is made from tubular steel, concrete, or steel lattice. Wind speed increases with height, so turbines on taller towers capture more energy and generate more electricity.

Wind direction—An upwind turbine operates facing into the wind; other turbines are designed to run downwind, facing away from the wind.

Wind vane—The wind vane measures wind direction and communicates with the yaw drive to orient the turbine properly with respect to the wind.

Yaw drive—The yaw drive of upwind turbines is used to keep the entire nacelle and thus the rotor facing into the wind as the wind direction changes. Downwind turbines do not require a yaw drive, as the wind blows the rotor downwind.

Yaw motor—The yaw motor powers the yaw drive.

> **DID YOU KNOW?**
>
> During the rotation of the nacelle, it is possible for the cables inside the tower to twist, with the cables becoming more and more twisted if the turbine keeps turning in the same direction. The wind turbine is therefore equipped with a cable twist counter, which notifies the controller that it is time to straighten the cables (Khaligh and Onar, 2010).

WIND ENERGY AND POWER CALCULATIONS

A wind turbine is a machine that converts the kinetic energy of wind into the mechanical energy of a shaft. Calculating the energy and power available in the wind relies on a knowledge of basic physics and geometry. The *kinetic energy* of an object is the extra energy it possesses because of its motion. It is defined as the work necessary to accelerate a body of a given mass from rest to its current velocity. Once in motion, a body maintains it kinetic energy unless its speed changes. The kinetic energy of a body is given by

$$\text{Kinetic Energy} = 0.5 \times m \times v^2 \tag{4.1}$$

where:

m = mass
v = velocity

■ Example 4.1. Determining Power in the Wind

Let's say we have a large packet of wind of thickness D that passes through the plane of a wind turbine's blades, which sweep out at cross-sectional area A (see Figure 4.6).

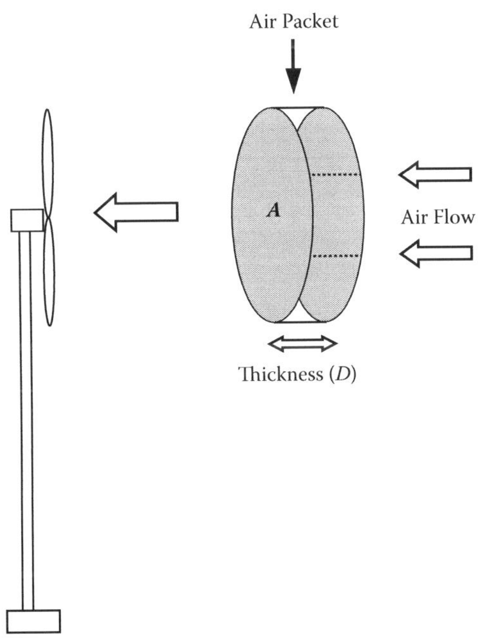

FIGURE 4.6 A packet of air passing through the plane of a wind turbine's blades, with the thickness (D) passing through the plane over a given time.

Step 1. To determine the power in the wind, we must first consider the kinetic energy of this packet of air, along with its mass (m) and velocity (v), as shown in Equation (4.1). We then divide by time to get power:

$$\text{Power through area } A = \frac{1}{2}\left(\frac{m \text{ passing through } A}{t}\right)v^2 \tag{4.2}$$

Step 3. The mass flow rate can be expressed as follows (where ρ is air density, the mass per unit volume of Earth's atmosphere):

$$m = \left(\frac{m \text{ passing through } A}{t}\right)\rho A v \tag{4.3}$$

Step 4. Combining Equations (4.2) and (4.3), we obtain:

$$\begin{aligned} \text{Power through area } A &= \frac{1}{2}(\rho A v)\,v^2 \\ \text{Power in the wind} = P_W &= \frac{1}{2}\left(\rho A v^3\right) \end{aligned} \tag{4.4}$$

where:

P_W = power in the wind (watts)
ρ = air density (2.225 kg/m^3 at 15°C and 1 atm)
A = cross-sectional area that wind passes through perpendicular to the wind (m^2)
v = wind speed normal to A (m/s; 1 m/s = 2.237 mph)

DID YOU KNOW?

Air density decreases with increasing altitude, as does air pressure. It also changes with variations in temperature or humidity.

From Equation (4.4), it is evident that wind power is a function of the cube of the wind speed; that is, doubling the wind speed increases the power by eight. As an example, the energy produced during 1 hour of 20-mph winds is the same as the energy produced during 8 hours of 10-mph winds. When we speak of power (W/m^2) and wind speed (mph), we cannot use average wind speed because the relationship is nonlinear. Power in the wind is also proportional to A. For a conventional horizontal-axis wind turbine, $A = (\pi/4)D^2$, so wind power is proportional to the blade diameter squared. Because cost is approximately proportional to blade diameter, larger wind turbines are more cost effective.

Air-Density Correction Factors

Earlier we pointed out that air density if affected by different temperature and pressure. Air-density correction factors can correct air density for temperature and altitude. Correction factors for both temperature and altitude correction can be found in standardized tables. Equation (4.5) is used to determine air density (ρ) for various temperatures and pressures:

$$\rho = \frac{P \times MW \times 10^{-3}}{RT} \tag{4.5}$$

where:

P = absolute gas pressure (atm)
MW = molecular weight of air (= 28.97 g/mol)
R = ideal gas constant (= 8.2056×10^{-5} $m^3 \cdot atm \cdot K^{-1} \cdot mol^{-1}$)
T = absolute temperature (degrees Kelvin = 273 + °C)

DID YOU KNOW?
Air density is greater at lower temperatures.

Elevation and Earth's Roughness

The speed of wind is affected by its elevation above the Earth and the roughness of Earth. Because power increases like the cube of wind speed, we can expect a significant economic impact from even a moderate increase in wind speed; thus, in the operation of wind turbines, wind speed is a very important parameter. The surface features of Earth cannot be ignored when deciding where to place wind turbines and calculating their output productivity. Natural obstructions such as mountains and forests and human-made obstructions such as buildings cause friction as winds flow over them. Generally, a lot of friction is encountered in the first few hundred meters above ground (although smooth surfaces such as water offer little resistance). For this reason, taller wind turbine towers are better.

When actual measurements are not possible or available, it is possible to characterize or approximate the impact of rough surfaces and height on wind speed:

$$\frac{v}{v_0} = \left(\frac{H}{H_0}\right)^{\alpha} \tag{4.6}$$

where:

α = friction coefficient, obtained from standardized tables (typical values of α are 0.10 for calm water or smooth, hard ground; 0.15 for open terrain; and 0.40 for a large city with tall buildings)
v = wind speed at height H
v_0 = wind speed at height H_0 (H_0 is usually 10 m)

Wind Turbine Rotor Efficiency

Generally, when we think or talk about efficiency we think about input vs. output and know that if we put 100% into something and get 100% output then we have a very efficient machine, operation, or process. In engineering, we can approximate efficiency (input vs. output) by performing mass balance calculations. We know that, according to the laws of conservation, mass cannot disappear or be destroyed; thus, it must continue to exist in one form or another. With regard to this input vs. output concept, we should address a couple views on maximum rotor efficiency that are misinformed and make no sense but are stated here to point out what really does make sense. The two wrong assumptions are that (1) it can be assumed that the downwind velocity is zero and the turbine

DID YOU KNOW?

The energy in wind is a cubed function of wind speed, which means that if the wind speed doubles there is eight times as much available energy, not twice as much as one might expect.

extracts 100% of the power from the wind, or (2) downwind velocity is the same as the upwind velocity and the turbine has not reduced the wind speed at all. Wrong! Albert Betz set the record straight in 1919 when he proposed that there exists a maximum theoretical efficiency for extracting kinetic energy from the wind.

Derivation of Betz's Law

To understand Betz's law, we must first understand the constraints on the ability of a wind turbine to convert kinetic energy in the wind into mechanical power. Visualize wind passing through a turbine (see Figure 4.7)—it slows down and the pressure is reduced so it expands. Equation (4.7) is used to determine the power extracted by the blades:

$$P_b = \frac{1}{2} m \left(v^2 - v_d^2\right) \tag{4.7}$$

where:

m = mass flow rate of air within stream tube
v = upwind undisturbed wind speed
v_d = downwind wind speed

Determining the mass flow rate is the next step. In making the determination, it is easiest to use the cross-sectional area A at the plane of the rotor because we know what this value is; thus, the mass flow rate is

$$m = p \times A \times v_b \tag{4.8}$$

Assume that the velocity through the rotor (v_b) is the average of the upwind velocity (v) and the downwind velocity (v_d):

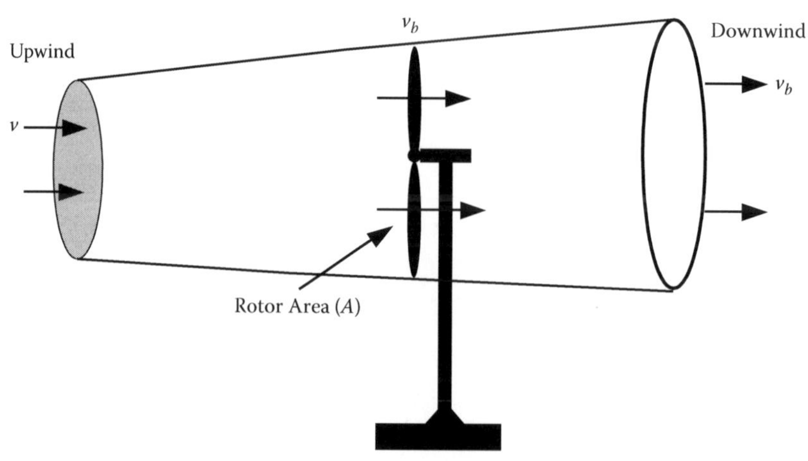

FIGURE 4.7 Wind passing through a turbine.

$$v_b = \frac{v + v_d}{2} \quad \rightarrow \quad m = \rho A\left(\frac{v + v_d}{2}\right)$$

Equation (4.7) becomes:

$$P_b = \frac{1}{2}\rho A\left(\frac{v + v_d}{2}\right)\left(v^2 - v_d^2\right) \tag{4.9}$$

Before moving on in the derivation process, it is important that we define operating speed ratio (λ):

$$\lambda = \frac{v_d}{v} \tag{4.10}$$

We can now rewrite Equation (4.9) as

$$P_b = \frac{1}{2}\rho A\left(\frac{v + \lambda v}{2}\right)\left(v^2 - \lambda^2 v^2\right) \tag{4.11}$$

$$\downarrow$$

$$\left(\frac{v+\lambda v}{2}\right)\left(v^2 - \lambda^2 v^2\right) = \frac{v^3}{2} - \frac{\lambda^2 v^3}{2} + \frac{\lambda v^3}{2} - \frac{\lambda^3 v^3}{2} = \frac{v^3}{2}\left[(1+\lambda) - \lambda^2(1+\lambda)\right] = \frac{v^3}{2}\left[(1+\lambda)(1-\lambda^2)\right]$$

$$P_b = \left(\frac{1}{2}\rho A v^3\right) \quad \times \quad \frac{1}{2}\left[(1+\lambda)(1-\lambda^2)\right]$$

$$\downarrow \qquad\qquad\qquad \downarrow$$

$$(P_W = \text{Power in the wind}) \quad (C_P = \text{Rotor efficiency})$$

The next step is to find the operating speed ratio (λ) that maximizes the rotor efficiency (C_P):

$$C_P = \frac{1}{2}\left[(1+\lambda)(1-\lambda^2)\right] = \frac{1}{2} - \frac{\lambda^2}{2} + \frac{\lambda}{2} - \frac{\lambda^3}{2}$$

Set the derivative of rotor efficiency to zero and solve for λ:

$$\frac{\partial C_P}{\partial \lambda} = -2\lambda + 1 - 3\lambda^2 = 0$$

$$\frac{\partial C_P}{\partial \lambda} = 3\lambda^2 + 2\lambda - 1 = 0$$

$$\frac{\partial C_P}{\partial \lambda} = (3\lambda - 1)(\lambda + 1) = 0$$

$$\vdots$$

$$\lambda = 1/3$$

We plug the optimal value for λ back into the equation for c_P to find the maximum rotor efficiency:

$$\left(\frac{v+\lambda v}{2}\right)\left(v^2 - \lambda^2 v^2\right) = \frac{v^3}{2} - \frac{\lambda^2 v^3}{2} + \frac{\lambda v}{2} \tag{4.12}$$

TABLE 4.2
Betz Limit for 80-M Rotor Turbines

Wind Speed (mph)	Wind Speed (m/s)	Wind Power (kW)	Power Based on 59.3% Betz Limit (kW)
5	2.2	36	21
10	4.5	285	169
15	6.7	962	570
20	8.9	2280	1352
25	11.2	4453	2641
28	12.5	6257	3710
30	13.4	7695	4563
35	15.6	12,220	7246
40	17.9	18,241	10,817
45	20.1	25,972	15,401
50	22.4	35,626	21,126
55	24.6	47,419	28,119
56 (cutoff speed)	25.0	50,053	29,681
60	26.8	61,563	36,507

Source: Adapted from Devlin, L., *Wind Turbine Efficiency*, 2007 (http://k0lee.com/turbineeff.htm).

The maximum efficiency of 59.3% occurs when air is slowed to 1/3 of its upstream rate. This value is called the *Betz efficiency* (Betz, 1966). In plain English, Betz's law states that all wind power cannot be captured by the rotor; otherwise, air would be completely still behind the rotor and not allow more wind to pass through. For illustrative purposes, in Table 4.2, we list wind speed, power of the wind, and power of the wind based on the Betz limit (59.3%).

Tip Speed Ratio

Efficiency is a function of how fast the rotor turns. The tip speed ratio (TSR), an extremely important factor in wind turbine design, is the ratio of the speed of the rotating blade tip to the speed of the free stream wind (see Figure 4.8). Stated differently, TSR is the speed of the outer tip of the blade divided by wind speed. There is an optimum angle of attack that creates the highest lift-to-drag ratio. If the rotor of the wind turbine spins too slowly, most of the wind will pass straight through the gap between the blades, thus giving it no power. But, if the rotor spins too fast, the blades will blur and act like a solid wall to the wind. Moreover, rotor blades create turbulence as they spin through the air. If the next blade arrives too quickly, it will hit that turbulent air. Thus, it is actually better to slow down the blades. Because the angle of attack is dependent on wind speed, there is an optimum tip speed ratio:

$$\text{TSR} = \Omega R/V \tag{4.13}$$

where:

Ω = rotational speed (rad/s)
R = rotor radius
V = wind "free stream" velocity

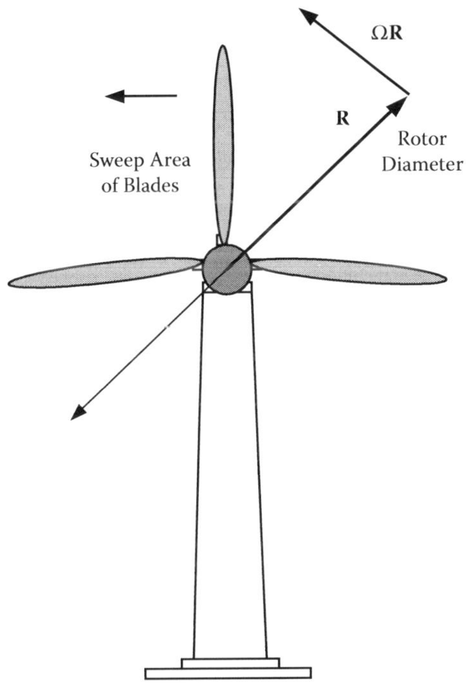

FIGURE 4.8 Tip speed ratio (TSR).

SMALL-SCALE WIND POWER

To meet the typical domestic user's need for electricity, a small-scale wind machine may be the answer. Such a wind turbine has rotors between 8 and 25 feet in diameter, stands around 30 feet, and can supply the power needs of an all-electric home or small business. Utility-scale turbines range in size from 50 to 750 kW. Single small turbines below 50 kW in size are used for homes, telecommunication dishes, or water pumping. Potential users of small wind turbines should ask many questions before installing such a system (EERE, 2009). Let's ask the obvious question first: "What are the benefits to homeowners from using wind turbines?" Wind energy systems provide a cushion against electricity price increases. Wind energy systems reduce U.S. dependence on fossil fuels, and they don't emit greenhouse gases. If you are building a home in a remote location, a small wind energy system can help you avoid the high cost of extending utility power lines to your site.

Although wind energy systems involve a significant initial investment, they can be competitive with conventional energy sources when you account for a lifetime of reduced or altogether avoided utility costs. The length of payback time—the time before the savings resulting from your system equal the system cost—depends on the system you choose, the wind resources at your site, the electric utility rates in your area, and how you use your wind system.

Another frequently asked question is "Is wind power practical for me?" Small wind energy systems can be used in connection with an electricity transmission and distribution system (*grid-connected systems*) or in stand-alone applications that are not connected to the utility grid. A grid-

DID YOU KNOW?

The difference between power and energy is that power (kilowatts) is the rate at which electricity is consumed and energy (kilowatt-hours) is the quantity consumed.

> **SAFETY NOTE**
>
> For safety, batteries should be isolated from living areas and electronics because they contain corrosive and explosive substances. Lead–acid batteries also require protection from temperature extremes.

connected wind turbine can reduce the consumption of utility-supplied electricity for lighting, appliances, and electric heat. If the turbine cannot deliver the amount of energy required, the utility makes up the difference. When the wind system produces more electricity than the household requires, the excess can be sold to the utility. With the interconnections available today, switching takes place automatically. Stand-alone wind energy systems can be appropriate for homes, farms, or even entire communities that are far from the nearest utility lines.

Stand-alone systems (systems not connected to the utility gird) require batteries to store excess power generated for use when the wind is calm. They also must have a charge controller to keep the batteries from overcharging. Deep-cycle batteries, such as those used for golf carts, can discharge and recharge 80% of their capacity hundreds of times, which makes them a good option for remote renewable energy systems. Automotive batteries are shallow-cycle batteries and should not be used in renewable energy systems because of their short life for deep-cycling operations.

Small wind turbines generate direct-current (DC) electricity. In very small systems, DC appliances operate directly off the batteries. To use standard appliances that use conventional household alternating current (AC), an inverter must be installed to convert DC electricity from the batteries to AC (see Figure 4.9). Although the inverter slightly lowers the overall efficiency of the system, it

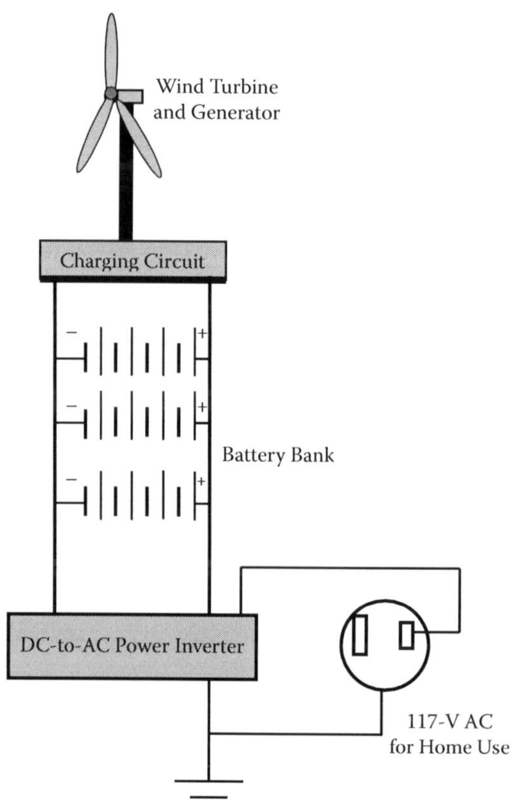

FIGURE 4.9 Stand-alone, small-scale wind power system.

allows the home to be wired for AC, a definite plus with lenders, electrical code officials, and future homebuyers. In grid-connected (or interactive) systems, the only additional equipment required is a power condition unit (inverter) that makes the turbine electrical output compatible with the utility grid. Usually, batteries are not needed. Either type of system can be practical if the following conditions exist:

- *Conditions for stand-alone systems*
  You live in an area with average annual wind speeds of at least 4.0 mps (9 mph).
  A grid connection is not available or can only be made through an expensive extension. The cost of running a power line to a remote site to connect with the utility grid can be prohibitive, ranging from $15,000 to more than $50,000 per mile, depending on the terrain.
  You have an interest in gaining energy independence from the utility.
  You would like to reduce the environmental impact of electricity production.
  You acknowledge the intermittent nature of wind power and have a strategy for using intermittent resources to meet your power needs.

- *Conditions for grid-connected systems*
  You live in an area with average annual wind speeds of at least 4.5 mps (10 mph).
  Utility-supplied electricity is expensive in your area (about 10 to 15¢ per kilowatt-hour).
  The utility's requirements for connecting your system to its grid are not prohibitively expensive.
  Local building codes or covenants allow you to erect a wind turbine on your property.
  You are comfortable with long-term investments.

After comparing stand-alone systems and grid-connected systems and determining which is best-suited for your particular circumstance, the next question to consider is whether your location is the appropriate site for installing a small-scale wind turbine system. Is it legal to install the system on your property? Are there any environmental or economic issues? Does the wind blow frequently and hard enough to make a small wind turbine system economically worthwhile? This is a key question that is not always easily answered. The wind resource can vary significantly over an area of just a few miles because of local terrain influences on the wind flow; yet, there are steps you can take that will go a long way toward answering the above question.

Wind resource maps like the ones included in the U.S. Department of Energy's *Wind Energy Resource Atlas of the United States* (RReDC, 2010) can be used to estimate the wind resources in your region. The highest average wind speeds in the United States are generally found along seacoasts, on ridge lines, and on the Great Plains; however, many other areas have wind resources strong enough to power a small wind turbine economically. The wind resource estimates provided by the *Wind Energy Resource Atlas* generally apply to terrain features that are well exposed to the wind, such as plains, hilltops, and ridge crests. Local terrain features may cause the wind resources at specific sites to differ considerably from these estimates.

Average wind speed information can be obtained from a nearby airport; however, caution should be used because local terrain and other factors may cause the wind speed to differ from that recorded at an airport. Airport wind data are generally measured at heights about 20 to 33 ft (6 to 10 m) above ground. Also, average wind speeds increase with height and may be 15 to 25% greater at a typical wind turbine hub height of 80 ft (24 m) than those measured at airport anemometer heights.

If wind speed data for a particular site are not available, it may be necessary to measure wind speeds at that location for a year. A recording anemometer generally costs $500 to $1500. The most accurate readings are taken at hub height (i.e., the elevation at the top of the wind turbine tower). This requires placing the anemometer high enough to avoid turbulence created by trees, buildings, and other obstructions.

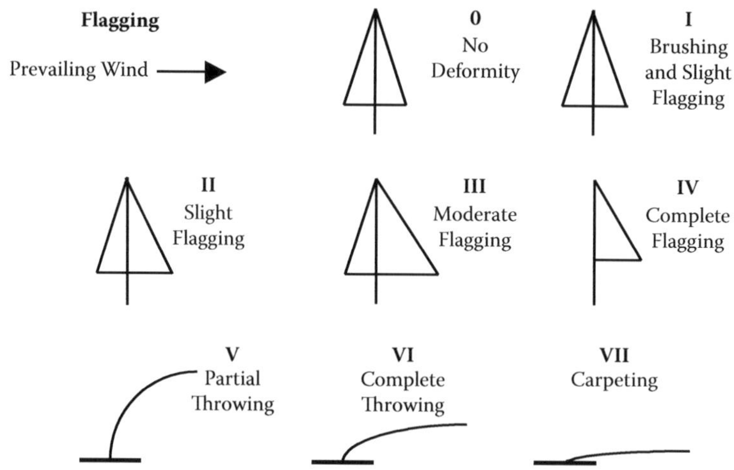

Griggs–Putnam Index of Deformity

Index	I	II	III	IV	V	VI	VII
Wind mph	7–9	9–11	11–13	13–16	15–18	16–21	22+
Speed m/s	3–4	4–5	5–6	6–7	7–8	8–9	10

FIGURE 4.10 A crude method of approximating average annual wind speed from the deformation of trees (and other foliage).

Within the same property it is not unusual to have varied wind resources. Those living in complex terrains should take care when selecting an installation site. A wind turbine installed on the top of a hill or on the windy side of a hill, for example, will have greater access to prevailing winds than if it were located in a gully or on the leeward (sheltered) side of a hill on the same property. Consider existing obstacles and plan for future obstructions, including trees and buildings, that could block the wind. Also recall that the power in the wind is proportional to its speed cubed. This means that the amount of power provided by a generator increases exponentially as the wind speed increases; for example, a site that has an annual average wind speed of about 5.6 mps (12.6 mph) has twice the energy available as a site with an average of 4.5 mps (10 mph) ($12.6/10^3$). Another useful indirect measurement of the wind resource is observation of an area's vegetation. Trees, especially conifers or evergreens, can be permanently deformed by strong winds. This deformity, known as *flagging*, has been used to estimate the average wind speed for an area (see Figure 4.10).

In addition to ensuring the proper siting of a small wind turbine system, other legal, environmental, and economic issues must be addressed:

- Research potential legal and environmental obstacles.
- Obtain cost and performance information from manufacturers.
- Perform a complete economic analysis that accounts for a multitude of factors.
- Understand the basics of a small wind system.
- Review possibilities for combining the system with other energy sources, backups, and energy efficiency improvements.

With regard to economic issues, because energy efficiency is usually less expensive than energy production, making your house more energy efficient first will likely result in spending less money on a wind turbine, because a smaller one may meet your needs. *A word of caution:* Before investing in a wind turbine, research potential legal and environmental obstacles to installing one. Some jurisdictions, for example, restrict the height of the structures permitted in residentially zoned areas, although variances are often obtainable. Neighbors might object to a wind machine that blocks their view, or they might be concerned about noise. Consider obstacles that might block the wind in the future (large planned developments or saplings, for example). Saplings that will grow into large trees can be a problem in the future. As mentioned, trees can affect wind speed (see Figure 4.10). If you plan to connect the wind generator to your local utility company's grid, find out its requirements for interconnections and buying electricity from small independent power producers. When you are convinced that a small wind turbine is what you want and there are no obstructions restricting its installation, approach buying a wind system as you would any major purchase.

THE BOTTOM LINE ON WIND POWER

Technology for utilizing our wind resources is much more advanced today, and the United States is home to one of the best wind resource areas in the world: North and South Dakota, Nebraska, Kansas, Montana, Iowa, and Oklahoma (Archer and Jacobson, 2004). As with any other source of energy, nonrenewable or renewable, the use of wind energy has its advantages and disadvantages. On the positive side, it should be noted that wind energy is a free, renewable resource, so no matter how much is used today, there will still be the same supply in the future. Wind energy is also a source of clean, nonpolluting, electricity. Wind turbines can be installed on farms or ranches, thus benefiting the economy in rural areas, where most of the best wind sites are found. Moreover, farmers and ranchers can continue to work the land because the wind turbines use only a fraction of the land—the distance between turbines allows the land to also be used for agriculture and grazing (only about 5% of the land in a wind farm is actually occupied by the turbines themselves). One huge advantage of wind energy is that it is a domestic, abundant source of energy.

Wind energy does have a few drawbacks. Wind projects face opposition. Wind power must compete with conventional generation sources on a cost basis; even though the cost of wind power has decreased dramatically in the past 10 years, the technology requires a higher initial investment than fossil-fueled generators. A challenge to using wind as a source of power is that the wind is intermittent and does not always blow when electricity is needed. Wind energy cannot be stored (unless batteries are being used), and not all winds can be harnessed to meet the timing of electricity demands. Another problem is that good sites are often located in remote locations, far from cities where the electricity is needed. Moreover, wind resource development may compete with other uses for the land, and those alternative uses may be more highly valued than electricity generation. Finally, with regard to the environment, wind power plants have relatively little impact on the environment compared to other conventional power plants, but there is some concern over the noise produced by the rotor blades and aesthetic (visual) impacts. Sometimes birds are killed when they fly into the rotors. Most of these problems have been resolved or greatly reduced through technological development or by properly siting wind plants, but the "not in my backyard" (NIMBY) point of view is still alive and strong in the United States and has succeeded in killing many projects, even renewable, clean energy projects.

We understand some of the issues related to the NIMBY point of view, although we do not agree with them. It is questionable and definitely arguable that the United States has reached its apogee in attaining the so-called good life for most of its citizenry and that our best days are behind us, but a brighter future depends on innovation and freeing ourselves from the bondage of fossil fuels and the accompanying economics, politics, and potential turmoil. With fossil-fuel supplies dwindling and world political pressures for accessibility ramping up to dangerous levels, we simply must find reliable alternative energy sources for the future. We may never be able to improve on the so-called "good life," but it would be beneficial for most of us if we are able to maintain the level of living that we have obtained to this point while at the same time reducing oil imports and carbon emissions. One thing is certain: Wind energy production continues to be one of the fastest growing energy technologies that is set to become a major generator of electricity throughout the world.

DISCUSSION AND REVIEW QUESTIONS

1. What factors must be taken into account when considering wind energy as an alternative renewable energy source?
2. Why should automotive batteries not be used in renewable energy systems?
3. ____________ are necessary to produce motion and changes in motion.
4. ____________ winds are horizontal air movements.
5. Localized air circulation gives rise to ____________ circulation.
6. Drag can cause ____________ of air movements.
7. Wind turbines convert the kinetic energy in the wind into ____________ power.
8. Downwind turbines don't require a yaw drive. Why?
9. Efficiency is based on input vs. ____________.
10. In your opinion, is wind power a viable source of renewable energy?

REFERENCES AND RECOMMENDED READING

Alternative Energy. (2010). *Wind Turbines*, http://www.alternative-energy-news.info/technology/wind-pworer/wind-turbines/.

Archer, C. and Jacobson, M.Z. (2004). *Evaluation of Global Wind Power*. Stanford, CA: Department of Civil and Environmental Engineering, Stanford University.

Burton, T., Sharpe, D., Jenkins, N., and Bossanyi, E. (2001). *Wind Energy Handbook*. New York: John Wiley & Sons.

Betz, A. (1966). *Introduction to the Theory of Flow Machines* (D.G. Randall, trans.). Oxford: Pergamon Press.

EERE. (2006a). *Wind & Water Power Program*. Washington, D.C.: Energy Efficiency & Renewable Energy, U.S. Department of Energy (http://www1.eere.energy.gov/windandhydro/).

EERE. (2006b). *How Wind Turbines Work*. Washington, D.C.: Energy Efficiency & Renewable Energy, U.S. Department of Energy (http://www1.eere.energy.gov/Windandhydro/wind_how.html).

Hanson, B.J. (2004). *Energy Power Shift*. Maple, WI: Lakota Scientific Press.

Khaligh, A. and Onar, O.C. (2010). *Energy Harvesting*. Boca Raton, FL: Taylor & Francis.

RReDC. (2010). *Wind Energy Resource Atlas of the United States*. Washington, D.C.: Renewable Resource Data Center, National Renewable Energy Laboratory (http://rredc.nrel.gov/wind/pubs/atlas/).

Serway, R.A. and Jewett, J.W. (2004). *Physics for Scientists and Engineers*, 6th ed. New York: Thomson/Brooks/Cole.

Tripler, P. and Llewellyn, R. (2002). *Modern Physics*, 4th ed. New York: W.H. Freeman.

White, F.M. (1988). *Fluid Mechanics*, 2nd ed. Singapore: McGraw-Hill.

Wind Energy Programmatic EIS. (2009). *Wind Energy Basics*. Washington, D.C.: Wind Energy Development Programmatic Environmental Impact Statement, U.S. Department of the Interior Bureau of Land Management and Argonne National Laboratory (http://windeis.anl.gov/guide/basics/index.cfm).

5 Hydropower

When we speak of water and its many manifestations we are speaking of that endless quintessential cycle that predates all other cycles. Water is our most precious natural resource; we can't survive without it. There is no more water today than there was yesterday; that is, there is no more this calendar year than 100 million years ago. The water present today is the same water used by all the animals that ever lived…by cave dwellers, Caesar, Cleopatra, Christ, Leonardo da Vinci, John Snow, Teddy Roosevelt, and the rest of us. Again, there is not one drop more nor one drop less of water than there has always been. Although it is a unique blend of thermal and mechanical aspects, the life-giving water cycle is dependent on solar energy and gravity (see Figure 5.1). Nothing on Earth is truly infinite in supply, but the energy available from water sources, in practical terms, comes closest to that ideal.

Grand Coulee Dam, Washington. (Photo by Frank R. Spellman.)

Cutaway view of turbine type used at Grand Coulee Dam, Washington. (Photo by Frank R. Spellman.)

Tops of turbines at Rocky Reach Dam, Washington. (Photo by Frank R. Spellman.)

Turbine blade from Rocky Reach Dam, Washington. (Photo by Frank R. Spellman.)

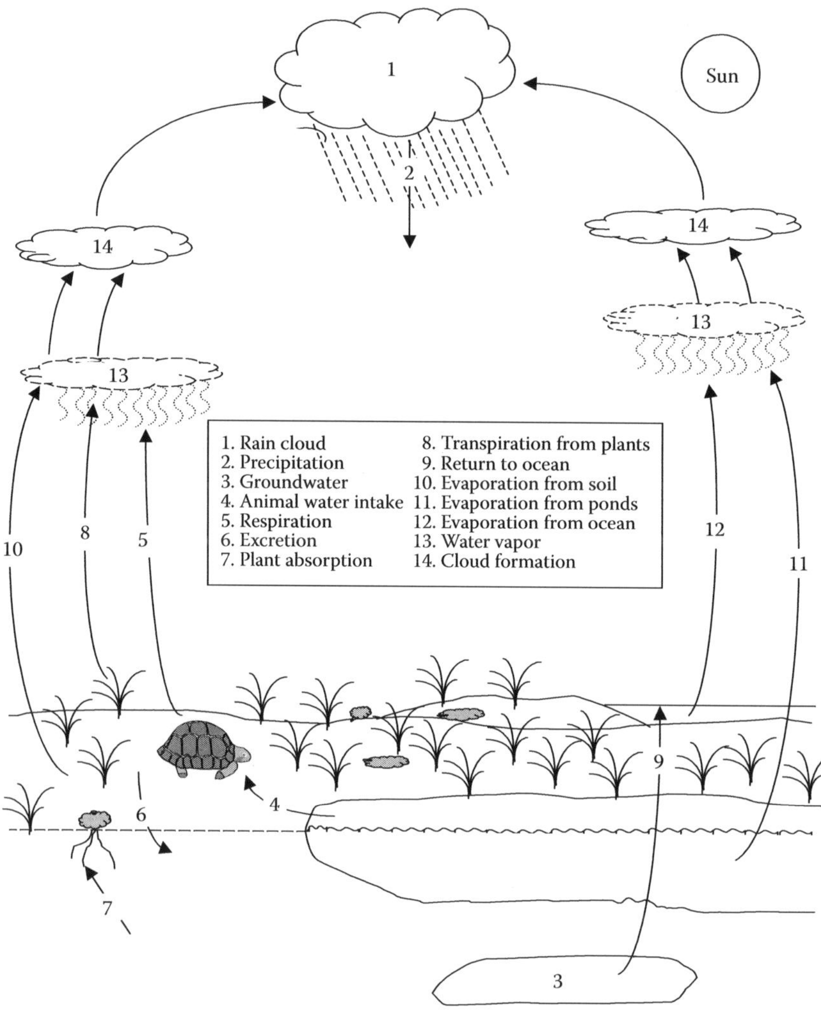

FIGURE 5.1 Water cycle. (Adapted from Carolina Biological Supply Co., Burlington, NC.)

INTRODUCTION

When we look at rushing waterfalls and rivers, we may not immediately think of electricity, but water-powered (hydroelectric) power plants are responsible for lighting many of our homes and neighborhoods. Hydropower is the harnessing of water to perform work. The power of falling water has been used in industry for thousands of years (see Table 5.1). The Greeks used water wheels to grind wheat into flour more than 2000 years ago. In addition to grinding flour, the power of water has been used to saw wood and power textile mills and manufacturing plants.

The technology for using falling water to create hydroelectricity has existed for more than a century. The evolution of the modern hydropower turbine began in the mid-1700s when a French hydraulic and military engineer, Bernard Forest de Belidor, wrote a four-volume work describing the use of a vertical-axis vs. horizontal-axis machine. Water turbine development continued into the 1800s. A Brush arc light dynamo driven by a water turbine provided theater and storefront lighting in Grand Rapids, Michigan, in 1880, and in 1881 a Brush dynamo connected to a turbine in a flour mill provided street lighting at Niagara Falls, New York. These two projects used direct-current (DC) technology.

TABLE 5.1
History of Hydropower

Date	Hydropower Event
B.C.E.	Hydropower was used by the Greeks more than 2000 years ago to turn water wheels for grinding wheat into flour.
Mid-1770s	French hydraulic and military engineer Bernard Forest de Belidor wrote a four-volume work describing vertical- and horizontal-axis machines.
1775	U.S. Army Corps of Engineers was founded, with establishment of a Chief Engineer for the Continental Army.
1880	Michigan's Grand Rapids Electric Light and Power Company, generating electricity by dynamo belted to a water turbine at the Wolverine Chair Factory, lit up 16 brush-arc lamps.
1881	Niagara Falls city street lamps were powered by hydropower.
1882	World's first hydroelectric power plant began operation on the Fox River in Appleton, Wisconsin.
1886	About 45 water-powered electric plants were operating in the United States and Canada.
1887	First hydroelectric plant in the west was opened in San Bernardino, California.
1889	Two hundred electric plants in the United States used water power for some or all of their power generation.
1901	First Federal Water Power Act was passed.
1902	Bureau of Reclamation was established.
1907	Hydropower provided 15% of U.S. electrical generation.
1920	Hydropower provided 25% of U.S. electrical generation. Federal Power Act established the Federal Power Commission with authority to issue licenses for hydro development on public lands.
1933	Tennessee Valley Authority was established.
1935	Federal Power Commission authority was extended to all hydroelectric projects built by utilities engaged in interstate commerce.
1937	Bonneville Dam, the first federal dam, began operation on the Columbia River in Oregon and Washington; the federal Bonneville Power Administration was established in the Pacific Northwest.
1940	Hydropower provided 40% of electrical generation. Conventional capacity had tripled in the United States since 1920.
1980	Conventional capacity had nearly tripled in the United States since 1900.
2003	About 10% of U.S. electricity came from hydropower. Capacity was about 80,000 MW of conventional capacity and 18,000 MW of pumped storage.

Source: EERE, *History of Hydropower*, Energy Efficiency & Renewable Energy, U.S. Department of Energy, Washington, D.C., 2008 (http://www1.eere.energy.gov/windandhydro/hydro_history.html).

Alternating current (AC) is used today. That breakthrough came when the electric generator was coupled to the turbine in 1882, which resulted in the world's first hydroelectric plant, located on the Fox River in Appleton, Wisconsin. The Appleton hydroelectric power plant is considered to be one of the major accomplishments of the Gilded Age (1878–1889) (Library of Congress, 2009). People across the United States soon enjoyed electricity in their homes, schools, and offices, where they were able to read by electric lamp instead of candlelight or kerosene. Today, we take electricity for granted and cannot imagine life without it.

Table 5.2 shows the energy consumption by energy source for 2008; in that year, hydropower power accounted for 2.512 quadrillion Btu. Ranging in size from small systems (100 kW to 30 MW) for a home or village to large projects (capacity greater than 30 MW) that produce electricity for utilities, hydropower plants are of three types: impoundment, diversion, and pumped storage. Some hydropower plants use dams, and some do not. Many dams were built for other purposes, and the hydropower function was added later. The United States has about 80,000 dams, of which only 2400 produce power. The other dams are for recreation, stock/farm ponds, flood control, water supply, and irrigation. The types of hydropower plants are described below.

TABLE 5.2
U.S. Energy Consumption by Energy Source, 2008

Energy Source	Energy Consumption (quadrillion Btu)
Total	99.438
Fossil fuels (coal, natural gas, petroleum)	83.532
Electricity net imports	0.113
Nuclear electric power	8.427
Renewable	7.367
Biomass (total)	3.852
Biofuels	1.372
Waste	0.436
Wood and wood-derived fuels	2.044
Geothermal energy	0.360
Hydroelectric conventional	**2.512**
Solar thermal/PV energy	0.097
Wind energy	0.546

Source: EIA, *U.S. Energy Consumption by Energy Source*, Energy Information Administration, Washington, D.C., 2007 (www.eia.doe.gov/cneaf/solar.renewables/page/trends/table1.html).

Impoundment

The most common type of hydroelectric power plant is an impoundment facility, typically a large hydropower system that uses a dam to store river water in a reservoir. This type of facility works best in mountainous or hilly terrain where high dams can be built and deep reservoirs can be maintained. Potential energy available in a reservoir depends on the mass of water contained in it, as well as on overall depth of the water. Water released from the reservoir flows through a turbine and makes it spin, which in turn activates a generator that produces electricity. The water may be released either to meet changing electricity needs or to maintain a constant reservoir level.

Key Point: The potential energy in a specific slug or parcel of water is expressed in newton-meters (N·m). The newton is the standard unit of force, equivalent to meters per second squared (1 m/s²). The potential energy is the product of the mass of the slug (kg), acceleration due to gravity (about 9.8 m/s²), and elevation of the parcel in meters (vertical distance it falls as its energy is harnessed). The equivalent kinetic energy unit is the joule (J), which is, in effect, equal to a watt-second (W·s).

Diversion

A diversion (sometimes called *run-of-river*) facility channels all or a portion of the flow of a river from its natural course through a canal or penstock, and the current is used to drive turbines. This approach may not require the use of a dam. This type of system is best suited for locations where a river drops considerably per unit of horizontal distance. The ideal location is near a natural waterfall or rapids. The chief advantage of a diversion system is the fact that, lacking a dam, it has far less impact on the environment than an impoundment facility (Gibilisco, 2007).

Pumped Storage

When the demand for electricity is low, a pumped storage facility stores energy by pumping water from a lower reservoir to an upper reservoir. During periods of high electrical demand, the water is released back to the lower reservoir to generate electricity.

HYDROPOWER BASIC CONCEPTS*

$$\text{Air pressure (at sea level)} = 14.7 \text{ pounds per square inch (psi)}$$

The relationship shown above is important because our study of hydropower basics begins with air. A blanket of air, many miles thick surrounds the Earth. The weight of this blanket on a given square inch of the Earth's surface will vary according to the thickness of the atmospheric blanket above that point. As shown above, at sea level the pressure exerted is 14.7 pounds per square inch (psi). On a mountaintop, air pressure decreases because the blanket is not as thick.

$$1 \text{ cubic foot (ft}^3\text{) of water} = 62.4 \text{ lb}$$

This relationship is also important; note that both cubic feet and pounds are used to describe a volume of water. A defined relationship exists between these two methods of measurement. The specific weight of water is defined relative to a cubic foot. One cubic foot of water weighs 62.4 lb. This relationship is true only at a temperature of 4°C and at a pressure of 1 atmosphere, conditions referred to as standard temperature and pressure (STP). One atmosphere equals 14.7 psi at sea level and 1 ft^3 of water contains 7.48 gal. The weight varies so little that, for practical purposes, this weight is used for temperatures ranging from 0 to 100°C. One cubic inch of water weighs 0.0362 lb. Water 1 ft deep will exert a pressure of 0.43 psi on the bottom area (12 in × 0.0362 lb/in.^{3}). A column of water 2 ft high exerts 0.86 psi (2 ft × 0.43 psi/ft), one 10 ft high exerts 4.3 psi (10 ft × 0.43 psi/ft), and one 55 ft high exerts 23.65 psi (55 ft × 0.43 psi/ft). A column of water 2.31 ft high will exert 1.0 psi. To produce a pressure of 50 psi requires a 115.5-ft water column (50 psi × 2.31 ft/psi). Two important points are being made here:

1. 1 ft^3 of water = 62.4 lb (see Figure 5.2).
2. A column of water 2.31 ft high will exert 1.0 psi.

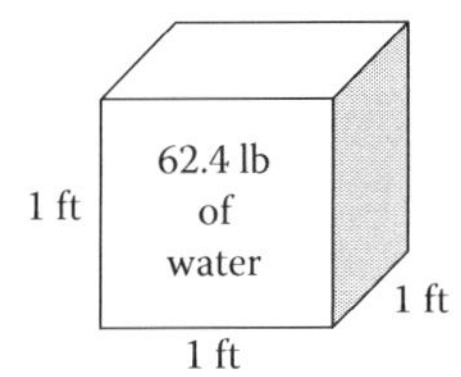

FIGURE 5.2 One cubic foot of water weights 62.4 lb.

Another relationship is also important. As noted above, at standard temperature and pressure, 1 ft^3 of water contains 7.48 gal and weighs 62.4 lb. Thus, we can determine the weight of 1 gal of water:

$$\text{Weight of 1 gal water} = 62.4 \text{ lb/ft}^3 \div 7.48 \text{ gal/ft}^3 = 8.34 \text{ lb/gal}$$

Thus,

$$1 \text{ gal water} = 8.34 \text{ lb}$$

■ Example 5.1

Problem: Find the number of gallons in a reservoir that has a volume of 855.5 ft^3.
Solution: Convert cubic feet to gallons by multiplying the number of cubic feet by 7.48 gal/ft^3.

$$855.5 \text{ ft}^3 \times 7.48 \text{ gal/ft}^3 = 6399 \text{ gal (rounded)}$$

Note: The term *head* is used to designate water pressure in terms of the height of a column of water in feet; for example, a 10-ft column of water exerts 4.3 psi. This can be referred to as 4.3-psi pressure or 10 feet of head.

STEVIN'S LAW

Stevin's law deals with water at rest. Specifically, it states: "The pressure at any point in a fluid at rest depends on the distance measured vertically to the free surface and the density of the fluid." Stated as a formula, this becomes:

* Much of the information in this section is from Spellman, F.R., *The Science of Water*, CRC Press, Boca Raton, FL, 2008.

$$p = w \times h \tag{5.1}$$

where:

p = pressure in pounds per square foot (psf)
w = density in pounds per cubic foot (lb/ft^3)
h = vertical distance in feet

■ Example 5.2

Problem: What is the pressure at a point 18 ft below the surface of a reservoir?
Solution: To calculate this, we must know that the density of the water (w) is 62.4 lb/ft^3.

$$p = w \times h = 62.4 \text{ lb/ft}^3 \times 18 \text{ ft} = 1123 \text{ lb/ft}^2 \text{ (psf)}$$

Water practitioners generally measure pressure in pounds per square *inch* rather than pounds per square *foot*; to convert, divide by 144 $in.^2/ft^2$ (12 in. × 12 in. = 144 $in.^2$):

$$p = \frac{1123 \text{ psf}}{144 \text{ in.}^2/\text{ft}^2} = 7.8 \text{ lb/in.}^2 \text{ or psi (rounded)}$$

DENSITY AND SPECIFIC GRAVITY

Table 5.3 shows the relationships among temperature, specific weight, and density of water. When we say that iron is heavier than aluminum, we say that iron has greater density than aluminum. In practice, what we are really saying is that a given volume of iron is heavier than the same volume of aluminum. A cubic foot of water, for example, weighs 62.4 lb, and a cubic foot of aluminum, weighs 178 lb; thus, aluminum is 2.7 times heavier than water.

Note: *Density* is the mass per unit volume of a substance. *Specific gravity* is the weight (or density) of a substance compared to the weight (or density) of an equal volume of water. The specific gravity of water is 1.

Suppose you had a tub of lard and a large box of cold cereal, each having a mass of 600 grams. The density of the cereal would be much less than the density of the lard because the cereal occupies a much larger volume than the lard occupies. The density of an object can be calculated by using the formula:

$$\text{Density} = \frac{\text{Mass}}{\text{Volume}} \tag{5.2}$$

Perhaps the most common measures of density are pounds per cubic foot (lb/ft^3) and pounds per gallon (lb/gal):

- 1 ft^3 of water weighs 62.4 lb; density = 62.4 lb/ft^3.
- 1 gal of water weighs 8.34 lb; density = 8.34 lb/gal.

The density of a dry material, such as cereal, lime, soda, or sand, is usually expressed in pounds per cubic foot. The density of a liquid, such as liquid alum, liquid chlorine, or water, can be expressed either as pounds per cubic foot or as pounds per gallon. The density of a gas, such as chlorine gas, methane, carbon dioxide, or air, is usually expressed in pounds per cubic foot. As shown in Table 5.3, the density of a substance such as water changes slightly as the temperature of the substance changes. This occurs because substances usually increase in volume (size) by expanding as they become warmer. Because of this expansion with warming, the same weight is spread over a larger volume, so the density is lower when a substance is warm than when it is cold.

TABLE 5.3
Water Properties (Temperature, Specific Weight, and Density)

Temperature (°F)	Specific Weight (lb/ft³)	Density (slugs/ft³)	Temperature (°F)	Specific Weight (lb/ft³)	Density (slugs/ft³)
32	62.4	1.94	130	61.5	1.91
40	62.4	1.94	140	61.4	1.91
50	62.4	1.94	150	61.2	1.90
60	62.4	1.94	160	61.0	1.90
70	62.3	1.94	170	60.8	1.89
80	62.2	1.93	180	60.6	1.88
90	62.1	1.93	190	60.4	1.88
100	62.0	1.93	200	60.1	1.87
110	61.9	1.92	210	59.8	1.86
120	61.7	1.92			

It is not that difficult to find the specific gravity of a piece of metal. All you have to do is to weigh the metal in air, then weigh it under water. The loss of weight is the weight of an equal volume of water. To find the specific gravity, divide the weight of the metal by its loss of weight in water:

$$\text{Specific gravity} = \frac{\text{Weight of substance}}{\text{Weight of equal volume of water}} \tag{5.3}$$

■ Example 5.3

Problem: If a piece of metal weighs 150 lb in air and 85 lb under water, what is the specific gravity?

Solution:

$$150\ \text{lb} - 85\ \text{lb} = 65\ \text{lb loss of weight in water}$$

$$\text{Specific Gravity} = \frac{150}{65} = 2.3$$

Key Point: When performing specific gravity calculations, it is essential to express densities in the same units.

As stated earlier, the specific gravity of water is 1, which is the standard, the reference to which all other liquid or solid substances are compared. Specifically, any object that has a specific gravity greater than 1 will sink in water (e.g., rocks, steel, iron, grit, floc, sludge). Substances with a specific gravity less than 1 will float (e.g., wood, scum, gasoline). Considering the total weight and volume of a ship, its specific gravity is less than 1, which is why it can float. The most common use of specific gravity in water operations is in gallon-to-pound conversions. In many cases, the liquids being handled have a specific gravity of 1 or very nearly 1 (between 0.98 and 1.02), so a value of 1 may be used in the calculations without introducing significant error. For calculations involving a liquid with a specific gravity of less than 0.98 or greater than 1.02, however, the conversions from gallons to pounds must consider specific gravity. The technique is illustrated in the following example.

■ Example 5.4

Problem: A basin contains 1455 gal of a liquid. If the specific gravity of the liquid is 0.94, how many pounds of liquid are in the basin?

Solution: Normally, for a conversion from gallons to pounds, we would use the factor 8.34 lb/gal (the density of water) if the specific gravity of the substance is between 0.98 and 1.02; in this instance, however, the substance has a specific gravity outside this range, so the 8.34 factor must be adjusted. Multiply 8.34 lb/gal by the specific gravity to obtain the adjusted factor:

$$8.34 \text{ lb/gal} \times 0.94 = 7.84 \text{ lb/gal (rounded)}$$

Then convert 1455 gal to pounds using the corrected factor:

$$1455 \text{ gal} \times 7.84 \text{ lb/gal} = 11{,}407 \text{ lb (rounded)}$$

FORCE AND PRESSURE

Water exerts force and pressure against the walls of its container, whether it is stored in a tank or flowing in a pipeline. Force and pressure are different, although they are closely related. Force and pressure are defined below. *Force* is the push or pull influence that causes motion. In the English system, force and weight are often used in the same way. The weight of 1 ft^3 of water is 62.4 lb. The force exerted on the bottom of a 1-ft cube is 62.4 lb (see Figure 5.2). If we stack two 1-ft cubes on top of one another, the force on the bottom will be 124.8 lb. *Pressure* is the force per unit of area. In equation form, this can be expressed as

$$P = \frac{F}{A} \tag{5.4}$$

where:

P = pressure
F = force
A = area over which the force is distributed

Key Point: Storing liquids in a hydraulic system can be a problem. Air is readily compressible and can be stored in large quantities in relatively small containers, but liquids such as water cannot be compressed. It is not possible to store a large amount of water in a small tank, as 62.4 lb of water occupies a volume of 1 ft^3, regardless of the pressure applied.

Pounds per square inch ($lb/in.^2$ or psi) or pounds per square foot (lb/ft^2) are commonly used to express pressure. The pressure on the bottom of our 1-ft cube is 62.4 lb/ft^2 (see Figure 5.2). It is normal to express pressure in pounds per square inch. This is easily accomplished by determining the weight of 1 $in.^2$ of a 1-ft cube. If we have a cube that is 12 in. on each side, the number of square inches on the bottom surface of the cube is $12 \times 12 = 144$ $in.^2$. Dividing the weight by the number of square inches determines the weight on each square inch:

$$\text{psi} = \frac{62.4 \text{ lb/ft}}{144 \text{ in.}^2} = 0.433 \text{ psi/ft}$$

This is the weight of a column of water 1-in. square and 1 ft tall. If the column of water were 2 ft tall, the pressure would be 2 ft × 0.433 psi/ft = 0.866.

Note: 1 ft of water = 0.433 psi.

With this information, feet of head can be converted to psi by multiplying the feet of head times 0.433 psi/ft.

■ Example 5.5

Problem: A tank is mounted at a height of 90 ft. Find the pressure at the bottom of the tank.
Solution:

$$90 \text{ ft} \times 0.433 \text{ psi/ft} = 39 \text{ psi (rounded)}$$

Note: To convert psi to feet, divide the psi by 0.433 psi/ft.

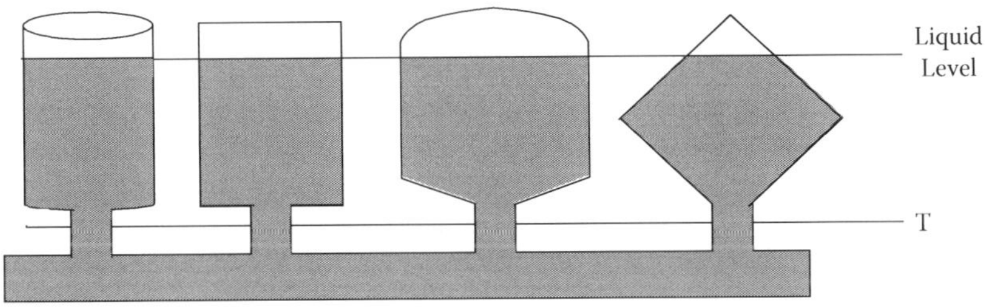

FIGURE 5.3 Hydrostatic pressure

■ Example 5.6

Problem: Find the height of water in a tank if the pressure at the bottom of the tank is 22 psi.
Solution:

$$\text{Height} = \frac{22 \text{ psi}}{0.433 \text{ psi/ft}} = 51 \text{ ft (rounded)}$$

Hydrostatic Pressure

Figure 5.3 shows a number of differently shaped, connected, open containers of water. Note that the water level is the same in each container, regardless of the shape or size of the container. This occurs because pressure is developed, within water (or any other liquid), by the weight of the water above. If the water level in any one container is momentarily higher than that in any of the other containers, the higher pressure at the bottom of this container would cause some water to flow into the container having the lower liquid level. In addition, the pressure of the water at any level (such as line T) is the same in each of the containers. Pressure increases because of the weight of the water. The farther down from the surface, the more pressure is created. This illustrates that the *weight*, not the volume, of water contained in a vessel determines the pressure at the bottom of the vessel. Important principles that always apply for hydrostatic pressure include the following (Nathanson, 1997):

1. The pressure depends only on the depth of water above the point in question (not on the water surface area).
2. The pressure increases in direct proportion to the depth.
3. The pressure in a continuous volume of water is the same at all points that are at the same depth.
4. The pressure at any point in the water acts in all directions at the same depth.

Head

Head is defined as the vertical distance the water must be lifted from the supply tank to the discharge, or as the height a column of water would rise due to the pressure at its base. A perfect vacuum plus atmospheric pressure of 14.7 psi would lift the water 34 ft. If the top of the sealed tube is open to the atmosphere and the reservoir is enclosed, the pressure in the reservoir is increased; the water will rise in the tube. Because atmospheric pressure is essentially universal, we usually ignore the first 14.7 psi of actual pressure measurements and measure only the difference between the water pressure and the atmospheric pressure; we call this *gauge pressure*. Water in an open reservoir, for example, is subjected to the 14.7 psi of atmospheric pressure, but subtracting this 14.7 psi leaves a gauge pressure of 0 psi. This shows that the water would rise 0 ft above the reservoir surface. If the gauge pressure in a water main is 120 psi, the water would rise in a tube connected to the main:

$$120 \text{ psi} \times 2.31 \text{ ft/psi} = 277 \text{ ft (rounded)}$$

Total Dynamic (System) Head

Total head includes the vertical distance the liquid must be lifted (*static head*), the loss to friction (*friction head*), and the energy required to maintain the desired velocity (*velocity head*):

$$\text{Total Dynamic Head} = \text{Static Head} + \text{Friction Head} + \text{Velocity Head} \tag{5.5}$$

Static Head

Static head is the actual vertical distance the liquid must be lifted:

$$\text{Static Head} = \text{Discharge Elevation} - \text{Supply Elevation} \tag{5.6}$$

■ Example 5.7

Problem: A supply tank is located at elevation 118 ft. The discharge point is at elevation 215 ft. What is the static head in feet?

Solution:

$$\text{Static Head} = 215\text{ ft} - 118\text{ ft} = 97\text{ ft}$$

Friction Head

Friction head is the equivalent distance of the energy that must be supplied to overcome friction. Engineering references include tables showing the equivalent vertical distances for various sizes and types of pipes, fittings, and valves. The total friction head is the sum of the equivalent vertical distances for each component. For friction head,

$$\text{Friction Head (ft)} = \text{Energy losses due to friction} \tag{5.7}$$

Velocity Head

Velocity head is the equivalent distance of the energy consumed in achieving and maintaining the desired velocity in the system:

$$\text{Velocity Head (ft)} = \text{Energy losses to maintain velocity} \tag{5.8}$$

Pressure and Head

The pressure exerted by water is directly proportional to its depth or head in the pipe, tank, or channel. If the pressure is known, the equivalent head can be calculated.

$$\text{Head (ft)} = \text{Pressure (psi)} \times 2.31\text{ (ft/psi)} \tag{5.9}$$

■ Example 5.8

Problem: The pressure gauge on the discharge line from the influent pump reads 72.3 psi. What is the equivalent head in feet?

Solution:

$$\text{Head} = 72.3 \times 2.31\text{ ft/psi} = 167\text{ ft}$$

Head and Pressure

If the head is known, the equivalent pressure can be calculated by

$$\text{Pressure (psi)} = \frac{\text{Head (ft)}}{2.31\text{ ft/psi}} \tag{5.10}$$

■ Example 5.9

Problem: A tank is 22 ft deep. What is the pressure in psi at the bottom of the tank when it is filled with water?

Solution:

$$\text{Pressure} = \frac{22 \text{ ft}}{2.31 \text{ ft/psi}} = 9.52 \text{ psi (rounded)}$$

Flow and Discharge Rates: Water in Motion

The study of fluid flow is much more complicated than that of fluids at rest, but it is important to have an understanding of these principles because the water used in hydropower applications is nearly always in motion (e.g., the water is used to propel turbine blades). *Discharge* (or flow) is the quantity of water passing a given point in a pipe or channel during a given period. Stated another way for open channels, the flow rate through an open channel is directly related to the velocity of the liquid and the cross-sectional area of the liquid in the channel:

$$Q = A \times V \tag{5.11}$$

where:

Q = flow, or discharge in cubic feet per second (cfs)
A = cross-sectional area of the pipe or channel (ft^2)
V = water velocity in feet per second (fps or ft/s)

■ Example 5.10

Problem: A channel is 6 ft wide, and the water depth is 3 ft. The velocity in the channel is 4 fps. What is the discharge or flow rate in cubic feet per second?
Solution:

$$\text{Flow} = 6 \text{ ft} \times 3 \text{ ft} \times 4 \text{ ft/s} = 72 \text{ cfs}$$

Discharge or flow can be recorded as gal/day (gpd), gal/min (gpm), or cubic feet per second (cfs). Flows treated by many hydropower systems are large and are often expressed in million gallons per day (MGD). The discharge or flow rate can be converted from cfs to other units such as gpm or MGD by using appropriate conversion factors.

■ Example 5.11

Problem: A 12-in.-diameter pipe has water flowing through it at 10 fps. What is the discharge in (a) cfs, (b) gpm, and (c) MGD?
Solution: Before we can use the basic formula, we must determine the area (A) of the pipe. The formula for the area of a circle is

$$\text{Area } (A) = \pi \times \frac{D^2}{4} = \pi \times r^2 \tag{5.12}$$

where:

π = the constant value 3.14159, or simply 3.14
D = diameter of the circle in feet
r = radius of the circle in feet

Therefore, the area of the pipe is

$$\text{Area } (A) = \pi \times \frac{D^2}{4} = 3.14 \times \frac{(1 \text{ ft})^2}{4} = 0.785 \text{ ft}^2$$

(a) Now we can determine the discharge in cfs:

$$Q = V \times A = 10 \text{ fps} \times 0.785 \text{ ft}^2 = 7.85 \text{ ft}^3\text{/s (cfs)}$$

(b) We need to know that 1 cfs is 449 gpm, so 7.85 cfs × 449 gpm/cfs = 3525 gpm (rounded).

(c) Finally, 1 million gallons per day is 1.55 cfs, so:

$$\frac{7.85 \text{ cfs}}{1.55 \text{ cfs/MGD}} = 5.06 \text{ MGD}$$

Area and Velocity

The *law of continuity* states that the discharge at each point in a pipe or channel is the same as the discharge at any other point (if water does not leave or enter the pipe or channel). That is, under the assumption of steady-state flow, the flow that enters the pipe or channel is the same flow that exits the pipe or channel. In equation form, this becomes:

$$Q_1 = Q_2 \quad \text{or} \quad A_1V_1 = A_2V_2 \tag{5.13}$$

■ Example 5.12

Problem: A pipe 12 inches in diameter is connected to a 6-in.-diameter pipe. The velocity of the water in the 12-in. pipe is 3 fps. What is the velocity in the 6-in. pipe?

Solution: Using the equation $A_1V_1 = A_2V_2$, we need to determine the area of each pipe:

$$\text{12-inch: Area } (A) = \pi \times \frac{D^2}{4} = 3.14 \times \frac{(1 \text{ ft})^2}{4} = 0.785 \text{ ft}^2$$

$$\text{6-inch: Area } (A) = \pi \times \frac{D^2}{4} = 3.14 \times \frac{(0.5 \text{ ft})^2}{4} = 0.196 \text{ ft}^2$$

The continuity equation now becomes:

$$0.785 \text{ ft}^2 \times 3 \text{ ft/s} = 0.196 \text{ ft}^2 \times V_2$$

Solving for V_2:

$$V_2 = \frac{(0.785 \text{ ft}^2) \times (3 \text{ ft/s})}{(0.196 \text{ ft}^2)} = 12 \text{ ft/s (fps)}$$

Pressure and Velocity

In a closed pipe flowing full (under pressure), the pressure is indirectly related to the velocity of the liquid:

$$\text{Velocity}_1 \times \text{Pressure}_1 = \text{Velocity}_2 \times \text{Pressure}_2 \quad \text{or} \quad V_1P_1 = V_2P_2 \tag{5.14}$$

Conservation of Energy

The volume of water flowing past any given point in the pipe or channel per unit time is called the *flow rate* or *discharge*—or just *flow*. The *continuity of flow* and the *continuity equation* have been discussed (see Equation 5.14). Along with the continuity of flow principle and continuity equation, the law of conservation of energy is also important to our study of water hydraulics.

Many of the principles of physics are important to the study of hydraulics. When applied to problems involving the flow of water, few of the principles of physical science are more important and useful to us than the law of conservation of energy. Simply, the law of conservation of energy states that energy can be neither created nor destroyed, but it can be converted from one form to another. In a given closed system, the total energy is constant.

Energy Head

Hydraulic systems have both kinetic and potential energy. The three forms of mechanical energy in hydraulic systems are the potential energy due to elevation, the potential energy due to pressure, and the kinetic energy due to velocity. Energy is expressed in the units of foot pounds (ft-lb). It is convenient to express hydraulic energy in terms of *energy head* in feet of water, equivalent to foot-pounds per pound of water (ft-lb/lb = ft).

Energy Available

Energy available is directly proportional to flow rate and to the hydraulic head. As mentioned, the head is equivalent to stored potential energy. This is shown as

$$\text{Head} = m \times g \times h$$

where:

m = mass of water
g = acceleration due to gravity (can be taken as 10 m/s^2 in most applications)
h = head difference

The diameter of a pipe must be large enough to handle the volume of water flowing. Friction in the pipes will reduce the effective head of water so larger diameters are used, although cost is a factor. Ideally, the pipes should narrow as they proceed downhill; however, friction losses are highest where the velocity is highest, so usually the pipe diameter changes very little. Friction losses in piping are classified as either major head loss or minor head loss (Tovey, 2005).

DID YOU KNOW?

Hydro (water) kinetic energy is the energy possessed by a body of water because of its motion (kinetic energy = $1.2 \times m \times v^2$). Hydrostatic (at rest) energy is the energy possessed by a body because of its position or location at an elevation or height above a reference or datum (potential energy = $m \times g \times h$).

Major Head Loss

Major head loss consists of pressure decreases along the length of pipe caused by friction created as water encounters the surfaces of the pipe. It typically accounts for most of the pressure drop in a pressurized or dynamic water system. The components that contribute to major head loss are roughness, length, diameter, and velocity:

- *Roughness*—Even when new, the interior surfaces of pipes are rough. The roughness varies, of course, depending on pipe material, corrosion (tuberculation and pitting), and age. Because normal flow in a water pipe is turbulent, the turbulence increases with pipe roughness, which, in turn, causes pressure to drop over the length of the pipe.

- *Pipe length*—With every foot of pipe length, friction losses occur. The longer the pipe, the more head loss. Friction loss because of pipe length must be factored into head loss calculations.
- *Pipe diameter*—Generally, small-diameter pipes have more head loss than large-diameter pipes. In large-diameter pipes, less of the water actually touches the interior surfaces of the pipe (thus encountering less friction) than in a small-diameter pipe.
- *Water velocity*—Turbulence in a water pipe is directly proportional to the speed (or velocity) of the flow; thus, the velocity head also contributes to head loss.

DID YOU KNOW?

For the same diameter pipe, when flow increases head loss increases.

Calculating Major Head Loss

Darcy, Weisbach, and others developed the first practical equation to determine pipe friction in about 1850. The equation or formula now known as the *Darcy–Weisbach* equation for circular pipes is

$$h_f = f\frac{LV^2}{D^2 g} \tag{5.15}$$

In terms of the flow rate (Q), the equation becomes:

$$h_f = \frac{8fLQ^2}{\pi^2 g D^5} \tag{5.16}$$

where:

h_f = head loss (ft)
f = coefficient of friction
L = length of pipe (ft)
V = mean velocity (ft/s)
D = diameter of pipe (ft)
g = acceleration due to gravity (32.2 ft/s^2)
Q = flow rate (ft^3/s)

The Darcy–Weisbach formula was meant to apply to the flow of any fluid, and into this friction factor was incorporated the degree of roughness and an element called the *Reynold's number*, which is based on the viscosity of the fluid and the degree of turbulence of flow. The Darcy–Weisbach formula is used primarily for determining head loss calculations in pipes. For open channels, the *Manning equation* was developed during the latter part of the 19th century. Later, this equation was used for both open channels and closed conduits.

In the early 1900s, a more practical equation, the *Hazen–Williams* equation, was developed for use in making calculations related to water pipes and wastewater force mains:

$$Q = 0.435 \times CD^{2.63} \times S^{0.54} \tag{5.17}$$

where:

Q = flow rate (ft^3/s)
C = coefficient of roughness (C decreases with roughness)
D = hydraulic radius R (ft)
S = slope of energy grade line (ft/ft)

DID YOU KNOW?

An alternative to calculating the Hazen–Williams formula, an *alignment chart*, has become quite popular for fieldwork. The alignment chart can be used with reasonable accuracy.

C Factor

C factor, as used in the Hazen–Williams formula, designates the coefficient of roughness. *C* does not vary appreciably with velocity, and by comparing pipe types and ages it includes only the concept of roughness, ignoring fluid viscosity and Reynold's number. Based on experience (experimentation), accepted tables of *C* factors have been established for pipe (see Table 5.4). Generally, the *C* factor decreases by one with each year of pipe age. Flow for a newly designed system is often calculated with a *C* factor of 100, based on averaging it over the life of the pipe system.

DID YOU KNOW?

A high *C* factor means a smooth pipe. A low *C* factor means a rough pipe.

TABLE 5.4
***C* Factors**

Type of Pipe	*C* Factor
Asbestos cement	140
Brass	140
Brick sewer	100
Cast iron (10 years old)	110
Cast iron (20 years old)	90
Ductile iron (cement lined)	140
Concrete or concrete-lined	
Smooth, steel forms	140
Wooden forms	120
Rough	110
Copper	140
Fire hose (rubber-lined)	135
Galvanized iron	120
Glass	140
Lead	130
Masonry conduit	130
Plastic	150
Steel	
Coal-tar enamel lined	150
New unlined	140
Riveted	110
Tin	130
Vitrified	120
Wood stave	120

Source: Adapted from Lindeburg, M.R., *Civil Engineering Reference Manual*, 4th ed., Professional Publications, San Carlos, CA, 1986.

Slope

Slope is defined as the head loss per foot. In open channels, where the water flows by gravity, slope is the amount of incline of the pipe and is calculated as feet of drop per foot of pipe length (ft/ft). Slope is designed to be just enough to overcome frictional losses, so the velocity remains constant, the water keeps flowing, and solids will not settle in the conduit. In piped systems, where pressure loss for every foot of pipe is experienced, slope is not provided by slanting the pipe but instead by pressure added to overcome friction.

Minor Head Loss

In addition to the head loss caused by friction between the fluid and the pipe wall, losses also are caused by turbulence created by obstructions (i.e., valves and fittings of all types) in the line, changes in direction, and changes in flow area.

DID YOU KNOW?

In practice, if minor head loss is less than 5% of the total head loss, it is usually ignored.

RESERVOIR STORED ENERGY

One of the major components of a hydroelectric dam is the area behind the dam, its reservoir. The water temporarily stored there is called *gravitational potential energy*. The water is stored above the rest of the dam facility so gravity can carry the water down to the turbines. Because this higher altitude is not where the water would naturally be, the water is considered to be at an altered equilibrium. The result is stored energy of position, or gravitational potential energy. The water has the potential to do work because of the position it is in. As shown in Figure 5.4 , gravity will force the water to fall to a lower position through the intake and the control gate. When the control gate is opened, the water from the reservoir moves through the intake and becomes translational kinetic energy as it falls through the next main part of the system, the penstock. *Translational kinetic energy* is the energy due to the motion from one location to another. The water moves (falls) from the reservoir through the long shaft of the penstock toward the turbines, where the kinetic energy becomes mechanical energy. The force of the water is used to turn the turbines, which turn the

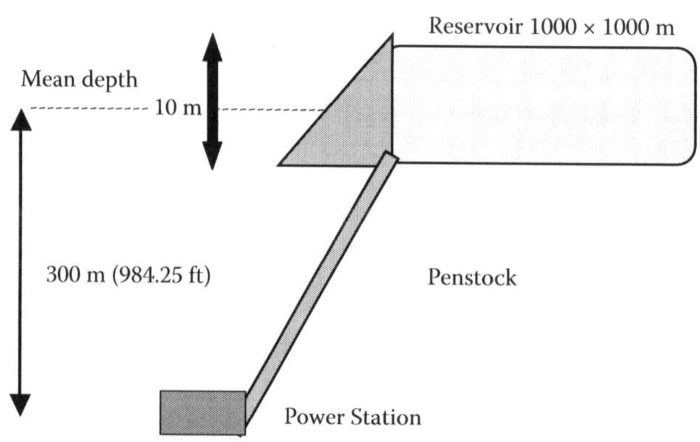

FIGURE 5.4 Schematic representation of a hydro scheme. (Adapted from Tovey, N.K., *ENV-2E02 Energy Resources 2004–2005 Lecture*, University of East Anglia, Norwich, U.K., 2005, pp. 108–109; http://www2.env.uea.ac.uk/gmmc/energy/env2e02/env2e02.htm.)

generator shaft. The generators then convert the energy of the water into electricity, and step-up transformers increase the voltage produced to higher voltage levels. Because the potential energy stored in the reservoir is converted into kinetic energy at the inlet to the water turbine, we can equate:

$$m \times g \times h = mV^2 \tag{5.18}$$

where:

m = mass of water
g = acceleration due to gravity (can be taken as 10 m/s^2 in most applications)
h = head difference
V = velocity of water at the inlet

■ Example 5.13*

Problem: A reservoir has an area of 1 km^2, and the difference between the crest of the dam and the inlet to a hydro station is 10 m (see Figure 5.4). The station runs at an overall efficiency of 80% and is situated 305 m below the crest of the dam. The rainfall is 1000 mm per annum, the catchment area of the reservoir is 10 times the area of the reservoir, and the run is 50%. What should the rated output of the turbine be if its maximum output is designed to be 5 times the mean output at the site? What is the maximum time the station could operate at full power during a sustained drought?

Solution:

1. Mean head between the maximum and minimum levels = 305 – 10/2 = 300 m.
2. Average annual flow into the reservoir = 50% of 10 × area × rainfall = 0.5 × 10 × (1000 m × 1000 m) × 1 m = 5,000,000 m^3.
3. Mean energy generated per annum at 80% efficiency = $m \times g \times h \times 0.8 = 5{,}000{,}000 \times 1000 \times 10 \times 300 \times 0.8$ = 12,000,000 MJ.
4. Rated output (mean power) = 12,000,000/(60 × 60 × 24 × 365) = 0.381 MW.
5. So, maximum power out = 5 × 0.381 = 190 MW, and time at maximum power assuming that the reservoir falls by 10 m is

$$\text{Days} = \frac{\text{Area} \times \text{Depth} \times \text{Density} \times g \times h \times 0.8}{\text{Maximum Power}}$$

$$= \frac{1000 \times 1000 \times 10 \times 1000 \times 300 \times 0.8}{1{,}900{,}000 \times 60 \times 60 \times 24} = 146.2 \text{ days}$$

WATER TURBINES

The two main types of water turbines are *impulse* and *reaction* (EERE, 2008). The type of hydropower turbine chosen for a project is based on the height of standing water—referred to as *head*—and the flow, or volume of water, at the site. Cost, efficiency desired, and how deep the turbine must be set are other deciding factors.

IMPULSE TURBINE

The impulse turbine uses the velocity of the water to move the runner and discharges to atmospheric pressure. The water stream hits each bucket on the runner. There is no suction on the down side of the turbine, and water flows out the bottom of the turbine housing after hitting the runner. An impulse turbine is generally suitable for high-head, low-flow applications.

* Adapted from Tovey, N.K., *ENV-2E02 Energy Resources 2004–2005 Lecture*, University of East Anglia, Norwich, U.K., 2005 (http://www2.env.uea.ac.uk/gmmc/energy/env2e02/env2e02.htm).

Reaction Turbine

A reaction turbine develops power from the combined action of pressure and moving water. The runner is placed directly in the water stream flowing over the blades rather than striking each individually. Reaction turbines are generally used for sites with lower head and higher flows than compared with impulse turbines.

ADVANCED HYDROPOWER TECHNOLOGY

The U.S. Department of Energy (USDOE, 2008) and its associated technical activities support the development of technologies that will enable existing hydropower power to generate more electricity with less environmental impact. This will be done by: (1) developing new turbine systems that have improved overall performances; (2) developing new methods to optimize hydropower operations at the unit, plant, and reservoir system levels; and (3) conducting research to improve the effectiveness of the environmental mitigation practices required at hydropower projects.

The main objective of its research into advanced hydropower technology is to develop new system designs and operation modes that will enable both better environmental performance and competitive generation of electricity. The products of DOE research will allow hydropower projects to generate cleaner electricity, and DOE-sponsored projects will develop new equipment and operational techniques to optimize water-use efficiency, increase generation, and improve environmental performance and mitigation practices at existing plants. Ongoing research efforts contributing to the success of these objectives will enable up to a 10% increase in the hydropower generation at existing dams; these objectives include the following:

- Test a new generation of large turbines in the field to demonstrate that these turbines are commercially viable, compatible with today's environmental standards, and capable of balancing environmental, technical, operational, and cost considerations.
- Develop new tools to improve water use efficiency and operations optimization within hydropower units, plants, and river systems with multiple hydropower facilities.
- Identify improved practices that can be applied at hydropower plants to mitigate for environmental effects of hydro development and operation.

HYDROPOWER GENERATION: DISSOLVED OXYGEN CONCERNS

The benefits derived from the use of hydropower include: (1) it is a clean fuel source; (2) it is a fuel source that is domestically supplied; (3) it relies on the water cycle and thus is a renewable power source; (4) it is generally available as needed; (5) it creates reservoirs that offer a variety of recreational opportunities, such as fishing, swimming, and boating; and (6) it creates a supply of water where needed and assists in flood control. Many of these benefits are well known and often taken for granted.

Coins are two sided, of course; that is, the good side of anything is generally accompanied by a bad side. Many view this to be the case with hydropower. The bad side or disadvantages of hydropower include the impact on fish populations, such as salmon, when the fish cannot migrate upstream past impoundment dams to their spawning grounds or migrate downstream to the ocean. Hydropower can also be impacted by drought, in that when water is not available the plant is not able to produce electricity. Hydropower plants also compete with other uses for the land.

Other lesser known negatives of hydropower plants concern their impact on water flow and quality; hydropower plants can cause low water levels that impact riparian (riverbank) habitats. Water quality is also affected by hydropower plants. Low dissolved oxygen levels in the water, which are harmful to riparian habitats, can result when reservoirs stratify; that is, they develop layers of water of different temperatures (see Figure 5.5). Stratification can affect the water temperature, thus

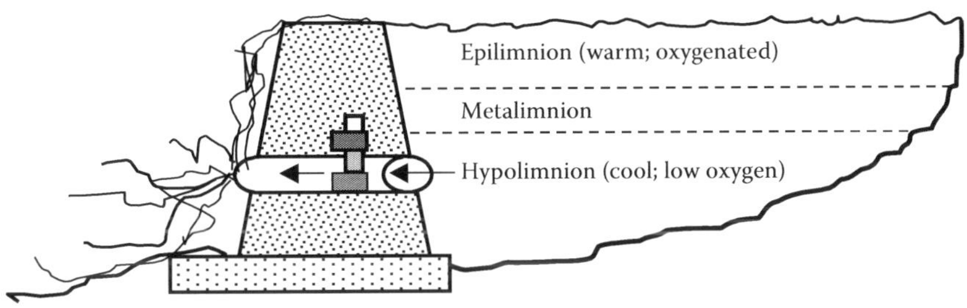

FIGURE 5.5 Thermal stratification of a hydropower reservoir.

causing changes in dissolved oxygen levels, nutrient levels, productivity, and the bioavailability of heavy metals. During the summer, the natural process of stratification can divide a reservoir into distinct vertical strata, such as a warm, well-mixed upper layer (epilimnion) over a cooler, relativity stagnant lower layer (hypolimnion). Plant and animal respiration, bacterial decomposition of organic matter, and chemical oxidation can all act to progressively remove dissolved oxygen from hypolimnetic waters. This decrease in hypolimnetic dissolved oxygen is not generally offset by the renewal mechanisms of atmospheric diffusion, circulation, and photosynthesis that operate in the epilmnion (Spellman, 1996, 2008). In temperate regions, the decline in hypolimnetic dissolved oxygen concentrations begins at the onset of stratification (spring or summer) and continues until either anaerobic conditions predominate or reoxygenation occurs during the fall turnover of the water body.

Numerous structural, operational, and regulatory techniques are available to resolve low dissolved oxygen issues. Levels of dissolved oxygen can be increased through modifications in dam operations that include such techniques as fluctuating the timing and duration of flow releases, spilling or sluicing water, increasing minimum flows, flow mixing, and turbine aeration; at some sites, injection of air or oxygen for weir aeration has proven effective. The most effective strategy for addressing the dissolved oxygen problem depends on the site.

THE BOTTOM LINE ON HYDROPOWER

Hydropower offers advantages over the other energy sources but faces unique environmental challenges. The advantages of using hydropower begin with that fact that hydropower does not pollute the air like power plants that burn fossil fuels, such as coal and natural gas. Moreover, hydropower can be produced locally and does not have to be imported into the country as foreign oil must be. Because hydropower relies on the water cycle, driven by the sun, it is a renewable resource that will be around for at least as long as humans. Hydropower is controllable; that is, engineers can control the flow of water through the turbines to produce electricity on demand. Finally, hydropower impoundment dams create huge lake areas for recreation, irrigation of farm lands, reliable supplies of potable water, and flood control.

Hydropower also has some disadvantages; for example, fish populations can be impacted if they cannot migrate upstream past impoundment dams to their spawning grounds or if they cannot migrate downstream to the ocean. Many dams have installed fish ladders or elevators to aid upstream fish passage. Downstream fish passage is aided by diverting fish from turbine intakes using screens or racks or even underwater lights and sounds and by maintaining a minimum spill flow past the turbine. Hydropower can also impact water quality and flow. Hydropower plants can cause low dissolved oxygen levels in the water, which can be harmful to riparian habitats; the problem of low dissolved oxygen levels can be addressed by using various aeration techniques to oxygenate the water. Maintaining minimum flows of water downstream of a hydropower installation is also critical for the survival of riparian habitats. Hydropower is also susceptible to drought.

When water is not available, the hydropower plants cannot produce electricity. Finally, construction of new hydropower facilities impact investors and others by competing with other uses of the land. Preserving local flora and fauna and historical or cultural sites is often more highly valued than electricity generation.

DISCUSSION AND REVIEW QUESTIONS

1. A few environmentalist groups advocate tearing down existing hydroelectric dams in the United States to allow affected rivers to run free. Do you support or oppose this stance? Why?
2. Do we have any more water on Earth today than yesterday or 1000 years ago?
3. The most common type of hydroelectric power plant is an ____________ facility.
4. Water pressure in terms of height is designated ____________.
5. The pressure at any point in a fluid at rest depends on the distance measured vertically to the free ____________ and the ____________ of the fluid.
6. ____________ is the mass per unit volume of a substance.
7. ____________ is the push or pull influence that causes motion.
8. Static Head + Friction Head + Velocity Head = ____________.
9. What states that energy can be neither created nor destroyed?
10. What designates the coefficient of roughness?

REFERENCES AND RECOMMENDED READING

Anon. (2006). *Urbanization and Global Change*. Flint: University of Michigan (http://globalchange.umich.edu/globalchange2/current/lectures/urban_gc/).

Bender, M.D., Hauser, G.E., and Shiao, M.S. (1991). *Modeling Boone Reservoir to Evaluate Cost-Effectiveness of Point and Nonpoint Source Pollutant Controls*, Report No. WR28-1-31-107. Norris, TN: Engineering Laboratory, Tennessee Valley Authority.

Berryman, A.A. (1999). *Principles of Population Dynamics and Their Application*. London: Garland Science.

Berryman, A.A. (2002). *Population Cycles: The Case for Trophic Interactions*. London: Oxford University Press.

Brahic, C. (2009). Fish 'an ally' against climate change. *New Scientist*, January 16 (http://www.newscientist.com/article/dn16432-fish-an-ally-against-climate-change.html).

Brookfield, H.C. (1989). *Sensitivity to Global Change: A New Task for Old/New Geographers*. Reading, U.K.: Norma Wilkinson Memorial Lecture, University of Reading.

Brundtland, G.H. (1987). *Our Common Future*. New York: Oxford University Press.

Caldeira, K. and Wickett, M.E. (2003). Anthropogenic carbon and ocean pH. *Nature*, 425(6956), 365.

Calder, W.A. (1983). An allometric approach to population cycles of mammals. *Journal of Theoretical Biology*, 100, 275–282.

Calder, W.A. (1996). *Size, Function and Life History*. Mineola, NY: Dover Publications.

Christensen, C. (1982). In *Making It Happen: A Positive Guide to the Future*, Richardson, J.M., Ed. Washington, D.C.: U.S. Association for the Club of Rome.

Commoner, B. (1971). *The Closing Circle: Nature, Man, and Technology*. New York: Knopf.

Damuth, J. (1991). Of size and abundance. *Nature* 351, 268–269.

Debres, K. (2005). Burgers for Britain: a cultural geography of McDonald's UK. *Journal of Cultural Geography*, March 22.

Easton, T.A. and Goldfarb, T.D. (1989). Is population control the key to preventing environmental deterioration? In *Taking Sides: Clashing Views on Controversial Environmental Issues*. Guilford, CT: Duskin Publishing Group.

EERE. (2008). *History of Hydropower*. Washington, D.C.: Energy Efficiency & Renewable Energy, U.S. Department of Energy (http://www1.eere.energy.gov/windandhydro/hydro_history.html).

Enger, E., Kormelink, J.R., Smith, B.F., and Smith, R.J. (1989). *Environmental Science: The Study of Interrelationships*. Dubuque, IA: William C. Brown Publishers.

Gibilisco, S. (2007). *Alternative Energy Demystified*. New York: McGraw-Hill.

Ginzburg, L.R. (1986). The theory of population dynamics. 1. Back to first principles. *Journal of Theoretical Biology*, 122, 385–399.

Ginzburg, L.R. and Colyvan, M. (2004). *Ecological Orbits: How Planets Move and Populations Grow*. New York: Oxford University Press.

Ginzburg, L.R. and Jensen C.X.J. (2004). Rules of thumb for judging ecological theories. *Trends in Ecology and Evolution*, 19, 121–126.

Hardin, G. (1986). Cultural carrying capacity: a biological approach to human problems. *BioScience*, 36, 599–606.

Hickman, C.P., Roberts, L.S., and Hickman, F.M. (1990). *Biology of Animals*. St. Louis, MO: Times Mirror/Mosby College Publishing.

Holdren, J.P. (1991). Population and the energy problem. *Population and Environment*, 12, 231–255.

Hubbell, S.P. and Johnson, L.K. (1977). Competition and next spacing in a tropical stingless bee community. *Ecology*, 58, 949–963.

Jarvie, M. and Solomon, B. (1998). Point–nonpoint effluent trading in water sheds: a review and critique. *Environmental Impact Assessment Review*, 18, 135–157.

Keller, E.A. (1988). *Environmental Geology*, Columbus, OH: Merrill.

Krebs, R.E. (2001). *Scientific Laws, Principles and Theories*. Westport, CT: Greenwood Press.

Kuhlken, R. (2002). Intensive agricultural landscapes of Oceania. *Journal of Cultural Geography*, 19(2), 161–195.

Library of Congress. (2009). *The World's First Hydroelectric Power Plant Began Operation September 30, 1882*. Washington, D.C.: Library of Congress (http://www.americaslibrary.gov/jb/gilded/jb_gilded_hydro_1.html).

Liebig, J. (1840). *Chemistry and Its Application to Agriculture and Physiology*. London: Taylor & Walton.

Lindeburg, M.R. (1986). *Civil Engineering Reference Manual*, 4th ed. San Carlos, CA: Professional Publications.

Lund, J.W. (2007). Characteristics, development and utilization of geothermal resources. *Geo-Heat Centre Quarterly Bulletin*, 28(2), 1–9.

March, P.A., Brice, T.A., Mobley, M.H., and Cybularz, J.M. (1992). Turbines for solving the DO dilemma. *Hydro Review*, 11(1), 30–36.

Masters, G.M. (1991). *Introduction to Environmental Engineering and Science*. Englewood Cliffs, NJ: Prentice Hall.

Miller, G.T. (1988). *Environmental Science: An Introduction*. Belmont, CA: Wadsworth.

Nathanson, J.A. (1997). *Basic Environmental Technology: Water Supply, Waste Management, and Pollution Control*, 2nd ed. Upper Saddle River, NJ: Prentice Hall, pp. 21–22.

NREL. (2009). *Concentrating Solar Power*. Washington, D.C.: National Renewable Energy Laboratory (http://www.nrel.gov/learning/re_csp.html).

Podar, M.K., Crossman, J.C., Burmaster, D.E., Ruane, R.J., Jaksch, J.A., Hauser, G., and Sessions, S.L. (1985). *Optimizing Point/Nonpoint Source Tradeoff in the Houston River Near Kingsport, Tennessee*, paper presented at the National Conference on Nonpoint Sources of Pollution, May 19–22, Kansas City, MO.

Reid, W. et al. (1988). *Bankrolling Successes*. Washington, D.C.: Environmental Policy Institute and National Wildlife Federation.

Searchinger, T. et al. (2008). Use of U.S. croplands for biofuels increases greenhouse gases through emissions from land-use change. *Science*, 319(5867), 1238–1240.

Sharov, A. (1992). Life-system approach: a system paradigm in population ecology. *Oikos*, 63, 485–494.

Simon, J.L. (1980). Resources, population, environment: an oversupply of false bad news. *Science*, 208, 1431–1437.

Sinha, A. (2006). Cultural landscape of Pavagadh: the abode of Mother Goddess Kalika; *Journal of Cultural Geography*, 23(2), 89–103.

Spellman, F.R. (1996). *Stream Ecology and Self-Purification*. Lancaster, PA: Technomic.

Spellman, F.R. (2008). *The Science of Water,* 2nd ed. Boca Raton, FL: CRC Press.

Tovey, N.K. (2005). *ENV-2E02 Energy Resources 2004–2005 Lecture*. Norwich, U.K.: University of East Anglia (http://www2.env.uea.ac.uk/gmmc/energy/env2e02/env2e02.htm).

Turchin, P. (2001). Does population ecology have general laws? *Oikos*, 94, 17–26.

Turchin, P. (2003). *Complex Population Dynamics: A Theoretical/Empirical Synthesis*. Princeton, NJ: Princeton University Press.

UCS. (2008). *Land Use Changes and Biofuels*. Cambridge, MA: Union of Concerned Scientists. (ucsusa.org/assets/documents/clean_vehicles/Indirect-Land-Use-Factsheet.pdf).

USDOE. (2006). *Geothermal Technologies Program: A History of Geothermal Energy in the United States*. Washington, D.C.: U.S. Department of Energy (http://www1.eere.energy.gov/geothermal/history.html).

USDOE. (2008). *Advanced Hydropower Technology*. Washington, D.C.: U.S. Department of Energy (http://www1.eere.energy.gov/windandhydro/hydro_advtech.html).

USDOE. (2009a). *Fossil Fuels*. Washington, D.C.: U.S. Department of Energy. (http://www.energy.gov/ energysources/fossilfuels.htm).

USDOE. (2009b). *Solar Energy Technologies Program*. Washington, D.C.: U.S. Department of Energy (http://www1.eere.Energy.gov/solar/printable_versions/about.html).

USEPA. (2009). *Ozone Science: The Facts Behind the Phase-Out*. Washington, D.C.: U.S. Environmental Protection Agency (http://www.epa.gov/ozone/science/sc_fact.html).

USGS. (1999). *Hawaiian Volcano Observatory*. Washington, D.C.: U.S. Geological Survey (http://hvo.wr.usgs.gov/).

Vitousek, P.M., Ehrlich, A.H., and Matson, P.A. (1986). Human appropriation of the products of photosynthesis. *BioScience*, 36, 368–373.

Volterra, V. (1926). Variazioni e fluttuazioni del numero d'indivudui in specie animali conviventi. *Memorie dell'Accademia dei Lincei*, 2, 31–113.

Wetzel, R.G. (1975). *Limnology*. Philadelphia, PA: W.B. Saunders.

Wilson, E.O. (2001). *The Future of Life*. New York: Knopf.

World Commission on Environment and Development. (1987). *Our Common Future*. New York: Oxford University Press.

Zelinsky, W. (2004). Globalization reconsidered: the historical geography of modern western male attire. *Journal of Cultural Geography*, 22(1), 83–134.

6 Bioenergy

The wretched and the poor look for water and find none,
their tongues are parched with thirst;
but I the Lord will give them an answer,
I, the God of Israel, will not forsake them.
I will open rivers among the sand-dunes
and wells in the valleys;
I will turn the wilderness into pools
and dry land into springs of water;
I will plant cedars in the wastes,
And acacia and myrtle and wild olive;
the pine shall grow on the barren heath
side by side with fir and box...

—Isaiah 41: 17–20

A nation that runs on oil can't afford to run short.

—Old oil industry slogan

INTRODUCTION

In a recent article in *The New York Times*, Friedman (2010) stated that, "The fat lady has sung." Specifically, he was speaking about America's transition from the "greatest generation" to what novelist Kurt Anderson has referred to as the "grasshopper generation." According to Friedman, we are "going from the age of government handouts to the age of citizen givebacks, from the age of companions fly free to the age of paying for each bag." Our parents were the greatest generation, but we have changed course and become the grasshopper generation, which is "eating through the prosperity that was bequeathed us like hungry locusts." Emphasizing again the major theme of this text, what we are eating through, among other things, is our readily available, relatively inexpensive source of energy. The point is we can, like the grasshopper, gobble it all up until it is all gone or we can find alternatives—renewable alternatives of energy.

One of the promising forms of alternative energy is *bioenergy*, a general term for energy derived from materials such as straw, wood, or animal wastes, which, in contrast to fossil fuels, were living matter relatively recently. Such materials can be burned directly as solids (*biomass*) to produce heat or power, but they can also be converted into liquid biofuels. Interest is growing in *biofuels*, liquid fuels (biodiesel and bioethanol) that can be used for transport. At the moment, transport has taken center stage in our search for renewable, alternative fuels to eventually replace hydrocarbon fuels. Unlike biofuels, solid biomass fuel is used primarily for electricity generation or heat supply.

Even though we have stated that bioenergy is a promising source of energy for the future, it is rather ironic that the experts (or anyone else for that matter) frequently make this point without qualification. The qualification? The reality? Simply, keep in mind that only 100 yeas ago our economy was based primarily on bioenergy from biomass, or carbohydrates, rather than from hydrocarbons. In the late 1800s, the largest selling chemicals were alcohols made from wood and grain, the first plastics were produced from cotton, and about 65% of the nation's energy came from wood (USDOE, 2003a). By the 1920s, the economy started shifting toward the use of fossil resources, and

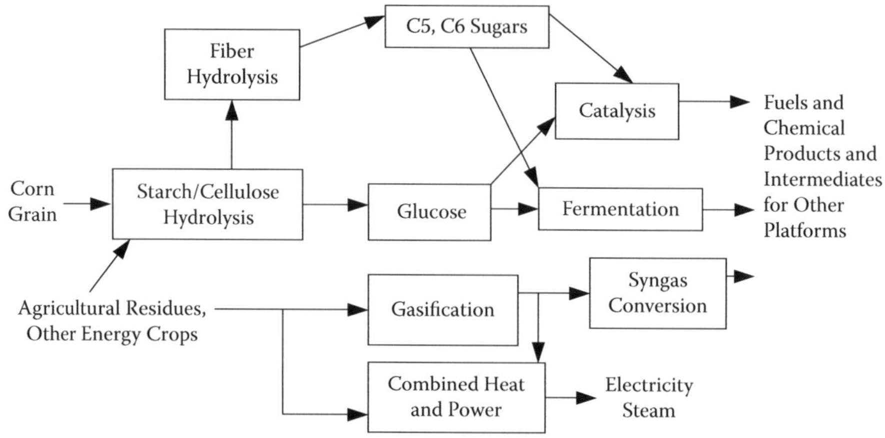

FIGURE 6.1 Biorefinery.

after World War II this trend accelerated as technology breakthroughs were made. By the 1970s, fossil energy was established as the backbone of the U.S. economy, and all but a small portion of the carbohydrate economy remained (Morris, 2002). In 1989, in the industrial sector, plants accounted for about 16% of input, compared with 35% in 1925 (USDOE, 2003b).

Processing costs and the availability of inexpensive fossil energy resources continue to be driving factors in the dominance of hydrocarbon resources. In many cases, it is still more economical to produce goods from petroleum or natural gas than from plant matter. This trend is about to shift dramatically as we reach peak oil production and as the world continues to demand unprecedented amounts of petroleum supplies from an ever dwindling supply.

Assisting in this trend shift are technological advances in the biological sciences and engineering, political change, and concern for the environment, which have begun to swing the economy back toward carbohydrates on a number of fronts. Consumption of biofuels in vehicles, for example, has risen from 0 in 1977 to nearly 1.5 billion gallons in 1999. The use of inks produced from soybeans in the United States increased fourfold between 1989 and 2000 and now represents more than 22% of total use (Morris, 2002).

Technological advances are also beginning to make an impact on reducing the costs of producing industrial products and fuels from biomass, making them more competitive with those produced from petroleum-based hydrocarbons. Developments in pyrolysis, ultracentrifuges, membranes, and the use of enzymes and microbes as biological factories are enabling the extraction of valuable components from plants at a much lower cost. As a result, industry is investing in the development of new bioproducts that are steadily gaining a share of current markets (USDOE 2003a).

New technologies are helping the chemical and food processing industries develop processes for more cost-effective production of all kinds of industrial products from biomass. One example is a plastic polymer derived from corn that is now being produced at a 300-million-pound-per-year plant in Nebraska, a joint venture between Cargill, one of the largest grain trading companies in the world, and Dow Chemical, one of the largest chemical producers (Fahey, 2001).

Other chemical companies are exploring the use of low-cost biomass processes to make chemicals and plastics that are now made from more expensive petrochemical processes (USDOE, 2003b). In this regard, new innovative processes such as biorefineries may become the foundation of the new bioindustry. A biorefinery is similar in concept to a petroleum refinery, except that it is based on conversion of biomass feedstocks rather than crude oil. Biorefineries in theory would use multiple forms of biomass to produce a flexible mix of products, including fuels, power, heat, chemicals, and materials (see Figure 6.1). In a biorefinery, biomass would be converted into high-value chemical

TABLE 6.1
U.S. Energy Consumption by Energy Source, 2008

Energy Source	Energy Consumption (quadrillion Btu)
Total	99.438
Fossil fuels (coal, natural gas, petroleum)	83.532
Electricity net imports	0.113
Nuclear electric power	8.427
Renewable	7.367
Biomass (total)	**3.852**
Biofuels	**1.372**
Waste	**0.436**
Wood and wood-derived fuels	**2.044**
Geothermal energy	0.360
Hydroelectric conventional	2.512
Solar thermal/PV energy	0.097
Wind energy	0.546

Source: EIA, *U.S. Energy Consumption by Energy Source*, Energy Information Administration, Washington, D.C., 2007 (www.eia.doe.gov/cneaf/solar.renewables/page/trends/table1.html).

products and fuels (both gas and liquid). Byproducts and residues, as well as some portion of the fuels produced, would be used to fuel on-site power generation or cogeneration facilities producing heat and power. The biorefinery concept has already proven successful in U.S. agricultural and forest products industries, where such facilities now produce food, feed fiber, or chemicals, as well as heat and electricity to run plant operations. Biorefineries offer the most potential for realizing the ultimate opportunities of the bioenergy industry. Table 6.1 shows the energy consumption by energy source for 2008; in that year, biomass (biofuels, waste, wood and wood-derived fuels) accounted for 3.852 quadrillion Btu.

DID YOU KNOW?

In 2004, it was estimated that in the United States alone more than 100 billion gallons of light fuel oil per year could be produced from waste and biomass, including municipal solid waste, municipal sewage sludge, hazmat waste, agricultural crop waste, feedlot manure, plastics, tires, heavy oil or tar sands, forestry waste, restaurant grease, and biomass crops (switchgrass) grown on idle land and cropland.

BIOMASS

Biomass (all of Earth's living matter) consists of the energy from plants and plant-derived organic-based materials; it is essentially stored energy from the sun. Biomass can be biochemically processed to extract sugars, thermochemically processed to produce biofuels or biomaterials, or combusted to produce heat or electricity. Biomass is also an input into other end-use markets, such as forestry products (pulpwood) and other industrial applications. This complicates the economics of biomass feedstock and requires that we differentiate between what is technically possible from what is economically feasible, taking into account relative prices and intermarket competition. The biomass industry is one of the fastest-growing industries in the United States. Biomass has been used since people began burning wood to cook food and keep warm. Trees have been the principal fuel of almost every society for over 5000 years, from the Bronze Age until the middle of the 19th century (Perlin, 2005). Wood is still the most significant biomass energy resource today, but other sources of biomass can also be used, including food crops grassy and woody plants, residues from agriculture or forestry, and the organic component of municipal and industrial wastes. Even the fumes (methane) from landfills can be used as a biomass energy source. Organic material that has been transformed by geological processes into substances such as coal or petroleum are excluded from this category.

Feedstock Types

A variety of biomass feedstocks can be used to produce transportation fuels, bioproducts, and power. Feedstock refers to the crops or products, such as waste vegetable oil, that can be used as or converted into biofuels and bioenergy. The advantages or disadvantages of one type of feedstock as compared to another are gauged in terms of how much usable material they yield, where they can grow, and how energy and water intensive they are. Feedstock types are listed as first-generation or second-generation feedstocks. First-generation feedstocks include those that are already widely grown and used for some form of bioenergy or biofuel production, which means that possible foods vs. fuel conflicts could arise. First-generation feedstocks include sugars (sugar beets, sugar cane, sugar palm, sweet sorghum, and *Nipa* palm), starches (cassava, corn, milo, sorghum, sweet potato, and wheat), waste feedstocks (e.g., whey and citrus peels), and oils and fats (coconut oil, oil palm, rapeseed, soybeans, sunflower seed, castor beans, *Jatropha*, jojoba, karanj, waste vegetable oil and animal fat). Second-generation feedstocks refer broadly to crops that have high potential yields of biofuels but are not widely cultivated or are not cultivated as an energy crop. These would include cellulosic feedstocks or conventional crops such as miscanthus, prairie grass, switchgrass, willows, and hybrid poplar trees. Algae and halophytes (saltwater plants) are other second-generation feedstocks.

Currently, a majority of the ethanol produced in the United States is made from corn or other starch-based crops. The current focus, however, is on developing cellulosic feedstocks (non-grain, non-food-based feedstocks such as switchgrass, corn stover, and wood material) and on technologies to convert cellulosic material into transportation fuels and other products. Using cellulosic feedstocks not only can alleviate the potential concern of diverting food crops to produce fuel but also offers a variety of environmental benefits (EERE, 2008). Because such a wide variety of cellulosic feedstocks can be used for energy production, potential feedstocks are grouped into categories, or pathways. Figure 6.2 shows some of the specific feedstocks in each of these areas.

Composition of Biomass

The ease with which biomass can be converted to useful products or intermediates is determined by the composition of the biomass feedstock. Biomass contains a variety of components, some of which are readily accessible and others that are much more difficult and costly to extract. The composition and subsequent conversion issues for current and potential biomass feedstock compounds are listed and described below.

Starch (Glucose)

Starch is readily recovered and converted from grain (corn, wheat, rice) into products. Starch from corn grain provides the primary feedstock for today's existing and emerging sugar-based bioproducts, such as polylactide, as well as the entire fuel ethanol industry. Corn grain serves as the primary feedstock for starch used to manufacture today's biobased products. Corn wet mills use a multistep

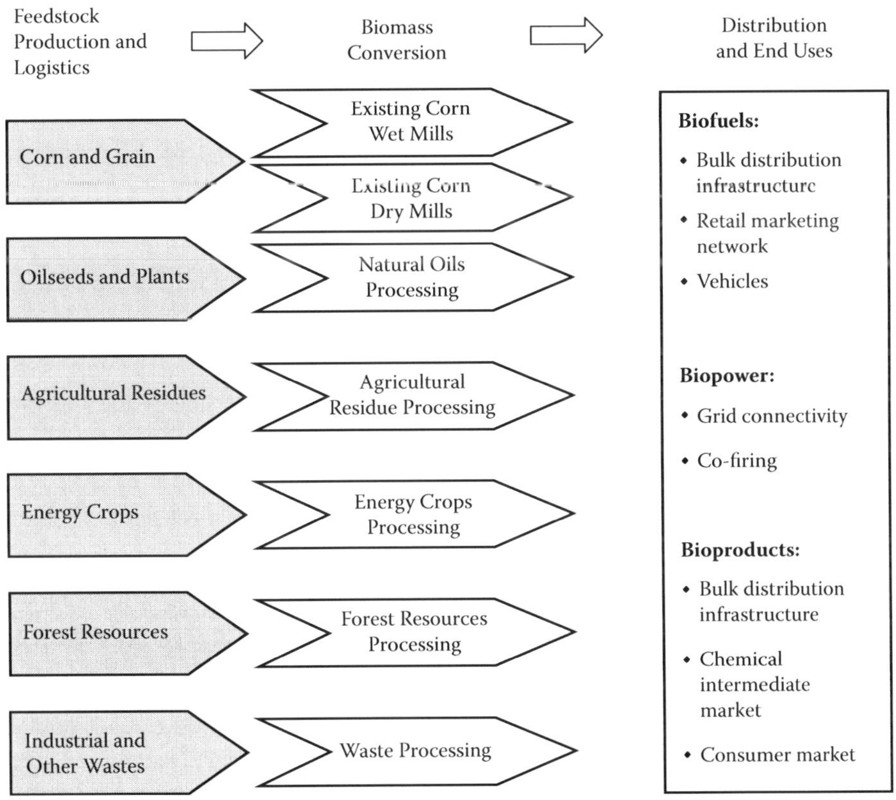

FIGURE 6.2 Resource-based biorefinery pathways.

process to separate starch from the germ, gluten (protein), and fiber components of corn grain. The starch streams generated by wet milling are very pure. Acid and enzymatic hydrolysis is used to break the glycosidic linkages of starch to yield glucose, which is then converted into a multitude of useful products.

Lignocellulosic Biomass

This includes the non-grain portion of biomass (e.g., cobs, stalks), often referred to as *agricultural stover* or *residues*. Energy crops such as switchgrass also contain valuable components, but they are not as readily accessible as starch. These lignocellulosic biomass resources (also called *cellulosic*) are comprised of cellulose, hemicellulose, and lignin. Generally, lignocellulosic material contains 30 to 50% cellulose, 20 to 30% hemicellulose, and 20 to 30% lignin. Some exceptions to this are cotton (98% cellulose) and flax (80% cellulose). Lignocellulosic biomass is perceived as a valuable and largely untapped resource for the future bioindustry; however, recovering the components in a cost-effective way represents a significant technical challenge.

Cellulose

Cellulose, one of nature's polymers, is composed of glucose, a six-carbon sugar. The glucose molecules are joined by glycosidic linkages that allow the glucose chains to assume an extended ribbon conformation. Hydrogen bonding between chains leads to the formation of flat sheets that lay on top of one another in a staggered fashion, similar to the way staggered bricks add strength and stability to a wall. As a result, cellulose is very chemically stable and insoluble and serves as a structural component in plant walls.

Hemicellulose

Hemicellulose is a polymer containing primarily 5-carbon sugars such as xylose and arabinose with some glucose and mannose dispersed throughout. It forms a short-chain polymer that interacts with cellulose and lignin to form a matrix in the plant wall, strengthening it. Hemicellulose is more easily hydrolyzed than cellulose. Much of the hemicellulose in lignocellulosic material is solubilized and hydrolyzed to pentose and hexose sugars.

Lignin

Lignin helps bind the cellulosic/hemicellulose matrix while adding flexibility to the mix. The molecular structure of lignin polymers is very random and disorganized and consists primarily of carbon ring structures (benzene rings with methoxyl, hydroxyl, and propyl groups) interconnected by polysaccharides (sugar polymers). The ring structures of lignin have great potential as valuable chemical intermediates; however, the process of lignin separation and recovery is difficult.

Oils and Proteins

The seeds of certain plants offer two families of compounds with great potential for bioproducts: oils and protein. Oils and protein are found in the seeds of certain plants (soybeans, castor beans) and can be extracted in a variety of ways. Plants raised for this purpose include soy, corn, sunflower, safflower, rapeseed, and others. A large portion of the oil and protein recovered from oilseeds and corn is processed for human or animal consumption, but they can also serve as raw materials for lubricants, hydraulic fluids, polymers, and a host of other products.

Vegetable Oils

Vegetable oils are composed primarily of triglycerides, also referred to as *triacylglycerols*. Triglycerides contain a glycerol molecule as the backbone with three fatty acids attached to the glycerol hydroxyl groups.

Proteins

Proteins are natural polymers with amino acids as the monomer unit. They are incredibly complex materials, and their functional properties depend on their molecular structure. Twenty amino acids can be differentiated by their side chain or R-group, and they can be classified as nonpolar and hydrophobic, polar uncharged, and ionizable. The interactions among the side chains, the amide protons, and the carbonyl oxygen help create the three-dimensional shape of the protein.

Plant Basics*

To optimize plant biomass for more efficient processing requires a better understanding of plants and plant cell-wall structure and function. The plant kingdom ranks second in importance only to the animal kingdom (at least from the human point of view). The importance of plants and plant communities to humans, bioenergy production, and their environment cannot be overstated. Some of the important things plants provide are listed below.

- *Aesthetics*—Plants add to the beauty of the places where we live.
- *Medicine*—Eighty percent of all medicinal drugs originate in wild plants.
- *Food*—Ninety percent of the world's food comes from only 20 plant species.
- *Industrial products*—Plants are very important for the goods they provide (e.g., plant fibers provide clothing), and wood is used to build homes.

* This section is adapted from Spellman, F.R., *Biology for the Nonbiologist*, Government Institutes Press, Lanham, MD, 2009.

TABLE 6.2
A Comparison of Plants and Animals

Plants	Animals
Plants contain chlorophyll and can make their own food.	Animals cannot make their own food and are dependent on plants and other animals for food.
Plants give off oxygen and take in the carbon dioxide given off by animals.	Animals breathe in oxygen and give off carbon dioxide which plants need to make food.
Plants generally are rooted in one place and do not move on their own.	Most animals have the ability to move fairly freely.
Plants have either no or a very basic ability to sense.	Animals have a much more highly developed sensory and nervous system.

- *Recreation*—Plants are the basis for many important recreational activities, including fishing, nature observation, hiking, and hunting.
- *Fish and wildlife habitats*—Plants provide the necessary habitats for wildlife and fish populations.
- *Air quality*—The oxygen in the air we breathe comes from the photosynthesis of plants.
- *Water quality*—Plants aid in maintaining healthy watersheds, streams, and lakes by holding soil in place, controlling stream flows, and filtering sediments from water.
- *Erosion control*—Plant cover helps prevent wind or water erosion of the top layer of soil.
- *Climate*—Regional climates are impacted by the amount and type of plant cover.
- *Ecosystem*—Every plant species serves an important role or purpose in its community.
- *Feedstock for bioenergy production*—Some important fuel chemicals come from plants; for example, ethanol comes from corn and soy diesel from soybeans.

Though both are important kingdoms of living things, plants and animals differ in many important aspects. Some of these differences are summarized in Table 6.2. Before discussing the basic specifics of plants, it is important to first define a few key plant terms.

Plant Terminology

- *Apical meristem*—Meristematic cells located at tip (apex) of a root or shoot.
- *Cambium*—Lateral meristem in plants.
- *Chloroplasts*—Eukaryotic disk-like organelles that have a double membrane.
- *Companion cells*—Specialized cells in the phloem that load sugars into the sieve elements.
- *Cork cambium*—Secondary growth meristem that produces cork, a tough substance that replaces the epidermis.
- *Cotyledons*—Leaf-like structure (sometimes referred to as *seed leaf*) that is present in the seeds of flowering plants.
- *Dicot*—One of the two main types of flowering plants, characterized by having two cotyledons.
- *Diploid*—Having two of each kind of chromosome.
- *Guard cells*—Specialized epidermal cells flanking the stomata and whose opening and closing regulate gas exchange and water loss.
- *Haploid*—Having only a single set of chromosomes.
- *Meristem*—Group of plant cells that can divide indefinitely, providing new cells for the plant.
- *Monocots*—One of two main types of flowering plants, characterized by having a single cotyledon.

TABLE 6.3
Main Divisions of Plants

Division	Examples
Bryophyta	Mosses, liverworts, and hornworts
Coniferophyta	Conifers such as redwoods, pines, and firs
Cycadophyta	Cycads (sago palms)
Gnetophyta	Shrub trees and vines
Ginkgophyta	*Ginkgo* is the only genus
Lycophyta	Lycopods (look like mosses)
Pterophyta	Ferns and tree-ferns
Anthophyta	Flowering plants including oak, corn, maize, and herbs

- *Periderm*—Outer layer of plant tissue derived from the cork cambium and tissues produced by it (e.g., cork).
- *Phloem*—Complex vascular tissue that transports carbohydrates throughout the plant.
- *Sieve tubes*—Conducting cells in the phloem of vascular plants.
- *Stomata*—Pores on the underside of leaves that can be opened or closed to control gas exchange and water loss.
- *Thallus*—Main plant body, not differentiated into a stem or leaves.
- *Tropism*—Plant behavior that controls the direction of plant growth.
- *Vascular tissue*—Tissues found in the bodies of vascular plants that transport water, nutrients, and carbohydrates; the two major kinds are *xylem* and *phloem*.
- *Xylem*—Vascular tissue of plants that transports water and dissolved minerals from the roots upward to other parts of plant; xylem often also provides mechanical support against gravity.

Although not typically acknowledged, plants are as intricate and complicated as animals. Plants evolved from photosynthetic protists (unicellular organisms) and are characterized by photosynthetic nutrition, cell walls made from cellulose and other polysaccharides, lack of mobility, and a characteristic life cycle involving an alternation of generations. The divisions of plants and examples are listed in Table 6.3.

The Plant Cell

A brief summary of plant cells is provided here (see Figure 6.3):

- *Plants have all the organelles that animal cells have* (e.g., nucleus, ribosomes, mitochondria, endoplasmic reticulum, Golgi apparatus).
- *Plants have chloroplasts.* Chloroplasts are special organelles that contain chlorophyll and allow plants to carry out photosynthesis.
- *Plant cells can sometimes have large vacuoles for storage.*
- *Plant cells are surrounded by a rigid cell wall made of cellulose*, in addition to the cell membrane that surrounds animal cells; these walls provide support.

Vascular Plants

Vascular plants, also called *tracheophytes*, have special vascular tissue for the transport of necessary liquids and minerals over long distances. Vascular tissues are composed of specialized cells that create tubes through which materials can flow throughout the plant body. These vessels are

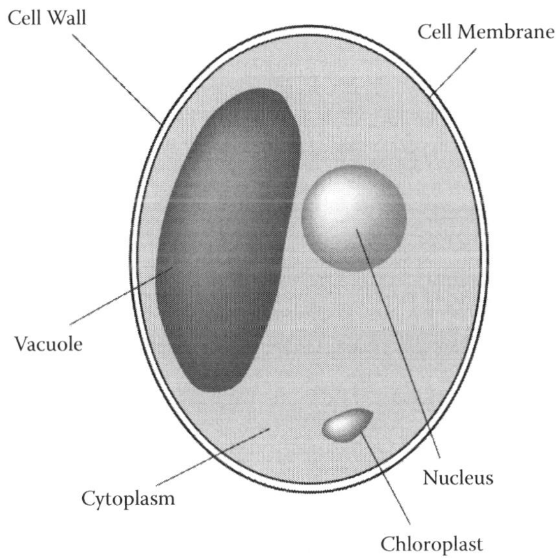

FIGURE 6.3 Plant cell.

continuous throughout the plant, allowing for the efficient and controlled distribution of water and nutrients. In addition to this transport function, vascular tissues also support the plant. The two types of vascular tissue are xylem and phloem:

- *Xylem* consists of a tube or a tunnel (pipeline) in which water and minerals are transported throughout the plant to leaves for photosynthesis. In addition to distributing nutrients, xylem (wood) provides structural support. After a time, the xylem at the center of older trees ceases its transport function and takes on a supportive role only.
- *Phloem* tissue consists of cells called *sieve tubes* and *companion cells*. Phloem tissue moves dissolved sugars (carbohydrates), amino acids, and other produces of photosynthesis from the leaves to other regions of the plant.

The two most important tracheophytes are gymnosperms (*gymno*, naked; *sperma*, seed) and angiosperms (*angio*, vessel or container):

- *Gymnosperms*—The plants we recognize as gymnosperms represent the sporophyte generation (i.e., the spore-producing phase in the life cycle of a plant that exhibits alternation of generation). Gymnosperms were the first tracheophytes to use seeds for reproduction. The seeds develop in protective structures called *cones*. A gymnosperm contains some cones that are female and some that are male. Female cones produce spores that, after fertilization, become eggs enclosed in seeds that fall to the ground. Male cones produce pollen, which is carried by the wind and fertilizes female eggs. Unlike flowering plants, the gymnosperm does not form true flowers or fruits. Coniferous tress such as firs and pines are good examples of gymnosperms.
- *Angiosperms*—The flowering plants are the most highly evolved plants and the most dominant today. They have stems, roots, and leaves. Unlike gymnosperms such as conifers and cycads, angiosperm seeds are found in a flower. Angiosperm eggs are fertilized and develop into a seed in an ovary that is usually in a flower.

The two types of angiosperms are monocots and dicots:

- *Monocots*—These angiosperms start with one seed leaf (cotyledon), thus their name, which is derived from the presence of a single cotyledon during embryonic development. Monocots include grasses, grains, and other narrow-leaved angiosperms. The main veins of their leaves are usually parallel and unbranched, the flower parts occur in multiples of three, and a fibrous root system is present. Monocots include orchids, lilies, irises, palms, grasses, and wheat, corn, and oats.
- *Dicots*—Angiosperms in this group grow two seed leaves (cotyledons). Most plants are dicots, including maples, oaks, elms, sunflowers, and roses. Their leaves usually have a single main vein or three or more branched veins that spread out from the base of the leaf.

Leaves

The principal function of leaves is to absorb sunlight for the manufacture of plant sugars by photosynthesis. The broad, flattened surfaces of leaves gather energy from sunlight, while apertures on their undersides bring in carbon dioxide and release oxygen. Leaves develop as a flattened surface in order to present a large area for efficient absorption of light energy. On its two exteriors, the leaf has layers of epidermal cells that secrete a waxy, nearly impermeable cuticle (chitin) to protect against water loss (dehydration) and fungal or bacterial attack. Gases diffuse in or out of the leaf through *stomata*, small openings on the underside of the leaf. The opening or closing of the stomata occurs through the swelling or relaxing of *guard cells*. If the plant wants to limit the diffusion of gases and the transpiration of water, the guard cells swell together and close the stomata. Leaf thickness is kept to a minimum so gases that enter the leaf can diffuse easily throughout the leaf cells.

Chlorophyll and Chloroplasts

The green pigment in leaves is *chlorophyll*. Chlorophyll absorbs red and blue light from the sunlight that falls on leaves; therefore, the light reflected by the leaves is diminished in red and blue and appears green. The molecules of chlorophyll are large. They are not soluble in the aqueous solution that fills plant cells. Instead, they are attached to the membranes of disc-like structures, called *chloroplasts*, inside the cells. Chloroplasts are the site of photosynthesis, the process in which light energy is converted to chemical energy. In chloroplasts, the light absorbed by chlorophyll supplies the energy used by plants to transform carbon dioxide and water into oxygen and carbohydrates. Chlorophyll is not a very stable compound; bright sunlight causes it to decompose. To maintain the amount of chlorophyll in their leaves, plants continuously synthesize it. The synthesis of chlorophyll in plants requires sunlight and warm temperatures; therefore, in the summer chlorophyll is continuously broken down and regenerated in the leaves of trees.

Photosynthesis

Because our quality of life, and indeed our very existence, depends on photosynthesis, it is essential to understand it. In photosynthesis, plants (and other photosynthetic autotrophs) use the energy from sunlight to create the carbohydrates necessary for cell respiration. More specifically, plants take water and carbon dioxide and transform them into glucose and oxygen:

$$6CO_2 + 6H_2O + \text{light energy} \rightarrow C_6H_{12}O_6 + 6O_2$$

This general equation of photosynthesis represents the combined effects of two different stages. The first stage is the light reaction and the second stage is the dark reaction. The *light reaction* is the photosynthesis process in which solar energy is harvested and transferred into the chemical

> **DID YOU KNOW?**
>
> Charles Darwin was the first to discuss how plants respond to light. He found that the new shoot of grasses bend toward the light because the cells on the dark side grow faster than the lighted side.

bonds of adenosine triphosphate (ATP); it can only occur in light. The *dark reaction* is the process in which food (sugar) molecules are formed from carbon dioxide from the atmosphere with the help of ATP; it can occur in the dark as long as ATP is present.

Roots

Roots absorb nutrients and water, anchor the plant in the soil, provide support for the stem, and store food. They are usually below ground and lack nodes, shoots, and leaves. Plants have two major types of root systems. *Taproot systems* have a stout main root with a limited number of side-branching roots. Examples of taproot system plants are nut trees, carrots, radishes, parsnips, and dandelions. Taproots make transplanting difficult. The *fibrous root system* has many branched roots. Examples of fibrous root plants are most grasses, marigolds, and beans. Radiating from the roots is a system of root hairs, which vastly increase the absorptive surface area of the roots. Roots also anchor the plant in the soil.

Growth in Vascular Plants

Vascular plants undergo two kinds of growth (growth is primarily restricted to meristems): primary and secondary growth. *Primary growth* occurs relatively close to the tips of roots and stems. It is initiated by apical meristems, and it is primarily involved in the extension of the plant body. The tissues that arise during primary growth are called *primary tissues*, and the plant body composed of these tissues is called the *primary plant body*. Most primitive vascular plants are made up entirely of primary tissues. *Secondary growth* that occurs in some plants thickens the stems and roots. Secondary growth results from the activity of lateral meristems, or *cambia*, of which there are two types:

- *Vascular cambia* give rise to secondary vascular tissues (secondary xylem and phloem), xylem to the inside and phloem to the outside.
- *Cork cambia* form the periderm (bark). The periderm replaces the epidermis in woody plants.

Plant Hormones

Plant growth is controlled by plant hormones, which influence cell differentiation, elongation, and division. Some plant hormones also affect the timing of reproduction and germination:

- *Auxins* affect cell elongation (tropism), apical dominance, and fruit drop or retention. Auxins are also responsible for root development, secondary growth in the vascular cambium, inhibition of lateral branching, and fruit development. Auxins are involved in the absorption of vital minerals and fall color. As a leaf reaches its maximum growth auxin, production declines. In deciduous plants, this triggers a series of metabolic steps that cause the reabsorption of valuable materials (such as chlorophyll) and their transport into the branch or stem for storage during the winter months. Once chlorophyll is gone, the other pigments typical of fall color become visible.

- *Kinins* promote cell division and tissue growth in the leaf, stem, and root. Kinins are also involved in the development of chloroplasts, fruits, and flowers. In addition, they have been shown to delay senescence (aging), especially in leaves, which is one reason why florists use cytokinins on freshly cut flowers; when treated with cytokinins the flowers last longer, protein synthesis continues, and carbohydrates do not breakdown.
- *Gibberellins* are produced in the root tips, and they act as messengers to stimulate growth, especially elongation of the stem. They can also end the dormancy period of seeds and buds by encouraging germination. Gibberellins play an additional role in root growth and differentiation.
- *Ethylene* controls the ripening of fruits. Ethylene may ensure that flowers are carpelate (female), and gibberellin confers maleness on flowers. Ethylene also contributes to the senescence of plants by promoting leaf loss and other changes.
- *Inhibitors* restrain growth and maintain the period of dormancy in seeds and buds.

Tropisms

Tropism is the movement (and growth in plants) of an organism in response to an external stimulus. Tropisms controlled by hormones are a unique characteristic of sessile organisms such as plants that enables them to adapt to different features of their environment—gravity, light, water, and touch—so they can flourish. The three main tropisms are

- *Phototropism*—The tendency of plants to grow or bend (move) in response to light. Phototropism results from the rapid elongation of cells on the dark side of the plant, which causes the plant to bend in the opposite direction; for example, the stems and leaves of a geranium plant growing on the windowsill always turn toward the light.
- *Gravitropism*—Refers to the tendency of a plant to grow toward or against gravity. A plant that displays positive gravitropism (plant roots) will grow downward, toward the center of the Earth. Gravity causes the roots of plants to grow down so the plant is anchored in the ground and has enough water to grow and thrive. Plants that display negative gravitropism (plant stems) will grow upward, away from the earth. Most plants are negatively gravitropic. Gravitropism is also controlled by auxin. In a horizontal root or stem, auxin is concentrated in the lower half, pulled by gravity. In a positively gravitropic plant, this auxin concentration will inhibit cell growth on the lower side, causing the stem to bend downward. In a negatively gravitropic plant, this auxin concentration will inspire cell growth on that lower side, causing the stem to bend upward.
- *Thigmotropism*—Some people notice that their houseplants respond to thigmotropism (i.e., growing or bending in response to touch), growing better when they touch them and pay attention to them. Touch causes parts of the plant to thicken or coil as they touch or are touched by environmental entities; for example, tree trunks grow thicker when exposed to strong winds, and vines tend to grow straight until they encounter a substrate to wrap around.

Photoperiodism

Photoperiodism is the response of an organism (such as a plant) to naturally occurring changes in light over a 24-hour period. Leaves are responsible for the photoperiodism in plants. Sunflowers are particularly known for their photoperiodism, as they open and close in response to the changing position of the sun throughout the day. All flowering plants have been placed in one of three categories with respect to photoperiodism: (1) short-day plants, (2) long-day plants, and (3) day-neutral plants:

1. *Short-day plants*—Flowering is promoted by day lengths shorter than a certain critical day length; includes poinsettias, chrysanthemums, goldenrod, and asters.
2. *Long-day plants*—Flowering is promoted by day lengths longer than a certain critical day length; includes spinach, lettuce, and most grains.
3. *Day-neutral plants*—Flowering is in response insensitive to day length; includes tomatoes, sunflowers, dandelions, rice, and corn.

Plant Reproduction

Plants can reproduce both sexually and asexually. Each type of reproduction has its benefits and disadvantages. A comparison of sexual and asexual plant reproduction is provided in the following text.

Sexual Reproduction

- Sexual reproduction occurs when a sperm nucleus from the pollen grain fuses with an egg cell from the ovary of a *pistil* (the female reproductive structures in flowers, consisting of the stigma, style, and ovary).
- Each brings a complete set of genes and produces genetically unique organisms.
- The resulting plant embryo develops inside the seed and grows when the seed is germinated.

Asexual Reproduction

- Asexual reproduction occurs when a vegetative part of a plant, root, stem, or leaf gives rise to a new offspring plant whose genetic content is identical to the parent plant. An example would be a plant reproducing by root suckers, shoots that come from the root system. The breadfruit tree is an example.
- Asexual reproduction is also called *vegetative propagation*. It is an important way for plant growers to get many identical plants from one very quickly.
- Plants (e.g., crabgrass) can spread and colonize an area quickly in this way.

Plant Cell Walls

Figure 6.4 shows the basic organelles within a standard plant cell, including the cell wall. It should be pointed out, however, that plants can have two types of cell walls: primary and secondary. Primary cell walls contain cellulose consisting of hydrogen-bonded chains of thousands of glucose molecules, in addition to hemicellulose and other materials all woven into a network. Certain types of cells, such as those in vascular tissues, develop secondary walls inside the primary wall

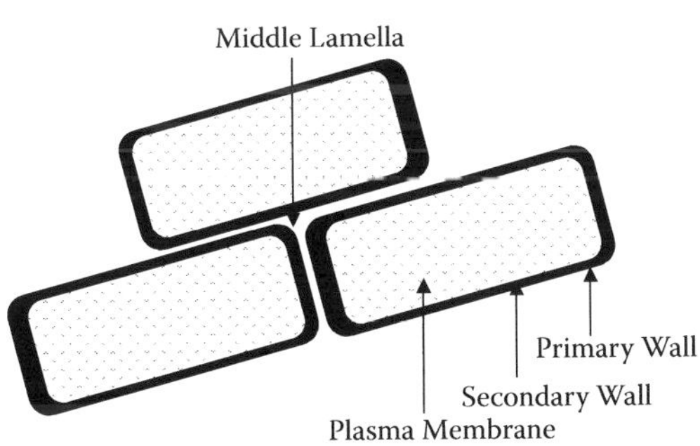

FIGURE 6.4 Plant cell walls.

> **DID YOU KNOW?**
>
> Cellulose microfibrils are composed of linear chains of glucose molecules that hydrogen binds to form the microfibrils.

after the cell has stopped growing. These cell-wall structures also contain lignin, which provides rigidity and resistance to compression. The area formed by two adjacent plant cells, the *middle lamella*, typically is enriched with pectin. Cellulose in higher plants is organized into microfibrils, each measuring about 3 to 6 nm in diameter and containing up to 36 glucan chains having thousands of glucose residues. Like steel girders stabilizing a skyscraper, the mechanical strength of the primary cell wall is due mainly to the microfibril scaffold (i.e., crystalline cellulose core) (USDOE, 2007).

Feedstocks*

First-Generation Feedstocks

For first-generation feedstocks, the type of major end-use product is easily categorized (see Table 6.4). Corn and sugarcane are the most commonly used feedstocks for ethanol production. Soybean and other vegetable oils and animal fats are used for biodiesel production (and bioproducts). Manure and landfill organic waste are used for methane production and the generation of electricity. Corn is used for ethanol production and currently is the leading feedstock used in United States. Several factors favor a positive outlook for further near-term growth in corn-ethanol production. Continued high oil prices will provide economic support for the expansion of all alternative fuel programs, including corn ethanol. Technology improvements that increase feedstock productivity and fuel conversion yields and positive spillovers from second-generation technologies (biomass gasification in ethanol refiners) will also help to lower production costs for corn ethanol. Among the factors likely to limit future growth of corn-ethanol production are increased feedstock and other production costs, increased competition from unconventional liquid fossil fuels (from oil sands, coal, heavy oil, and shale), the emergence of cellulosic ethanol as a low-cost competitor, and new policies to reduce greenhouse gas emissions that could favor advanced biofuels over corn ethanol.

Biodiesel is another type of biofuel experiencing expansion. Although its production costs are higher than ethanol, biodiesel does offer some environmental advantages, including biodegradability and lower sulfur and carbon dioxide emissions when burned. Biodiesel production in the United States has increased rapidly, from less than 2 million gallons in 2000 to about 500 million gallons reported in 2007. Policy incentives in the Energy Independence and Security Act of 2007 are expected to sustain demand of 1 billion gallons per year for this fuel after 2011. A variety of oil-based feedstocks are converted to biodiesel using a process known as *transesterification*. The transesterification process involves the reaction of a triglyceride (fat or oil) with an alcohol to form esters and glycerol. These reactions are often catalyzed by the addition of an acid or base. The oil-based feedstocks include vegetable oils (mostly soy oil), recycled oils and yellow grease, and animal fats such as beef tallow. It takes 3.4 kg of oil or fat to produce 1 gallon of biodiesel (Baize, 2006). Biodiesel production costs are high compared to ethanol, with feedstocks accounting for 80% or more of total costs.

* Much of this section is from Biomass Research and Development Board, *The Economics of Biomass Feedstocks in the United States: A Review of the Literature*, Occasional Paper No. 1, Biomass Research & Development Initiative, Washington, D.C., 2008.

TABLE 6.4
Biomass Feedstocks and Major Bioenergy End-Use Applications

		First-Generation Fuels			Second-Generation Fuels	
Feedstock		**Ethanol**	**Biodiesel**	**Methane**	**Cellulosic Ethanol**	**Thermoch Fuels**[a]
First Generation						
Starch and sugar feedstock for ethanol	Corn	X				
	Sugarcane	X				
	Molasses	X				
	Sorghum	X				
Vegetable oil and fats for biodiesel	Vegetable oils		X			
	Recycled fats and grease		X			
	Beef tallow		X			
Second Generation (Short-Term)						
Agricultural residues and livestock byproducts	Corn stover				X	
	Wheat straw				X	
	Rice straw				X	
	Bagasse				X	
	Manure			X		
Forest biomass	Logging residues				X	X
	Fuel treatments				X	X
	Conventional wood				X	X
Urban woody waste and landfills	Primary wood products				X	X
	Secondary mill residues				X	X
	Municipal solid waste			X	X	X
	Construction/demolition wood					
	Landfills			X	X	
Second-Generation (Long-Term)						
Herbaceous energy crops	Switchgrass				X	X
	Miscanthus				X	X
	Reed canary grass				X	X
	Sweet sorghum					
	Alfalfa				X	X
Short-rotation woody crops	Willow				X	X
	Hybrid poplar				X	X
	Cottonwood				X	X
	Sycamore				X	X
	Eucalyptus				X	X

[a] Ethanol, diesel, butanol, etc.

The biodiesel industry consists of many small plants that are highly dispersed geographically. Decisions about plant location are primarily determined by local availability and access to the feedstock. Recent expansion in biodiesel production is affecting the soybean market. Achieving a nationwide target of a 2% biodiesel blend in diesel transportation fuel, for example, would require 2.8 million metric tons (MT) of vegetable oil, or about 30% of current U.S. soybean oil production. One of the key biodiesel byproducts is glycerin, which is a concern because there are no existing markets for the product; however, recent technological developments include an alternative chemical process that would produce biodiesel without glycerin. In addition, new processes are being tested that further transform glycerin into propylene glycol, which is used in the manufacture of antifreeze (Zeman, 2007).

Second-Generation Feedstocks: Short-Term Availability

Agricultural residues, a second-generation biomass feedstock, offer a potentially large and readily available biomass resource, but sustainability and conservation constraints could place much of it out of reach. Given current U.S. cropland use, corn and wheat offer the most potentially recoverable residues; however, these residues play an important role in recycling nutrients in the soil and maintaining long-term fertility and land productivity. Removing too much residue could aggravate soil erosion and deplete the soil of essential nutrients and organic matter. Safe removal methodologies, based on soil erosion, have been developed. Methodologies to determine removal rates while safeguarding soil fertility and meeting conservation objectives still must be developed. Studies have shown that under current tillage practices the national average safe removal rate based on soil erosion for corn stover is less than 30%. Actual rates vary widely depending on local conditions. In other words, much of the generated crop residues may be out of reach for biomass use if soil conservation goals are to be achieved (Graham et al., 2007). The estimated delivery costs for agricultural residues vary widely depending on crop type, load resource density, storage and handling requirements, and distance and transportation costs. Moreover, existing estimates are largely derived from engineering models, which may not account for economic conditions. Agricultural residue feedstocks (such as corn stover) have a significant advantage in that they can be readily integrated into the expanding corn-ethanol industry; however, dedicated energy crops (such as switchgrass) may have more benign environmental impacts.

Another significant biomass source is *forest biomass*, which can be immediately available should the bioenergy market develop. Logging residues are associated with timber industry activities and are significant biomass resources in many states, particularly in the Northeast, North Central, Pacific Northwest, and Southeast states. In the Western states, the predominance of public lands and environmental pressure reduce the supply potential for logging residues, but there is a vast potential for biomass from thinning undertaken to reduce the risk of forest fires. The few analyses that have examined the recoverability of logging residues cite the need to account for factors such as the scale and location of biorefineries and biopower plants, as well as regional resource density. The potential for forest residues may be large, but the actual quantities available for biomass conversion may be low due to the economics of harvesting, handling, and transporting the residues from forest areas to locations where they could be used. It is not clear how these residues compete with fossil fuels in the biopower and co-firing industries. In addition, competing uses for these products can be found in the pulp and paper industry, as well as various bioenergy end uses. Economic studies of logging residues suggest a current lack of competitiveness with fossil fuels (coal, gas), but logging residues could become more cost competitive with further improvements in harvesting and transportation technologies and with policies that require a fuller accounting of the social and environmental benefits from converting forest residues to biopower or biofuels.

Another source of forest residues that could be recovered in significant quantities is biomass from fuel treatments and thinnings. Fuel treatment residues are the byproduct of efforts to reduce the risk of loss from fire, insects, and disease; therefore, they present substantially different challenges than logging residues. The overall value of forest health benefits such as clean air and water is generally believed to exceed the cost of treatment; however, treated forests are often distant from end-use markets, resulting in high transportation costs to make use of the harvested material. Road or trail access, steep terrain, and other factors commonly limit thinning operations in Western forests. Transportation costs can be a significant factor in the cost of recovering biomass. As much as half the cost of the material delivered to a manufacturing facility may be attributed to transportation. The offset to high-cost transportation of forest thinnings is onsite densification of the biomass. This could entail pelletization, fast pyrolysis (to produce bio-oil), or baling. The economics of transporting thinned woody residues vs. onsite densification depend on the distance to end-use markets. Densification may be more economical if power generation facilities are far away. In addition to co-firing or co-generation facilities, improvements in thermochemical conversion efficiency and establishment of small-scale conversion facilities using gasification or pyrolysis may favor the use of forest residues for biofuel production (Polagye et al., 2007).

A third major category of immediately available second-generation biomass is wood residues from *secondary mill products* and *urban wood waste*. Urban wood waste provides a relatively inexpensive feedstock to supplement other biomass resources (Wiltsee, 1998). Urban wood waste encompasses the biomass portion of commercial, industrial, and municipal solid waste, while secondary mill residues include sawdust, shavings, wood trims, and other byproducts generated from processing lumber, engineered wood products, or wood particles. Both urban wood waste and secondary mill residues have several primary uses and disposal methods. Urban wood waste not used in captive markets (such as the pulpwood industry) could be used as a biomass to generated electricity or to produce cellulosic ethanol when it becomes commercially viable. The amount of urban wood waste produced in the United States is significant, and its use as biomass could be economically viable, particularly in large urban centers (Wiltsee, 1998). Several national availability estimates exist for various types of urban wood waste, but estimates vary depending on methodology, product coverage, and assumptions about alternative uses (McKeever, 2004; Wiltsee, 1998). One of the challenges facing the potential availability of urban woody waste is to sort out the operation that is available (not currently used) and determine alternative uses, including those used by captive markets (not likely to be diverted to bioenergy). One assessment of urban wood waste found that 36% of total biomass generated is currently sold to noncaptive markets, and 50% of the unused residues are not available due to contamination, quality, or recoverability.

One source of recurring and potentially available carbon feedstock is municipal solid waste (MSW). In 2005, the USEPA estimated that 245.7 million tons of MSW were generated in the United States, of which 79 million tons were recycled, 33.4 millions tons were diverted to energy recovery, and 133.3 million tons were disposed of in landfills. As such, landfilled material represents a potentially significant source of renewable carbon that could be used for fuel or energy production or in support of biofuel production.

Second-Generation Feedstocks: Long-Term Availability

Large-scale biofuels production, in the long run, will require other resources, including dedicated energy crops. Dedicated feedstocks are perennial grasses and trees grown as crops specifically to provide the required raw materials to bioenergy producers. A steady supply of low-cost, uniform, and consistent-quality biomass feedstock will be critical for the economic viability of cellulosic ethanol production. During the late 1980s, the Department of Energy sponsored research on perennial herbaceous (grassy) biomass crops, particularly switchgrass, considered a model energy crop

ATTRIBUTES OF THE PERFECT ENERGY CROP

The perfect energy crop is a perennial that has:

- High biomass
- Improved composition and structure
- Disease and pest resistance
- Optimized architecture
- Salt, pH, and aluminum tolerance
- Rapid and cost-effective propagation
- Stand establishment (cold germination and cold growth)
- Deep roots

DID YOU KNOW?

With regard to establishing switchgrass, no single method can be suggested for all situations (Parrish et al., 2008). No-till and conventional tillage can be used to establish the crop. When seeded as part of a diverse mixture, planting guidelines for warm-season grass mixtures for conservation plantings should be followed. Several key factors can increase the likelihood of success for establishing switchgrass, including:

- Plant switchgrass after the soil is well warmed during the spring.
- Use seeds that are highly germinable and plant them 0.6 to 1.2 cm deep, or up to 2 cm deep in sandy soils.
- Pack the soil both before and after seeding.
- Provide no fertilization at planting to minimize competition.
- Control weeds with chemical or cultural control methods.

because of its many perceived advantages: It is native to North America, has a high biomass yield per acre, has wide regional coverage, and is adaptable to marginal land conditions. An extensive research program on switchgrass in the 1990s generated a wealth of information on high-yielding varieties, regional adaptability, and management practices. Preliminary field trials show that the economic viability of switchgrass cultivation depends critically on the initial establishment success. During this phase, seed dormancy and seedling sensitivity to soil and weed conditions require that recommended practices be closely followed by growers. Viable yields require fertilization rates at about half the average for corn.

Switchgrass (*Panicum virgatum*) is a hardy, deep-rooted, perennial, rhizomatous, warm-season grass native to North America. It is believed to be most suitable for cultivation in marginal lands, low-moisture lands, and lands with lower opportunity costs such as pastures, including lands under the Conservation Reserve Program (CRP), where the federal government pays landowners annual rent for keeping land out of production. Additionally, a large amount of highly erodible land in the Corn Belt is unsuitable for straw or stover removal but potentially viable for dedicated energy crops such as switchgrass. Factors favoring adoption of switchgrass include selection of suitable lands, environmental benefits (carbon balances, improved soil nutrients and quality), and use of existing hay production techniques to grow the crop. Where switchgrass is grown on CRP lands, payments help to offset production costs. Factors discouraging switchgrass adoption include no possibility for crop rotation; farmers' risk aversion for producing a new crop because of lack of information, skills, and know-how; potential conflict with on-farm and off-farm scheduling activities; and a lack of compatibility with long-term land tenure. Overall, production budget and delivery cost assessments suggest that switchgrass is a high-cost crop (under current technology and price conditions) that may not compete with established crops, except in areas with low opportunity costs (e.g., pasture land, marginal lands).

Substantial variability is apparent in the economics of switchgrass production and assessments of production budgets and delivered costs. Factors at play include methods for storage and handling, transport distances, yields, and types of land used (cropland vs. grassland). When delivered costs of switchgrass are translated into break-even prices (compared with conventional crops), it becomes apparent that cellulosic ethanol or biopower plants would have to offer relatively high prices for switchgrass to induce farmer to grow it (Rinehart, 2006). The economics of switchgrass could improve, however, if growers benefited from CRP payments and other payments tied to environmental services (such as carbon credits). In the long run, the viability of an energy crop such as switchgrass hinges on continued reductions in cellulosic ethanol conversion costs and sustained improvements in switchgrass yield and productivity through breeding, biotechnology, and agronomic research.

DID YOU KNOW?

The main advantages of using switchgrass over corn as an ethanol feedstock are that its cost of production is generally about half that of grain corn and more biomass energy per hectare can be captured in the field (Samson et al., 2008).

Although it is a potentially important biofuels crop, switchgrass does have limitations. It is not optimally grown everywhere; for example, in the upper Midwest, under wet soils, reed canary grass is more suitable, while semitropical grass species are better adapted to the Gulf Coast region. State and local efforts are testing alternatives to switchgrass, including reed canary grass (*Phalaris arundinacea*), which is a tall, perennial grass that forms extensive single-species stands along the margins of lakes and streams, and *Miscanthus* (often confused with elephant grass), a rapidly growing plant with low mineral content and high biomass yield.

With regard to long-run sustainability, the ecology of perennial grassy crops favors a multiplicity of crops or even a mix of species within the same area. Both ecological and economic sustainability favor the development of a range of herbaceous species for optimal use of local soil and climatic conditions. A mix of several energy crops in the same region would help reduce the risk of epidemic pests and disease outbreak and optimize the supply of biomass to an ethanol or biopower plant because the various types of grasses mature and can be harvested at different times. Moreover, development of future energy crops must be evaluated from the standpoint of their water use efficiency, impact on soil nutrient cycling, effect on crop rotations, and environmental benefit (improved energy use efficiency and reduced greenhouse gas emissions, nutrient runoff, pesticide runoff, and land-use impacts). In the long run, developing a broad range of grassy corps for energy use is compatible with both sustainability and economic viability criteria.

Short-rotation woody crops (SRWCs) represent another important category of future dedicated energy crops. Among the SRWCs, hybrid poplar, willow, American sycamore, sweetgum, and loblolly have been extensively researched for their very high biomass yield potential. Breeding programs and management practices continue to be developed for these species. SRWCs are based on a high-density plantation system and more frequent harvesting (every 3 to 4 years for willow and 7 years for hybrid poplar). Following is a summary of short-rotation woody crops:

- *Hybrid poplar* is very site specific and has a limited growing niche. It requires an abundant and continuous supply of moisture during the growing season. Soils should be moist but not continually saturated and should have good internal drainage. The hybrid poplar prefers damp, well-drained, fine sandy-loam soils located near streams, where coarse sand is first deposited as flooding occurs. Hybrid poplars are among the fastest growing trees in North America—in just six growing seasons, hybrid poplars can reach 60 feet or more in height. They are well suited for the production of bioenergy, fiber, and other bio-based products.
- In folklore and myth, a *willow tree* (*Salix*) is believed to be quite sinister, capable of uprooting itself and stalking travelers. The reality is that willows are used for biofiltration, constructed wetlands, ecological wastewater treatment systems, hedges, land reclamation, landscaping, phytoremediation, stream bank stabilization (bioengineering), slope stabilization, soil erosion control, shelterbelts and windbreaks, soil building, soil reclamation, tree bog compost toilets, and wildlife habitat. Willow is grown for biomass or biofuel in energy forestry systems because of its high energy-in/energy-out ratio, large carbon mitigation potential, and fast growth (Aylott, 2008). Willow is also grown and harvested for making charcoal.
- *American sycamore* prefers alluvial soils along streams in bottomlands. Sycamore growth and yield are less than the hybrid poplar and willow.

- *Sweetgum (Liquidambar styraciflua)* is a species tolerant of a variety of soils, but it grows best on rich, moist, alluvial clay and loam soils of river bottoms.
- *Loblolly pine* (*Pinus taeda*) is quite adaptable to a variety of sites. It performs well on both poorly drained bottomland flats and modestly arid uplands. Biomass for energy is currently being obtained from precommercial thinnings and from logging residue in loblolly pine stands. Utilization will undoubtedly increase, and loblolly pine energy plantations may become a reality.

In many parts of the country, plantations of willow, poplar, pines, and cottonwood have been established and are being commercially harvested. Willow plants are being planted in New York, particularly following the enactment of renewable portfolio standards and other state incentives. Over 30,000 hectares of poplar are grown in Minnesota, and several thousand hectares are also grown as part of a DOE-funded project to provide biomass for a power utility company in southern Minnesota. The Pacific Northwest has large plantations of hybrid poplars, estimated recently at 60,000 hectares. Most of these plantations are currently used for pulp wood, with little volume being used for bioenergy. Because SRWCs can be used for biomass or as feedstock for pulp and other products, pulp demand will influence the cost of using it for bioenergy production.

An important consideration for energy crops (e.g., switchgrass, poplar, willow) is the potential for increasing yields and developing other desirable characteristics. Most energy crops are unimproved or have been bred only recently for biomass yield, whereas corn and other commercial food crops have undergone substantial improvements in yield, disease resistance, and other agronomic traits. A more complex understanding of biological systems and application of the latest biotechnological advances would accelerate the development of new biomass crops with desirable attributes. These attributes include increased yields and processability, optimal growth in specific microclimates, better pest resistance, efficient nutrient use, and greater tolerance to moisture deficits and other sources of stress. Agronomic and breeding improvements of these new crops could provide a significant boost to future energy crop production.

Bioethanol Production by Dry Corn Mill Process

Ethanol production from dry corn milling follows a seven-step process: milling, liquefaction, saccharification, fermentation, distillation, dehydration, and denaturing (Figure 6.5) (USDOE, 2010). Corn feedstock is cleaned and milled (ground into corn meal), and the milled corn is then converted to slurry. The slurry is liquefied (slurried with water—mash), and enzymes are added to the mash to convert starch into simple sugar (saccharification). The mash is cooked in a saccharification tank to reduce bacterial levels and is then placed in a fermenter, where sugar is converted to ethanol by yeast (fermentation). The resulting "beer" containing 2 to 12% ethanol is then distilled into ethanol at 95% alcohol and 5% water. The remaining solids (stillage) are collected during distillation, dried, and sold as an animal feed called dried distillers' grain (DDG). The final removal of water from the ethanol to less than 1% (dehydration) allows it to be blended with gasoline.

DID YOU KNOW?

Fermentation is a series of chemical reactions that convert sugars to ethanol. The fermentation reaction is caused by yeast or bacteria, which feed on the sugars. Ethanol and carbon dioxide are produced as the sugar is consumed. The simplified fermentation reaction equation for the 6-carbon sugar, glucose, is

$$\underset{\text{Glucose}}{C_6H_{12}O_6} \rightarrow \underset{\text{Ethanol}}{2CH_3CH_2OH} + \underset{\text{Carbon Dioxide}}{2CO_2}$$

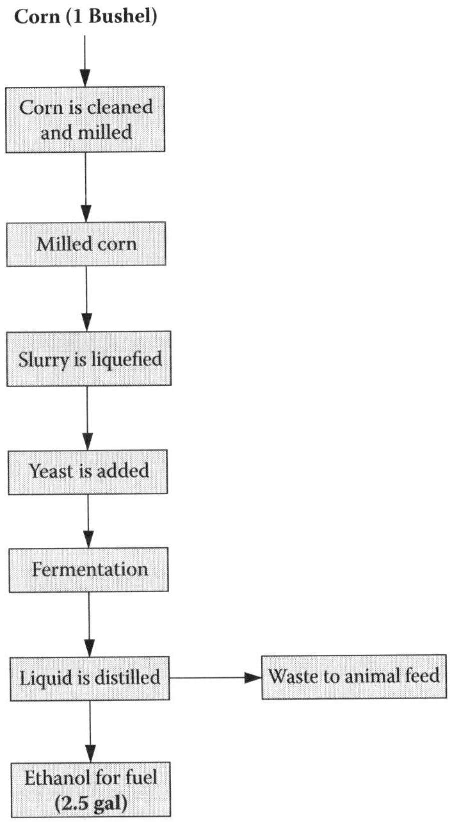

FIGURE 6.5 Flow diagram for production of ethanol from corn.

Gasoline Gallon Equivalent

It is important for the reader to have a fundamental understanding of the difference between conventional gasoline and diesel fuel energy output as compared to nonconventional renewable products. Typically, this comparison is made utilizing a standard engineering parameter known as the gasoline gallon equivalent (GGE), which is the ratio of the number of British thermal units (Btu) available in 1 U.S. gallon (1 gal) of gasoline to the number of British thermal units available in 1 gal of the alternative substance in question. In 1994, NIST (1994) defined a gasoline gallon equivalent as 5660 pounds of natural gas. The GGE parameter allows consumers to compare the energy content of competing fuels against a commonly known fuel—gasoline. Table 6.5 lists GGE and Btu/unit values for various fuels.

Pros and Cons of Bioethanol

Ethanol can be used as an extender and octane enhancer in conventional gasoline. Its primary advantage is that its feedstock (plants) is renewable. It can also be used as a primary fuel (in E85), thus reducing dependence on petroleum products. In some locations, E85 and gasohol are lower in cost per gallon than conventional gasoline. Ethanol can be used to prevent gas-line freeze in extremely cold weather, and, because it is not as flammable as gasoline, it is less likely to cause accidental explosions. Ethanol is an important player in the ongoing effort to reduce environmental pollution from carbon monoxide (CO) gas and global carbon dioxide (CO_2). Farmers will benefit from increasing production and demand for ethanol as a fuel.

TABLE 6.5
Gasoline Gallon Equivalents (GGE)

Fuel	GGE	Btu/Unit
Gasoline (base)	1.00 U.S. gal	114,000 Btu/gal
Gasoline (conventional, summer)	0.996 U.S. gal	114,500 Btu/gal
Gasoline (conventional, winter)	1.013 U.S. gal	112,500 Btu/gal
Gasoline (reformulated gasoline, ethanol)	1.019 U.S. gal	111,836 Btu/gal
Gasoline (reformulated gasoline, ETBE)	1.019 U.S. gal	111,811 Btu/gal
Gasoline (reformulated gasoline, MTBE)	1.020 U.S. gal	111,745 Btu/gal
Gasoline (10% MBTE)	1.02 U.S. gal	112,000 Btu/gal
Gasoline (regular unleaded)	1.00 U.S. gal	114,100 Btu/gal
Diesel #2	0.88 U.S. gal	129,500 Btu/gal
Biodiesel (B100)	0.96 U.S. gal	118,300 Btu/gal
Biodiesel (B20)	0.90 U.S. gal	127,250 Btu/gal
Liquid natural gas (LNG)	1.52 U.S. gal	75,000 Btu/gal
Compressed natural gas	1.26 ft^3 (3.587 m^3)	900 Btu/ft^3
Hydrogen at 101.325 kPa	357.37 ft^3	319 Btu/ft^3
Hydrogen by weight	2.198 lb (0.997 kg)	51,500 Btu/lb (119.9 MJ/kg)
Liquefied petroleum gas (LPG)	1.35 U.S. gal	84,300 Btu/gal
Methanol fuel (M100)	2.01 U.S. gal	56,800 Btu/gal
Ethanol fuel (E100)	1.500 U.S. gal	76,100 Btu/gal
Ethanol (E85)	1.39 U.S. gal	81,800 Btu/gal
Jet fuel (naphtha)	0.97 U.S. gal	118,700 Btu/gal

Note: ETBE, ethyl *tert*-butyl ether; MBTE, methyl *tert*-butyl ether.
Source: USEPA, *Fuel Economy Impact Analysis of RFG*, U.S. Environmental Protection Agency, Washington, D.C., 2007.

Ethanol use does have a few drawbacks. Probably the most obvious shortcoming is its limited availability at local gas stations. Gas stations selling E85 as part of their normal product line are not as abundant as conventional gas stations. In some locations, E85 and gasohol are more expensive per gallon than conventional gasoline. E85 can damage vehicles that are not designed to burn it as a fuel. Also, ethanol contains less energy than gasoline; this means that a car will not go as far on a gallon of E85, and fuel economy will decrease by 20 to 30%. Ethanol is produced from plant matter that could otherwise be consumed as human and animal food products. Critics argue that food products used for fuel instead of to feed people and animals will contribute indirectly to mass world hunger and could lead to higher food prices. Although higher corn prices increase animal feed costs for farmers and ingredient costs for food manufacturers, these costs are passed through to retail at a rate less than 10% of the change in corn price.

Corn for Ethanol Production or Food Supply?*

In 2007, record U.S. trade driven by economic growth in developing countries and favorable exchange rates, combined with tight global grain supplies, resulted in record or near-record prices for corn, soybeans, and other food and feed grains. For corn, these factors, along with increased demand for ethanol, helped push prices from under $2 per bushel in 2005 to $3.40 per bushel in 2007. By the end of the 2006/07 crop year, over 2 billion bushels of corn (19% of the harvested crop)

* Information for this section was obtained from Coyle, W., The future of biofuels: a global perspective, *Amber Waves*, November, 2007 (http://www.ers.usda.gov/AmberWaves/november07/features/biofuels.htm).

were used to produce ethanol, a 30% increase from the previous year. Higher corn prices motivated farmers to increase corn acreage at the expense of other crops, such as soybeans and cotton, raising their prices as well.

The pressing and pertinent question becomes: What effect do these higher commodity costs have on retail food prices? In general, retail food prices are much less volatile than farm-level prices and tend to rise by a fraction of change in farm prices. The magnitude of response depends on both the retailing costs beyond the raw food ingredients and the nature of competition in retail food markets. The impact of ethanol on retail food prices depends on how long the increased demand for corn drives up farm corn prices and the extent to which higher corn prices is passed through to retail.

Retail food prices adjust as the cost of inputs into retail food production change and the competitive environment in a given market evolves. Strong competition among three to five retail store chains in most U.S markets has had a moderating effect on food price inflation. Overall, retail food prices have been relatively stable over the past 20 years, with prices increasing an average of 3.0% per year from 1987 through 2007, just below the overall rate of inflation. Since then, food price inflation has averaged just 2.5% per year.

Field corn is the predominant corn type grown in the United States, and it is primarily used for animal feed. Currently, less than 10% of the U.S. field corn crop is used for direct domestic human consumption in corn-based foods such as corn meal, corn starch, and corn flakes, while the remainder is used for animal feed, exports, ethanol production, seed, and industrial uses. Sweet corn, both white and yellow, is usually consumed as immature whole-kernel corn by humans and also as an ingredient in other corn-based foods, but it makes up only about 1% of the total U.S. corn production.

Because U.S. ethanol production uses field corn, the most direct impact of increased ethanol production should be on field corn prices and on the price of food products based on field corn; however, even for those products heavily based on field corn, the effect of rising corn prices is dampened by other market factors. An 18-ounce box of corn flakes, for example, contains about 12.9 ounces of milled field corn. When field corn is priced at \$2.28 per bushel (the 20-year average), the actual value of corn represented in the box of corn flakes is about 3.3¢ (1 bushel = 56 pounds). (The remainder is packaging, processing, advertising, transportation, and other costs.) At \$3.40 per bushel, the average price in 2007, the value is about 4.9¢. That 49% increase in corn prices would be expected to raise the price of a box of corn flakes by about 1.6¢, or 0.5%, assuming no other cost increases.

In the 1980s, Coca Cola made the shift from sugar to corn syrup in most of its U.S.-produced soda, and many other beverage makers followed suit. Currently, about 4.1% of U.S.-produced corn is made into high-fructose corn syrup. A 2-liter bottle of soda contains about 15 ounces of corn in the form of high-fructose corn syrup. At \$3.40 per bushel, the actual value of corn represented is 5.7¢, compared with 3.8¢ when corn is priced at \$2.28 per bushel. Assuming no other cost increases, the higher corn price in 2007 would be expected to raise soda prices by 1.9¢ per 2-liter bottle, or 1%. These are notable changes in terms of price measures and inflation but relatively minor changes in the average household budget.

In addition to the impact on corn flakes and soda beverage prices, livestock prices are also impacted. This stands to reason when you consider that livestock rations traditionally contain a large amount of corn; thus, a bigger impact would be expected in meat and poultry prices due to higher feed costs than in other food products. Currently, 55% of corn produced in the United States is used as animal feed for livestock and poultry; however, estimating the actual corn used as feed to produce retail meat is a complicated calculation. Livestock producers have many options when deciding how much corn to include in a feed ration; for example, at one extreme, grass-fed cattle consume no corn, while other cattle may have a diet consisting primarily of corn. For hog and poultry producers, ration variations may be less extreme, but they can still vary quite a bit. To estimate the impact of higher corn prices on retail meat prices, it is necessary to make a series of assumptions about feeding practices and grain conversion rates from animal to final retail meat products.

To avoid downplaying potential impacts, this analysis uses upper-bound conversion estimates of 7 pounds of corn to produce 1 pound of beef, 6.5 pounds of corn to produce 1 pound of pork, and 2.6 pounds of corn to produce 1 pound of chicken.

Using these ratios and data from the Bureau of Labor Statistics, a simple pass-through model provides estimates of the expected increase in meat prices given the higher corn prices. The logic of this model is illustrated by an example using chicken prices. Over the past 20 years, the average price of a bushel of corn in the United States has been $2.28, implying that a pound of chicken at the retail level uses 8¢ worth of corn, or about 4% of the $2.05 average retail price for chicken breasts. Using the average price of corn for 2007 ($3.40 per bushel) and assuming producers do not change their animal-feeding practices, retail chicken prices would rise 5.2¢, or 2.5%. Using the same corn data, retail beef prices would go up 14¢ per pound, or 8.7%, while pork prices would rise 13¢ per pound, or 4.1%.

These estimates for meat, poultry, and corn-related foods, however, assume that the magnitude of the corn price change does not affect the rate at which cost increases are passed through to retail prices. It could be the case that corn price fluctuations have little impact on retail food prices until corn prices rise high enough for a long enough time to elicit a large price adjustment by food produces and notably higher retail food prices.

On the other hand, these estimates may be overstating the effect of corn price increases on retail food prices since they do not account for potential substitution by producers from more expensive to less costly inputs. Such substitution would dampen the effect of higher corn prices on retail meat prices. Even assuming the upper-bound effects outlined above, the impact of rising corn commodity prices on overall food prices is limited. Given that less than a third of retail food contains corn as a major ingredient, these rising prices for corn-related products would raise overall U.S. retail food prices less than 1 percentage point per year above the normal rate of inflation.

Continuing elevated prices for corn will depend on the extent to which corn remains the most efficient feedstock for ethanol production and ethanol remains a viable source of alternative energy. Both of these conditions may change over time as other crops and biomass are used to produce ethanol and other alternative energy sources develop.

DID YOU KNOW?

While higher commodity costs may have a relatively modest impact on U.S. retail food prices, there may be greater effect on retail food prices in lower income developing countries. As a relatively low-priced food, grains have historically accounted for a large share of the diet in less developed countries. Even with incomes rising, consumers in such countries consume a less processed diet than is typical in the United States and other industrialized countries, so food prices are more closely tied to swings in both domestic and global commodity prices.

Even if these conditions do not change in the near term, market adjustments may dampen long-run impacts. In 1996, when field corn prices reached an all-time high of $3.55 per bushel due to drought-related tighter supplies in the United States and strong demand for corn from China and other parts of Asia, the effect on food prices was short lived. At that time, retail prices rose for some foods, including pork and poultry, but these effects did not extend beyond the middle of 1997. For the most part, food markets adjusted to the higher corn prices and corn producers increased supply, bringing down price.

Food producers, manufacturers, and retailers may also adjust to the changing market conditions by adopting more efficient production methods and improved technologies to counter higher costs. For example, soft drink manufacturers may consider substituting sugar for corn syrup as a sweetener if corn prices remain high, while livestock and poultry produces may develop alternative feed rations that minimize corn needed for animal feed. Adjustments by producers, manufacturers, and retailers, along with continued strong retail competition, imply that U.S. retail food prices will remain relatively stable.

Brazilian Example*

Brazil has the world's second largest ethanol program and is capitalizing on plentiful soybean supplies to expand into biodiesel. More than half of the nation's sugarcane crop is processed into ethanol, which now accounts for about 20% of the country's fuel supply. Initiated in the 1970s after the OPEC oil embargo, Brazil's policy program was designed to promote the nation's energy independence and to create an alternative and value-added market for sugar producers. The government has spent billions to support sugarcane producers, develop distilleries, build up a distribution infrastructure, and promote production of pure-ethanol-burning and, later, flex-fuel vehicles able to run on gasoline, ethanol–gasoline blends, or pure hydrous ethanol (wet; 7 to 4% water content). Advocates contend that, while the costs were high, the program saved far more in foreign exchange from reduced petroleum imports.

In the mid- to late 1990s, Brazil eliminated direct subsidies and price setting for ethanol. It pursued a less intrusive approach with two main elements—a blending requirement (now about 25%) and tax incentives favoring ethanol use and the purchase of ethanol-using or flex-fuel vehicles. Today, more than 80% of Brazil's newly produced automobiles have flexible fuel capability, up from 30% in 2004. With ethanol widely available at almost all of Brazil's 32,000 gas stations, Brazilian consumers currently choose primarily between 100% hydrous ethanol and a 2% ethanol–gasoline blend on the basis of relative prices.

Approximately 20% of current fuel use (alcohol, gasoline, and diesel) in Brazil is ethanol, but it may be difficult to raise the share as Brazil's fuel demand grows. Brazil is a middle-income economy with per capita energy consumption only 15% that of the United States and Canada. Current ethanol production levels in Brazil are not much higher than they were in the later 1990s. Production of domestic off- and onshore petroleum resources has grown more rapidly than ethanol and accounts for a larger share of expanding fuel use than does ethanol in the last decade.

Biomass for Biopower

In addition to advanced fuels, biomass can also be used for the production of biopower. This can be done in several ways, including direct combustion of biomass in dedicated power plants, co-firing biomass with coal, biomass gasification in a combined cycle plant to produce steam and electricity, or via anaerobic digestion (EERE and EPRI, 1997). Combustion is the burning of biomass in air. This involves the conversion of chemical energy stored in biomass into heat, mechanical power, or electricity (McKendry, 2002). Although it is possible to use all types of biomass, combustion is preferable when biomass is more than 50% dry. High-moisture biomass is better suited for biological conversion processes. Net bioenergy conversion efficiencies for biomass combustion power plants range from 20 to 40%. Higher efficiencies are obtained with combined heat and power (CHP) facilities or large power-only systems (over 100 megawatt electrical), or when the biomass is co-fired with coal in power plants.

Co-firing biomass with coal is a straightforward and inexpensive way to diversify the fuel supply, reduce coal plant air emissions (NO_x, SO_2, CO_2), divert biomass from landfills, and stimulate the biomass power industry (Hughes, 2000). Moreover, biomass is the only renewable energy technology that can directly displace coal. Given the dominance of coal-based power plants in U.S. electricity production, co-firing with biomass fuel is the most economical way to reduce greenhouse

* This example was adapted from Coyle, W., The future of biofuels: a global perspective, *Amber Waves*, November, 2007 (http://www.ers.usda.gov/AmberWaves/november07/features/biofuels.htm).

gas emissions. Possible biomass fuel for co-firing includes wood waste, short-rotation woody crops, switchgrass, alfalfa stems, various types of manure, landfill gas, and wastewater treatment gas (Tillman, 2000). In addition, agricultural residues such as straw can also be used for co-firing.

A promising technology development currently at the demonstration stage is biomass integrated gasification/combined cycle (BIG/CC), where a gas turbine converts the gaseous fuel to electricity with a high conversion efficiency, reaching 40 to 50% of the heating value of the incoming gas (McKendry, 2002). An important advantage of gasification is the ability to work with a wider variety of feedstocks, such as high-alkali fuels that are problematic with direct combustion. High-alkali fuels such as switchgrass, straw, and other agricultural residues can cause corrosion, but gasification systems can easily remove the alkali species from the fuel gas before it is combusted. High silica, also a problem with grasses, can cause slagging in the reactor. The slagging problem is not unique to any particular form of biomass but instead is common with many different types of biomass fuels (Miles et al., 1993). Slagging deposits can reduce heat transfer and combustion efficiency, as well as damage combustion chambers when large particles break off. Research has focused on two alkali metals, potassium and sodium, and silica, all elements commonly found in living plants. In general, it appears that faster growing plants (or faster growing plant components such as seeds) tend to have higher concentrations of alkali metal and silica. Thus, material such as straw, nut hulls, fruit pits, weeds, and grasses tend to create more problems when burned than wood from a slow-growing tree.

Potassium and sodium metals, whether in the form of oxides, hydroxides, or metalloorganic compounds, tend to lower the melting point of ash mixtures containing various other minerals such as silica (SiO_2). The high alkali content (up to 35%) in the ash from burning annual crop residues lowers the fusion, or sticky temperature, of these ashes from 2200°F for wood ash to as low as 1300°F. This results in serious slagging on the boiler grate or in the bed and fouling of convection heat transfer surfaces. Even small percentages (10%) of some of these high-alkali residues burned with wood in conventional boilers will cause serious slagging and fouling in a day to two, necessitating combustion system shutdown. A method to predict slagging and fouling from the combustion of biomass fuels has been adapted from the coal industry. It involves calculating the weight in pounds of alkali ($K_2O + Na_2O$) per million Btu in the fuel as follows:

$$\frac{1\times10^6}{\text{Btu/lb}}\times\%\ \text{Ash}\times\%\ \text{Alkali of the Ash}=\frac{\text{lb Alkali}}{\text{MM Btu}}$$

This method combines all the pertinent data into one index number. A value below 0.4 lb/MM Btu (MM Btu = thousand thousand Btu) is considered a fairly low slagging risk. Values between 0.4 and 0.8 lb/MM Btu will probably slag with increasing certainty of slagging as 0.8 lb/MM Btu is approached. Above 0.8 lb/MM Btu, the fuel is virtually certain to slag and foul (see Table 6.6).

Another process for biomass is the application of anaerobic digestion to produce biogas (methane) for electricity generation (see later discussion on biogas). Anaerobic digestion involves the controlled breakdown of organic wastes by bacteria in the absence of oxygen. Major agricultural feedstocks for anaerobic digestion include food-processing wastes and manure from livestock operations. A significant increase in the generation of electricity from municipal waste and landfill gas, to about 0.5% of U.S. electricity consumption, is predicted (EIA, 2007).

Biomass for Bioproducts*

Bioproducts are industrial and consumer goods manufactured wholly or in part from renewable biomass (plant-based resources). Today's industrial bioproducts are amazingly diverse, ranging from solvents and paints to pharmaceuticals, soaps, cosmetics, and building materials (see Table

* Information for this section was obtained from USDOE, *Industrial Bioproducts: Today and Tomorrow*, U.S. Department of Energy, Washington, D.C. (http://www.brdisolutions.com/pdfs/BioProductsOpportunitiesReportFinal.pdf).

TABLE 6.6
Alkali Content and Slagging Potential of Various Biofuels

Fuel	Total Alkali (lb/MM Btu)	Slagging Potential
Wood		
Pine chips	0.07	Minimal
White oak	0.14	Minimal
Hybrid poplar	0.46	Probable
Urban wood waste	0.46	Probable
Clean tree trimmings	0.73	Probable
Pits, nuts, shells		
Almond shells	0.97	Certain
Refuse derived fuel	1.60	Certain
Grasses		
Switchgrass	1.97	Certain
Wheat straw (average)	2.00	Certain
Wheat straw (high alkali)	5.59	Certain
Rice straw	3.80	Certain

Source: Adapted from Miles, T.R. et al., in *Proceedings of the First Biomass Conference of the Americas: Energy, Environment, Agriculture, and Industry*, Burlington, VT, August 30–September 2, 1993.

6.7). Industrial bioproducts are integral to our way of life—few sectors of the economy do not rely in some way on products made from biomass. Corn, wood, soybeans, and plant oils are the primary resources used to create this remarkable diversity of industrial and consumer goods. In some cases, it is not readily apparent that a product is derived in part from biomass. Biomass components are often combined with other materials such as petrochemicals and minerals to manufacture the final product. Soybean oil, for example, is blended with other components to produce paints, toiletries, solvents, inks, and pharmaceuticals. Some products, such as starch adhesives, are derived entirely from biomass. The many derivatives of corn illustrate the diversity of products that can be obtained from a single biomass resource (see Figure 6.6). In addition to being an important source of food and feed, corn serves as a feedstock for ethanol and sorbitol (a sweetish, crystalline alcohol),

TABLE 6.7
Common Products from Biomass

Biomass Resource	Uses
Corn	Solvents, pharmaceuticals, adhesives, starch, resins, binders, polymers cleaners, ethanol
Vegetable oils	Surfactants in soaps and detergents, pharmaceuticals (inactive ingredients), inks, paints, resins, cosmetics, fatty acids, lubricants, biodiesel
Wood	Paper, building materials, cellulose for fibers and polymers, resins, binders, adhesives, coatings, paints, inks, fatty acids, road and roofing pitch

Source: USDOE, *Industrial Bioproducts: Today and Tomorrow*. Energy Efficiency & Renewable Energy, U.S. Department of Energy, Washington, D.C., 2003.

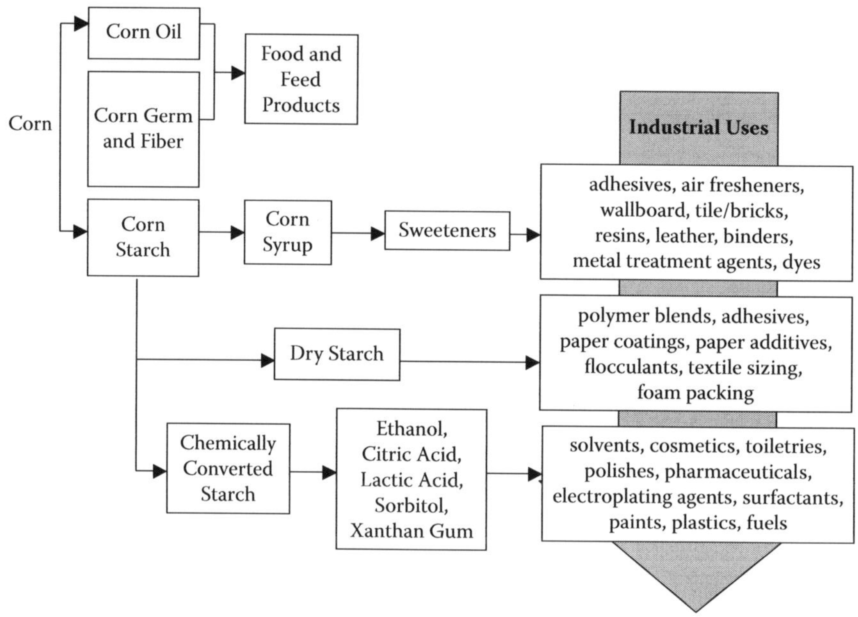

FIGURE 6.6 Bioproducts from corn grain.

industrial starches and sweeteners, citric and lactic acid, and many other products. Biomass, which is comprised of carbohydrates, can be used to produce some of the products that are commonly manufactured from petroleum and natural gas, or hydrocarbons. Both resources contain the essential elements of carbon and hydrogen. In some cases, both resources have captured a portion of market share (see Table 6.8).

TABLE 6.8
Products from Hydrocarbons versus Carbohydrates

Product	Total Production (millions of tons)	Percent Derived from Plants (%)
Adhesives	5.0	40
Fatty acids	2.5	40
Surfactants	3.5	35
Acetic acid	2.3	17.5
Plasticizers	0.8	15
Activated carbon	1.5	12
Detergents	12.6	11
Pigments	15.5	6
Dyes	4.5	6
Wall paint	7.8	3.5
Inks	3.5	3.5
Plastics	30	1.8

Source: Morris, D., *Accelerating the Shift to a Carbohydrate Economy: The Federal Role*, Institute for Local Self-Reliance, Washington, D.C., 2002.

TABLE 6.9
Biodiesel Physical Characteristics

Specific gravity	0.87 to 0.89
Kinematic viscosity at 40°C	3.7 to 5.8
Cetane number	46 to 70
Higher heating value (Btu/lb)	16,928 to 17,996
Sulfur (wt%)	0.0 to 0.0024
Cloud point temperature (°C)	–11 to 16
Pour point temperature (°C)	–15 to 13
Iodine number	60 to 135
Lower heating value (lb/lb)	15,700 to 16,735

Classes of Bioproducts

The thousands of industrial bioproducts produced today can be categorized into five major areas:

1. *Sugar and starch bioproducts* derived through fermentation and thermochemical processes include alcohols, acids, starch, xanthum gum, and other products derived from biomass sugars. Primary feedstocks include sugarcane, sugarbeets, corn, wheat, rice, potatoes, barley, sorghum grain, and wood.
2. *Oil- and lipid-based bioproducts* include fatty acids, oils, alkyd resins, glycerine, and a variety of vegetable oils derived from soybeans, rapeseed, or other oilseeds.
3. *Gum and wood chemicals* include tall oil (liquid rosin), alkyd resins, rosins, pitch, fatty acids, turpentine, and other chemicals derived from trees.
4. *Cellulose derivatives, fibers, and plastics* include products derived from cellulose, such as cellulose acetate (cellophane) and triacetate, cellulose nitrate, alkali cellulose, and regenerated cellulose. The primary sources of cellulose are bleached wood pulp and cotton linters.
5. *Industrial enzymes* are used as biocatalysts for a variety of biochemical reactions in the production of starch and sugar, alcohols, and oils. They are also used in laundry detergents, leather tanning, and textile sizing (Uhlig, 1998).

BIODIESEL

The diesel engine is the workhorse of heavy transportation and industrial processes; it is widely used to power trains, tractors, ships, pumps, and generators. Powering the diesel engine is conventional diesel fuel or biodiesel fuel. Biodiesel is a rather viscous liquid fuel made up of fatty acid alkyl esters, fatty acid methyl esters (FAME), or long-chain mono-alkyl esters. It is produced from renewable sources such as new and used vegetable oils and animal fats and is a cleaner-burning replacement for petroleum-based diesel fuel. It is nontoxic and biodegradable. This fuel is designed to be used in compression-ignition (diesel) engines similar or identical to those that burn petroleum diesel. Biodiesel has physical properties similar to those of petroleum diesel (see Table 6.9). In the United States, most biodiesel is made from soybean oil or recycled cooking oils. Animal fats, other vegetable oils, and other recycled oils can also be used to produce biodiesel, depending on the costs and availability. In the future, blends of all kinds of fats and oils may be used to produce biodiesel. Before providing a basic description of the vegetable and recycled grease-to-biodiesel process, it is important to review a few of the key technological terms associated with the conversion process:

- *Acid esterification*—Oil feedstocks containing more than 4% free fatty acids go through an acid esterification process to increase the yield of biodiesel. These feedstocks are filtered and preprocessed to remove water and contaminants and then fed to the acid esterification process. The catalyst, sulfuric acid, is dissolved in methanol and then mixed with the pretreated oil. The mixture is heated and stirred, and the free fatty acids are converted to biodiesel. Once the reaction is complete, it is dewatered and then fed to the transesterification process.

- *Transesterification*—Oil feedstocks containing less than 4% free fatty acids are filtered and preprocessed to remove water and contaminants and then fed directly to the transesterification process along with any products of the acid esterification process. The catalyst, potassium hydroxide, is dissolved in methanol and then mixed with the pretreated oil. If an acid esterification process is used, then extra base catalyst must be added to neutralize the acid added in that step. Once the reaction is complete, the major coproducts, biodiesel and glycerin, are separated into two layers.
- *Methanol recovery*—The methanol is typically removed after the biodiesel and glycerin have been separated, to prevent the reaction from reversing itself. The methanol is cleaned and recycled back to the beginning of the process.
- *Biodiesel refining*—Once separated from the glycerin, the biodiesel goes through a clean-up or purification process to remove excess alcohol, residual catalyst, and soaps. This consists of one or more washings with clear water. It is then dried and sent to storage. Sometimes the biodiesel goes through an additional distillation step to produce a colorless, odorless, zero-sulfur biodiesel.
- *Glycerin refining*—The glycerin byproduct contains unreacted catalyst and soaps that are neutralized with an acid. Water and alcohol are removed to produce 50 to 80% crude glycerin. The remaining contaminants include unreacted fats and oils. In large biodiesel plants, the glycerin can be further purified, to 99% or higher purity, for sale to the pharmaceutical and cosmetic industries.

The most popular biodiesel production process is transesterification (production of the ester) of vegetable oils or animal fats, using alcohol in the presence of a chemical catalyst. Each gallon of biodiesel produced requires about 3.4 kg of oil or fat (Baize, 2006). The transesterification of degummed soybean oil produces ester and glycerin. The reaction requires heat and a strong base catalyst such as sodium hydroxide or potassium hydroxide. The simplified transesterification reaction is shown below.

$$\text{Triglycerides} + \text{Free Fatty Acids } (<4\%) + \text{Alcohol} \rightarrow \text{Alkyl Esters} + \text{Glycerin}$$

Some feedstocks must be pretreated before they can go through the transesterification process. Feedstocks with less than 4% free fatty acids, including vegetable oils and some food-grade animal fats, do not require pretreatment. Feedstocks with more than 4% free fatty acids, which include inedible animal fats and recycled greases, must be pretreated in an acid esterification process. In this step, the feedstock is reacted with an alcohol (such as methanol) in the presence of a strong acid catalyst (sulfuric acid), converting the free fatty acids into biodiesel. The remaining triglycerides are converted to biodiesel in the transesterification reaction:

$$\text{Triglycerides} + \text{Free Fatty Acids } (>4\%) + \text{Alcohol} \rightarrow \text{Alkyl Esters} + \text{Triglycerides}$$

Figure 6.7 illustrates the basic technology for processing vegetable oils (such as soybeans) and recycled greases (used cooking oil and animal fat). When the feedstock is vegetable oil, the extracted oil is processed to remove all traces of water, dirt, and other contaminants. Free fatty acids are also removed. A combination of methyl alcohol and a catalyst, usually sodium hydroxide or potassium hydroxide, breaks the oil molecules apart in the esterification process. The resulting esters are then refined into usable biodiesel.

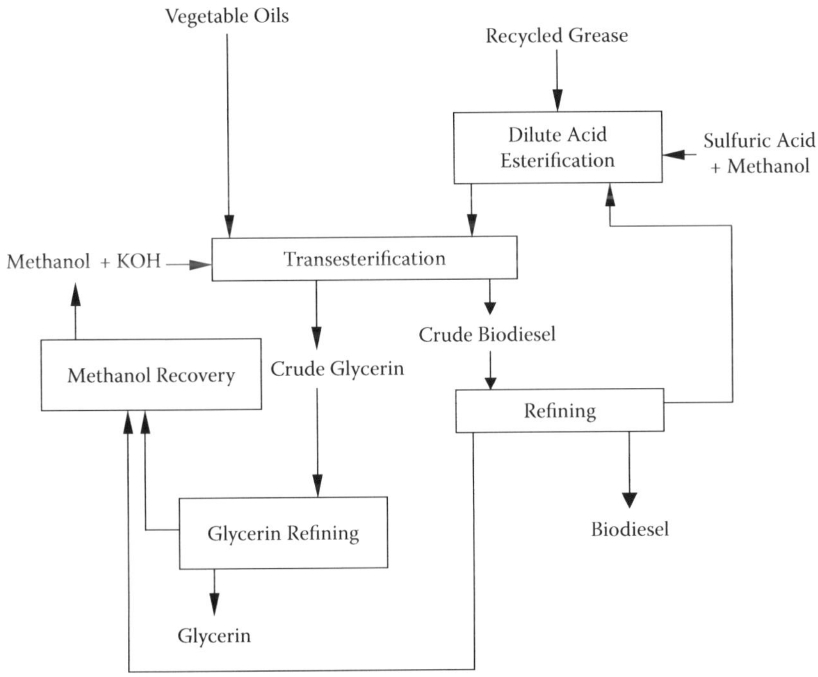

FIGURE 6.7 Biodiesel process.

When the feedstock is used-up cooking oil and animal fats refined to produce biodiesel, the process is similar to the way biodiesel is derived from vegetable oil, except that an additional step is involved (Figure 6.7). Methyl alcohol and sulfur are used in a process called *dilute acid esterification* to obtain a substance resembling fresh vegetable oil, which is then processed in same way as vegetable oil to obtain the final product.

DID YOU KNOW?

Biodiesel is routinely blended with petroleum diesel. The percentage of biodiesel in the blend is written following an uppercase letter B to denote the proportion. For example, a mixture of 20% biodiesel and 80% petroleum diesel is designated as B20, a mixture of equal parts biodiesel and petroleum diesel is B50, and neat (or pure) biodiesel is B100. Most conventional diesel engines can burn blends from pure petroleum diesel up to B20 without modification. With minor modification, most diesel engines built since 1994 can use blends from B20 to B100. The diesel vehicle owner's manual and vehicle warranty should be checked before using any alternative fuel.

Algae to Biodiesel

The focus of our discussion of biofuels to this point has been on terrestrial sources of biomass fuels. This section is concerned with photosynthetic organisms that grow in aquatic environments—namely, macroalgae, microalgae, and emergents. Macroalgae, more commonly known as seaweed, are fast-growing marine and freshwater plants that can grow to considerable size (up to 60 m in length). Emergents are plants that grow partially submerged in bogs and marshes. Microalgae are, as the name suggests, microscopic photosynthetic organisms. The focus of our discussion here is primarily on *microalgae*, organisms capable of photosynthesis that are less than 0.4 mm in diameter, including the diatoms and cyanobacteria, as opposed to *macroalgae* (seaweed). Microalgae are the preferred choice because they are less complex in structure, have a fast growth rate, and typically (for some species) contain a high concentration of oil. It is important to point out, however, that recent research efforts also are focusing on using seaweeds for algae fuel (biofuels), probably due to their high availability (Lewis, 2005).

The following algal species are currently being studied for their suitability as a mass oil-producing crop across various locations worldwide (Anon., 2008):

- *Ulva*
- *Botryococcus braunii*
- *Chlorella*
- *Dunaliella tertiolecta*
- *Gracilaria*
- *Pleurochrysis carterae*
- *Sargassum* (has 10 times the output volume of *Gracilaria*)

DID YOU KNOW?

Algae fuel, also called algal fuel, algaeoleum, or second-generation biofuel, is a biofuel derived from the natural oils of algae which are used to produce biodiesel.

The algae-to-biodiesel alternative renewable fuel source program got its start during the Carter administration in response to the 1970s Arab fuel embargo. At the same time, the Carter administration consolidated all federal energy activities under the auspices of the newly established U.S. Department of Energy (DOE). Among its various programs established to develop all forms of alternative energy (related to solar energy), the DOE initiated research on the use of plant life as a source of transportation fuels (today called the Biofuels Program).

Before discussing the algae-to-biofuels process in detail, it is important for the reader to be well-grounded in basic algal concepts. In many ways, the study of microalgae is a relatively limited field of study. Algae are not nearly as well understood as other organisms that have found a role in today's biotechnology industry. The study of microalgae represents an area of higher risk and high gains; it also presents an opportunity for the curious and ambitious to break new ground in science by conducting research in this growing area of interest.

Algae can be both a nuisance and an ally; we might say they possess Jekyll and Hyde-like characteristics (good vs. bad). Many ponds, lakes, rivers, streams, and bays (e.g., Chesapeake Bay) in the United States (and elsewhere) are undergoing *eutrophication*, the killing off (albeit very slowly in many cases) of calm-water environments, especially ponds, marshes, and lakes, by enrichment

DID YOU KNOW?

In the 1980s, the DOE gradually shifted its focus to technologies that could have large-scale impacts on national consumption of fossil energy. Much of the DOE publications from this period reflect a philosophy of energy research that might, somewhat pejoratively, be called the *quads mentality.* A quad is a shorthand name for the unit of energy often used by the DOE to describe the amounts of energy that a given technology might be able to displace. Quad is short for "quadrillion Btus"—a unit of energy representing 10^{15} (1,000,000,000,000,000) Btus of energy. This perspective led the DOE to focus on the concept of immense algae farms.

of the environment with inorganic substances (phosphorous and nitrogen). When eutrophication occurs, filamentous algae such as *Caldophora* break loose in a pond, lake, stream, or river and wash ashore, making their noxious presence known. More important than the offensive odor dying algae give off is that the dying algae initiate a biodegradation process whereby instead of adding oxygen to their watery environment they begin to use it up. Algae have their good side, too; for example, algae are allies in many wastewater treatment operations, and they can be valuable in long-term oxidation ponds, where they aid in the purification process by producing oxygen. Before discussing the specifics and different types of algae, it is important to be familiar with algal terminology:

- *Algae*—Large and diverse assemblages of eukaryotic organisms that lack roots, stems, and leaves but have chlorophyll and other pigments for carrying out oxygen-producing photosynthesis.
- *Algology (phycology)*—The study of algae.
- *Antheridium*—Male reproductive structures where sperm are produced.
- *Aplanospore*—Nonmotile spores produced by sporangia.
- *Benthic*—Algae attached to and living on the bottom of a body of water.
- *Binary fission*—Nuclear division followed by division of the cytoplasm.
- *Chloroplasts*—Packets that contain chlorophyll *a* and other pigments.
- *Chrysolaminarin*—The carbohydrate reserve in Chrysophyta organisms.
- *Diatoms*—Photosynthetic, circular, or oblong chrysophyte cells.
- *Dinoflagellates*—Unicellular, photosynthetic protistan algae.
- *Dry mass factor*—Percentage of dry biomass in relation to the fresh biomass; for example, if the dry mass factor is 5%, one would need 20 kg of wet algae (algae in the media) to obtain 1 kg of dry algae cells.
- *Epitheca*—The larger part of the frustule (diatoms).
- *Euglenoids*—Algae (*Euglena*) that contain chlorophylls *a* and *b* in their chloroplasts.
- *Fragmentation*—A type of asexual algal reproduction in which the thallus breaks up and each fragmented part grows to form a new thallus.
- *Frustule*—The distinctive two-piece wall of silica in diatoms.
- *Hypotheca*—The small part of the frustule (diatoms).
- *Lipid content*—The percentage of oil in relation to the dry biomass required to get it; for example, if the algae lipid content is 40%, one would need 2.5 kg of dry algae to obtain 1 kg of oil.
- *Neustonic*—Algae that live at the water–atmosphere interface.
- *Oogonia*—Vegetative cells that function as female sexual structures in the algal reproductive system.
- *Pellicle*—A *Euglena* structure that allows for turning and flexing of the cell.
- *Phytoplankton*—Includes algae and small plants.
- *Plankton*—Free-floating, primarily microscopic aquatic organisms.
- *Planktonic algae*—Algae suspended in water as opposed to attached and living on the bottom (benthic).
- *Protothecosis*—Disease in humans and animals caused by the green algae *Prototheca moriformis.*
- *Thallus*—Vegetative body of algae.

Algae are autotrophic, contain the green pigment chlorophyll, and are a form of aquatic plant. Algae differ from bacteria and fungi in their ability to carry out photosynthesis, the biochemical process requiring sunlight, carbon dioxide, and raw mineral nutrients. Photosynthesis takes place in the chloroplasts. The chloroplasts are usually distinct and visible. They vary in size, shape, distribution, and numbers. In some algal types, the chloroplast may occupy most of the cell space. Algae usually grow near the surface of the water because light cannot penetrate very far down into the water.

Although in mass (multicellular forms like marine kelp) they can easily be seen by the unaided eye, many of them are microscopic. Algal cells may be nonmotile or motile by means of one or more flagella, or they may exhibit a gliding motility, as in diatoms. They occur most commonly in water (fresh and polluted water, as well as in saltwater), where they may be suspended (planktonic) or attached and living on the bottom (benthic). A few algae that live at the water–atmosphere interface are neustonic. Within the freshwater and saltwater environments, they are important primary producers (at the start of the food chain for other organism). During their growth phase, they are important oxygen-generating organisms and constitute a significant portion of the plankton in water.

Algae belong to seven divisions (phylums) distributed between two different kingdoms (Plantae and Protista) in the standard biological five-kingdom system. Only five of these divisions are discussed in this text:

- Chlorophyta (green algae)
- Euglenophyta (euglenoids)
- Chrysophyta (golden-brown algae and diatoms)
- Phaeophyta (brown algae)
- Pyrrophyta (dinoflagellates)

The primary classification of algae is based on their cellular properties. Several characteristics are used to classify algae, including: (1) cellular organization and cell-wall structure; (2) nature of chlorophylls present; (3) type of motility, if any; (4) carbon polymers that are produced and stored; and (5) reproductive structures and methods. Table 6.10 summarizes the properties of the five divisions discussed in this text.

Algae show considerable diversity in the chemistry and structure of their cells. Some algal cell walls are thin, rigid structures composed of cellulose modified by the addition of other polysaccharides. In some algae, the cell wall is strengthened by the deposition of calcium carbonate. Other forms have chitin present in the cell wall. Complicating the classification of algal organisms are the euglenoids, which lack cell walls. The cell walls of diatoms are composed of silica; the frustules (shells) of diatoms are extremely resistant to decay and remain intact for long periods of time, as fossil records indicate.

TABLE 6.10
Comparative Summary of Algal Characteristics

Algal Group	Common Name	Structure	Pigments	Carbon Reserve	Motility	Reproduction
Chlorophyta	Green algae	Unicellular to multicellular	Chlorophylls *a* and *b*, carotenes, xanthophylls	Starch, oils	Most are nonmotile	Asexual and sexual
Euglenophyta	Euglenoids	Unicellular	Chlorophylls *a* and *b*, carotenes	Fats	Motile	Asexual
Chrysophyta	Golden brown algae, diatoms	Multicellular	Chlorophylls *a* and *b*, special carotenoids, xanthophylls	Oils	Gliding by diatoms or by flagella	Asexual and sexual
Phaeophyta	Brown algae	Unicellular	Chlorophylls *a* and *b*, carotenoids, xanthophylls	Fats	Motile	Asexual and sexual
Pyrrophyta	Dinoflagellates	Unicellular	Chlorophylls *a* and *b*, carotenes, xanthophylls	Starch	Motile	Asexual and sexual (rare)

The principal feature used to distinguish algae from other microorganisms (e.g., fungi) is the presence of chlorophyll and other photosynthetic pigments in algae. All algae contain chlorophyll *a*. Some, however, contain other types of chlorophylls. The presence of these additional chlorophylls is characteristic of a particular algal group. In addition to chlorophyll, other pigments encountered in algae include fucoxanthin (brown), xanthophylls (yellow), carotenes (orange), phycocyanin (blue), and phycoerythrin (red). Many algae have flagella (a threadlike appendage), locomotor organelles that may be of the single polar or multiple polar type. Euglenophyta (euglenoids) have a simple flagellate form with a single polar flagellum. Golden algae have either two or four polar flagella, but diatoms have a gliding motility rather than flagella. Pyrrophyta (dinoflagellates) have two flagella of different lengths. In some cases, algae are nonmotile until they form motile gametes (a haploid cell or nucleus) during sexual reproduction.

Algae can be either autotrophic or heterotrophic. Most are photoautotrophic; they require only carbon dioxide and light as their principal source of energy and carbon. In the presence of light, algae carry out oxygen-producing photosynthesis; in the absence of light, algae use oxygen. Chlorophyll and other pigments are used to absorb light energy for photosynthetic cell maintenance and reproduction. One of the key characteristics used in the classification of algal groups is the nature of the reserve polymer synthesized as a result of utilizing carbon dioxide present in water.

Algae may reproduce asexually or sexually. Three types of asexual reproduction occur: binary fission, spores, and fragmentation. Binary fission in some unicellular algae occurs when the parent organism divides to form two new individuals that are like the parent. Some algae reproduce through spores. These spores are unicellular and germinate without fusing with other cells. In fragmentation, the thallus breaks up and each fragment grows to form a new thallus. Sexual reproduction can involve the union of cells, where eggs are formed within vegetative cells called *oogonia* (which function as female structures) and sperm are produced in a male reproductive organs called *antheridia*. Algal reproduction can also occur through a reduction of chromosome number or the union of nuclei.

Characteristics of Algal Divisions

Chlorophyta (Green Algae)

The majority of algae found in ponds belong to this group; they also can be found in saltwater and soil. Several thousand species of green algae are known today. Many are unicellular; others are multicellular filaments or aggregated colonies. The green algae have chlorophylls *a* and *b*, along with specific carotenoids, and they store carbohydrates at starch. Few green algae are found at depths greater than 7 to 10 meters, largely because sunlight does not penetrate to that depth. Some species have a holdfast structure that anchors them to the bottom of the pond and to other submerged inanimate objects. Green algae reproduce by both sexual and asexual means. Multicellular green algae display some division of labor in their production of various reproductive cells and structures.

Euglenophyta (Euglenoids)

The Euglenophyta are a small group of unicellular microorganisms that have a combination of animal and plant properties. Euglenoids lack a cell wall, possess a gullet, have the ability to ingest food, have the ability to assimilate organic substances, and, in some species, have no chloroplasts. They occur in fresh, brackish, and salt waters and on moist soils. A typical *Euglena* cell is elongated and bound by a plasma membrane; the absence of a cell wall makes them very flexible in movement. Inside the plasma membrane is a structure known as the pellicle that gives the organisms a definite form and allows the cell to turn and flex. Euglenoids that are photosynthetic contain chlorophylls *a* and *b*, and they always have a red eyespot (stigma) that is sensitive to light (photoreceptive). Some euglenoids move about by means of flagella; others move about using contracting and expanding motions. The characteristic food supply for euglenoids is lipopolysaccharides. Reproduction in euglenoids is by simple cell division.

Note: Some autotrophic species of *Euglena* become heterotrophic when light levels are low.

Chrysophyta (Golden-Brown Algae)

The Chrysophyta group is quite large, having several thousand diversified members. They differ from green algae and euglenoids in that: (1) chlorophylls *a* and *c* are present; (2) fucoxanthin, a brownish pigment, is present; and (3) they store food in the form of oils and leucosin, a polysaccharide. The combination of yellow pigments, fucoxanthin, and chlorophylls causes most of these algae to appear golden-brown. The Chrysophyta are also diversified in their cell wall chemistry and flagellation. The division is divided into three major classes: golden-brown, yellow-brown algae, and diatom. Some Chrysophyta lack cell walls; others have intricately patterned coverings external to the plasma membrane, such as walls, plates, and scales. The diatoms are the only group having hard cell walls of pectin, cellulose, or silicon, constructed in two halves (the epitheca and the hypotheca) called a *frustule*. Two anteriorly attached flagella are common among Chrysophyta; others have no flagella. Most Chrysophyta are unicellular or colonial. Asexual cell division is the usual method of reproduction in diatoms; other forms of Chrysophyta can reproduce sexually. Diatoms have direct significance for humans. Because they make up most of the phytoplankton of the cooler ocean parts, they are the ultimate source of food for fish. Water and wastewater operators understand the importance of their ability to function as indicators of industrial water pollution. As water quality indicators, their specific tolerances to environmental parameters such as pH, nutrients, nitrogen, concentration of salts, and temperature have been compiled.

Note: Diatoms secrete a silicon dioxide shell (frustule) that forms the fossil deposits known as diatomaceous earth, which is used in filters and as abrasives in polishing compounds.

Phaeophyta (Brown Algae)

With the exception of a few freshwater species, all algal species of this division exist in marine environments as seaweed. They are a highly specialized group, consisting of multicellular organisms that are sessile (attached and not free moving). These algae contain essentially the same pigments seen in the golden-brown algae, but they appear brown because of the predominance of and the masking effect of a greater amount of fucoxanthin. Brown algal cells store food as the carbohydrate laminarin and some lipids. Brown algae reproduce asexually. They are used in foods, animal feeds, and fertilizers and as a source for alginate, a chemical emulsifier added to ice cream, salad dressing, and candy.

Pyrrophyta (Dinoflagellates)

The principal members of this division are the dinoflagellates, a diverse group of biflagellated and nonflagellated unicellular, eukaryotic organisms. The dinoflagellates occupy a variety of aquatic environments, although the majority live in marine habitats. Most of these organisms have a heavy cell wall composed of cellulose-containing plates. They store food as starch, fats, and oils. These algae have chlorophylls *a* and *c* and several xanthophylls. The most common form of reproduction in dinoflagellates is by cell division, but sexual reproduction has also been observed.

Algal Biomass

Algal biomass contains three main components:

1. Carbohydrates
2. Protein
3. Natural oils

DID YOU KNOW?

Cell division in dinoflagellates differs from that of most protistans; chromosomes attach to the nuclear envelope and are pulled apart as the nuclear envelope stretches. During cell division in most other eukaryotes, the nuclear envelope dissolves.

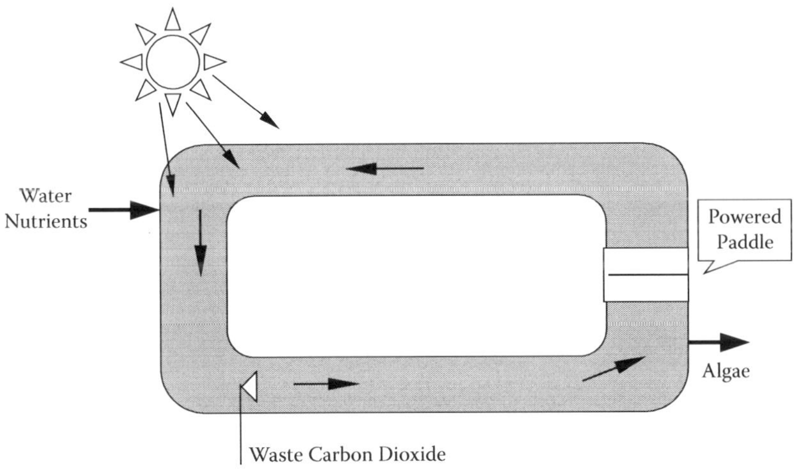

FIGURE 6.8 Raceway design algae pool.

Biodiesel production applies exclusively to the natural oil fraction, the main product of interest to us in this section. The bulk of natural oil made by oilseed crops is in the form of triacylglycerols (TAGs). TAGs consist of three long chains of fatty acids attached to a glycerol backbone. The oil-producing algae species of concern (*Ulva*, *Botryococcus braunii*, *Chlorella*, *Dunaliella tertiolecta*, *Gracilaria*, *Pleurochrysis carterae*, and *Sargassum*) can produce up to 60% of their body weight in the form of TAGs. Thus, algae represent an alternative source of biodiesel, one that does not compete with the existing oilseed market.

Algae can produce up to 300 times more oil per acre than conventional crops, such as rapeseed, palms, soybean, or *Jatropha* (Christi, 2007). Moreover, algae has a harvesting cycle of 1 to 10 days, permitting several harvests in a very short time frame, a strategy quite different from yearly crops. Algae can also be grown on land that is not suitable for other established crops, such as arid land, land with excessively saline soil, or drought-stricken land. This minimizes the issue of removing areas of land from the cultivation of food crops (Schenk et al., 2008). Algae can grow 20 to 30 times faster than food crops (McDill, 2009). Algae can be produced and harvested for biofuel using various technologies, including photobioreactors (plastic tubes full of nutrients exposed to sunlight), closed-loop systems that are not exposed to open air, and open ponds. Although open ponds have many objectors and dissenters, we have chosen to discuss open ponds configured as algae farms because they represent a relatively low-cost system and are easily understood.

Open-Pond Algae Farms

Algae farms consist of open, shallow ponds in which some source of waste carbon dioxide (CO_2) can be bubbled into the ponds and captured by the algae. As shown in Figure 6.8, the ponds in an algae farm are "raceway" designs, in which the algae, water and nutrients circulate around a racetrack. Paddlewheels provide the flow. The algae suspended in the water circulates back up to the surface at a regular frequency. The ponds are kept shallow because of the need to keep the algae exposed to sunlight and the limited depth to which sunlight can penetrate the pond water. The ponds operate continuously; that is, water and nutrients are constantly fed to the pond, while algae-containing water is removed at the other end. Some kind of harvesting system is required to recover the algae, which contains substantial amounts of natural oil.

Figure 6.9 illustrates the concept of an algae farm. The size of these ponds is measured in terms of surface area (as opposed to volume), because surface area is so critical to capturing sunlight. Their productivity is measured in terms of biomass produced per day per unit of available surface area. Even at levels of productivity that would stretch the limits of an aggressive research and development

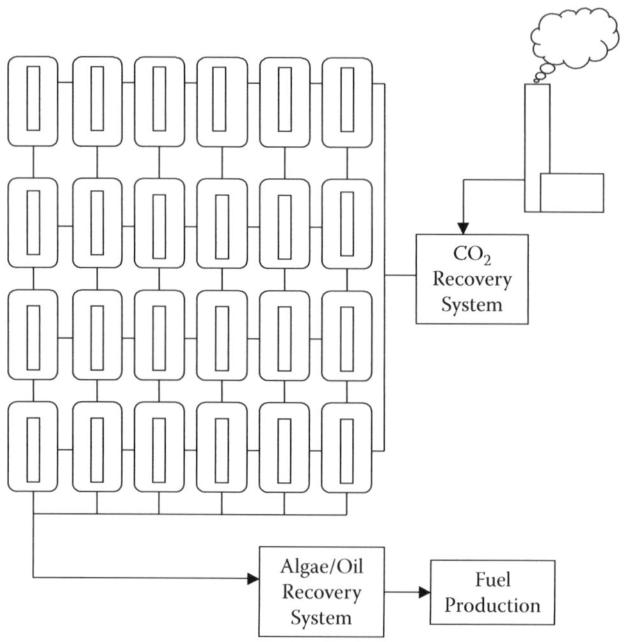

FIGURE 6.9 Algae farm.

program, such systems would require acres of land. At such large sizes, it is more appropriate to think of these operations on the scale of a farm. Waste carbon dioxide is readily available from a number of sources. Every operation that involves combustion of fuel for energy is a potential source. Generally, coal and other fossil fuel-fired power plants are targeted as the main sources of carbon dioxide. Typical coal-fired power plants emit flue gas from their stacks containing up to 13% carbon dioxide. This high concentration of carbon dioxide enhances the transfer and uptake of carbon dioxide in the ponds. The concept of coupling a coal-fired power plant with an algae farm offers a win–win approach to recycling the carbon dioxide resulting from coal combustion into a usable liquid fuel.

Jatropha to Biodiesel

The uninformed or misinformed might question that a byproduct of the plant *Jatropha curcas* (yes, the same plant known in some places as the "black vomit nut" and in others as the "bellyache bush" or "tuba-tub") could be of any value. *Jatropha* shrubs are inedible plants that grow mostly in countries such as the Philippines. The shrub is resistant to drought and can easily be planted or propagated through seeds or cuttings. It starts producing seeds within 14 months, but reaches its maximum productivity level after 4 to 5 years. The *Jatropha* plant can produce an oil content of 30 to 58%, depending on the quality of the soil where it is planted. The seeds yield an annual equivalent of 0.75 to 2 tons or 1892 liters of biodiesel per hectare. The plant remains useful for around 30 to 40 years and can be planted in harsh climates where it does not compete for the resources required to grow food. Along with having the advantage of being a renewable fuel source, the *Jatropha* plant also reduces greenhouse gas emissions and our dependence on oil imports.

Pros and Cons of Biodiesel

The greatest advantage of biodiesel over conventional petroleum diesel is that biodiesel comes from renewable resources. The supply can be grown, over and over again. Biodiesel combustion also produces fewer emissions (except for nitrous oxides) than combustion of an equal amount of petroleum diesel. The widespread use of biodiesel can also reduce the dependency on imported oil. From a safety standpoint, biodiesel is safer than petroleum diesel because it is less combustible. From an environmental standpoint, when accidentally spilled, biodiesel is not persistent within environmental media (air, water, soil) because it is biodegradable. Biodiesel can also be produced from waste products such as cooking oils and grease.

Probably the most pressing disadvantages or shortcomings of biodiesel, at least at the present time, are its lack of availability or accessibility and high cost relative to petroleum diesel. This lack of availability is bound to change as the currently less expensive and more accessible petroleum diesel becomes more expensive and difficult to find. Additional disadvantages to consider are that biodiesel requires special handling, storage, and transportation management as compared to petroleum diesel. When combusted, biodiesel produces more nitrous oxide emissions than an equal amount of petroleum diesel. Another potential problem with the production of biodiesel is its current dependence on soybeans as its primary feedstock; many are concerned that the widespread use of biodiesel as fuel will contribute to higher food prices and indirectly to world hunger. Performance and mileage per gallon are slightly reduced with biodiesel as compared with petroleum diesel. Biodiesel can also act as a solvent in some diesel engines, loosening deposits which may clog filters.

BIOGAS (METHANE, CH_4)

Primarily known as a fuel for interior heating systems, methane or biogas can also be used as a replacement for natural gas, a fossil fuel, for electricity generation and for cooking and heating, as well as an alternative fuel to gasoline. Methane is a natural gas produced by the breakdown of organic material in the absence of oxygen in termite mounds and wetlands and by some animals. Humans are also responsible for the release of methane through biomass burning, rice production, cattle, and releases from gas exploration. Methane can also be obtained directly from the earth; however, other methods of production have been developed, most notably the fermentation or composting of plant and animal waste. The reasons for considering biogas (methane) as a possible biofuel include:

- It is viable because of its potential use as an alternative fuel source.
- Its use will improve air quality.
- It can be produced locally, thus reducing reliance on imported natural gas.

Methane is produced under anaerobic (no oxygen) conditions where organic material is biodegraded or broken down by a group of microorganisms. The three main sources of feedstock material for anaerobic digestion are shown in Figure 6.10 and described in the following.

Anaerobic Digestion

Anaerobic digestion is traditionally used to manage waste (e.g., sludge stabilization) and to release energy. It involves using bacteria that thrive in the absence of oxygen; it is slower than aerobic digestion but has the advantage that only a small percentage of the wastes is converted into new bacterial cells. Instead, most of the organics are converted into carbon dioxide and methane gas.

Key Point: Allowing air to enter an anaerobic digester should be prevented because the mixture of air and gas produced in the digester can be explosive.

Stages of Anaerobic Digestion

Anaerobic digestion (see Figure 6.11) has four key biological and chemical stages (Spellman, 2009; USEPA, 1979, 2006):

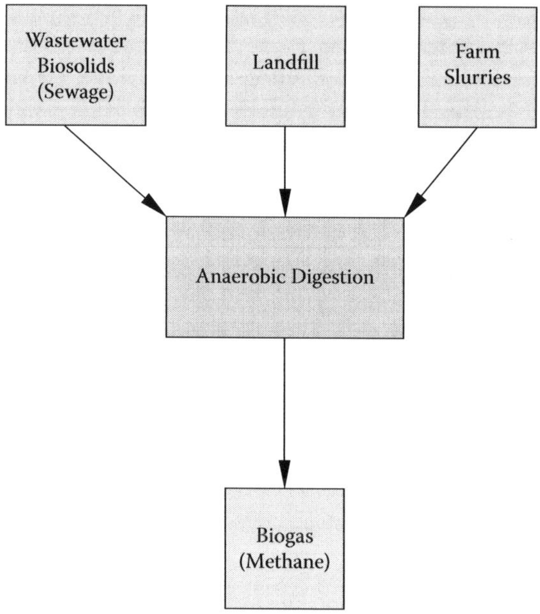

FIGURE 6.10 Production of biogas (methane CH_4).

1. *Hydrolysis* occurs when the proteins, cellulose, lipids, and other complex organics are broken down into smaller molecules and become soluble by utilizing water to split the chemical bonds of the substances.
2. *Acidogenesis* occurs when the products of hydrolysis are converted into organic acids (where monomers are converted to fatty acids).
3. *Acetogenesis* occurs when the fatty acids are converted to acetic acid, carbon dioxide, and hydrogen.
4. *Methanogenesis* occurs when the organic acids produced during the fermentation step are converted to methane and carbon dioxide.

The efficiency of each phase is influenced by the temperature and the amount of time the process is allowed to react. The organisms that perform hydrolysis and volatile acid fermentation (*acidogenic bacteria*) are fast-growing microorganisms that prefer a slightly acidic environment and

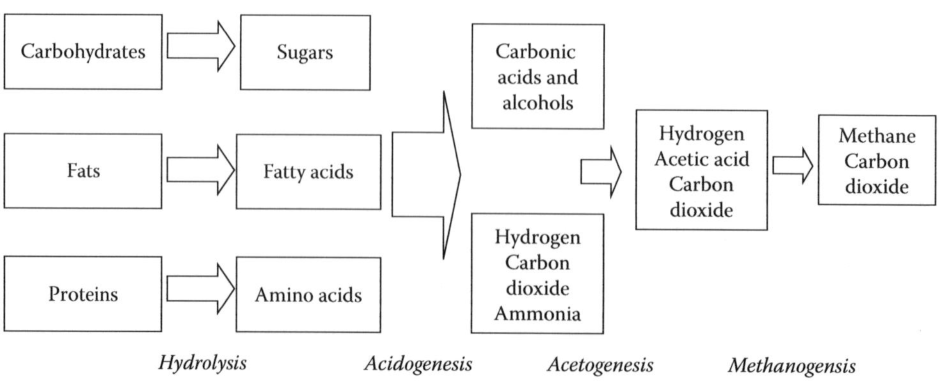

FIGURE 6.11 Key stages of anaerobic digestion.

TABLE 6.11
Typical Contents of Biogas

Matter	Percentage (%)
Methane (CH_4)	50–75
Carbon dioxide (CO_2)	25–50
Nitrogen (N_2)	0–10
Hydrogen (H_2)	0–1
Hydrogen sulfide (H_2S)	0–3
Oxygen (O_2)	0–2

higher temperatures than the organisms that perform the methane formation step (*methanogenic bacteria*). A simplified generic chemical equation for the overall processes outlined above is as follows:

$$C_6H_{12}O_6 \rightarrow 3CO_2 + 3CH_4$$

Biogas is the ultimate waste product of the bacteria feeding off the input biodegradable feedstock and is mostly methane and carbon dioxide, with a small amount of hydrogen and trace hydrogen sulfide (see Table 6.11). Keep in mind that the ultimate output from a wastewater digester is water; biogas (methane) is more of an off-gas that can be used as an energy source. Wastewater digestion and the production of biogas are discussed in the next section.

Anaerobic Digestion of Sewage Biosolids (Sludge)

Equipment used in anaerobic digestion typically includes a sealed digestion tank with either a fixed or a floating cover or an inflatable gas bubble, heating and mixing equipment, gas storage tanks, solids and supernatant withdrawal equipment, and safety equipment (e.g., vacuum relief, pressure relief, flame traps, explosion proof electrical equipment).

Caution: Biosolids are inherently dangerous as potential sources of explosive gases, and such facilities should never be entered without following OSHA's confined-space entry permit requirements; only fully trained personnel should enter permit-required confined spaces.

In operation, process residual (thickened or unthickened sludge) is pumped into the sealed digester. The organic matter digests anaerobically by a two-stage process. Sugars, starches, and carbohydrates are converted to volatile acids, carbon dioxide, and hydrogen sulfide. The volatile acids are then converted to methane gas. This operation can occur in a single tank (single stage) or in two tanks (two stages). In a single-stage system, supernatant and digested solids must be removed whenever flow is added. In a two-stage operation, solids and liquids from the first stage flow into the second stage each time fresh solids are added. Supernatant is withdrawn from the second stage to provide additional treatment space. Periodically, solids are withdrawn for dewatering or disposal. The methane gas produced in the process may be used for many plant activities.

Various performance factors affect the operation of the anaerobic digester. The percent volatile matter in raw sludge, digester temperature, mixing, volatile acids/alkalinity ratio, feed rate, percent solids in the raw sludge, and pH are all important operational parameters that the operator must

DID YOU KNOW?

The primary purpose of a secondary digester is to allow for solids separation.

TABLE 6.12
Anaerobic Digester Sludge Parameters

Raw Sludge Solids	Impact
<4%	Loss of alkalinity
	Decreased sludge retention time
	Increased heating requirements
	Decreased volatile acid/alkalinity ratio
4–8%	Normal operation
>8%	Poor mixing
	Organic overloading
	Decreased volatile acid/alkalinity ratio

monitor (see Table 6.12). In addition to being able to recognize normal and abnormal anaerobic digester performance parameters, digester operators must also know and understand normal operating procedures, including sludge additions, supernatant withdrawal, sludge withdrawal, pH control, temperature control, mixing, and safety requirements.

Caution: Keep in mind that in fixed-cover operations additions must be balanced by withdrawals. If not, structural damage occurs.

Sludge must be pumped (in small amounts) several times each day to achieve the desired organic loading and optimum performance, and supernatant withdrawal must be controlled for maximum sludge retention time. Digested sludge is withdrawn only when necessary, always leaving at least 25% seed. pH should be maintained at 6.8 to 7.2 by adjusting the feed rate, sludge withdrawal, or alkalinity additions. Anaerobic digesters must be continually monitored to ensure proper operation, including determining supernatant pH, volatile acids, alkalinity, BOD or COD, total solids, and temperature. Sludge (in and out) should be routinely tested for percent solids and percent volatile matter. Normal operating parameters are listed in Table 6.13.

TABLE 6.13
Normal Operating Ranges for Anaerobic Digester

Parameter	Normal Range
Sludge retention time	
Heated	30–60 days
Unheated	180+ days
Volatile solids loading	0.04–0.1 lb/day/ft^3
Operating temperature	
Heated	90–95°F
Unheated	Varies with season
Mixing	
Heated (primary)	Yes
Unheated (secondary)	No
Methane in gas	60–72%
Carbon dioxide in gas	28–40%
pH	6.8–7.2
Volatile acids/alkalinity ratio	≤0.1
Volatile solids reduction	40–60%
Moisture reduction	40–60%

Note: The buffer capacity of an anaerobic digester is indicated by the volatile acid/alkalinity relationship. Decreases in alkalinity cause a corresponding increase in ratio.

If the digester is heated, the temperature must be controlled to a normal temperature range of 90 to 95°F (never adjusting the temperature by more than 1°F per day). If the digester is equipped with mixers, mixing should be accomplished to ensure that organisms are exposed to food materials. Anaerobic digesters are inherently dangerous, and several catastrophic failures have been recorded. To prevent such failures, safety equipment such as pressure-relief and vacuum-relief valves, flame traps, condensate traps, and gas collection safety devices must be installed. It is important that these critical safety devices be checked routinely and maintained for proper operation.

Key Point: Because of the inherent danger involved with working inside anaerobic digesters, they are automatically classified as permit-required confined spaces; therefore, all operations involving internal entry must be made in accordance with OSHA's confined-space entry permit regulations. If safety questions concerning safe entry into confined spaces (of any type) arise, consult with a Certified Safety Professional (CSP), Certified Industrial Hygienist (CIH), or Professional Engineer (PE).

Process control calculations involved with anaerobic digester operation include determining the required seed volume, volatile acids/alkalinity ratio, sludge retention time, estimated gas production, volatile matter reduction, and percent moisture reduction in the digester sludge. Examples of how to make these calculations are provided in the following text.

Required Seed Volume in Gallons

$$\text{Seed Volume (gal)} = (\text{Digester Volume}) \times (\%\ \text{Seed}) \tag{6.1}$$

■ Example 6.1

Problem: A new digester requires a 25% seed to achieve normal operation within the allotted time. If the digester volume is 266,000 gal, how many gallons of seed material will be required?

Solution:

$$\text{Seed Volume} = 266{,}000 \times 0.25 = 66{,}500 \text{ gal}$$

Volatile Acids/Alkalinity Ratio

The volatile acids/alkalinity ratio can be used to control operation of an anaerobic digester.

$$\text{Ratio} = \frac{\text{Volatile Acids Concentration}}{\text{Alkalinity Concentration}} \tag{6.2}$$

■ Example 6.2

Problem: A digester contains 240 mg/L volatile acids and 1860 mg/L alkalinity. What is the volatile acids/alkalinity ratio?

Solution:

$$\text{Ratio} = \frac{240 \text{ mg/L}}{1860 \text{ mg/L}} = 0.13$$

Increases in the ratio normally indicate a potential change in the operating condition of the digester as shown in Table 6.14.

Sludge Retention Time

Sludge retention time is the length of time the sludge remains in the digester.

$$\text{Sludge Retention Time (days)} = \frac{\text{Digester Volume (gal)}}{\text{Sludge Volume Added per Day (gpd)}} \tag{6.3}$$

TABLE 6.14
Operating Conditions of Digester

Condition	Volatile Acid/Alkalinity Ratio
Optimum	≤0.1
Acceptable range	0.1–0.3
Carbon dioxide in gas increases	≥0.5
pH decreases	≥0.8

■ Example 6.3

Problem: Sludge is added to a 525,000-gal digester at the rate of 12,250 gpd. What is the sludge retention time?
Solution:

$$\text{Sludge Retention Time} = \frac{525{,}000 \text{ gal}}{12{,}250 \text{ gpd}} = 42.9 \text{ days}$$

Estimated Gas Production in Cubic Feet/Day

The rate of gas production is normally expressed as the volume of gas (ft^3) produced per pound of volatile matter (VM) destroyed. The total cubic feet of gas a digester will produce per day can be calculated by:

$$\text{Gas Production (ft}^3) = \text{VM}_{in} \text{ (lb/day)} \times \%\text{VM Reduction} \times \text{Production Rate (ft}^3\text{/lb)} \quad (6.4)$$

■ Example 6.4

Problem: A digester receives 11,450 lb of volatile matter per day. Currently, the volatile matter reduction achieved by the digester is 52%. The rate of gas production is 11.2 ft^3 of gas per pound of volatile matter destroyed.
Solution:

$$\text{Gas Production} = 11{,}450 \text{ lb/day} \times 0.52 \times 11.2 \text{ ft}^3\text{/lb} = 66{,}685 \text{ ft}^3\text{/day}$$

Percent Volatile Matter Reduction

Because of changes occurring during sludge digestion, the calculation used to determine percent volatile matter (%VM) reduction is more complicated.

$$\%\text{VM Reduction} = \frac{(\%\text{VM}_{in} - \%\text{VM}_{out}) \times 100}{\%\text{VM}_{in} - (\%\text{VM}_{in} \times \%\text{VM}_{out})} \quad (6.5)$$

■ Example 6.5

Problem: Using the data provided below, determine the percent volatile matter reduction for the digester.

Raw sludge volatile matter = 74%
Digested sludge volatile matter = 55%

Solution:

$$\%\text{VM Reduction} = \frac{(0.74 - 0.55) \times 100}{0.74 - (0.74 \times 0.55)} = 57\%$$

Percent Moisture Reduction in Digested Sludge

$$\% \text{ Moisture Reduction} = \frac{(\% \text{ Moisture}_{in} - \% \text{ Moisture}_{out}) \times 100}{\% \text{ Moisture}_{in} - (\% \text{ Moisture}_{in} \times \% \text{ Moisture}_{out})} \tag{6.6}$$

■ Example 6.6

Problem: Using the digester data provide below, determine the percent moisture reduction and percent volatile matter reduction for the digester.

Raw sludge percent solids = 6%
Digested sludge percent solids = 14%

Solution:

Note: % Moisture = 100% – Percent Solids

$$\% \text{ Moisture Reduction} = \frac{(0.94 - 0.86) \times 100}{0.94 - (0.94 \times 0.86)} = 61\%$$

Anaerobic Digestion of Agricultural Wastes

Animal waste accounts for 10% of methane emissions in the United States (Spellman and Whiting, 2007). When not correctly managed, farm waste slurries can also seriously pollute local watercourses. Small anaerobic digesters have been installed on farms to treat excess animal slurries that cannot be placed on the land. The biogas formed is normally used for heat but can also be used to fuel engines and other onsite energy needs such as electricity and heating. Onsite biogas production and management can also reduce offensive odors from overloaded or improperly managed manure storage facilities. These odors impair air quality and may be a nuisance to nearby communities. Anaerobic digestion of animal waste reduces these offensive odors because the volatile organic acids (the odor-causing compounds) are consumed by biogas-producing bacteria. In addition to biogas, another important byproduct of anaerobic digestion is ammonium, which is the major constituent of commercial fertilizer and is readily available to and utilized by crops. The bottom line on the production of biogas on the farm: Biogas recovery can improve profitability while improving environmental quality.

Landfill Biogas

Landfills can be a source of energy. Landfills produce methane as organic waste decomposes in the same anaerobic digestion process used in converting wastewater and farm waste slurries into biogas. Most landfill gas results from the degradation of cellulose contained in municipal and industrial solid waste. Unlike animal manure digesters, which control the anaerobic digestion process, the digestion occurring in landfills is an uncontrolled process of biomass decay. To be technically feasible, a landfill must be at least 40 feet deep and have at least 1 million tons of waste in place for landfill gas collection.

The efficiency of the process depends on the waste composition and moisture content of the landfill, cover material, temperature, and other factors. The biogas released from landfills, commonly referred to as *landfill gas*, is typically 50% methane, 45% carbon dioxide, and 5% other gases. The energy content of landfill gas is 400 to 550 Btu/ft^3. Figure 6.12 shows a landfill energy system. Such a system consists of a series of wells drilled into the landfill. A piping system connects the wells and collects the gas. Dryers remove moisture from the gas and filters remove impurities. The gas typically fuels an engine-generator set or gas turbine to produce electricity. The gas can also fuel a

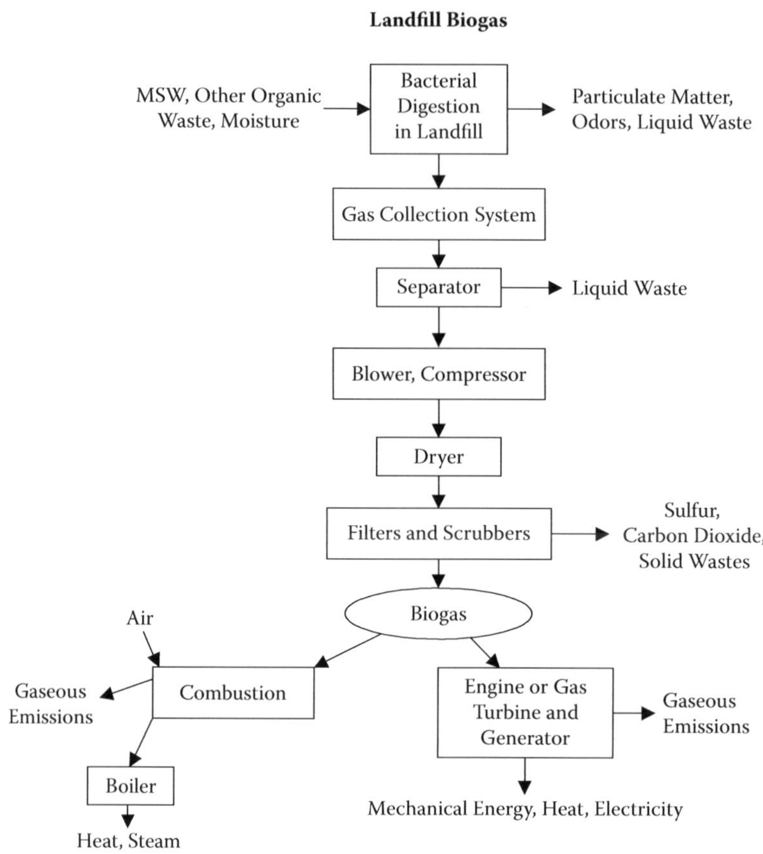

FIGURE 6.12 Landfill biogas system flow diagram.

boiler to produce heat or steam. Because waste-generated biogas is considered to be a "dirty gas" as compared to natural gas, further gas cleanup is required to improve biogas to pipeline quality, or the equivalent of natural gas. Reforming the gas to hydrogen would make possible the production of electricity using fuel cell technology.

THE BOTTOM LINE ON BIOFUELS

With near-record oil prices, the future of biofuel—made from plant material (biomass)—is of keen interest worldwide. Using biofuels to power industry, private vehicles, and personal appliances offers several benefits over the use of conventional fuels; for example, environmental benefits include the use of biomass energy to greatly reduce greenhouse gas emissions. Burning biomass releases about the same amount of carbon dioxide as burning fossil fuels; however, fossil fuels release carbon dioxide captured by photosynthesis millions of years ago—an essentially "new" greenhouse gas. Biomass, on the other hand, releases carbon dioxide that is largely balanced by the carbon dioxide captured during its own growth (depending on how much energy was used to grow, harvest, and process the fuel). Another benefit of biomass use for fuel is that it can reduce dependence on foreign oil because biofuels are renewable liquid transportation fuels.

Moreover, biomass energy supports U.S. agricultural and forest-product industries. The main biomass feedstocks for power are paper mill residue, lumber mill scrap, and municipal waste. For biomass fuels, the feedstocks are corn (for ethanol) and soybeans (for biodiesel), both surplus crops. In the near future—and with developed technology—agricultural residues such as corn stover (the stalks, leaves, and husks of the plant) and wheat straw will also be used. Long-term plans include growing and using dedicated energy crops, such as fast-growing trees and grasses that can grow sustainably on land that will not support intensive food crops.

Despite all of the benefits of using biomass fuel, the reality is that the future role of biofuels depends on profitability and new technologies. Technological advances and efficiency gains (higher biomass yields per acre and more gallons of biofuel per ton of biomass) could steadily reduce the economic cost and environmental impact of biofuel production. Biofuel production will likely be most profitable and environmentally benign in tropical areas where growing seasons are longer, per-acre biofuel yields are higher, and fuel and other input costs are lower. Brazil, for example, uses bagasse, which is a byproduct from sugar production, to power ethanol distilleries, whereas the United States uses natural gas or coal.

Biofuels will most likely be part of a portfolio of solutions to high oil prices, including conservation and the use of other alternative fuels. The role of biofuels in global fuel supplies is likely to remain modest because of its land intensity. In the United States, replacing all current gasoline consumption with ethanol would require an increase in land devoted to corn production. Technology will be central to boosting the role of biofuels. If the energy of widely available, cellulose materials could be economically harnessed around the world, biofuel yields per acre could more than double, reducing land requirements significantly (Coyle, 2007).

DISCUSSION AND REVIEW QUESTIONS

1. Do you agree with Kurt Anderson's assertion that the current generation is the grasshopper generation? Explain.
2. Solid biomass fuel is used primarily for ________, ________, and/or heat supply.
3. The biorefinery is similar in concept the petroleum refinery, except that it is based on conversion of ________ feedstocks rather than crude oil.
4. What are feedstocks?
5. Why do the authors characterize algae as Jekyll and Hyde-like organisms?
6. What is characterized by having two cotyledons?
7. What absorbs red and blue light from the sunlight that falls on leaves?
8. Plant growth is controlled by ________.
9. What is considered a model energy crop?
10. In folklore and myth, a ________ tree is believed to be quite sinister.
11. ________ is the ratio of the number Btus available in 1 gal of gasoline to the number of Btus available in 1 gal of the alternative substance in question.
12. Do you think using food corn as a feedstock for ethanol production is a good practice? Explain.

REFERENCES AND RECOMMENDED READING

Allen, H.L., Fox, T.R., and Campbell, R.G. (2005). What's ahead for intensive pine plantation silviculture? *Southern Journal of Applied Forestry*, 29, 62–69.

Anon. (2008). Algae eyed as biofuel alternative. *Taipei Times*, January 12, p. 2 (http://www.taipeitimes.com/News/taiwan/archives/2008/01/12/2003396760).

Ashford, R.D. (2001). *Ashford's Dictionary of Industrial Chemicals*, 2nd ed. London: Wavelength Publications.

Aylott, M.J. (2008). Yield and spatial supply of bioenergy poplar and willow short-rotation coppice in the U.K. *New Phytologist*, 178(2), 358–370.

Baize, J. (2006). Bioenergy and biofuels. *Agricultural Outlook Forum*, February 17.

Baker, J.B. and Broadfoot, W.M. (1979). *A Practical Field Method of Site Evaluation for Commercially Important Southern Hardwoods*, Gen. Tech. Rep. SO-26. New Orleans, LA: Forest Service, Southern Forest Experiment Station, U.S. Department of Agriculture.

Bender, M. (1999). Economic feasibility review for community-scale farmer cooperatives for biodiesel. *Bioresource Technology*, 70, 81–87.

Christi, Y. (2007). Biodiesel from microalgae. *Biotechnology Advances*, 25, 294–306.

Coyle, W. (2007). The future of biofuels: a global perspective. *Amber Waves*, November (http://www.ers.usda.gov/AmberWaves/november07/features/biofuels.htm).

Dickman, D. (2006). Silviculture and biology of short-rotation wood crops in temperate regions: then and now. *Biomass and Bioenergy*, 30, 696–705.

EERE. (2008). *Feedstock Types*. Washington, D.C.: Energy Efficiency & Renewable Energy, U.S. Department of Energy (http://www1.eere.energy.gov/biomass/feedstocks_types.html).

EERE and EPRI. (1997). *Renewable Energy Technology Characterizations*, TR-109496. Washington, D.C.: Energy Efficiency & Renewable Energy, U.S. Department of Energy (http://www1.eere.energy.gov/ba/pba/tech_characterizations.html).

EIA. (2007). *Annual Energy Outlook 2007 with Projections to 2030*, DOE/EIA-0383. Washington, D.C.: Energy Information Administration, U.S. Department of Energy.

Elliot, D.C. et al. (1999). Production of levulinic acid and use as a platform chemical for derived products, in *Biomass, a Growth Opportunity in Green Energy and Value-Added Products: Proceedings of the 4th Biomass Conference of the Americas* (Vol. 1, pp. 595–600). Oxford: Elsevier.

Fahey, J. (2001). Shucking petroleum. *Forbes Magazine*, 168(13), 206–208.

Farrell, A.E. and Gopal, A.R. (2008). Bioenergy research needs for heat, electricity, and liquid fuels. *MRS Bulletin*, 33, 373–387.

Friedman, T.L. (2010). The fat lady has sung. *The New York Times*, February 21, p. WK8.

Gentille, S.B. (1996). *Reinventing Energy: Making the Right Choices*. Washington, D.C.: Diane Publishing.

Graham, R., Nelson, R., Sheehan J., Perlack, R., and Wright L. (2007). Current and potential U.S. corn stover supplies. *Agronomy Journal*, 99(1), 1–11.

Haas, M., McAloon, A., Yee, W., and Foglia, T. (2006). A process model to estimate biodiesel production costs. *Bioresource Technology*, 97, 671–678.

Harrar, E.S. and Harrar, J.G. (1962). *Guide to Southern Trees*, 2nd ed. New York: Dover Publications.

Hileman, B. (2003). Clashes over agbiotech. *Chemical Engineering News*, 81, 25–33.

Hoffman, J. (2001). BDO outlook remains healthy. *Chemical Market Reporter*, April 2, p. 5.

Hughes, E. (2000). Biomass cofiring: economics, policy and opportunities. *Biomass and Bioenergy*, 19, 45–465.

Kantor, S.L., Lipton, K., Manchester, A., and Oliveira, V. (1997). Estimating and addressing America's food losses. *Food Review*, January–April, pp. 2–12.

Klass, D.I. (1998). *Biomass for Renewable Energy, Fuels, and Chemicals*. San Diego, CA: Academic Press.

Lee, S. (1996). *Alternative Fuels*. Boca Raton, FL: Taylor & Francis.

Lewis, L. (2005). Seaweed to breathe new life into fight against global warming. *The Times & The Sunday Times*, May 14 (http://www.timesonline.co.uk/tol/news/world/article522203.ece).

Little, A. (2001). *Aggressive Growth in the Use of Bioderived Energy and Products in the United States by 2010: Final Report*. Washington, D.C.: U.S. Department of Energy.

Markarian, J. (2003). New additives and basestocks smooth way for lubricants. *Chemical Market Reporter*, April 28.

McDill, S. (2009). Can algae save the world—again? *Reuters*, February 10 (http://www.reuters.com/article/environmentNews/idUSTRE5196HB20090210?pageNumber=2&virtualBrandChannel=0).

McGraw, L. (1999). Three new crops for the future. *Agricultural Research Magazine*, 47(2), 17.

McKeever, D. (2004). Inventories of woody residues and solid wood waste in the United States, 2002, in *Proceedings of the Ninth International Conference, Inorganic-Bonded Composite Materials*, Vancouver, BC, October 10–13.

McKendry, P. (2002). Energy production from biomass. Part I. Overview of biomass. *Bioresource Technology*, 83, 37–46.

McLaughlin, S.B. and Kzos, L.A. (2005). Development of switchgrass as a bioenergy feedstock in the United States. *Biomass & Bioenergy*, 28, 515–535.

Miles, T.R., Miles, Jr., T.R., Baxter, L.L., Jenkins, B.M., and Oden, L.L. (1993). Alkali slagging problems with biomass fuels, in *Proceedings of the First Biomass Conference of the Americas: Energy, Environment, Agriculture, and Industry*, Burlington, VT, August 30–September 2.

Morey, R.V., Tiffany, D.G., and Hartfield D.L. (2006). Biomass for electricity and process heat at ethanol plants. *Applied Engineering in Agriculture*, 22, 723–728.

Morris, D. (2002). *Accelerating the Shift to a Carbohydrate Economy: The Federal Role*. Washington, D.C.: Institute for Local Self-Reliance.

NIST. (1994). *Handbook 44: Appendix D: Definitions*. Washington, D.C.: National Institute of Standards and Technology.

Parrish, D.J. Fike, J.H. Bransby, D.I., and Samson, R. (2008). Establishing and managing switchgrass as an energy crop. *Forage and Grazinglands*, doi:10.1094/FG-2008-0220-01-RV.

Perlin, J. (2005). *A Forest Journey: The Story of Wood and Civilization*. Woodstock, VT: The Countryman Press.

Polagye, B., Hodgson, K., and Malte, P. (2007). An economic analysis of bioenergy options using thinning from overstocked forests. *Biomass and Bioenergy*, 31, 105–125.

Rinehart, L. (2006). *Switchgrass as a Bioenergy*. Butte, MT: National Center for Appropriate Technology.

Rossell, J.B. and Pritchard, J.L.R., Eds. (1991). *Analysis of Oilseeds, Fats, and Fatty Foods*. New York: Elsevier.

Samson, R. et al. (2008). Developing energy crops for thermal applications: optimizing fuel quality, energy security and GHG mitigation, in *Biofuels, Solar and Wind as Renewable Energy Systems: Benefits and Risks*, Pimental, D., Ed., pp. 395–423. Berlin: Springer Science.

Sauer, P. (2000). Domestic spearmint oil producers face flat pricing. *Chemical Market Reporter*, Vol. 258.

Schenk, P. et al. (2008). Second generation biofuels: high-efficiency microalgae for industrial production. *BioEnergy Research*, 1(1), 20–45.

Sedjo, R. (1997). The economics of forest-based biomass supply. *Energy Policy*, 25(6), 559–566.

Silva, B. (1998). Meadowfoam as an alternative crop. *AgVentures*, 2(4), 28.

Sioru, B. (1999). Process converts trash into oil. *Waste Age*, 30(11), 20.

Spellman, F.R. (2009). *Handbook of Water and Wastewater Treatment Plant Operations*, 2nd ed. Boca Raton, FL: CRC Press.

Spellman, F.R. and Whiting, N. (2007). *Concentrated Animal Feeding Operations (CAFOs)*. Boca Raton, FL: CRC Press.

Tillman, D. (2000). Biomass cofiring: the technology, the experience, the combustion consequences. *Biomass and Bioenergy*, 19, 365–384.

Uhlig, H. (1998). *Industrial Enzymes and their Applications*. New York: John Wiley & Sons.

USDOE. (2003a). *Industrial Bioproducts: Today and Tomorrow*. Washington, D.C.: U.S. Department of Energy (http://www.brdisolutions.com/pdfs/BioProductsOpportunitiesReportFinal.pdf).

USDOE. (2003b). *The Bioproducts Industry: Today and Tomorrow*. Washington, D.C.: U.S. Department of Energy.

USDOE. (2006). *Breaking the Biological Barriers to Cellulosic Ethanol: A Joint Research Agenda*. Washington, D.C.: U.S. Department of Energy.

USDOE. (2007). *Understanding Biomass: Plant Cell Walls*. Washington, D.C.: U.S. Department of Energy (http://genomicscience.energy.gov/biofuels/cellwall_placemat.pdf.).

USDOE. (2010). *Fuel Ethanol Production*. Washington, D.C.: Genomic Science Program, U.S. Department of Energy (http://genomicscience.energy.gov/biofuels/ethanolproduction.shtml).

USEPA. (1979). *Process Design Manual for Sludge Treatment and Disposal*, EPA 625/1-79-011. Washington, D.C.: Office of Research and Development, U.S. Environmental Protection Agency.

USEPA. (2006). *Biosolids Technology Fact Sheet: Multi-Stage Anaerobic Digestion*, EPA 832-F-06-031. Washington, D.C.: U.S. Environmental Protection Agency.

USEPA. (2007). *Fuel Economy Impact Analysis of RFG*. Washington, D.C.: U.S. Environmental Protection Agency (http://www.epa.gov/oms/rfgecon.htm)

USEPA. (2010). *Overview of Biogas Technology*. Washington, D.C.: U.S. Environmental Protection Agency (www.epa.gov/agstar/documents/chapter1.pdf).

Valigra, L. (2000). Tough as soybeans. *The Christian Science Monitor*, January 20.

Wilhelm, W.W., Johnson, J.M.F., Karlen, D.L., and Lightle, D.T. (2002). Corn stover to sustain soil organic carbon further constrains biomass supply. *Agronomy Journal*, 99, 1665–1667.

Wiltsee, G. (1998). *Urban Wood Waste Resource Assessment*, NREL/SR-570-25918. Golden, CO: National Renewable Energy Laboratory, U.S. Department of Energy.

Wood, M. (2002). Desert shrub may help preserve wood. *Agricultural Research Magazine*, 50(4), 10–11.

World Agricultural Outlook Board. (2010). *World Agricultural Supply and Demand Estimates*. Washington, D.C.: U.S. Department of Agriculture (http://www.usda.gov/oce/commodity/wasde/).

Zeman, N. (2007). The preventative chemist. *Biodiesel Magazine*, February 2007.

7 Geothermal Energy

The U.S. Geological Survey has calculated that the heat energy in the upper 10 kilometers of the Earth's crust in the U.S. is equal to over 600,000 times the country's annual non-transportation energy consumption. Probably no more than a tiny fraction of this energy could ever be extracted economically. However, just one hundredth of 1% of the total is equal to half the country's current non-transportation energy needs for more than a century, with only a fraction of the pollution from fossil-fueled energy sources.

—McLarty et al. (2000)

Geothermal heat is the only renewable energy source created naturally by the Earth itself.

—Kimberly K. Smith, Carlton College

If we utilize waste biomass, solar (passive and thermal), wind (on shore), photovoltaic, geothermal, and other renewable resources available to us in the United States, we would exceed the demand (what we need) by at least five times as much energy as we need, all from clean, renewable sources.

—Frank R. Spellman

INTRODUCTION*

Approximately 4000 miles below the Earth's surface is the core, where temperatures can reach 9000°F. This heat—geothermal energy (*geo* meaning "earth" and *thermos* meaning "heat")—flows outward from the core, heating the surrounding area, which can form underground reservoirs of hot water and steam. These reservoirs can be tapped for a variety of uses, such as to generate electricity or heat buildings. The geothermal energy potential in the uppermost 6 miles of the Earth's crust amounts to 50,000 times the energy of all oil and gas resources in the world. In the United States, most geothermal reservoirs are located in the western states, Alaska, and Hawaii; however, geothermal heat pumps (GHPs), which take advantage of the shallow ground's stable temperature for heating and cooling buildings, can be used almost anywhere. Again, it is important to point out that there is nothing new about renewable energy. From solar power to burning biomass (wood) in caves and elsewhere, humans have taken advantage of renewable resources from time immemorial. Hot springs have been used for bathing since Paleolithic times or earlier (USDOE, 2010), and the early Romans used hot springs to supply public baths and for underfloor heating systems. The world's oldest geothermal district heating system has been operating in France since the 14th century (Lund, 2007). The history of geothermal energy use in the United States is interesting and lengthy; following is a brief chronology of major geothermal events in this country (EERE, 2006).

DID YOU KNOW?

Scientists estimate that geothermal potential could be as large as 100 million kW.

* Based on information from USDOE, *Renewable Energy: An Overview*, National Renewable Energy Laboratory, U.S. Department of Energy, Washington, D.C., 2001.

Geothermal Timeline

8000 B.C. (and Earlier)

Paleo-Indians used hot springs for cooking and for refuge and respite. Hot springs were neutral zones where members of warring nations would bathe together in peace. Native Americans have a history with every major hot spring in the United States.

1807

As European settlers moved westward across the continent, they gravitated toward these springs of warmth and vitality. In 1807, the first European visited the Yellowstone area; John Colter (*c.* 1774–1813), widely considered to be the first mountain man, probably encountered hot springs leading to the designation "Colter's Hell." Also that year, settlers founded the city of Hot Springs in Arkansas, where, in 1830, Asa Thompson charged $1 each for the use of three spring-fed baths in a wooden tub—the first known commercial use of geothermal energy.

1847

William Bell Elliot, a member of John C. Fremont's survey party, stumbled upon a steaming valley just north of what is now San Francisco, California. Elliot called the area *The Geysers*—a misnomer—and thought he had found the gates of Hell.

1852

The Geysers was developed into a spa called The Geysers Resort Hotel. Guests include J. Pierpont Morgan, Ulysses S. Grant, Theodore Roosevelt, and Mark Twain.

1862

At springs located southeast of The Geysers, businessman Sam Brannan poured an estimated half million dollars into an extravagant development dubbed "Calistoga," replete with hotel, bathhouse, skating pavilion, and racetrack. Brannan's was one of many spas reminiscent of those of Europe.

1864

Homes and dwellings had been built near springs through the millennia to take advantage of their natural heat, but the construction of the Hot Lake Hotel near La Grande, Oregon, marked the first time that the energy from hot springs was used on a large scale.

1892

Boise, Idaho, provided the world's first district heating system as water was piped from hot springs to town buildings. Within a few years, the system was serving 200 homes and 40 downtown businesses. Today, there are four district heating systems in Boise that provide heat to over 5 million square feet of residential, business, and governmental space. The United States has 17 district heating systems, and dozens more can be found around the world.

1900

Hot springs water was piped to homes in Klamath Falls, Oregon.

1921

John D. Grant drilled a well at The Geysers with the intention of generating electricity. This effort was unsuccessful, but a year later Grant met with success across the valley at another site, and the United States' first geothermal power plant went into operation. Grant used steam from the first well to build a second well, and, several wells later, the operation was producing 250 kilowatts, enough electricity to light the buildings and streets at the resort. The plant, however, was not competitive with other sources of power, and it soon fell into disuse.

1927
Pioneer Development Company drilled the first exploratory well at Imperial Valley, California.

1930
The first commercial greenhouse use of geothermal energy was undertaken in Boise, Idaho. The operation used a 1000-foot well drilled in 1926. In Klamath Falls, Charlie Lieb developed the first downhole heat exchanger (DHE) to heat his house. Today, more than 500 DHEs are in use around the country.

1940
The first residential space heating in Nevada became available in the Moan area in Reno.

1948
Geothermal technology moved east when Carl Nielsen developed the first ground-source heat pump for use at his residence. J.D. Krocker, an engineer in Portland, Oregon, pioneered the first commercial building use of a groundwater heat pump.

1960
The country's first large-scale geothermal electricity-generating plant began operation. Pacific Gas and Electric operated the plant located at The Geysers. The first turbine produced 11 megawatts (MW) of net power and operated successfully for more than 30 years. Today, 69 generating facilities are in operation at 18 resource sites around the country.

1978
Geothermal Food Processors, Inc., opened the first geothermal food-processing (crop-drying) plant in Brady Hot Springs, Nevada. The Loan Guaranty Program provided $3.5 million for the facility.

1979
The first electrical development of a water-dominated geothermal resource occurred at the east Mesa field in the Imperial Valley in California. The plant was named for B.C. McCabe, the geothermal pioneer who, with his Magma Power Company, did field development work at several sites, including The Geysers.

1980
TAD's Enterprises of Nevada pioneered the use of geothermal energy for the cooking, distilling, and drying processes associated with alcohol fuel production. UNOCAL built the country's first flash plant, generating 10 MW at Brawley, California.

1982
Economical electrical generation began at California's Salton Sea geothermal field through the use of crystallizer-clarifier technology. The technology resulted from a government/industry effort to manage the high-salinity brines at the site.

1984
A 20-MW plant began generating power at Utah's Roosevelt Hot Springs. Nevada's first geothermal electricity was generated when a 1.3-MW binary power plant went into operation.

1987
Geothermal fluids were used in the first geothermal-enhanced heap leaching project for gold recovery near Round Mountain, Nevada.

1989

The world's first hybrid (organic Rankine/gas engine) geopressure–geothermal power plant began operation at Pleasant Bayou, Texas, using both the heat and the methane of a geopressured resource.

1992

Electrical generation began at the 25-MW geothermal plant in the Puna field of Hawaii.

1993

A 23-MW binary power plant was completed at Steamboat Springs, Nevada.

1995

In Empire, Nevada, Integrated Ingredients dedicated a food-dehydration facility capable of processing 15 million pounds of dried onions and garlic per year. A DOE low-temperature resource assessment of 10 western states identified nearly 9000 thermal wells and springs and 271 communities collocated with a geothermal resource greater than 50.

2002

Organized by GeoPowering the West, geothermal development working groups were active in five states—Nevada, Idaho, New Mexico, Oregon, and Washington. Group members represented all stakeholder organizations. The working groups began identifying barriers to geothermal development in their states and bringing together all interested parties to arrive at mutually beneficial solutions.

2003

The Utah Geothermal Working Group was formed.

Geothermal Energy as a Renewable Energy Source

Table 7.1 shows geothermal energy's ranking as a renewable energy source. The 0.353 geothermal quadrillion Btu figure is expected to steadily increase, which should be reflected in later figures when they are released.

GEOTHERMAL ENERGY: THE BASICS

As mentioned, geothermal energy processes utilize the natural heat of the Earth for beneficial purposes when the heat is collected and transported to the surface. To gain a proper understanding of geothermal energy as it is used at present, it is important to define *enthalpy*, the heat content of a substance per unit mass. Temperature alone is not sufficient to define the useful energy content of a steam/water mixture. A mass of steam at a given temperature and pressure can provide much more energy than the same mass of water under the same conditions. Enthalpy is a function of

$$\text{Pressure} + \text{Volume} + \text{Temperature} = \text{Enthalpy}$$

Geothermal practitioners usually classify geothermal resources as high enthalpy (water and steam at temperatures above about 180 to 200°C), medium enthalpy (about 100 to 180°C), and low enthalpy (<100°C). For the purpose of this text, though, it is sufficient to think of temperature and enthalpy as the same.

TABLE 7.1
U.S. Energy Consumption by Energy Source, 2008

Energy Source	Energy Consumption (quadrillion Btu)
Total	99.438
Fossil fuels (coal, natural gas, petroleum)	83.532
Electricity net imports	0.113
Nuclear electric power	8.427
Renewable	7.367
Biomass (total)	3.852
Biofuels	1.372
Waste	0.436
Wood and wood-derived fuels	2.044
Geothermal energy	**0.360**
Hydroelectric conventional	2.512
Solar thermal/PV energy	0.097
Wind energy	0.546

Source: EIA, *U.S. Energy Consumption by Energy Source*, Energy Information Administration, Washington, D.C., 2007 (www.eia.doe.gov/cneaf/solar.renewables/page/trends/table1.html).

Earth's Layers

Earth is made up of three main compositional layers: crust, mantle, and core (see Figure 7.1). The crust has variable thickness and composition; the continental crust is 25 to 70 km thick, while the oceanic crust is 7 to 10 km thick. The elements silicon, oxygen, aluminum, and ion make up the Earth's crust. Like the shell of an egg, the Earth's crust is brittle and can break. Based on seismic (earthquake) waves that pass through the Earth, we know that below the crust is the mantle, a dense, hot layer of semisolid rock approximately 2900 km thick. The mantle might be thought of as the white of a boiled egg. It contains silicon, oxygen, aluminum, and more iron, magnesium, and

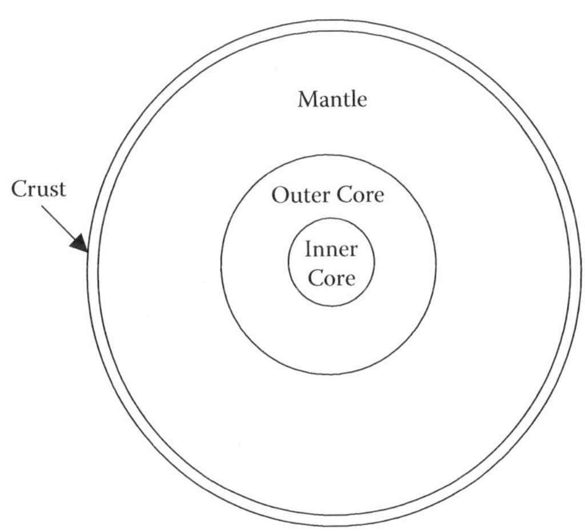

FIGURE 7.1 Earth's layers.

calcium than the crust. The mantle is hotter and more dense than the crust because temperature and pressure inside the Earth increase with depth. The temperature at the top of the mantle is 870°C; the temperature at the bottom is 2200°C. The 30-km thick transitional layer between the mantle and crust is called the Moho layer.

At the center of the Earth lies the core, which is nearly twice as dense as the mantle because its composition is metallic (comprised of iron–nickel alloy) rather than stony. Unlike the yolk of an egg, however, the Earth's core is actually made up of two distinct parts: a 2200-km-thick liquid outer core and a 1250-km-thick solid inner core. As the Earth rotates, the liquid outer core spins, creating the Earth's magnetic field. Several important points related to Earth's structure and geothermal properties include the following:

- Heat flows outward from the center as a result of radioactive decay.
- The crust insulates us from the interior heat.
- The temperature at the base of crust is about 1000°C and increases slowly deeper into the core.
- Hot spots are located 2 to 3 km from the surface.

Crustal Plates

Geologists have developed the theory of *plate tectonics* (from the Greek word for "builder"). The theory of plate tectonics deals with the formation, destruction, and large-scale motions of great segments of the Earth's surface (crust), called *plates.* This theory relies heavily on the older concept of continental drift (developed during the first half of the 20th century) and newer understanding of seafloor spreading, both of which help to explain earthquakes and volcanic eruptions, as well as the origin of fold mountain systems.

Large quantities of heat (that are economically extractable) tend to be concentrated in places where hot or even molten (magma) rock exists at relatively shallow depths in the Earth's outmost layer (the crust). Such hot zones generally are near the boundaries of the dozen or so slabs of rigid rock (or plates) that form the Earth's lithosphere. These crustal plates are composed of great slabs of rock (lithosphere) about 100 km thick and cover many thousands of square miles (they are thin in comparison to their length and width). Geologists recognize at least seven main plates:

- African Plate (covering Africa)
- Antarctic Plate (covering Australia)
- Eurasian Plate (covering Asia and Europe)
- Indo-Australian Plate (covering the Indian subcontinent, part of the Indian Ocean, and Australia)
- North American Plate (covering North America and northeast Siberia)
- Pacific Plate (covering the Pacific Ocean)
- South American Plate (covering south America)

as well as several minor plates, including:

- Arabian Plate
- Caribbean Plate
- Cocos Plate
- Juan de Fuca Plate
- Nazea Plate
- Philippine Plate
- Scotia Plate

The plates literally ride on the *asthenosphere*, which is the ductile, soft, plastic-like zone in the upper mantle. Crustal plates move in relation to one another at one of three types of plate boundaries: convergent (collision) boundaries, divergent (spreading) boundaries, and transform boundaries. These boundaries between plates are typically associated with deep-sea trenches, large faults, fold mountain ranges, and mid-oceanic ridges.

Convergent Boundaries

Convergent boundaries (or active margins) develop where two plates slide toward each other, commonly forming either a subduction zone (if one plate subducts or moves underneath the other) or a continental collision (if the two plates contain continental crust). To relieve the stress created by the colliding plates, one plate deforms and slips below the other.

Divergent Boundaries

Divergent boundaries occur where two plates slide apart from each other. Oceanic ridges, which are examples of divergent boundaries, occur where new oceanic, melted lithosphere materials well up, resulting in basaltic magmas, which intrude and erupt at the oceanic ridge, in turn creating new oceanic lithosphere and crust (new ocean floor). Along with volcanic activity, the mid-oceanic ridges are also areas of seismic activity.

Transform Boundaries

Transform boundaries do not separate or collide; rather, they slide past each other in a horizontal manner with a shearing motion. Most transform boundaries occur where oceanic ridges are offset on the sea floor. The San Andreas Fault in California is an example of a transform fault.

DID YOU KNOW?

- Plates are in constant motion, moving several centimeters a year.
- Collisions or grinding can create mountains, volcanoes, and geysers and cause earthquakes.
- Near the junction of these plates is where heat travels rapidly from the interior of the planet.

ENERGY CONVERSION

The conversion of heat to electricity is common to most power plants. This is the case whether the energy source is coal, gas, nuclear power, wind power, solar power, water power, or geothermal power. Powering the nontransportation section of our economy is important, of course, so converting any fuel source to electrical power for industrial use is a prime objective in our constant and insatiable appetite for energy. It is tempting to think that we should focus solely on the production of electricity to power not only our industries and homes but also everything else—one genie in one bottle to accomplish everything. Liquid fuels are the fuels of choice right now because they are accessible, available, and relatively inexpensive. Thus, although our ongoing research for other energy sources persists, we still do not have that absolute pressing need to come up with a liquid fuel replacement—at least not yet. Anyway, when our energy-needs focus shifts due to necessity, absolute or otherwise, geothermal energy will be available, and we need to continue our research in this important area. Energy conversion occurs routinely today, with a variety of types of energy, and geothermal conversion is nothing new.

Geothermal energy conversion refers to the power-plant technology that converts the hot geothermal fluids into electric power. Although geothermal power plants have much in common with traditional power-generating stations—turbines, generators, heat exchangers, and other standard power-generating equipment—there are important differences between geothermal and other power-generating technologies. Each geothermal site, for example, has its own unique set of characteristics and operating conditions that must be taken into account. The fluid produced from a geothermal well can be steam, brine, or a mixture or the two, and the temperature and pressure of the resource can vary substantially from site to site. The chemical composition of the resource can contain dissolved minerals, gases, and other substances that are difficult to manage. These site-specific conditions can have a profound affect on efficiency, productivity, and economic viability, so engineers must fine tune geothermal conversion technology, precisely matching plant design to the site-specific conditions.

DID YOU KNOW?

The temperature of the Earth increases, on average, by about 28°C for every kilometer (80°F for every mile) of depth below the surface for the first several kilometers down.

GEOTHERMAL POWER PLANT TECHNOLOGIES

Geothermal power plants fall into one of three conversion categories: *dry steam*, *flash*, or *binary cycle*. The type of conversion used depends on the state of the fluid (whether steam or water) and its temperature. The first geothermal power generation plant built was a dry-steam power plant; this type of plant uses the steam from the geothermal reservoir as it comes from wells and routes it directly through turbine/generator units to produce electricity. Flash steam plants, the most common type of geothermal power generation plants in operation today, pump water at temperatures greater than 360°F (182°C) and under high pressure to generation equipment at the surface. Binary cycle geothermal power generation plants differ from dry-steam and flash-steam systems in that the water or steam from the geothermal reservoir never comes in contact with the turbine/generator units.

DRY-STEAM POWER PLANTS

Location-specific, dry-steam geothermal power plants take advantage of subterranean rocks that are so hot that the water vaporizes on its way up through the production well (see Figure 7.2). This type of geothermal power plant is able to use steam directly from the ground to drive the turbine. This is the oldest type of geothermal power plant. It was first used in Italy in 1904 and is still very effective. These geothermal plants emit only excess steam and very minor amount of gases (EERE, 2010a).

FLASH-STEAM POWER PLANTS

Figure 7.3 illustrates a flash-steam geothermal power plant. Fluid is sprayed into a tank held at a much lower pressure than the fluid, causing some of the fluid to rapidly vaporize, or "flash." The vapor then drives a turbine, which drives a generator. If any liquid remains in the tank, it is returned to the groundwater pump to be forced down into the Earth again so it can be flashed again to extract more energy (EERE, 2010a).

BINARY-CYCLE POWER PLANTS

In a binary-cycle geothermal power plant (see Figure 7.4), water is pumped into the Earth and comes back up hot, just as it does in the flash-steam system. Instead of going into a flash tank, however, the hot water enters a *heat exchanger*, where most of its energy is transferred to another fluid, the binary

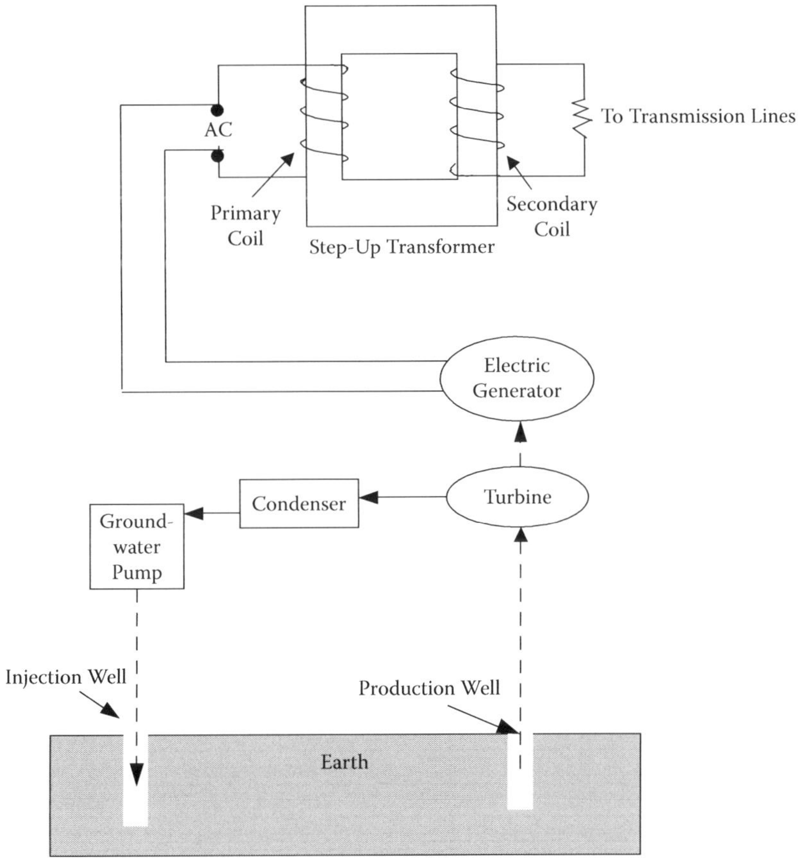

FIGURE 7.2 Simplified functional diagram of a dry-steam geothermal power plant.

liquid. This fluid can be water, but more often it is a volatile liquid resembling refrigerant that boils easily into vapor at a lower temperature than the water. The liquid-to-vapor conversion occurs in a special low-temperature boiler. Because this is a closed-loop system, virtually nothing is emitted to the atmosphere. Moderate temperature water is by far the more common geothermal resource, and most geothermal power plants in the future will be binary-cycle plants (EERE, 2010a).

ENHANCED GEOTHERMAL SYSTEMS

The great potential for dramatically expanding the use of geothermal energy can be realized by using enhanced geothermal systems (EGSs), also sometimes called *engineered geothermal systems*. Current geothermal power generation comes from hydrothermal reservoirs and is somewhat limited in geographic application to specific ideal places in the western United States. This represents the "lower hanging fruit" of geothermal energy potential (EERE, 2010b). Enhanced geothermal systems offer the ability to extend the use of geothermal resources to large areas of the western United States, as well as into new geographic areas of the entire country. More than 100,000 MWe (megawatt electrical) of economically viable capacity may be available in the continental United States, representing a 40-fold increase over current geothermal power generating capacity. This potential is about 10% of the overall U.S. electrical capacity today, and such systems represent a domestic energy source that is clean, reliable, and proven. The EGS concept is to extract heat by creating a subsurface fracture system, a reservoir, to which water can be added from injection wells, which are drilled into hot basement rock that has limited permeability and fluid content. This type of

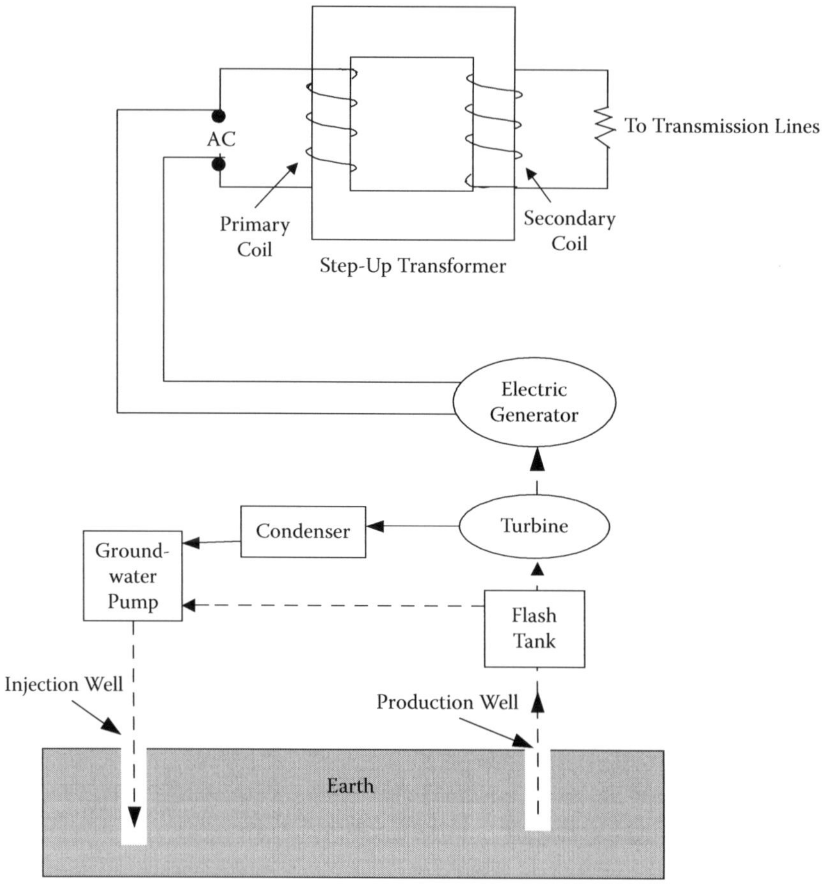

FIGURE 7.3 Simplified functional diagram of a flash-steam geothermal power plant.

geothermal resource is sometimes referred to as "hot, dry rock" and represents an enormous potential energy resource. Creating an enhanced, or engineered, geothermal system requires improving the natural permeability of rock. Rocks are permeable due to minute fractures and pore spaces between mineral grains. Water is injected at sufficient pressure to ensure fracturing; the water is heated by contact with the rock and returns to the surface through production wells, as in naturally occurring hydrothermal systems. Additional production wells are drilled to extract heat from large volumes of rock mass to meet power generation requirements. A previously unused but large energy resource is now available for clean, geothermal power generation.

GEOTHERMAL HEAT PUMPS

Geothermal heat pumps (GHPs), sometimes referred to as geoexchange, earth-coupled, ground-source, or water-source heat pumps, have been in use since the later 1940s. Geothermal heat pumps use the constant temperature of the Earth as the exchange medium instead of the outside air temperature. This allows the system to reach fairly high efficiencies (300 to 600%) on the coldest of winter nights, compared to 175 to 250% for air-source heat pumps on cool days. Geothermal heat pumps are used for space heating and cooling, as well as water heating. Their great advantage is that they work by concentrating naturally existing heat, rather than by producing heat through the combustion of fossil fuels. The system includes three principal components (see Figure 7.5):

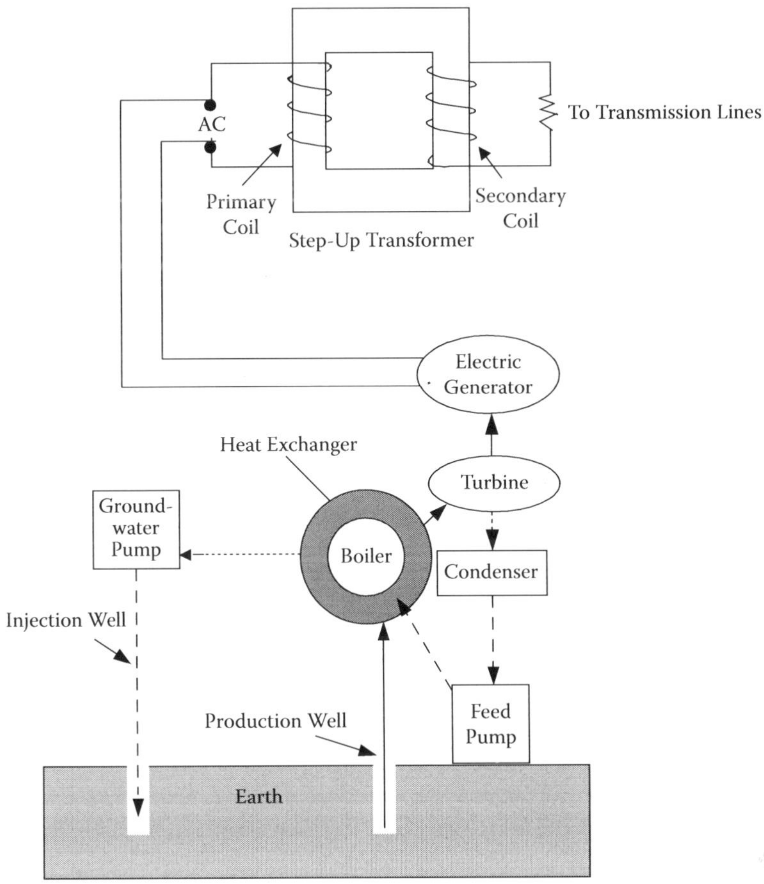

FIGURE 7.4 Simplified functional diagram of a binary-cycle geothermal power plant.

- *Geothermal earth connection subsystems*—Using the Earth as a heat source (sink), a series of pipes, commonly called a *loop*, is buried in the ground near the building to be conditioned. The loop can be buried either vertically or horizontally. It circulates a fluid (water, or a mixture of water and antifreeze) that absorbs heat from, or relinquishes heat to, the surrounding soil, depending on whether the ambient air is colder or warmer than the soil.
- *Geothermal heat pump subsystem*—For heating, a geothermal heat pump removes the heat from the fluid in the Earth connection, concentrates it, and then transfers it to the building. For cooling, the process is reversed.
- *Geothermal heat distribution subsystem*—Conventional ductwork is generally used to distribute heated or cooled air from the geothermal heat pump throughout the building.

In addition to space conditioning, geothermal heat pumps can be used to provide domestic hot water while the system is operating. Many residential systems are now equipped with *desuperheaters*, which transfer excess heat from the compressor of the geothermal heat pump compressor to the hot water tank. A desuperheater provides no hot water during the spring and fall when the geothermal heat pump system is not operating; however, because the geothermal heat pump is so much more efficient than other means of water heating, manufacturers are beginning to offer "full demand" systems that use a separate heat exchanger to meet all of a household's hot water needs. These units cost-effectively provide hot water as quickly as any competing system (EERE, 2009).

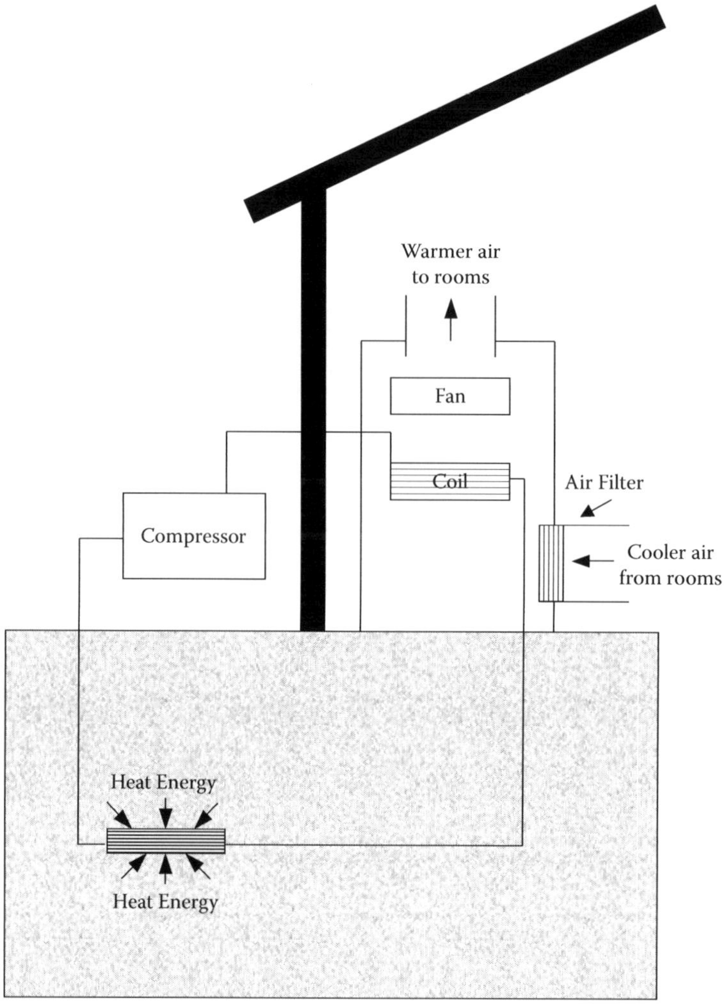

FIGURE 7.5 Geothermal heat pump system. (Adapted from Gibilisco, S., *Alternative Energy Demystified*, McGraw Hill, New York, 2007, p. 54.)

Types of Geothermal Heat Pumps

There are four basic types of ground loop geothermal heat pump systems. Three of these—horizontal, vertical, and pond/lake—are closed-loop systems. The fourth type of system is an open-loop option. Which one of these is best depends on the climate, soil conditions, available land, and local installation costs at the site. All of these approaches can be used for residential and commercial building applications (EERE, 2010c).

DID YOU KNOW?

Geothermal system life is estimated at 25 years for the inside components and 50+ years for the ground loop. Approximately 50,000 geothermal heat pumps are installed in the United States each year. Even with the large number of annual installations, it is important to keep in mind that heat pumps are relatively expensive to install new. This is especially true of the deep ground source type. It may take a long time for a new system to pay for itself.

Closed-Loop Systems

- *Horizontal*—This type of installation is generally most cost effective for residential installations, particularly for new construction where sufficient land is available. It requires trenches at least 4 feet deep. The most common layouts use either two pipes, one buried at 6 feet and the other at 4 feet, or two pipes placed side-by-side 5 feet in the ground in a 2-foot-wide trench. The Slinky™ method of looping pipe allows more pipe in a short trench, which cuts down on installation costs and makes horizontal installation possible in areas where conventional horizontal applications would not be possible.
- *Vertical*—Large commercial buildings and schools use vertical systems when the land area required for horizontal loops would be prohibitive. Vertical loops are also used where the soil is too shallow for trenching, and they minimize the disturbance to existing landscaping. For a vertical system, holes (approximately 4 inches in diameter) are drilled about 20 feet apart and 100 to 400 feet deep. Into these holes go two pipes that are connected at the bottom with a U-bend to form a loop. The vertical loops are connected with horizontal pipe (i.e., manifold), placed in trenches, and connected to the heat pump in the building.
- *Pond/lake*—If the site has an adequate water body, this may be the lowest cost option. A supply line pipe is run underground from the building to the water and coiled into circles at least 8 feet under the surface to prevent freezing. The coils should only be placed in a water source that meets minimum volume, depth, and quality criteria.

Open-Loop System

This type of system uses well or surface body water as the heat-exchange fluid that circulates directly through the GHP system. Once it has circulated through the system, the water returns to the ground through the well, a recharge well, or surface discharge. This option is obviously practical only where there is an adequate supply of relatively clean water and all local codes and regulations regarding groundwater discharge are met.

THE BOTTOM LINE ON GEOTHERMAL ENERGY

The supply of geothermal energy is vast and can be considered renewable, as long as each site is properly engineered and operated to ensure that excessive water is not pumped into the Earth in one location over too short a period of time. Geothermal energy can be and already is accessed by drilling water or steam wells in a process similar to drilling for oil. Geothermal energy is an enormous, underused heat and power resource that is clean (emits little or no greenhouse gases), reliable (average system availability of 95%), and home grown (making us less dependent on foreign oil). Geothermal resources range from shallow ground to hot water and rock several miles below the Earth's surface, and even farther down to the extremely hot molten rock called magma. Wells a mile or more deep can be drilled into underground reservoirs to tap steam and very hot water, which can be brought to the surface for use in a variety of applications. In the United States, most geothermal reservoirs are located in the western states, Alaska, and Hawaii; however, before geothermal electricity can be considered a key element of the U.S. energy infrastructure, it must become cost competitive with traditional forms of energy.

DISCUSSION AND REVIEW QUESTIONS

1. The Earth's transitional layer between the mantle and crust is called the ____________.
2. The ductile, soft, plastic-like zone in Earth's upper mantle is known as the ____________.
3. Enhanced geothermal systems are also called ____________.
4. In the United States, most geothermal reservoirs are located in the ____________.
5. Enthalpy is a function of ____________.

6. The Earth's core is nearly twice as ____________ as the mantle.
7. Convergent plates are known as ____________.
8. What type of geothermal system is able to use steam directly from the ground to drive a turbine?
9. The ____________ concept is to extract heat by creating a subsurface fracture zone.
10. What technology uses the constant temperature of the Earth as the exchange medium instead of the outside air temperature?
11. Is geothermal energy a viable alternative renewable energy source that is capable of replacing fossil fuels in the United States? Explain.

REFERENCES AND RECOMMENDED READING

Anon. (1996). Heat from the earth: geothermal heat pumps. *The Family Handyman*, 46 (9), 9–12.

Atkinson, L. and Sancetta, C. (1993). Hail and farewell. *Oceanography*, 6(34).

Bonner, J.T. (1965). *Size and Cycle.* Princeton, NJ: Princeton University.

Cataldo, S. (1999). Ground-source heat pumps dig in. *Home Energy Magazine Online*, September/October.

EERE. (2006). *A History of Geothermal Energy in the United States*. Washington, D.C.: Energy Efficiency & Renewable Energy, U.S. Department of Energy (http://www1.eere.energy.gov/geothermal/history.html).

EERE. (2008). *History of Hydropower*. Washington, D.C.: Energy Efficiency & Renewable Energy, U.S. Department of Energy (http://www1.eere.energy.gov/windandhydro/hydro_history.html).

EERE. (2009). *Geothermal Heat Pumps*. Washington, D.C.: Energy Efficiency & Renewable Energy, U.S. Department of Energy (http://www1.eere.energy.gov/geothermal/heatpumps.html).

EERE. (2010a). *Hydrothermal Power Systems*. Washington, D.C.: Energy Efficiency & Renewable Energy, U.S. Department of Energy (http://www1.eere.energy.gov/geothermal/powerplants.html).

EERE. (2010b). *How an Enhanced Geothermal System Works*. Washington, D.C.: Energy Efficiency & Renewable Energy, U.S. Department of Energy (http://www1.eere.energy.gov/geothermal/egs_animation.html).

EERE. (2010c). *Types of Geothermal Heat Pump Systems*. Washington, D.C.: Energy Efficiency & Renewable Energy, U.S. Department of Energy (http://www.energysavers.gov/your_home/space_heating_cooling/index.cfm/mytopic=12650).

Enger, E., Kormelink, J.R., Smith, B.F., and Smith, R.J. (1989). *Environmental Science: The Study of Interrelationships*. Dubuque, IA: William C. Brown Publishers.

Gibilisco, S. (2007). *Alternative Energy Demystified*. New York: McGraw Hill.

Holmes, A. (1978). *Principles of Physical Geology*, 3rd ed. New York: John Wiley & Sons.

Lund, J.W. (2007). Characteristics, development and utilization of geothermal sources. *Geo-Heat Centre Quarterly Bulletin*, 28(2), 1–9.

Lyman, J. and Fleming, R.H. (1940). Composition of seawater. *Journal of Marine Research*, 3, 134–146.

McKnight, T. (2004). *Geographica: The Complete Illustrated Atlas of the World*. New York: Barnes & Noble Books.

McLarty, L., Grabowski, P., Entingh, D., and Robertson-Tait, A. (2000). Enhanced geothermal systems R&D in the United States, in *Proceedings World Geothermal Congress 2000,* pp. 3793–3796.

Oreskes, N., Ed. (2003). *Plate Tectonics: An Insiders History of the Modern Theory of the Earth*. New York: Westview.

Rafferty, K. (2001). *An Information Survival Kit for the Prospective Geothermal Heat Pump Owner*. Portland, OR: Oregon Institute of Technology, Geo-Heat Center.

Stanley, S.M. (1999). *Earth System History*. New York: W.H. Freeman, pp. 211–222.

Sverdrup, H.U., Johnson, M.W., and Fleming, R.H. (1942). *The Oceans: Their Physics, Chemistry, and General Biology*. New York: Prentice-Hall.

Turcotte, D.L. and Schubert, G. (2002). *Geodynamics*, 2nd ed. New York: John Wiley & Sons.

USDOE. (2010). *Fossil Fuels*. Washington, D.C.: U.S. Department of Energy (http://www.energy.gov/ energysources/fossilfuels.htm).

USEPA. (1993). *Space Conditioning: The Next Frontier.* Washington, D.C.: U.S. Environmental Protection Agency.

8 Ocean Energy

One day King Canute told his followers to carry him and his throne to the seashore at low tide. "Set me down right there at the water's edge," he said. "And I will command the sea to stay away. I will order the tide not to rise." "Oh, wow!" said his followers as they obeyed the King. "This ought to be something to see!" King Canute sat. He shouted: "Sea, stay away! Tide, do not rise!" But slowly, slowly, the tide came in and the sea rose, over his feet, past his knees, up to his waist. The crowd of followers pulled back a little to keep their own feet dry. They were puzzled. Why did the sea not obey the King? The water rose higher and higher, up to the King's shoulders, over his chin. "Loyal [glub] subjects," he gurgled. "Now you see that [glub] there are some things no man, be he [glub] king or commoner, can [glub] do! Now pull me the [glub] out of here!"

—Based on Viking legend

INTRODUCTION

As King Canute reportedly discovered, the rise and fall of the seas represent a vast and relentless natural phenomenon—certainly beyond the absolute control of all earthly subjects. The ocean can produce two types of energy: *thermal energy* from the sun's heat and *mechanical energy* from the tides and waves. Generating technologies for deriving electrical power from the ocean include tidal power, wave power, ocean thermal energy conversion, ocean currents, ocean winds, and salinity gradients. Of these, the three most well-developed technologies are *tidal power*, *wave power*, and *ocean thermal energy conversion* (OTEC). Ocean thermal energy conversion is limited to tropical regions, such as Hawaii, and to a portion of the Atlantic coast. Ocean thermal energy can be used for many applications, including electricity generation, which utilizes either warm surface water or boiled seawater to turn a turbine, which activates a generator. Tidal power requires large tidal differences which, in the United States, occur only in Maine and Alaska. Wave energy has a more general application. The western coastline of the United States has the highest wave potential; in California, the greatest potential is along the state's northern coast.

It is important to distinguish tidal energy from hydropower. Recall that hydropower is derived from the hydrological climate cycle, powered by solar energy, which is usually harnessed via hydroelectric dams. In contrast, tidal energy is the result of the interaction of the gravitational pull of the moon and, to a lesser extent, the sun on the seas. Processes that use tidal energy rely on the twice-daily tides and the resultant upstream flows and downstream ebbs in estuaries and lower reaches of some rivers. The conversion of tidal and wave energy into electricity usually involves mechanical devices. A dam is typically used to convert tidal energy into electricity by forcing the water through turbines that activate a generator. The mechanical power created from these systems either directly activates a generator or transfers to a working fluid, water, or air, which then drives a turbine or generator. Before discussing the thermal and mechanical energy potential of the ocean, we will provide a basic understanding of oceans and especially their margins, where most, if not all, ocean energy is harnessed using current technology. The following section provides a foundation for better understanding the ocean energy concepts presented later in this chapter.

OCEANS AND THEIR MARGINS*

Oceans are a principal component of the hydrosphere and the storehouse of Earth's water. Oceans cover about 71% of Earth's surface. The average depth of the oceans is about 3800 m, but the greatest ocean depth of 11,036 m was recorded in the Mariana Trench. The volume of all of the oceans is about 1.35 billion cubic kilometers, representing 96.5% of Earth's total water supply; however, the volume fluctuates with the growth and melting of glacial ice. The composition of ocean water has remained constant throughout geologic time, with the major constituents dissolving in ocean water from rivers and precipitation, as well as from weathering and degassing of the mantle by volcanic activity. Seawater is approximately 3.5% salt and 96.5% water, by weight. The dissolved salts include chloride (55.07%), sodium (30.62%), sulfate (7.72%), magnesium (3.68%), calcium (1.17%), potassium (1.10%), bicarbonate (0.40%), bromine (0.19%), and strontium (0.02%). The most significant factor related to ocean water that everyone is familiar with is the salinity of the water—how salty it is. *Salinity*, a measure of the amount of dissolved ions in the water, ranges between 33 and 37 parts per thousand. The concentration is the amount (by weight) of salt in water, usually expressed as parts per million (ppm); 1 ppm is equivalent to a shot glass of water taken from an Olympic-sized swimming pool. Water is saline if it has a concentration of more than 1000 ppm of dissolved salts; ocean water contains about 35,000 ppm of salt (USGS, 2007). Chemical precipitation, absorption onto clay minerals, and plants and animals prevent seawater from having higher salinity concentrations; however, salinity does vary in the oceans because surface water evaporates, rain and stream water is added, and ice forms or thaws. Another important property of seawater is temperature. The temperature of surface seawater varies with latitude, from near 0°C close to the poles to 29°C near the equator. Some isolated areas can have temperatures as high as 37°C. Temperature decreases with ocean depth.

THE OCEAN FLOOR

The bottoms of the ocean basins (ocean floors) are marked by mountain ranges, plateaus, and other relief features similar to (although not as rugged as) those on the land. As shown in Figure 8.1, the floor of the ocean can be divided into four divisions: continental shelf, continental slope, continental rise, and deep-sea floor or abyssal plain:

- The *continental shelf* is the flooded, nearly flat true margins of the continents. Varying in width to about 40 miles and a depth of approximately 650 feet, continental shelves slope gently outward from the shores of the continents. Continental shelves occupy approximately 7.5% of the ocean floor.
- The *continental slope* is a relatively steep slope descending from the continental shelf; it descends rather abruptly to the deeper parts of the ocean. These slopes represent about 8.5% of the ocean floor.
- The *continental rise* is a broad gentle slope below the continental slope containing sediment that has accumulated along parts of the continental slope.
- The *abyssal plain* is a sediment-covered deep-sea plain about 12,000 to 18,000 feet below sea level. This plain accounts for about 42% of the ocean floor.

The deep ocean floor does not consist exclusively of the abyssal plain. Some areas have considerable relief, which can be described as follows:

- *Seamounts* are isolated mountain-shaped elevations more than 3000 feet high.
- *Mid-oceanic ridges* are submarine mountains, extending more than 37,000 miles through the oceans, generally 10,000 feet above the abyssal plain.

* Material in this section is adapted from Spellman, F.R., *Geology for Nongeologists*, Government Institutes Press, Lanham, MD, 2009.

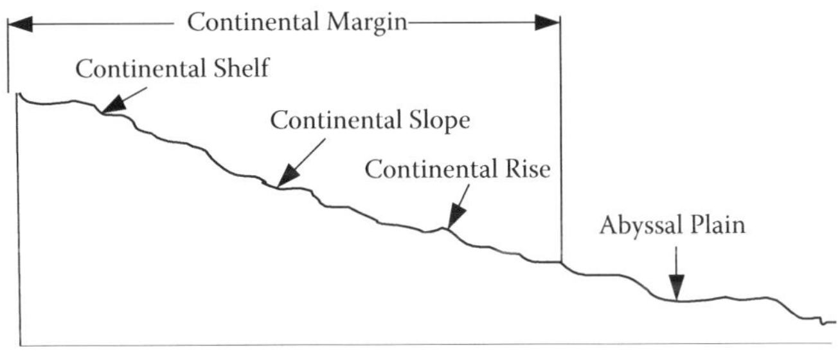

FIGURE 8.1 Cross section of ocean floor showing major elements of topography.

- *Trenches* are deep, steep-sided troughs in an abyssal plain.
- *Guyots* are flat-topped volcanic mountains that rise from the ocean bottom and usually are covered by 3000 to 6000 feet of water.

OCEAN TIDES, CURRENTS, AND WAVES

The Earth's coastline, including sea cliffs, shores, and beaches, is primarily sculpted by water—tides, currents, and waves. The ceaseless, restless motion of the sea is an extremely effective geologic agent. Seawater in motion erodes cliffs, transports debris along shores, and dumps it on beaches. Most coasts retreat or advance over time.

Tides

Tides are the periodic rise and fall of the sea, once every 12 hours and 26 minutes. The gravitational pull of the moon and, to a lesser extent, the sun causes water on Earth to bulge toward them. It is interesting to note that a similar bulge occurs on the opposite side of the Earth due to inertial forces (further explanation is beyond the scope of this text). The effect of the tides is not too noticeable in the open sea, as the difference between high and low tide amounts to only about 2 feet. The tidal range may be considerably greater near shore, however. It may range from less than 2 feet to as much as 50 feet. The tidal range will vary according to the phase of the moon and the distance of the moon from the Earth. The type of shoreline and the physical configuration of the ocean floor will also affect the tidal range.

Currents

Ocean currents are localized movements of masses of seawater. They are the result of drift of the upper 50 to 100 m of the ocean due to drag by wind. Thus, surface ocean currents generally follow the same patterns as atmospheric circulation, with the exception that atmospheric currents continue over the land surface while ocean currents are deflected by the land. Along with wind action, currents may also be caused by tides, variations in the salinity of the water, rotation of the Earth, and concentrations of turbid or muddy water. In addition to this surface circulation, seawater also circulates vertically due to differences in temperature and salinity that affect water density.

Waves

Waves are produced by the friction of wind on open water. Wave height and power depend on wind strength and fetch, the amount of unobstructed ocean over which the wind has blown. In a wave, water travels in loops (essentially up-and-down movements), with the diameter of the loops decreasing with depth. The diameter of loops at the surface is equal to the wave height. Breakers are formed when waves come into shallow water near the shore. The lower part of the wave is retarded by the ocean bottom, and the top, having greater momentum, is hurled forward, causing the wave to break. These breaking waves may do great damage to coastal property as they race across coastal lowlands driven by high winds.

COASTAL EROSION, TRANSPORTATION, AND DEPOSITION

The geologic work of the sea consists of erosion, transportation, and deposition. The sea accomplishes its work of coastal landform sculpting largely by means of waves and wave-produced currents; their effect on the seacoast may be quite pronounced. The coast and accompanying coastal deposits and landform development represent a balance between wave energy and sediment supply.

Wave Erosion

Waves attack shorelines and erode by a combination of several processes. The resistance of the rocks composing the shoreline and the intensity of wave action to which they are subjected are factors that determine how rapidly the shore will erode. Wave erosion works chiefly by hydraulic action, corrosion, and attrition. As waves strike a sea cliff, *hydraulic action* crams air into rock crevices, putting tremendous pressure on the surrounding rock; as waves retreat, the explosively expanding air enlarges cracks and breaks off chunks of rock (known as *scree*). Chunks hurled by waves against the cliff wear off more scree in a process called *corrasion*. In a similar process known as *attrition*, rocks collide with each other in the waves and become smaller and smoother until they are reduced to pebbles and sand grains (Lambert, 2007). Several coastal features are formed by marine erosion due to various combinations of wave action, rock types, and rock beds:

- *Sea cliffs or wave-cut platforms* are formed by wave erosion of underlying rock followed by caving-in of the overhanging rocks. As waves eat farther back inland, they leave a wave-cut beach or platform. Such cliffs are essentially vertical and are common at certain localities along the New England and Pacific coasts of North American.
- *Wave-cut benches* are the result of wave action not having enough time to lower the coastline to sea level. Because of the resistance to erosion, a relatively flat wave-cut bench develops. If subsequent uplift of the wave-cut bench occurs, it may be preserved above sea level as a marine terrace.
- *Headlands* are finger-like projections of resistant rock extending out into the water. Indentations between headlands are termed *coves*.
- *Sea caves, sea arches, and stacks* are formed by continued wave action on a sea cliff. Wave action hollows out cavities or caves in the sea cliffs. Eventually, waves may cut completely through a headland to form a sea arch; if the roof of the arch collapses, the rock left separated from the headland is called a *stack*.

Marine Transportation

Waves and currents are important transporting agents. Rip currents and undertow carry rock particles back to the sea, and long-shore currents will pick up sediments (some of it in solution), moving them out from shore into deeper water. Materials carried in solution or suspension may drift seaward for great distances and eventually be deposited far from shore. During the transportation process sediments undergo additional erosion, becoming reduced in size.

Marine Deposition

Marine deposition takes place whenever the velocity of currents and waves is reduced. Some rocks are thrown up on the shore by wave action. Most of the sediments thus deposited consist of rock fragments derived from the mechanical weathering of the continents, and they differ considerably from terrestrial or continental deposits. Due to input of sediments from rivers, deltas may form; due to beach drift such features as spits and hooks, bay barriers, and tombolos may form. Depositional features along coasts are discussed below.

- *Beaches* are transitory coastal deposits of debris that lie above the low-tide limit in the shore zone.
- *Barrier islands* are long, narrow accumulations of sand lying parallel to the shore and separated from the shore by a shallow lagoon.
- *Spits and hooks* are elongated, narrow embankments of sand and pebble extending out into the water but attached by one end to the land.
- *Tombolos* are bars of sand or gravel connecting an island with the mainland or another island.
- *Wave-built terraces* are structures built up from sediments deposited in deep water beyond a wave-cut terrace.

WAVE ENERGY

Waves are caused by the wind blowing over the surface of the ocean. In many areas of the world, the wind blows with enough consistency and force to provide continuous waves. Wave energy does not have the tremendous power of tidal fluctuations, but the regular pounding of the waves should not be underestimated because there is tremendous energy in the ocean waves. The total power of waves breaking on the world's coastlines is estimated at up to 2 terawatts. In optimal wave areas (such as the Pacific Northwest), 40 to 70 kilowatts of electricity could be produced per meter of coastline (EERE, 2010a). Because the wind is originally derived from the sun, we can consider the energy in ocean waves to be a stored, moderately high-density form of solar energy. According to certain estimates, wave technologies could feasibly fulfill 10% of the global electricity supply if fully developed (WEC, 2007). The west coasts of the United States and Europe and the coasts of Japan and New Zealand are good sites for harnessing wave energy.

Wave Energy: Facts, Parameters, and Equations

Three main processes create waves: (1) air flowing over the sea exerts a tangential stress on the water surface, resulting in the formation and growth of waves; (2) turbulent air flow close to the water surface creates rapidly varying shear stresses and pressure fluctuations (when these oscillations are in phase with existing waves, further wave development occurs); and (3) when waves have reached a certain size, the wind can exert a stronger force on the upwind face of the wave, resulting in additional wave growth. Waves located within or close to the areas where they are generated are called *storm waves*; waves known as *swell waves* can develop at great distances from their point of origin. The distance over which wind energy is transferred into the ocean to form waves is called the *fetch*. *Sea state* is the general condition of the free surface of a large body of water, with respect to wind waves and swell, at a certain location and moment (see Table 8.1).

The shape of a typical wave is described as *sinusoidal* (i.e., it has the form of a mathematical sine function) (see Figure 8.2). The difference in height between the peaks and troughs is height H, and the distance between successive peaks (or troughs) of the wave is wavelength λ. The time in seconds required for successive peaks (or troughs) to pass a given fixed point is period T. The *frequency* (ν) of the wave describes the number of peak-to-peak (or trough-to-trough) oscillations of the wave surface per second, as seen by a fixed observer, and is the reciprocal of the period; that is, $\nu = 1/T$.

If a wave is traveling at velocity v past a given fixed point, it will travel a distance equal to its wavelength (λ) in a time equal to the wave period T (i.e., $v = \lambda/T$). The power (P) of an idealized ocean wave is approximately equal to the square of height H multiplied by wave period T. The exact expression is

TABLE 8.1
World Meteorological Organization (WMO) Sea State Codes

WMO Sea State Code	Wave Height (meters)	Characteristics
0	0	Calm (glassy)
1	0–0.1	Calm (rippled)
2	0.1–0.5	Smooth (wavelets)
3	0.5–1.25	Slight
4	1.25–2.5	Moderate
5	2.5–4	Rough
6	4–6	Very rough
7	6–9	High
8	9–14	Very high
9	Over 14	Phenomenal

$$P = \frac{g^2 H T^2}{32}$$

where P is in units of watts per meter, and g is the acceleration due to gravity (9.81 m/s^2) (Phillips, 1977).

Wave Height

Wave height increases with increases in

- Wind speed
- Time duration of the wind blowing
- Fetch
- Water depth

Deep Water Waves

For deep water waves, the velocity of a long ocean wave can be shown to be proportional to the period (if the depth of water is greater than about half of wavelength λ) as follows:

$$v = \frac{gT}{2\pi}$$

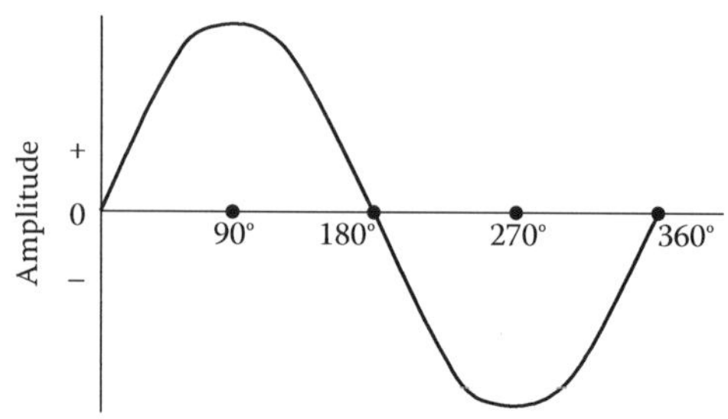

FIGURE 8.2 Sinusoidal wave showing wavelength and amplitude.

The velocity in meters per second is approximately 1.5 times the wave period in seconds. It is interesting to note that in the deep ocean long waves travel faster than shorter waves. Moreover, if the above relationships hold, we can find the deep water wavelength (λ) for any given wave period:

$$\lambda = \frac{gT^2}{2^\pi}$$

Intermediate Depth Waves

As the water becomes shallower, the properties of the waves become increasingly dominated by water depth. When waves reach shallow water, their properties are completely governed by the water depth, but in intermediate depths (i.e., between $d = \lambda/2\pi$ and $d = \lambda/4\pi$), the properties of the waves will be influenced by both water depth d and wave period T (Phillips, 1977).

Shallow Water Waves

As waves approach the shore, the seabed begins to have an effect on their speed, and it can be shown that if the water depth (d) is less than a quarter of the wavelength, the velocity is given by

$$v = \sqrt{gd}$$

Group Velocity

As waves propagate, their energy is transported. The energy transport velocity is the group velocity. The wave energy flux, through a vertical plane of unit width perpendicular to the wave propagation direction, is equal to

$$P = E \times c_g$$

where c_g is the group velocity (m/s).

Wave Energy Conversion Technology

In the early 1970s, the harnessing of wave power focused on using floating devices such as Cockerell rafts (a wave power hydraulic device), Salter's Duck (curved cam-like device that can capture 90% of waves for energy conversion), rectifiers (to convert AC to DC electricity), and the clam (a floating, rigid, doughnut-shaped device that converts wave energy to electrical energy). Wave energy converters can be classified in terms of their location: fixed to the seabed, generally in shallow water; floating offshore in deep water; or tethered in intermediate depths. These floating devices are not cost effective, and significant mooring problems remain to be solved, so current practice is to move closer to shore, sacrificing some energy. Fixed devices do have several advantages, including (Tovey, 2005):

- Easier maintenance
- Easier to land on device
- No mooring problems
- Easier power transmission
- Enhanced productivity
- Better design life

Key Point: Hydro (water) kinetic energy is the energy possessed by a body of water due to its motion (kinetic energy = $1.2 \times m \times v^2$). Hydrostatic (at rest) energy is the energy possessed by a body of water due to its position or location at an elevation or height above a reference or datum (potential energy = $m \times g \times h$).

Wave energy devices can be classified by means of their reaction system, but it is often more instructive to discuss how they interact with the wave field. In this context, each moving body may be labeled as either a displacer or a reactor:

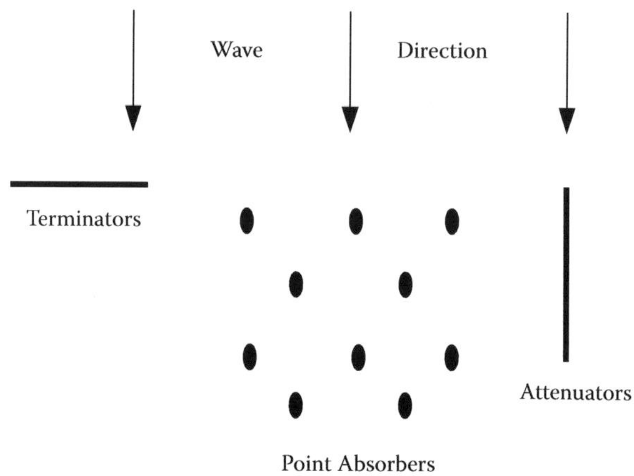

FIGURE 8.3 Types of wave energy converters. (Adapted from Tovey, N.K., *ENV-2E02 Energy Resources 2004–2005 Lecture*, University of East Anglia, Norwich, U.K., 2005; http://www2.env.uea.ac.uk/gmmc/energy/env2e02/env2e02.htm.)

- A *displacer* is a body moved by the waves. It might be a buoyant vessel or a mass of water. If buoyant, the displacer may pierce the surface of the waves or be submerged.
- A *reactor* is a body that provides reaction to the displacer. It could a body fixed to the seabed, or the seabed itself. It could also be another structure or mass that is not fixed but moves in such a way that reaction forces are created (e.g., by moving by a different amount or at different times). A degree of control over the forces acting on each body or acting between the bodies (particularly stiffness and damping characteristics) is often required to optimize the amount of energy captured.

In some designs, the reactor is actually inside the displacer, while in others it is an external body. Internal reactors are not subject to wave forces, but external reactors may experience loads that cause them to move in ways similar to a displacer; thus, some devices do not have dedicated reactors at all but rather a system of displacers whose relative motion creates a reaction system (INL, 2005). The three types of well-known wave energy conversion devices are point absorbers, terminators, and attenuators (see Figure 8.3):

- A *point absorber* is a floating structure that absorbs energy in all directions by virtue of its movements at or near the water surface (see Figure 8.4). It may be designed so as to resonate—that is, move with larger amplitudes than the waves themselves. This feature is useful to maximize the amount of power that is available for capture. The power take-off system may take a number of forms, depending on the figuration of displacers/reactors.
- A *terminator* is also a floating structure that moves at or near the water surface, but a terminator absorbs energy in only a single direction (see Figure 8.5). The device extends in the direction normal to the predominant wave direction, so that as waves arrive the device restrains them. Again, resonance may be employed and the power take-off system may take a variety of forms.

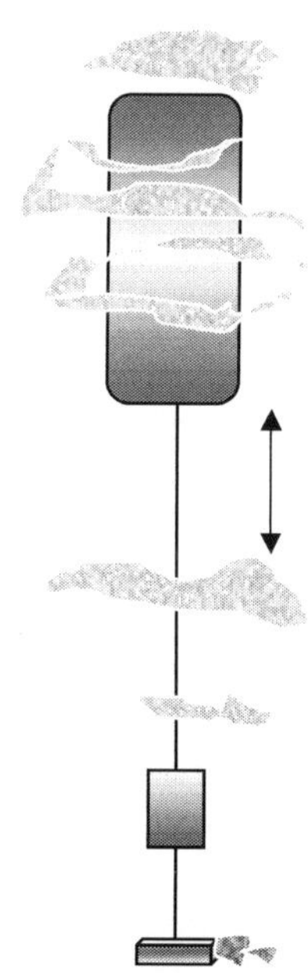

FIGURE 8.4 Point absorber.

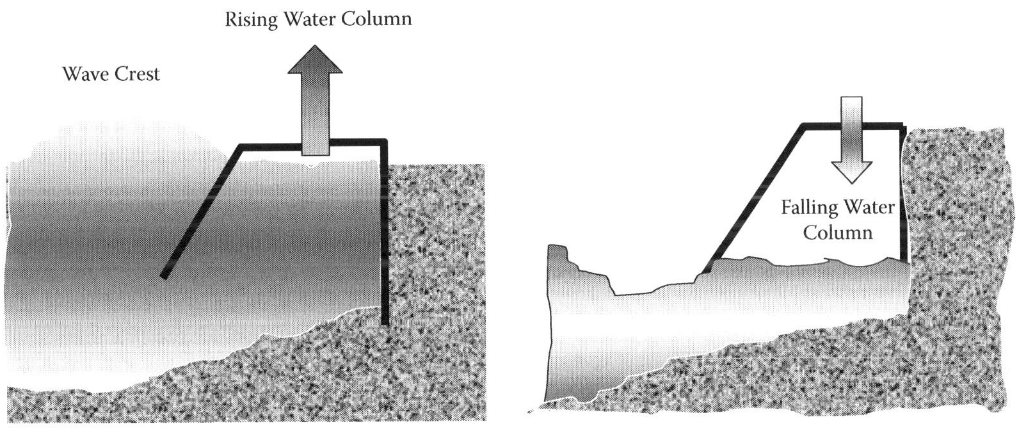

FIGURE 8.5 Terminator.

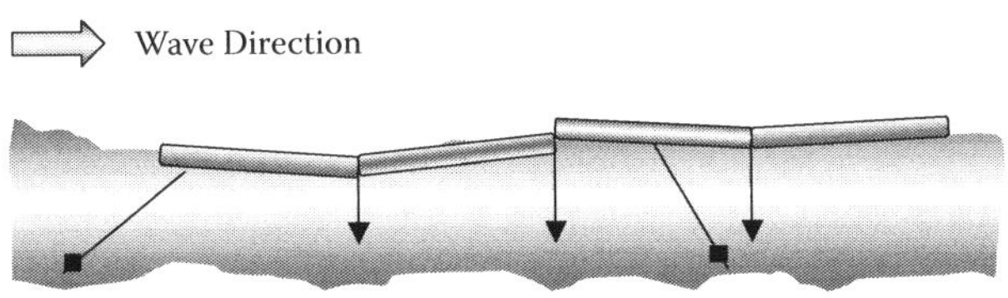

FIGURE 8.6 Attenuator.

- An *attenuator* is a long floating structure like the terminator, but it is orientated parallel to the waves rather than normal to them (see Figure 8.6). It rides the waves like a ship, and movements of the device at its bow and along its length can be restrained to extract energy. A theoretical advantage of the attenuator over the terminator is that its area normal to the waves is small so the forces it experiences are much lower.

TIDAL ENERGY

The tides rise and fall in eternal cycles. Tides are changes in the level of the oceans caused by the gravitational pull of the moon and sun and the rotation of the Earth. The relative motions of these cause several different tidal cycles, including a semidiurnal cycle (with a period of 12 hours and 25 minutes); a semimonthly cycle (spring or neap tides corresponding with the position of the moon, with the highest tides occurring in March and September); a semiannual cycle (with a period of about 178 days which is associated with the moon's orbit); and longer term cycles (e.g., a 19-year moon cycle). Nearshore water levels can vary up to 40 feet, depending on the season and local factors. Only about 20 locations have good inlets and a large enough tidal range—about 10 feet—to produce energy economically (USDOI, 2010). The tide ranges have been classified as follows (Masselink and Short, 1993):

DID YOU KNOW?

Spring tides have a range about twice that of neap tides; the other cycles can cause further variations of up to 15%. The tidal range is amplified in estuaries, and in some situations the shape of the estuary is such that near resonance occurs.

- *Micromareal*, when the tidal range is less than 2 meters
- *Mesomareal*, when the tidal range is between 2 meters and 4 meters
- *Macromareal*, when the tidal range is higher than 4 meters

Tidal Energy Technologies

Some of the oldest ocean energy technologies use tidal power. Tidal power is more predictable than solar power and wind energy. All coastal areas consistently experience two high and two low tides over a period of slightly greater than 24 hours. For those tidal differences to be harnessed into electricity, the difference between high and low tide must be at least 5 meters, or more than 16 feet. There only about 40 sites on the Earth with tidal ranges of this magnitude. Currently, there are no tidal power plants in the United States, although conditions are good for tidal power generation in both the Pacific Northwest and the Atlantic Northeast regions of the country. Tidal energy technologies include the following:

- *Tidal barrages*—A barrage is a simple generation system for tidal plants that involves installing a dam, or barrage, across an inlet. Sluice gates (gates commonly used to control water levels and flow rates) on the barrage allow the tidal basin to fill on the incoming high tides and to empty through the turbine system on the outgoing tide, also known as the ebb tide. Two-way systems generate electricity on both the incoming and outgoing tides. A potential disadvantage of a barrage tidal power system is the effect a tidal station can have on plants and animals in estuaries. Tidal barrages can change the tidal level in the basin and increase the amount of matter in suspension in the water (turbidity). They can also affect navigation and recreation.
- *Tidal fences*—These look like giant turnstiles. A tidal fence has vertical-axis turbines mounted in a fence. All the water that passes is forced through the turbines. Some of these currents run at 5 to 8 knots (5.6 to 9 miles per hour) and generate as much energy as winds of much higher velocity. Tidal fences can be used in areas such as channels between two land masses. They are less expensive to install than tidal barrages and have less impact on the environment, although they can disrupt the movement of large marine animals.
- *Tidal turbines*—These are basically wind turbines in the water that can be located anywhere with strong tidal flow; they function best where coastal currents run at between 3.6 and 4.9 knots (4 to 5.5 mph). Because water is about 800 times more dense than air, tidal turbines have to be much sturdier than wind turbines. Tidal turbines are heaver and more expensive to build but capture more energy.

OCEAN THERMAL ENERGY CONVERSION

The most plentiful renewable energy source in our planet by far is solar radiation: 170,000 TW ($170{,}000 \times 10^{12}$ W) fall on Earth. Because of its dilute and erratic nature, however, it is difficult to harness. To capture this energy, we must employ the use of large collecting areas and large storage capacities, requirements satisfied on Earth by only the tropical oceans. We are all taught at an early age that water covers about 71% (or two-thirds) of Earth's surface. With regard to the vast oceans covering the majority of Earth, it is fitting that Ambrose Bierce (1842–1914) referred to them as "a body of water occupying about two-thirds of the world made for man who has no gills." True, we have no gills, so those who look out upon those vast bodies of water that cover the surface might ask: "What is their purpose?" This is a good question with several possible answers. With regard to renewable energy, we could look out upon those vast seas and wonder how we might use this massive storehouse of energy for our own needs; it is so vast and deep that it absorbs much of the heat and light that come from the sun. One thing seems certain: The secret to our origin, past, present, and future lies within those massive wet confines we call oceans.

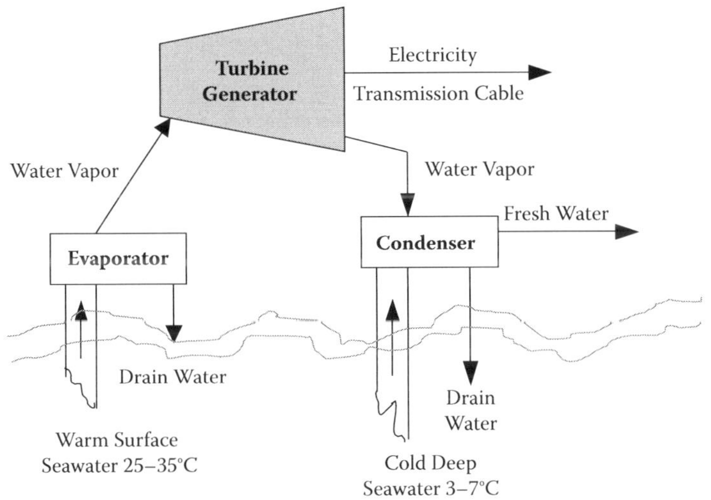

FIGURE 8.7 Schematic of ocean thermal energy conversion system.

Ocean Energy Conversion Process

The ocean is essentially a gigantic solar collector. The energy from the sun heats the surface water of the ocean. In tropical regions, the surface water can be 40 or more degrees warmer than the deep water. This temperature difference can be used to produce electricity. Ocean thermal energy conversion (OTEC) has the potential to produce more energy than tidal, wave, and wind energy combined. The OTEC systems can be open or closed. In a closed system, an evaporator turns warm surface water into stream under pressure (see Figure 8.7 and Figure 8.8). This steam spins a turbine generator to produce electricity. Water pumps bring cold deep water through pipes to a condenser on the surface. The cold water condenses the steam, and the closed cycle begins again. In an open system,

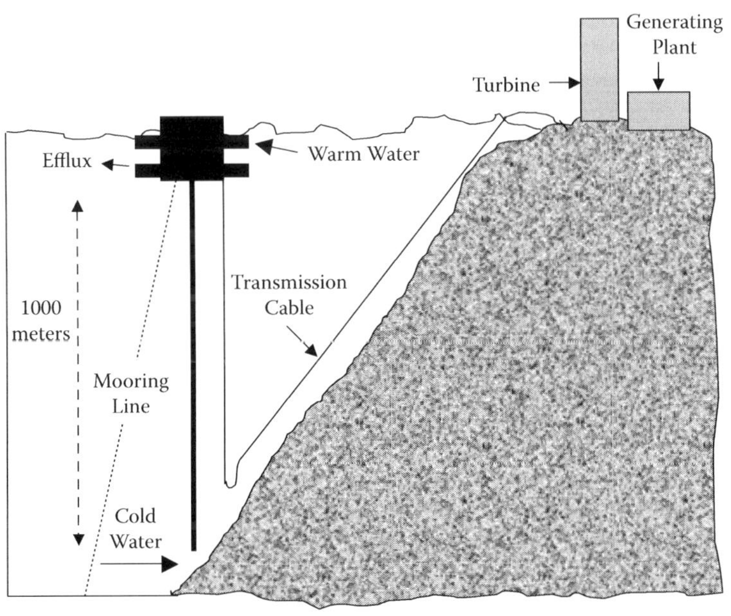

FIGURE 8.8 Ocean thermal energy conversion (OTEC) floating platform.

the steam is turned into fresh water, and new surface water is added to the system. A transmission cable carries the electricity to the shore. The OTEC systems must have a temperature difference of about 25°C to operate. This limits the use of OTEC to tropical regions where the surface waters are very warm and there is deep cold water. Hawaii, with its tropical climate, has experimented with OTEC systems since the 1970s. Because of the many challenges to their widespread use, no large or major OTEC systems are in operation today, but several experimental OTEC plants have been built. Pumping the water is a giant engineering challenge. Because of this, OTEC systems are not very energy efficient. It will probably be 10 to 20 years before the technology is available to produce and transmit electricity economically from OTEC systems.

Types of OTEC Technologies

The types of OTEC systems include the following:

- *Closed-cycle* systems use fluid with a low-boiling point, such as ammonia, to rotate a turbine to generate electricity. Warm surface seawater is pumped through a heat exchanger where the lower boiling point fluid is vaporized. The expanding vapor turns the turbogenerator. Cold, deep seawater pumped through a second heat exchanger condenses the vapor back into a liquid, which is then recycled through the system.
- *Open-cycle* systems use the warm surface water of tropical oceans to make electricity. When warm seawater is placed in a low-pressure container, it boils. The expanding steam drives a low-pressure turbine attached to an electrical generator. The steam, which has left its salt behind in the lower pressure container, is almost pure fresh water. It is condensed back into a liquid by exposure to cold temperatures from deep ocean water.
- *Hybrid* systems combine the features of both the closed-cycle and open-cycle systems. In a hybrid system, warm seawater enters a vacuum chamber where it is flash-evaporated into steam, similar to the open-cycle evaporation process. The steam vaporizes a lower boiling point fluid (in a closed-cycle loop) that drives a turbine to produce electricity.

THE BOTTOM LINE ON OCEAN ENERGY

The three forms of ocean energy—tidal, wave, and OTEC—are all renewable. This is a significant advantage over fossil fuels and other energy forms that pollute the environment. Although all three forms of ocean energy show promise for future development, the OTEC systems appear to be the most beneficial currently. OTEC systems offer both economic and noneconomic benefits. Economic benefits include the following:

- Help to produce fuels such as hydrogen, ammonia, and methanol
- Provide moderate-temperature refrigeration
- Produce baseload electrical energy
- Produce desalinated water for industrial, agricultural, and residential uses
- Serve as a resource for on-shore and near-shore mariculture operations
- Provide air conditioning to buildings
- Have significant potential to provide clean, cost-effective electricity for the future

Noneconomic benefits include:

- Enhance energy independence and energy security
- Promote competitiveness and international trade
- Promote international sociopolitical stability
- Have potential to mitigate greenhouse gas emissions resulting from burning fossil fuels

DISCUSSION AND REVIEW QUESTIONS

1. What are the two types of energy produced by oceans?
2. The composition of ocean water has ____________ throughout geologic time.
3. As waves strike sea cliffs, ____________ crams air into rock crevices.
4. What do we call bars of sand or gravel that connect an island with the mainland or other island?
5. Point absorbers may be designed to ____________ to maximize the amount of power that is available for capture.
6. An ____________ is a device that is oriented parallel to the waves and ride the waves like a ship to extract energy.
7. A ____________ tide is one where the tidal range is between 2 and 4 meters.
8. What uses a vertical axis turbine mounted in a fence?
9. The ocean is essentially a gigantic ____________ collector.
10. Is ocean energy a viable replacement alternative for fossil fuels? Explain.

REFERENCES AND RECOMMENDED READING

Berinstein, P. (2001). *Alternative Energy: Facts, Statistics, and Issues*. Westport, CT: Oryx Press.

Butti, K. and Perlin, J. (1980). *A Golden Thread: 2500 Years of Solar Architecture and Technology*. New York: Van Nostrand Reinhold.

EERE. (2010a). *Ocean Wave Power*. Washington, D.C.: Energy Efficiency & Renewable Energy, U.S. Department of Energy (http://www.energysavers.gov/renewable_energy/ocean/index.cfm/mytopic=50009).

EERE. (2010b). *Ocean Thermal Energy Conversion*. Washington, D.C.: Energy Efficiency & Renewable Energy, U.S. Department of Energy (http://www.energysavers.gov/renewable_energy/ocean/index.cfm/mytopic=50010).

Flavin, C. (1990). *Beyond the Petroleum Age: Designing a Solar Economy*. Washington, D.C.: Worldwatch Institute.

Garrett, C. and Cummins, P. (2004). Generating power from tidal currents. *Journal of Waterway, Port, Coastal and Ocean Engineering*, 130, 114–118.

Goda, Y. (2000). *Random Seas and Designs of Maritime Structures*. Singapore: World Scientific.

Gross, G.M. (1995). *Oceanography: A View of the Earth*, 7th ed. Englewood Cliffs, NJ: Prentice Hall.

INL. (2005). *Hydropower: Hydrokinetic & Wave Technologies*. Idaho Falls: Idaho National Laboratory (http://hydropower.inl.gov/hydrokinetic_wave/).

Lambert D. (2007). *The Field Guide to Geology*. New York: Checkmark Books.

Loch, S. and Norman, R.S. (1975). Osmotic power plants. *Science*, 189 654–655.

Masselink, G. and Short, A.D. (1993). The effect of tidal range on beach morphodynamics: a conceptual beach model. *Journal of Coastal Research*, 9(3), 785–800.

Phillips, O.M. (1977). *The Dynamics of the Upper Ocean*, 2nd ed. Cambridge, U.K.: Cambridge University Press.

Pinet, P.R. (1996). *Invitation to Oceanography*. St. Paul, MN: West Publishing.

Shaw, R. (1982). *Wave Energy: A Design Challenge*. West Sussex, U.K.: Ellis Horwood Ltd.

Sorensen, B. (2000). *Renewable Energy: Its Physics, Engineering, Use, Environmental Impact, Economy and Planning Aspects*. San Diego, CA: Academic Press.

Tovey, N.K. (2005). *ENV-2E02 Energy Resources 2004–2005 Lecture*. Norwich, U.K.: University of East Anglia (http://www2.env.uea.ac.uk/gmmc/energy/env2e02/env2e02.htm).

USDOI. (2010). *Ocean Energy*. Washington, D.C.: Minerals Management Service, U.S. Department of Interior (www.boemre.gov/mmsKids/PDFs/OceanEnergyMMS.pdf).

USGS. (2007). *The Water Cycle: Water Storage in Oceans*. Washington, D.C.: U.S. Geological Survey (http://ga.water.usgs.gov/edu/watercycleoceans.html).

WEC. (2010). *Survey of Energy Resources 2010*. London: World Energy Council (http://www.worldenergy.org/publications/3040.asp).

9 Fuel Cells

I believe fuel cell vehicles will finally end the hundred-year reign of the internal combustion engine as the dominant source of power for personal transportation. It's going to be a winning situation all the way around—consumers will get an efficient power source, communities will get zero emissions, and automakers will get another major business opportunity—a growth opportunity.

—William C. Ford, Jr., Ford chairman, International Auto Show, January 2000

Renewable energy innovation is to our energy future what sunshine is to plant growth.

—Frank R. Spellman

FUEL CELLS: THE BASICS

Depending on your education level or social, cultural, or economic background, the term *cell* may conjure up images as diverse in variety as the colors, sizes, and shapes of lightning bolts (which, by the way, are a huge source of renewable energy). Some may think of plant cells, animal cells, cell structure, cell biology, cell diagrams, cell membranes, human memory, cell theory, cell walls, cell parts, cell functions, honeycomb cells, prison cells, electrolytic cells (for producing electrolysis), aeronautic gas cells (contained in a balloon), ecclesiastical cells, or cell phones.

You may have noticed that nowhere in this particular list are terms related to the subject of this book: renewable energy. Some would argue that any kind of biological cell mechanism and cell phones are types of energy devices, producers or consumers that can be renewed unless destroyed. It is not our intention to argue this point either way; instead, it is our point that an important type of cell was not included in the above list, but it should be. We are referring to the electric cell, electrochemical cell, galvanic cell, or voltaic cell—often referred to as a *battery*. This type of cell, as discussed earlier in Chapter 2, is a device that generates electrical energy from chemical energy, usually consisting of two different conducting substances placed in an electrolyte. (Note that we could also include the solar cell in this discussion, but we discussed these earlier in the text.) The bottom line on this type of electric cell is that it is similar to other unmentioned cells: *fuel cells*.

Before moving on to our basic discussion of fuel cells and their associated terminology and applications, it is important to point out that, although we do not consider fuel cells as frequently as we do other types of cells (e.g., cell phones), we predict that the day is coming when we will refer to our fuel cells just as commonly as we mention our cell phones—for it will be the fuel cell that will power our lives, just as the cell phone powers our communication.

OPEN CELLS VS. CLOSED CELLS

Batteries store electrical energy chemically (contrary to popular belief, they do not make electrical energy), so they are considered to be thermodynamically *open* systems. By contrast, fuel cells differ from conventional electrochemical cell batteries in that they consume reactant from an external source, which must be replenished; thus, they are thermodynamically *closed* systems (Anon., 2007). A fuel cell is an electrochemical cell that converts a source fuel, such as hydrogen gas or a hydrogen-rich liquid fuel (in addition to hydrocarbons, alcohols, chlorine, chlorine dioxide, and others) into an electrical current (Meibuhr, 1966). The hydrogen gas or hydrogen-rich fuel is reacted or triggered in the presence of an electrolyte (an oxidant), usually oxygen from the air, to produce electrons, heat,

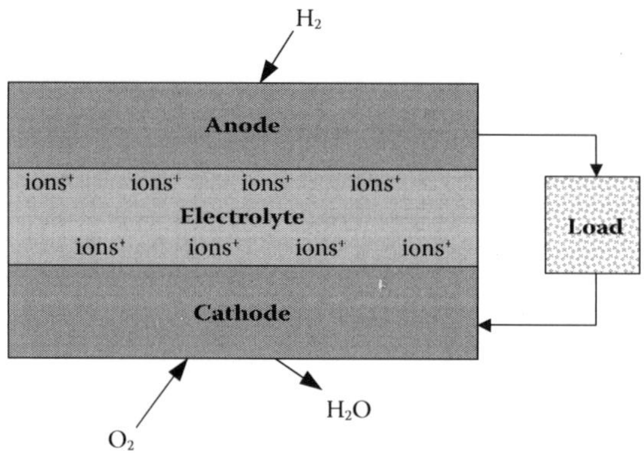

FIGURE 9.1 Generalized block diagram of the fuel cell process.

and water. The reactants flow into the cell and the reaction products flow out of it, while the electrolyte remains within the cell. Fuel cells can operate continuously as long as the reactant and oxidant flows are maintained. Figure 9.1 provides a generalized representation of the fuel cell process. Keep in mind that fuel cells come in many varieties (see Table 9.1); however, they all work in the same general manner. The principles of hydrogen-type fuel cell operation are described in the next section.

DID YOU KNOW?

To deliver the desired amount of energy, fuels cells can be combined (stacked) in series and parallel circuits. Recall from Chapter 2 that when cells are connected in series they yield higher voltage and in parallel they allow a higher current to be supplied. Although this reaction gives off a lot of heat (exothermic), the heat can be recovered for uses such as space heating, process heating, and domestic hot water.

HYDROGEN FUEL CELLS: A REALISTIC VIEW

Containing only one electron and one proton, hydrogen (chemical symbol H) is the simplest element on Earth. Hydrogen gas is a diatomic molecule, which means that each molecule has two atoms of hydrogen (pure hydrogen is commonly expressed as H_2). Hydrogen combines readily with other elements and is almost always found as part of another substance, such as water, hydrocarbons, or alcohols. Hydrogen is also found in biomass, which includes all plants and animals. Hydrogen can store and deliver usable energy, but, because it typically does not exist by itself in nature, it must be produced from compounds that contain it. Hydrogen can be produced using diverse, domestic resources, including nuclear, natural gas and coal, and biomass, as well as other renewables such as solar, wind, hydroelectric, or geothermal energy. This diversity of domestic energy sources makes hydrogen a promising energy carrier and important to our nation's energy security (USDOE, 2008a).

The U.S. Department of Energy is focusing on energy-efficient hydrogen-production technologies that result in near-zero net greenhouse gas emissions and use renewable energy sources, nuclear energy, and coal when combined with carbon sequestration. The processes used to produce hydrogen can be *thermal* (natural gas reforming, renewable liquid and bio-oil processing, biomass and coal gasification), *electrolytic* (splitting water using a variety of energy resources), or *photolytic* (splitting water using sunlight via biological and electrochemical materials). Hydrogen can be produced in large, central facilities located within 50 to 300 miles from the point of use or in smaller facilities located within 25 to 100 miles of use. For hydrogen to be successful in the marketplace,

TABLE 9.1
Types of Fuel Cells

Fuel Cell Type	Electrolyte	Current Uses
Metal hydride	Aqueous alkaline solution	Commercial/research
Electrogalvanic	Aqueous alkaline solution	Commercial/research
Direct formic acid	Polymer membrane (ionomer)	Commercial/research
Zinc–air battery	Aqueous alkaline solution	Mass production
Microbial	Polymer membrane/humic acid	Research
Upflow microbial	—	Research
Regenerative	Polymer membrane (ionomer)	Commercial/research
Direct borohydride	Aqueous alkaline solution	Commercial/research
Alkaline	Aqueous alkaline solution	Commercial/research
Direct methanol	Polymer membrane (ionomer)	Commercial/research
Reformed methanol	Polymer membrane (ionomer)	Commercial/research
Direct-ethanol	Polymer membrane (ionomer)	Research
Proton exchange membrane	Polymer membrane (ionomer)	Commercial/research
RFC–redox	Liquid electrolytes with redox shuttle and polymer membrane (ionomer)	Research
Phosphoric acid (H_3PO_4)	Molten phosphoric acid	Commercial/research
Molten carbonate	Molten alkaline carbonate	Commercial/research
Tubular solid oxide	O^{2-}-conducting ceramic oxide	Commercial/research
Protonic ceramic	H^+-conducting ceramic oxide	Research
Direct carbon	Several different	Commercial/research
Planar solid oxide	O^{2-}-conducting ceramic oxide	Commercial/research
Enzymatic biofuel	Any that will not denature the enzyme	Research
Magnesium air	Saltwater	Commercial/research

it must be cost competitive with the available alternatives. In the light-duty vehicle transportation market, this competitive requirement means that hydrogen must be available untaxed at \$2 to \$3/GGE (gasoline gallon equivalent). At this price, hydrogen fuel cell vehicles would cost the same on a cost-per-mile-driven basis compared to conventional vehicles. Some hydrogen production technologies are further along in development than others; some promise to be cost competitive for the transition period beginning in 2015, but others are considered long-term technologies that will not be cost competitive until after 2030 (USDOE, 2008a).

Infrastructure is required to move hydrogen from the location where it is produced to the dispenser at a refueling station or stationary power site. Infrastructure includes the pipelines, trucks, railcars, ships, and barges that deliver fuel, as well as the facilities and equipment needed to load and unload them. Some of the infrastructure is already in place because hydrogen has long been used in industrial applications, but it is not sufficient to support widespread consumer use of hydrogen as an energy carrier. Because hydrogen has a relatively low volumetric energy density, its transportation, storage, and final delivery to the point of use represent significant costs and result in some of the energy inefficiencies associated with using it as an energy carrier (USDOE, 2008b).

Options and trade-offs for hydrogen delivery from the production facilities to the point of use are complex. The choice of a hydrogen production strategy greatly affects the cost and method of delivery; for example, larger, centralized facilities can produce hydrogen at relatively low costs due to economies of scale, but the delivery costs are higher than for smaller, localized production facilities. Although these smaller production facilities would have relatively lower delivery costs, their hydrogen production costs are likely to be higher, as lower volume production means higher equipment costs on a per-unit-of-hydrogen basis (USDOE, 2008b).

Key challenges to hydrogen delivery include reducing delivery costs, increasing energy efficiency, maintaining hydrogen purity, and minimizing hydrogen leakage. Further research is needed to analyze the trade-offs between the hydrogen production options and the hydrogen delivery options taken together as a system. Building a national hydrogen delivery infrastructure is a big challenge. Such an infrastructure will take time to develop and will likely include combinations of various technologies. Infrastructure options will evolve as the demand for hydrogen grows and as delivery technologies develop and improve (USDOE, 2008b).

Hydrogen Storage

For hydrogen to be a competitive fuel for vehicles, the hydrogen vehicle must be able to travel a distance comparable to that of conventional hydrocarbon-fueled vehicles. Storing enough hydrogen on board a vehicle for a driving range of greater than 300 miles is a significant challenge. On a weight basis, hydrogen has nearly three times the energy content of gasoline (120 MJ/kg for hydrogen vs. 44 MJ/kg for gasoline); however, on a volume basis, the situation is reversed (8 MJ/L for liquid hydrogen vs. 32 MJ/L for gasoline). On-board hydrogen storage in the range of 5 to 13 kg is required to encompass the full platform of light-duty vehicles. Hydrogen can be stored as either a gas or a liquid. Storage as a gas typically requires high-pressure tanks (5000- to 10,000-psi tank pressure). Storage of hydrogen as a liquid requires cryogenic temperatures, because the boiling point of hydrogen at 1 atmosphere pressure is –252.8°C. Hydrogen can also be stored on the surfaces of solids (by adsorption) or within solids (by absorption). In adsorption, hydrogen is attached to the surface of a material as either hydrogen molecules or hydrogen atoms. In absorption, hydrogen is dissociated into H atoms, and then the hydrogen atoms are incorporated into the solid lattice framework (USDOE, 2008c).

Hydrogen storage in solids may make it possible to store large quantities of hydrogen in smaller volumes at low pressures and at temperatures close to room temperature. It is also possible to achieve volumetric storage densities greater than those of liquid hydrogen because the hydrogen molecule is dissociated into atomic hydrogen within a metal hydride lattice structure. Finally, hydrogen can be stored through the reaction of hydrogen-containing materials with water (or other compound such as alcohols). In this case, the hydrogen is effectively stored in both the material and the water. The terms *chemical hydrogen storage* and *chemical hydride* are used to describe this form of hydrogen storage. It is also possible to store hydrogen in the chemical structures of liquids and solids (USDOE, 2008c).

DID YOU KNOW?

Hydrogen fuel cell vehicles (FCVs) emit approximately the same amount of water per mile as vehicles using gasoline-powered internal combustion engines (ICEs).

How a Hydrogen Fuel Cell Works

A fuel cell uses the chemical energy of hydrogen to cleanly and efficiently produce electricity. Fuel cells have a variety of potential applications; for example, they can provide energy for systems as large as utility power stations or as small as laptop computers. Fuel cells offer several benefits over conventional combustion-based technologies currently used in many power plants and passenger vehicles. They produce much smaller quantities of greenhouse gases and none of the air pollutants that create smog and cause health problems. If pure hydrogen is used as a fuel, fuel cells emit only heat and water as byproducts (USDOE, 2010). A hydrogen fuel cell is a device that uses hydrogen (or hydrogen-rich fuel) and oxygen to create electricity through an electrochemical process. A single fuel cell consists of an electrolyte and two catalyst-coated electrodes (a porous anode and cathode). The various types of fuel cells all work similarly:

- Hydrogen (or hydrogen-rich fuel) is fed to an anode, where a catalyst separates the negatively charged hydrogen electrons from positively charged ions.
- At the cathode, oxygen combines with electrons and, in some cases, with species such as protons or water, resulting in water or hydroxide ions, respectively.
- For polymer electrolyte membrane and phosphoric acid fuel cells, protons move through the electrolyte to the cathode, where they combine with oxygen and electrons to produce water and heat.
- For alkaline, molten carbonate, and solid oxide fuel cells, negative ions travel through the electrolyte to the anode, where they combine with hydrogen to generate water and electrons.
- The electrons from the anode cannot pass through the electrolyte to the positively charged cathode; they must travel around it via an electrical circuit to reach the other side of the cell. This movement of electrons is an electrical current.

DISCUSSION AND REVIEW QUESTIONS

1. Why does society need fuel cells, and what can they be used for?
2. Are fuel cells a viable form of renewable energy? Explain.
3. What is the simplest element on Earth?
4. What three technologies are used to produce hydrogen?

REFERENCES AND RECOMMENDED READING

Anon. (2007). *Batteries, Supercapacitors, and Fuel Cells*. Washington, D.C.: Science Reference Services, Library of Congress (http://www.loc.gov/rr/scitech/tracer-bullets/batteriestb.html).

Barbir, F. (2005). *PEM Fuel Cells: Theory and Practice*. New York: Academic Press.

EG&G Technical Services, Inc. (2004). *Fuel Cell Handbook,* 7th ed. Morgantown, WV: National Energy Technology Laboratory, Office of Fuel Energy, U.S. Department of Energy.

Eisenberg, A. and Kim, J.S. (1998). *Introduction to Ionomers*. New York: John Wiley & Sons.

Hoogers, G. (2003). *Fuel Cell Technology Handbook*. Boca Raton, FL: CRC Press.

Meibuhr, S.G. (1966). *Electrochimica Acta*, 11, 1301.

Larminie, J. and Dicks A. (2003). *Fuel Cell Systems Explained*, 2nd ed. New York: John Wiley & Sons.

NREL. (2009). *Ultracapacitors*. Washington, D.C.: National Renewable Energy Laboratory (http://www.nrel.gov/vehiclesandfuels/energystorage/ultracapacitors.html?print).

Singhal, S.C. and Kendall, K. (2003). *High-Temperature Solid Oxide Fuel Cells: Fundamentals, Design, and Applications*. New York: Elsevier.

USDOE. (2008a). *Hydrogen Production*. Washington, D.C.: Energy Efficiency & Renewable Energy, U.S. Department of Energy (http://www1.eere.energy.gov/hydrogenandfuelcells/production/basics.html).

USDOE. (2008b). *Fuel Cell Technologies Program—Hydrogen Delivery*. Washington, D.C.: Energy Efficiency & Renewable Energy, U.S. Department of Energy (http://www1.eere.energy.gov/hydrogenandfuelcells/delivery/m/basics.html).

USDOE. (2008c). *Fuel Cell Technologies Program—Hydrogen Storage*. Washington, D.C.: Energy Efficiency & Renewable Energy, U.S. Department of Energy (http://www1.eere.energy.gov/hydrogenandfuelcells/storage/m/basics.html).

USDOE. (2010). *Fuel Cell Technologies Program—Fuel Cells*. Washington, D.C.: Energy Efficiency & Renewable Energy, U.S. Department of Energy (https://www1.eere.energy.gov/hydrogenandfuelcells/fuelcells/basics.html).

Vielstich, W. et al., Eds. (2009). *Handbook of Fuel Cells: Advances in Electrocatalysis, Materials, Diagnostics, and Durability*. New York: John Wiley & Sons.

Afterword: *Is It the Beginning of the End or the Beginning of the Beginning?*

When we consider that we are probably at or near peak oil natural production and already absorbing the higher costs associated with it (i.e., if we are able to obtain a reliable and consistent supply of it in the first place), renewable energy alternatives begin to look a lot more viable and necessary to us. On the other hand, we are currently suffering severe economic conditions; if we are lucky enough to have reliable and consistent employment with accompanying benefits, the advantages of switching from oil or natural gas supplies to renewable energy sources may not be that readily apparent or pressing. Until the summer of 2008, when gasoline prices at the pump exceeded $4.00/gal, many of us had not given much thought to the issue. After the spike in gasoline prices, however, people (those with and without employment) began to look at hybrid cars and domestic renewable energy alternatives in a very new light. Using Interstate 40 as our example, it is not that difficult or unusual to pick out numerous hybrid automobiles along with those ubiquitous heavy trucks, RVs, and other standard vehicles sharing the roadway. In addition, along the interstate it is not unusual to find giant wind towers with their turbine blades slashing through the air, turning turbine-generator shafts and cranking out electricity. Along this same superhighway, especially at rest stops (the ones not closed due to budget cuts) and state visitor centers, small arrays of solar cells can be spotted here and there. These solar panels supply electricity to fans, signal lights, natural gas metering systems, alarm systems, pump monitoring equipment, digital signs, and many other small appliances and industrial devices.

It is not unusual for drivers and passengers traveling the interstates to overlook these wind turbines, solar panels, and hydroelectric dams, these generators of renewable energy, because they are focused on getting from point A to point B and maybe safely back again to point A. These folks are missing the point that someday we will be driving high-capacity, highly efficient electric or hydrogen-like fuel-cell vehicles that will be as common as the gas- and diesel-powered vehicles of today. Much of the electricity required to recharge these electricity- or fuel-cell-powered vehicles will be provided by these same wind turbines, solar panels, and hydroelectric dams.

In light of today's current economic situation, which is driven by many factors, including the high price of crude oil and its dwindling accessibility, we can ask ourselves this question: Is it the beginning of the end or the beginning of the beginning? And the reader might wonder what we mean by that. Is doomsday around the corner, or is it already here and we are too ignorant to recognize the signs? Maybe we just don't give a damn, or our judgment is clouded by what is going on in the world today. The beginning of the end or the beginning of the beginning—do we anticipate *the* end, or *an* end, whatever that end might be?

The end of what you might ask? (Geez, at least that's what we *hope* you are asking.) We are referring to the end of the so-called good life as we know it—that is, the good life that our parents (the greatest generation) and we, the so-called grasshopper generation, knew very briefly, yesterday, in the not so distant past, and that, we can only hope, we will know again tomorrow.

Here, "yesterday" refers to the era just before that infamous dot.com bust in 1999 or thereabouts. Remember those days? When young computer whiz kids were driving around in all those BMWs and living the good life in million-dollar homes (mortgaged to the hilt, of course) with portfolios filled with promissory notes? Not to worry; the dot.com wizards were the new rich…until the bust, that is. Suddenly their paper was just paper and almost worthless, certainly not as utilitarian as toilet paper; it was just paper with print on it that promised untold fortunes that weren't meant to be. More recently, dysfunctional, asleep-at-the-switch government regulators and congressional officials allowed just about anyone and everyone to buy a home—even those with bad credit or no credit or no job or no future. These nonqualified buyers have brought us to the sharp edge of the abyss we are now straddling, and if we are not careful we will fall over.

So, is it all doom and gloom? Short answer: We don't have a clue. Long answer: Not necessarily, if we make the right moves soon. Right moves? What are they? How do we maintain the good life? How do we improve on our lives to the point where we are living as well as or better than our parents did?

We do not profess to understanding all of the problems facing our country today nor do we know all the solutions. We know what we know from what we know—from what we see and feel. Our great country is at a crossroads. We can maintain our current lifestyles, fall on our swords, or progress through innovation. We hope that it is progression, moving forward to maintain the good life, that lies before us. This is our hope.

So, how do we progress—how do we maintain that so-called good life? In our opinion, the first step is that we must achieve energy self-sufficiency through innovation and a fundamental shift from fossil fuels to renewable, nonpolluting energy supplies. Energy is the lifeblood of America. We must get off our knees and stop begging those who hate us to supply us with the energy we need. We must put America first. Using American genius we must innovate and come up with a solution to the pending energy crisis. Remember, every problem has a solution; we must find the solution to the energy crisis. Moreover, when we do innovate, we must make sure that our country holds onto its trade secrets regarding how to produce renewable energy sources; otherwise, other countries will simply steal our ideas and manufacture energy at rates against which we cannot compete. As an example, $9 of production costs (mainly for labor, raw materials, and taxes) in the United States cannot possibly compete with $1 production costs in other countries for the same product. If we are going to innovate, let's innovate for America. We have given away too much, and now we are paying for it at the pump and elsewhere.

Appendix: Answers to Chapter Discussion and Review Questions

CHAPTER 1

1. Energy products such as oil that, once used, are depleted
2. Alternative energy is intended to replace fuel sources without the undesired consequences of the replaced fuel; renewable energy is a form of energy capable of being constantly regenerated by natural processes at reasonable rates.
3. No; nuclear energy is derived from radioactive resources such as uranium; when the radioactivity is depleted the material is no longer usable.
4. Answers will vary.
5. Answers will vary.
6. Answers will vary.
7. Any 15 of the following: air conditioners, ammonia, antihistamine, antiseptics, artificial turf, balloons, bandages, basketballs, boats, bras, bubble gum, cameras, candles, car batteries, car bodies, carpet, caulking, chewing gum, cold cream, combs/brushes, compact discs, computers, contacts, cortisone, crayons, denture adhesives, deodorant, dice, dishwasher liquid, dresses, electric blankets, electrician's tape, eyeglasses, fertilizers, fiberglass, fishing lures, fishing rods, floor wax, floor wax, footballs, glue, golf balls, hair coloring, hair curlers, hair dryers, hearing aids, heart valves, house paint, ice chests, inks of all types, insect repellent, insulation, lens cleaner, lip balm, lipstick, loudspeakers, mops, motorcycle helmets, movie film, nail polish, oil filters, paint brushes, panty hose, parachutes, pens, perfumes, petroleum jelly, plastic chairs, plastic cups, plastic forks, plastic toys, plastic wrap, plywood adhesive, purses, refrigerator seals, roofing paper, rubber bands, rubber boots, rubber cement, running shoes, shirts (non-cotton), shoe polishes, shoes, shower curtains, solvents, stereos, synthetic clothing, tape recorders, telephones, thermos bottle, tires, toilet seats, toothpaste, transparent tape, trash bags, umbrellas, upholstery, vaporizers, volley balls, water pipes, water skis
8. Simply, the *Hirsch Report* reports on peaking world oil supplies.
9. It makes the point that we may be forced back to subsistence levels once population growth outpaces agricultural production; can be used generally for economics and world energy supplies.
10. Answers will vary.

CHAPTER 2

1. A series circuit has only one path for current flow, whereas a parallel circuit has more than one path.
2. Source voltage
3. Voltage drop
4. Counterclockwise
5. 2 amps
6. 12 volts

7. 16 volts
8. 80 watts
9. Less, more
10. Zero excessive
11. 120 ohms
12. One
13. Individual, total
14. $R_T = 10$ ohms
15. $I_T = 2$ amps
16. $P_T = 40$ watts
17. 12 ohms
18. 1 amp (rounded)
19. 1.5 volts
20. 12 watts (rounded)

CHAPTER 3

1. Answers will vary.
2. No; these terms are used interchangeably.
3. Some people can survive 5 minutes without air, 5 days without water, and 5 weeks without food.
4. How they collect solar energy—linear concentrators, dish engines, or power tower systems
5. Parabolic trough systems; linear Fresnel reflector systems
6. They produce zero emissions when operating on solar energy, operate more quietly than diesel or gasoline engines, are easier to operate and maintain than conventional engines, start up and shut down automatically, and operate for long periods with minimal maintenance.
7. The measure of the heat energy required to increase the temperature of a unit quantity of a substance by a unit of temperature
8. A passive solar design used to illuminate a living space, an industrial space, or a commercial working space
9. Ratio of the light output of a lamp operated by a specific ballast to the light output of the same lamp operated by a reference ballast
10. Answers will vary.

CHAPTER 4

1. Land requirement, access to generators, aesthetics
2. They are shallow-cycle batteries.
3. Forces
4. Advective
5. Thermal
6. Deflection
7. Mechanical
8. Wind blows the rotor downwind.
9. Output
10. Answers will vary.

CHAPTER 5

1. Answers will vary.
2. Answers will vary.
3. Impoundment
4. Head
5. Surface; density
6. Density
7. Force
8. Total head
9. Law of conservation of energy
10. *C* factor

CHAPTER 6

1. Answers will vary.
2. Electricity generation
3. Biomass
4. Crops or products, such as waste vegetable oil
5. Have both a good and bad side
6. Dicot
7. Chlorophyll
8. Plant hormones
9. Switchgrass
10. Willow
11. Gasoline gallon equivalent (GGE)
12. Answers will vary.

CHAPTER 7

1. Moho layer
2. Asthenosphere
3. Enhanced geothermal systems
4. Western states
5. Pressure + volume + temperature
6. Dense
7. Collision boundaries
8. Dry steam power plant
9. EGS
10. Geothermal heat pump (GHP)
11. Answers will vary.

CHAPTER 8

1. Mechanical, thermal
2. Remained constant
3. Hydraulic action
4. Tombolos
5. Resonate
6. Attenuator
7. Mesomareal
8. Tidal fence
9. Solar
10. Answers will vary.

CHAPTER 9

1. Answers will vary.
2. Answers will vary.
3. Hydrogen
4. Thermal, electrolytic, photolytic

Glossary*

Absorber: In a photovoltaic device, the material that readily absorbs photons to generate charge carriers (free electrons or holes).

Acid hydrolysis: A chemical process in which acid is used to convert cellulose or starch to sugar.

Albedo: The ratio of light reflected by a surface to the light falling on it.

Alcohol: A general class of hydrocarbons that contains a hydroxyl group (OH); the many types of alcohol include butanol, ethanol, and methanol.

Alcohol fuels: Alcohol can be blended with gasoline for use as transportation fuel. It may be produced from a wide variety of organic feedstock. The common alcohol fuels are methanol and ethanol. Methanol may be produced from coal, natural gas, wood, and organic waste; ethanol is commonly made from agricultural plants, primarily corn, containing sugar.

Alkaline fuel cell (AFC): A type of hydrogen/oxygen fuel cell in which the electrolyte is concentrated potassium hydroxide (KOH) and the hydroxide ions (OH) are transported from the cathode to the anode.

Alternating current (AC): An electric current that reverses its direction at regularly recurring intervals, usually 50 or 60 times per second.

Alternator: A device that turns the rotation of a shaft into alternating current (AC).

Ambient: Natural condition of the environment at any given time.

Amorphous silicon: An alloy of silica and hydrogen, with a disordered, noncrystalline internal atomic arrangement, that can be deposited in thin layers (a few micrometers in thickness) by a number of a deposition methods to produce thin-film photovoltaic cells on glass, metal, or plastic substrates.

Ampere (amp): A unit of electrical current flowing in a circuit; can be considered to be analogous to water flowing through a pipe being measured in liters per minute.

Ampere-hour: A measure of the flow of current (in amperes) over one hour; used to measure energy production over time and battery capacity.

Anaerobic digestion: A biochemical process by which organic matter is decomposed by bacteria in the absence of oxygen, producing methane and other byproducts.

Anemometer: Wind-speed measurement device used to send data to the controller; also used to conduct wind site surveys.

Angle of attack: In wind turbine operation, the angle of the airflow relative to the blade.

Angle of incidence: The angle that a ray of sun makes with a line perpendicular to the surface; for example, a surface that directly faces the sun has a solar angle of incidence of 0, but if the surface is parallel to the sun (e.g., sunrise striking a horizontal rooftop), the angle of incidence is 90°.

Anion: A negatively charged ion; an ion that is attracted to the anode.

Annual removals: The net volume of growing stock trees removed from the inventory during a specified year by harvesting, cultural operations such as timber land improvement, or land clearing.

Annualized growth rate: Calculated as follows: $(x_n/x_1)1/n$, where x is the value under consideration and n is the number of periods.

Anode: The electrode at which oxidation (a loss of electrons) takes place. For fuel cells and other galvanic cells, the anode is the negative terminal; for electrolytic cells, the anode is the positive terminal.

* Some definitions were retrieved from USDOE, *Glossary of Energy-Related Terms*, Energy Efficiency & Renewable Energy, U.S. Department of Energy, Washington, D.C., 2009 (http://www1.eere.energy.gov/site_administration/glossary.html).

Aquifer: Water-bearing stratum of permeable sand, rock, or gravel.

Asexual reproduction: The naturally occurring ability of some plant species to reproduce asexually through seeds, meaning the embryos develop without a male gamete. This ensures the seeds will produce plants identical to the mother plant.

Availability factor: A percentage representing the number of hours a generating unit is available to produce power (regardless of the amount of power) in a given period, compared to the number of hours in the period.

Azimuth angle: The angle between true south and the point on the horizon directly below the sun.

Bagasse: The fibrous material remaining after the extraction of juice from sugarcane; often burned by sugar mills as a source of energy.

Baseload plants: Electricity-generating units that are operated to meet the constant or minimum load on the system. The cost of energy from such units is usually the lowest available to the system.

Binary cycle: Combination of two power plant turbine cycles utilizing two different working fluids for power production. The waste heat from the first turbine cycle provides the heat energy for the operation of the second turbine, thus providing higher overall system efficiencies.

Biobased product: As defined by the Farm Security and Rural Investment Act (FSRIA), a product determined by the U.S. Secretary of Agriculture to be a commercial or industrial product (other than food or feed) that is composed, in whole or in significant part, of biological products or renewable domestic agriculture materials (including plant, animal, and marine materials) or forestry materials.

Biochemical conversion: The use of fermentation or anaerobic digestion to produce fuels and chemicals from organic sources.

Biodiesel: Fuel derived from vegetable oils or animals fats; it is produced when a vegetable oil or animal fat is chemically reacted with an alcohol.

Bioenergy: Useful, renewable energy produced from organic matter that may be used directly as a fuel or processed into liquids and gases.

Biofuels: Liquid fuels and blending components produced from biomass (plant) feedstocks; used primarily for transportation.

Biogas: Combustible gas derived from decomposing biological waste; normally consists of 50 to 60% methane.

Biomass: Any organic nonfossil material of biological origin constituting a renewable energy source; produced from organic matter that is available on a renewable or recurring basis, including agricultural crops and trees, wood and wood residues, plants (including aquatic plants), grasses, animal manure, municipal residues, and other residue materials. Biomass is generally produced in a sustainable manner from water and carbon dioxide by photosynthesis. The three main categories of biomass are primary, secondary, and tertiary.

Biomass gas (biogas): A medium-Btu gas containing methane and carbon dioxide; it is produced by the action of microorganisms on organic materials such as a landfill.

Biomaterials: Products derived from organic-based (as opposed to petroleum-based) products.

Bio-oil: Intermediate fuel derived from fast pyrolysis.

Biopower: The use of biomass feedstock to produce electric power through direct combustion of the feedstock, gasification, and then combustion of the resultant gas, or through other thermal conversion processes. Power is generated with engines, turbines, fuel cells, or other equipment.

Biorefinery: A facility that processes and converts biomass into value-added products, ranging from biomaterials to fuels such as ethanol or important feedstocks for the production of chemicals and other materials. Biorefineries can utilize mechanical, thermal, chemical, and biochemical processes.

Blackbody: An ideal substance that absorbs all radiation falling on it, reflecting nothing.

Black liquor (pulping liquor): The alkaline spent liquor removed from the digesters in the process of chemically pulping wood. After evaporation, the liquor is burned as a fuel in a recovery furnace that permits the recovery of certain basic chemicals.

Bone dry: Having zero moisture content.

Borehole breakouts: Failure of the borehole wall that occurs because of stress in the rock surrounding the borehole. The breakout is generally located symmetrically in the wellbore perpendicular to the direction of greatest horizontal stress on a vertical wellbore.

Brine: A geothermal solution containing appreciable amounts of sodium chloride or other salts.

British thermal unit (Btu): A basic measure of thermal (heat) energy. A Btu is defined as the amount of energy required to increase the temperature of 1 pound of water by 1°F at normal atmospheric pressure. 1 Btu = 1055 joules.

Bulk density: Weight per unit of volume, usually specified in pounds per cubic foot.

Cap rocks: Rocks of low permeability that overlie a geothermal reservoir.

Capacity factor: The ratio of the electrical energy produced by a generating unit for the period of time considered to the electrical energy that could have been produced at continuous full-power operation during the same period.

Capacity, gross: The full-load continuous rating of a generator, prime mover, or other electric equipment under specified conditions as designated by the manufacturer. It is usually indicated on a nameplate attached to the equipment.

Capitol cost: The cost of field development and plant construction and the equipment required for the generation of electricity.

Carbon dioxide (CO_2): A product of combustion; the most common greenhouse gas.

Carbon monoxide (CO): A colorless, odorless gas produced by incomplete combustion; carbon monoxide is poisonous when inhaled.

Carbon sequestration: The absorption and storage of carbon dioxide from the atmosphere by naturally occurring plants.

Carnot cycle: An ideal heat engine (conceived by Sadi Carnot) in which the sequence of operations forming the working cycle consist of isothermal expansion, adiabatic expansion, isothermal compression, and adiabatic compression back to its initial state.

Cascading heat: A process that uses a stream of geothermal hot water or steam to perform successive tasks requiring lower and lower temperatures.

Casing: Pipe placed in a wellbore as a structural interface between the wellbore and the surrounding formation. It typically extends from the top of the well and is cemented in place to maintain the diameter of the wellbore and provide stability.

Cast silicon: Crystalline silicon obtained by pouring pure molten silicon into a vertical mold and adjusting the temperature gradient along the mold during cooling to achieve slow, vertically advancing crystallization of the silicon. The resulting polycrystalline ingot is composed of large, relatively parallel, interlocking crystals. The cast ingots are sawed into wafers for further fabrication into photovoltaic cells. Cast-silicon wafers and ribbon-silicon sheets fabricated into cells are usually referred to as *polycrystalline photovoltaic cells.*

Cathode: The electrode at which reduction (a gain of electrons) occurs. For fuel cells and other galvanic cells, the cathode is the positive terminal; for electrolytic cells (where electrolysis occurs), the cathode is the negative terminal.

Cation: A positively charged ion.

Cellulose: The main carbohydrate in living plants; cellulose forms the skeletal structure of the plant cell wall.

Cellulosic ethanol: Ethanol derived from cellulosic and hemicellulosic portions of plant biomass.

Chips: Woody material cut into short, thin wafers. Chips are used as a raw material for pulping and fiberboard or as biomass fuel.

Closed-loop biomass: Crops grown in a sustainable manner for the purpose of optimizing their value for bioenergy and bioproduct uses. These include annual crops, such as maize and wheat, and perennial crops, such as trees, shrubs, and grasses (e.g., switchgrass).

Coarse materials: Wood residues suitable for chipping, such as slabs, edgings, and trimmings.

Co-firing: Practice of introducing biomass into the boilers of coal-fired power plants.

Combined cycle: An electric generating technology in which electricity is produced from otherwise lost waste heat exiting one or more gas (combustion) turbines. The exiting heat is routed to a conventional boiler or to a heat-recovery steam generator for utilization by a steam turbine in the production of electricity. Such designs increase the efficiency of the electric generating unit.

Combined heat and power (CHP) plant: A plant designed to produce both heat and electricity from a single heat source. This term has replaced the term "cogenerator." CHP better describes the facilities because some of the plants included do not produce heat and power in a sequential fashion; as a result, they do not meet the legal definition of cogeneration specified in the Public Utility Regulatory Polices Act (PURPA).

Commercial sector: An energy-consuming sector that consists of service-providing facilities and equipment of nonmanufacturing businesses; federal, state, and local governments; and other private and public organizations, such as religious, social, or fraternal groups. The commercial sector includes institutional living quarters, as well as sewage treatment facilities. Common uses of energy associated with this sector include space heating, water heating, air conditioning, lighting, refrigeration, cooking, and running a wide variety of other equipment. This sector include generators that produce electricity or useful thermal output primarily to support the activities of the above-mentioned commercial establishments.

Commercial species: Tree species suitable for industrial wood products.

Compressed natural gas (CNG): Mixture of hydrocarbon gases and vapors, consisting principally of methane in gaseous form that has been compressed.

Concentrator: A reflective or refractive device that focuses incident insolation onto an area smaller than the reflective or refractive surface, resulting in increased insolation at the point of focus.

Condensate: Water formed by condensation of steam.

Condenser: Equipment that condenses turbine exhaust steam into condensate.

Conservation Reserve Program (CRP): CRP provides farm owners or operators with an annual per-acre rental payment and half the cost of establishing a permanent land cover in exchange for retiring environmentally sensitive cropland from production for 10 to 15 years. In 1996, Congress reauthorized CRP for an additional round of contracts, limiting enrollment to 36.4 million acres at any time. The 2002 Farm Act increased the enrollment limit to 39 million acres. Producers can offer land for competitive bidding based on an Environmental Benefits Index (EBI) during periodic signups, or they can automatically enroll more limited acreages in such practices as riparian buffers, field windbreaks, and grass strips on a continuous basis. CRP is funded through the Commodity Credit Corporation (CCC).

Cooling tower: A structure in which heat is removed from hot condensate.

Conventional hydroelectric (hydropower) plant: A plant in which all the power is produced from natural streamflow as regulated by available storage.

Core: A cylinder of rock recovered from a well by a special coring drill bit.

Crop failure: Acreage on which crops have failed due to weather, insects, or diseases, but can also include some land not harvested due to lack of labor, lower market prices, or other factors. Acreage planted with cover crops or soil improvement crops not intended for harvest is excluded and is considered idle.

Cropland: Total cropland includes cropland harvested, crop failure, cultivated summer fallow, cropland used only for pasture, and idle cropland.

Cropland harvested: Includes row crops and closely sown crops, hay and silage crops, tree fruits, small fruits, berries, tree nuts, vegetables, melons, and miscellaneous other minor crops and hay. In recent years, farmers have double-cropped about 4% of this acreage.

Cropland used for crops: Cropland used for crops includes cropland harvested, crop failure, and cultivated summer fallow.

Cropland used for pasture: Land used for long-term crop rotation; however, some cropland pasture is marginal for crop uses and may remain in pasture indefinitely. This category also includes land that was used for pasture before crops reached maturity and some land used for pasture that could have been cropped without additional improvement.

Crust: Earth's outer layer of rock; also called the *lithosphere.*

Cryogenic liquefaction: The process through which gases such as nitrogen, hydrogen, helium, and natural gas are liquefied under pressure at very low temperatures.

Cull tree: A live tree, 5 inches in diameter at breast height or larger, that is not merchantable for saw logs now or prospectively because of rot, roughness, or species.

Cut-in speed: The speed at which a shaft must turn to generate electricity and send it over a wire.

Daylighting: The use of direct, diffuse, or reflected sunlight to provide supplemental lighting for building interiors.

d.b.h: The diameter measured at approximately breast height from the ground.

Densification: A mechanical process to compress biomass (usually wood waste) into pellets, briquettes, cubes, or densified logs.

Depletion factor: Annual percentage of the depletion of the thermal resource.

Digester gas: Biogas that is produced using a digester, an airtight vessel or enclosure in which bacteria decomposes biomass in water to produce biogas.

Diode: A solid-state device that acts as a one-way valve for electricity.

Direct current (DC): An electric current that flows in a constant direction; the magnitude of the current does not vary or has only a slight variation.

Direct methanol fuel cell (DMFC): A fuel cell in which the fuel is methanol (CH_3OH) in gaseous or liquid form. The methanol is oxidized directly at the anode instead of first being reformed to produce hydrogen. The electrolyte is typically polymer electrolyte membrane (PEM).

Direct use: Use of geothermal heat without first converting it to electricity, such as for space heating and cooling, food preparation, industrial processes, etc.

Distributed generation (distributed energy resources): Refers to electricity provided by small, modular power generators (typically ranging in capacity from a few kilowatts to 50 megawatts) located at or near customer demand.

District heating: A type of direct use in which a utility system supplies multiple users with hot water or steam from a central plant or well field.

Downwind turbine: A turbine that does not face into the wind and whose direction is controlled directly by the wind.

Drag bit: Drilling bit that drills by scraping or shearing rock with fixed hard surfaces known as cutters.

Drilling: Boring into the ground to access geothermal resources, usually with oil and gas drilling equipment that has been modified to meet geothermal requirements.

Dry steam: Very hot steam that occurs without liquid.

Dump load: A device that allows excess energy to be safely disposed of.

E10: A mixture of 10% ethanol and 90% gasoline based on volume.

E85: A mixture of 85% ethanol and 15% gasoline based on volume.

Efficiency: The ratio of the useful energy output of a machine or other energy-converting plant to the energy input.

Electric power sector: An energy-consuming sector that consists of electricity-only and combined heat and power (CHP) plants whose primary business is to sell electrical, or electricity and heat, to the public (i.e., North American Industry Classification System 22 plants).

Electric utility: A corporation, person, agency, authority, or other legal entity or instrumentality aligned with distribution facilities for delivery of electric energy for use primarily by the public. Included are investor-owned electric utilities, municipal and state utilities, federal electric utilities, and rural electric cooperatives. A few entities that are tariff based and corporately aligned with companies that own distribution facilities are also included.

Emissions: Anthropogenic releases of gases to the atmosphere. In the context of global climate change, they consist of radiatively important greenhouse gases (e.g., the release of carbon dioxide during fuel combustion).

Endothermic: A chemical reaction that absorbs or requires energy (usually in the form of heat).

Energy: The ability to do work.

Energy crops: Crops grown specifically for their fuel value. They include food crops such as corn and sugarcane and nonfood crops such as poplar trees and switchgrass. Low-energy crops that are under development include short-rotation woody crops (fast-growing hardwood trees) harvested in 5 to 8 years and herbaceous energy crops (e.g., perennial grasses) harvested annually after taking 2 to 3 years to reach full productivity.

Enhanced geothermal systems (EGS): Engineered reservoirs that can extract economic amounts of heat from geothermal resources. Rock fracturing, water injection, and water circulation technologies sweep heat from unproductive areas of existing geothermal fields or new fields lacking sufficient production capacity.

Enthalpy: A thermodynamic property of a substance, defined as the sum of its internal energy plus the pressure of the substance times its volume, divided by the mechanical equivalent of heat. The total heat content of air; the sum of enthalpies of dry air and water vapor, per unit weight of dry air; measured in Btu per pound (or calories per kilogram).

Environmental Impact Statement: A document created from a study of the expected environmental effects of a new development or installation.

Ethanol (CH_3CH_2OH; also known as ethyl alcohol or grain alcohol): A clear, colorless flammable oxygenated hydrocarbon with a boiling point of 173.5°F in the anhydrous state. It readily forms a binary azetrope with water, with a boiling point of 172.67°F at a composition of 95.57% by weight ethanol. It is used in the United States as a gasoline octane enhancer and oxygenate (maximum 10% concentration). Ethanol can be used in higher concentrations (E85) in vehicles designed for its use. Ethanol is typically produced chemically from ethylene or biologically from fermentation of various sugars from carbohydrates found in agricultural crops and cellulosic residues from crops or wood. The lower heating value, equal to 76,000 Btu per gallon, is assumed for most estimates in this text.

Evacuated tube: In a solar thermal collector, an absorber tube contained in an evacuated glass cylinder through which collector fluids flows.

Exothermic: A chemical reaction that gives off heat.

Externality: A cost or benefit not accounted for in the price of goods and services; often refers to the cost of pollution and other environmental impacts.

Fast pyrolysis: Thermal conversion of biomass by rapid heating to 450 to 600°C in the absence of oxygen.

Fault: A fracture in rock exhibiting relative movement between the adjoining surfaces.

Feller buncher: A self-propelled machine that cuts trees with giant shears near ground level and then stacks the trees into piles to await skidding.

Feedstock: A product used as the basis for the manufacture of another product.

Fenestration: Openings in a building (e.g., windows, doors, skylights). The whole-building design approach determines what type of fenestration products should be used. It is desirable to select products with characteristics that accommodate the local climate and address insulating, daylighting, heating and cooling, and natural ventilation needs.

Fermentation: Conversion of carbon-containing compounds by microorganisms for the production of fuels and chemicals such as alcohols, acids, or energy-rich gases.

Fiber products: Products derived from the fibers of herbaceous and woody plant materials. Examples include pulp, composition-board products, and wood chips for export.

Fine materials: Wood residues not suitable for chipping, such as planer shavings and sawdust.

Fischer–Tropsch fuels: Liquid hydrocarbon fuels produced by a process that combines carbon monoxide and hydrogen. The process is used to convert coal, natural gas, and low-value refinery products into high-value diesel substitutes.

Flash steam: Steam produced when the pressure on a geothermal liquid is reduced; also called *flashing.*

Flat plate pumped: A medium-temperature solar thermal collector that typically consists of a metal frame, glazing, absorbers (usually metal), and insulation and that uses a pump liquid as the heat-transfer medium; predominant use is in water heating applications.

Flexible-fuel vehicle: A vehicle with a single fuel tank designed to run on varying blends of unleaded gasoline with either ethanol or methanol.

Flow battery: An electrochemical energy storage device that utilizes tanks of rechargeable electrolyte to refresh the energy-producing reaction. Because its capacity is limited only by the size of its electrolyte tanks, it is useful for large-scale backup systems to supplement other forms of generation that may be intermittent in nature.

Forest land: Land at least 10% stocked by forest trees of any size, including land that formerly had such tree cover and that will be naturally or artificially regenerated. Forest land includes transition zones, such as areas between heavily forested and nonforested lands that are at least 10% stocked with forest trees and forested areas adjacent to urban or built-up lands. Also included are pinyon–juniper and chaparral areas in the West and afforested areas. The minimum area for classification of forest land is 1 acre. Roadside, streamside, and shelterbelt strips of trees must have a crown width of at least 120 feet to qualify as forest land. Unimproved roads and trails, streams, and clearings in the forest areas are classified as forest if less than 120 feet wide.

Fracture: Natural or induced break in rock.

Fracturing treatments: Fracturing treatments are performed by pumping fluid into the subsurface at pressures above the fracture pressure of the reservoir formation to create a highly conductive flow path between the reservoir and the wellbore.

Fuel cell: Cell capable of generating an electrical current by converting the chemical energy of a fuel directly into electrical energy. Fuel cells differ from conventional electrical cells in that the active materials such as fuel and oxygen are not contained within the cell but are supplied from outside.

Fuel cell poisoning: The lowering of a fuel cell's efficiency due to impurities in the fuel binding to the catalyst.

Fuel cell stack: Individual fuel cells connected in a series; fuel cells are stacked to increase voltage.

Fuel Treatment Evaluator (FTE): A strategic assessment tool capable of aiding the identification, evaluation, and prioritization of fuel treatment opportunities.

Fuelwood: Wood and wood products, possibly including coppices, scrubs, branches, etc., bought or gathered and used by direct combustion.

Full sun: The amount of power density in sunlight received at the Earth's surface at noon on a clear day (about 1000 W/m^2).

Fumarole: A vent or hole in the Earth's surface, usually in a volcanic region, from which steam, gaseous vapors, or hot gases issue.

Gallon: A volumetric measure equal to 4 quarts (231 cubic inches). One gallon equals 3785 liters; 1 barrel of fuel oil equals 42 gallons.

Gasification: A chemical or heat process to convert a solid fuel to a gaseous form.

Gasohol: A motor vehicle fuel that is a blend of 90% unleaded gasoline and 10% ethanol (by volume).

Gearbox: Increases the rpm of a low-speed shaft, transferring its energy to a high-speed shaft to provide enough speed to generate electricity.

Generation (electricity): The process of producing electric energy from other forms of energy; the amount of electric energy produced is expressed in watt-hours (Wh).

Geothermal: Of or relating to the Earth's interior heat.

Geothermal energy: The Earth's interior heat made available by extracting it from hot water or rocks. Electric power plants extract hot water or steam from geothermal reservoirs in the Earth's crust that is supplied to steam turbines to drive generators to produce electricity.

Geothermal gradient: The rate of temperature increase in the Earth as a function of depth. Temperature increases an average of 1°F for every 75 feet in descent.

Geothermal heat pumps: Devices that take advantage of the relatively constant temperature of the Earth's interior, using it as a source and sink of heat for both heating and cooling. For cooling purposes, heat is extracted from the space and dissipated into the Earth; for heating purposes, heat is extracted from the Earth and pumped into the space.

Geothermal plant: A plant in which a turbine is driven by either hot water or natural steam that derives its energy from heat found in rocks or fluids at various depths beneath the surface of the Earth. The fluids are extracted by drilling or pumping.

Geothermal resources: The natural heat of the Earth that can be used for beneficial purposes when the heat is collected and transported to the surface.

Geyser: A spring that shoots jets of hot water and steam into the air.

Glazing: Transparent or translucent material (glass or plastic) used to admit light or to reduce heat loss; used for windows, skylights, or greenhouses or for covering the aperture of a solar collector.

Global positioning system (GPS): A navigation system using satellite signals to fix the location of a radio receiver on or above the Earth's surface.

Grassland pasture and range: All open land used primarily for pasture and grazing, including shrub and brush land types of pasture, grazing land with sagebrush and scattered mesquite, and all tame and native grasses, legumes, and other forage used for pasture or grazing. Because of the diversity in vegetative composition, grassland pasture and range are not always clearly distinguishable from other types of pasture and range. At one extreme, permanent grassland may merge with cropland pasture, or grassland may often be found in transitional areas with forested grazing land.

Gravimetry: The use of precisely measured gravitational force to determine mass differences that can be correlated to subsurface geology.

Green pricing/marketing: In the case of renewable electricity, green pricing represents a market solution to the various problems associated with regulatory valuation of the non-market benefits of renewables. Green pricing programs allow electricity customers to express their willingness to pay for renewable energy development through direct payments on their monthly utility bills.

Greenhouse gas: Gases that trap the heat of the sun in the Earth's atmosphere, producing the greenhouse effect. The two major greenhouse gases are water vapor and carbon dioxide. Other greenhouse gases include methane, ozone, chlorofluorocarbons, and nitrous oxide.

Grid: The layout of an electrical distribution system.

Gross generation: The total amount of electric energy produced by the generating units at a generation station or stations, measured at the generator terminals.

Growing stock: A classification of timber inventory that includes live trees of commercial species meeting specified standards of quality or vigor. Cull tress are excluded. When associated with volume, includes only trees 5.0 inches in d.b.h. and larger.

HDR (hot dry rock): Subsurface geologic formations of abnormally high heat content that contain little or no water.

Hardwoods: Usually broad-leaved and deciduous trees.

Heat exchanger: A device for transferring thermal energy from one fluid to another.

Heat flow: Movement of heat from within the Earth to the surface, where it is dissipated into the atmosphere, surface water, and space by radiation.

Heat pump: A year-round heating and air-condition system employing a refrigeration cycle. In the refrigeration cycle, a refrigerant is compressed (as a liquid) and expanded (as a vapor) to absorb and reject heat. The heat pump transfers heat to a space to be heated during the winter period and, by reversing the operation, extracts (absorbs) heat from the same space to be cooled during the summer period. The refrigerant within a heat pump in the heating mode absorbs heat from an outside medium (air, ground, or ground water); in the cooling mode, it absorbs heat from the space to be cooled and disperses it to the outside medium.

Heat pump, air-source: The most common type of heat pump; it absorbs heat from the outside air and transfers the heat to the space to be heated in the heating mode. In the cooling mode, the heat pump absorbs heat from the space to be cooled and rejects the heat to the outside air. When the outside air approaches 32°F or less, air-source heat pumps in the heating mode lose efficiency and generally require a back-up (resistance) heating system.

Heat pump efficiency: The electrical energy required to operate a heat pump is directly related to the temperatures between which it operates. Geothermal heat pumps are more efficient than conventional heat pumps or air conditioners that use the outdoor air, as the temperature of the ground or ground water that is a few feet below the Earth's surface remains relatively constant throughout the year. It is more efficient in the winter to draw heat from the relatively warm ground than from the atmosphere where the air temperature is much colder, and in summer to transfer waste heat to the relatively cool ground rather than to hotter air. Geothermal heat pumps are generally more expensive to install than outside air heat pumps; however, depending on the location, geothermal heat pumps can reduce energy consumption (operating costs) and emissions by more than 20% compared to high-efficiency outside air heat pumps. Geothermal heat pumps also use the waste heat from air-conditioning to provide free hot water heating in the summer.

Heat pump, geothermal: A heat pump in which the refrigeration exchanges heat (in a heat exchanger) with a fluid circulating through an earth connection medium (ground or ground water). The fluid is contained in a variety of loop (pipe) configurations depending on the temperature of the ground and the ground area available. Loops may be installed horizontally or vertically in the ground or submersed in a body of water.

Heating value: The maximum amount of energy available from burning a substance.

Herbaceous: Nonwoody type of vegetation, usually lacking permanent strong stems, such as grasses, cereals, and canola (rapeseed).

High-speed shaft: Transmits force from the gearbox to the generator.

High-temperature collector: A solar thermal collector designed to operate at a temperature of 180°F or higher.

Hub: The center part of the rotor assembly which connects the blades to the low-speed shaft.

Hydraulic stimulation: A stimulation technique performed using fluid.

Hydrothermal: Pertaining to hot water.

Hydrothermal reservoir: An aquifer, or subsurface water, that has sufficient heat, permeability, and water to be exploited without stimulation or enhancement.

Idle cropland: Land in cover and soil improvement crops and cropland on which no crops were planted. Some cropland is idle each year for various physical and economic reasons. Acreage diverted from corps to soil-conserving uses (if not eligible for and used as cropland pasture) under federal farm programs is included in this component; cropland enrolled in the Federal Conservation Reserve Program (CRP) is included in idle cropland.

Incident light: Light that shines onto the face of a solar cell or module.

Induced seismicity: Induced seismicity refers to typically minor earthquakes and tremors that are caused by human activity that alters the stresses and strains on the Earth's crust. Most induced seismicity is of an extremely low magnitude, and in many cases human activity is merely the trigger for an earthquake that would have occurred naturally in any case.

Industrial wood: All commercial roundwood products except fuelwood.

Injection: The process returning spent geothermal fluids to the subsurface; sometimes referred to as *reinjection.*

Interferometric synthetic aperture radar (InSAR): A remote sensing technique that uses radar satellite images to determine movement of the surface of the Earth.

Internal collector storage: A solar thermal collector in which incident solar radiation is absorbed by the storage medium.

Ionomer: A polyelectrolyte comprised of copolymers containing both electrically neutral repeating units and a fraction of ionized units.

Irradiance: The direct, diffuse, and reflected solar radiation that strikes a surface.

Joule: The basic energy unit for the metric system or, in a later more comprehensive formulation, the International System of Units (SI). It is defined in terms of the meter, kilogram, and second.

Kaplan turbine: A type of turbine that has two blades whose pitch is adjustable. The turbine may have gates to control the angle of the fluid flow into the blades.

Kilowatt (kW): One thousand watts of electricity.

Kilowatt-hour (kWh): One thousand watt-hours.

Landfill gas: Gas that is generated by the decomposition of organic material at landfill disposal sites; landfill gas is approximately 50% methane.

Langley (L): Unit of solar irradiance; 1 gram-calorie per square centimeter; 1 L = 85.93 kWh/m^2.

Leeward: Away from the direction of the wind; opposite of windward.

Life-cycle analysis: Analysis focused on the environmental impact of a product during the entirety of its life cycle, from resource extraction to postconsumer waste disposal. It is a comprehensive approach to examining the environmental impacts of a product or package.

Lignin: Structural constituent of wood and (to a lesser extent) other plant tissues that encrusts the cell walls and cements the cells together.

Line shaft pump: Fluid pump that has the pumping mechanism in the wellbore and that is driven by a shaft connected to a motor on the surface.

Liner: A casing string that does not extend to the top of wellbore but instead is anchored or suspended from inside the bottom of the previous casing string.

Liquid collector: A medium-temperature solar thermal collector, employed predominately in water heating, which uses pumped liquid as the heat-transfer medium.

Lithology: The study and description of rocks, in terms of their color, texture, and mineral composition.

Live cull: A classification that includes live cull trees. When associated with volume, it is the net volume in live cull trees that are 5.0 inches in d.b.h. and larger.

Load: The simultaneous demand of all customers required at any specified point in an electric power system.

Load balancing: Keeping the amount of electricity produced (supply) equal to the consumption (demand). This is one of the challenges of wind energy production, which produces energy on a less predicable schedule than other methods.

Local solar time: A system of astronomical time in which the sun crosses the true north–south meridian at 12 noon and which differs from local time according to longitude, time zone, and equation of time.

Logging residues: The unused portions of growing stock and nongrowing stock trees cut or killed by logging and left in the woods.

Lost circulation: Zones in a well that imbibe drilling fluid from the wellbore, thus causing a reduction in the flow of fluid returning to the surface. This loss causes drilled rock particles to build up in the well and can cause problems in cementing the casing in place.

Low-speed shaft: Connects the rotor to the gearbox.

Low-temperature collectors: Metallic or nonmetallic solar thermal collectors that generally operate at temperatures below 110°F and use pumped liquid or air as the heat-transfer medium. They usually contain no glazing and no insulation, and they are often made of plastic or rubber, although some are made of metal.

Magma: Molten rock within the Earth from which igneous rock is formed by cooling.

Magnetic survey: Measurements of the Earth's magnetic field that are then mapped and used to determine subsurface geology.

Magneto–tellurie: An electromagnetic method of determining structures below the Earth's surface using electrical currents and the magnetic field.

Mantle: The Earth's inner layer of molten rock, lying beneath the Earth's crust and above the Earth's core of liquid iron and nickel.

Matrix treatments: Treatments performed below the reservoir fracture pressure; they are generally designed to restore the natural permeability of the reservoir following damage to the near-wellbore area. Matrix treatments typically use hydrochloric or hydrofluoric acids, to remove mineral material that reduces flow into the well.

Medium-temperature collectors: Solar thermal collectors designed to operate in the temperature range of 140 to 180°F but that can also operate at a temperature as low as 110°F. The collector typically consists of a metal frame, metal absorption panel with integral flow channels (attached tubing for liquid collectors or integral ducting for air collectors), and glazing and insulation on the sides and back.

Megawatt (WM): One million watts of electricity.

Methane: A colorless, flammable, odorless hydrocarbon gas (CH_4) that is the major component of natural gas. It is also an important source of hydrogen in various industrial processes. Methane is a greenhouse gas.

Methanol (CH_3OH; also known as methyl alcohol or wood alcohol): Methanol is usually produced by chemical conversion at high temperatures and pressures. Although normally produced from natural gas, methanol can be produced from gasified biomass.

Micro-seismicity: Small movements of the Earth causing fracturing and movement of rocks. Such seismic activity does not release sufficient energy for the events to be recognized except with sensitive instrumentation.

Mini-frac: A small fracturing treatment performed before the main hydraulic fracturing treatment to acquire stress data and to test prestimulation permeability.

MSW (municipal solid waste): Residential solid waste and some nonhazardous commercial, institutional, and industrial wastes.

MTBE (methyl *tert*-butyl ether): A fuel oxygenate that is produced by reacting methanol with isobutylene.

Multiplier effect: Sometimes called the *ripple effect* because a single expenditure in an economy can have repercussions throughout the entire economy. The multiplier is a measure of how much additional economic activity is generated from an initial expenditure.

Nacelle (or cowling): Contains and protects the gearbox and generator, sometimes large enough for an engineer or technician to stand in while doing maintenance.

Net metering: Arrangement that permits a facility (using a meter that reads inflows and outflows of electricity) to sell any excess power it generates over its load requirement back to the electrical grid to offset consumption.

Net photovoltaic cell shipment: The difference between photovoltaic cell shipments and photovoltaic cell purchases.

Net photovoltaic module shipment: The difference between photovoltaic cell shipments and photovoltaic cell purchases.

Net summer capacity: The maximum output, commonly expressed in megawatts, that generating equipment can supply to system load, as demonstrated by a multi-hour test, at the time of summer peak demand. This output reflects a reduction in capacity due to electricity use for station service or auxiliaries.

Nocturnal cooling: The effect of cooling by the radiation of heat from a building to the night sky.

Nonforest land: Land that has never supported forests and lands formerly forested where timber production is precluded by development for other uses; includes area used for crops, improved pasture, residential areas, city parks, improved roads of any width and adjoining clearings, powerline clearings of any width, and 1- to 4.5-acre areas of water classified by the Bureau of the Census as land. If intermingled in forest areas, unimproved roads and nonforest strips must be more than 120 feet wide, and clearings, etc., must be more than 1 acre in area to qualify as nonforest land.

Nonindustrial private: An ownership class of private lands where the owner does not operate wood-using processing plants.

Nonutility generation: Electric generation by nonutility power producers to supply electric power for industrial, commercial, and military operations or for sales to electric utilities.

Nonutility power producer: A corporation, person, agency, authority, or other legal entity or instrumentality that owns electric generating capacity and is not an electrical utility. Nonutility power produces include qualifying cogenerators, qualifying small power produces, and other nonutility generators without a designated, franchised service area that do not file forms listed in the Code of Federal Regulations, Title 18, Part 141.

Occupied space: The space within a building or structure that is normally occupied by people and that may be conditioned (heated, cooled, or ventilated).

Other biomass: Includes agricultural byproducts/crops (agricultural byproducts, straw), other biomass gas (digester waste alcohol), other biomass liquids (fish oil liquid acetonitrite, waste, tall oil, waste alcohol), and other biomass solids (medical waste, solid byproducts, sludge waste, tires).

Other forest land: Forest land other than timberland and reserved forest land; includes available forest land incapable of annually producing 20 ft^3/acre of industrial wood under natural conditions because of adverse site conditions such as sterile soils, dry climate, poor drainage, high elevation, steepness, or rockiness.

Other removals: Wood volume not utilized from cut or otherwise killed growing stock, from cultural operations such as precommercial thinnings, or from timberland clearing; does not include volume removed from inventory through reclassification of timber land to productive reserved forest land.

Other sources of roundwood: Sources of roundwood products that are not growing stock. Include available dead, rough, and rotten trees; trees of noncommercial species; trees less the 5.0 inches d.b.h.; tops; and roundwood harvested from nonforest land (e.g., fence rows).

Packer: Device that can be placed in the wellbore to block vertical fluid flow so as to isolate zones.

Paper pellets: Paper compressed and bound into uniform diameter pellets to be burned in a heating stove.

Parabolic dish: A high-temperature (above 180°F) solar thermal concentrator, generally bow shaped, with two-axis tracking.

Passive solar: A system in which solar energy alone is used for the transfer of thermal energy. Pumps, blowers, or other heat-transfer devices that use energy other than solar are not used.

Peak watt: A manufacturer's unit indicating the amount of power a photovoltaic cell or module will produce at standard test conditions (normally 1000 W/m^2 and 25°C).

Peaking plants: Electricity generating plants that are operated to meet the peak or maximum load on the system. The cost of energy from such plants is usually higher than from baseload plants.

Peat: Peat consists of partially decomposed plant debris. It is considered an early stage in the development of coal. Peat is distinguished from lignite by the presence of free cellulose and a high moisture content (exceeding 70%). The heat content of air-dried peat (about 50% moisture) is about 9 million Btu/ton. Most U.S. peat is used as a soil conditioner.

Permeability: The ability of a rock to transmit fluid through its pores or fractures when subjected to a difference in pressure. Typically measured in darcies or millidarcies.

Photovoltaic (PV) cell: An electronic device consisting of layers of semiconductor materials fabricated to form a junction (adjacent layers of materials with different electronic characteristics) and electrical contacts and being capable of converting incident light directly into electricity (direct current).

Photovoltaic (PV) module: An integrated assembly of interconnected photovoltaic cells designed to deliver a selected level of working voltage and current at its output terminals, packaged for protection against environmental degradation, and suited for incorporation in photovoltaic power systems.

Plate tectonics: A theory of global-scale dynamics involving the movement of many rigid plates of the Earth's crust. Tectonic activity is evident along the margins of the plates where buckling, grinding, faulting, and vulcanism occur as the plates are propelled by the forces of deep-seated mantle convection current. Geothermal resources are often associated with tectonic activity that allows groundwater to come in contact with deep subsurface heat sources.

Poletimber trees: Live trees at least 5.0 inches in d.b.h. but smaller than sawtimber trees.

Polycrystalline diamond compact (PDC) drilling bit: A drilling bit that uses polycrystalline diamond compact cutters to shear rock with a continuous scraping motion.

Porosity: The ratio of the aggregate volume of pore spaces in rock or soil to its total volume, usually stated as a percent.

Primary wood-using mill: A mill that converts roundwood products into other wood products. Common examples are sawmills that convert logs into lumber and pulp mills that convert pulpwood roundwood into wood pulp.

Process heating: The direct process end use in which energy is used to raise the temperature of substances involved in the manufacturing process.

Proppant: Suspended particles in the fracturing fluid that are used to hold fractures open after a hydraulic fracturing treatment, thus producing a pathway that fluids can easily flow along.

Protonic ceramic fuel cell: Fuel cell based on a ceramic electrolyte material that exhibits high protonic conductivity at elevated temperatures.

Public Utility Regulatory Policies Act of 1978 (PURPA): Contains measures designed to encourage the conservation of energy, more efficient use of resources, and equitable rates. Principal among these are suggested retail rate reforms and new incentives for production of electricity by cogenerators and uses of renewable resources.

Pulpwood: Roundwood, whole tree chips, or wood residues that are used for the production of wood pulp.

Pumped-storage hydroelectric plant: A plant that usually generates electric energy during peak load periods by using water previously pumped into an elevated storage reservoir during off-peak periods when excess generating capacity is available to do so. When additional generating capacity is needed, the water can be released from the reservoir through a conduit to turbine generators located in a power plant at a lower level.

Pyrolysis: Thermal decomposition of biomass at high temperatures (greater than 400°F, or 200°C) in the absence of air. The end product is a mixture of solids (char), liquid (oxygenated oils), and gases (methane, carbon monoxide, and carbon dioxide), the proportions of which are determined by operating temperature, pressure, oxygen content, and other conditions.

Quadrillion Btu (quad): Equivalent to 10^{15} Btu.

Qualifying facility (QF): A cogeneration or small power production facility that meets certain ownership, operating, and efficiency criteria established by the Federal Energy Regulatory Commission (FERC) pursuant to the Public Utility Regulatory Polices Act of 1978 (PURPA).

Rankine cycle: The thermodynamic cycle that is an ideal standard for comparing the performance of heat engines, steam power plants, steam turbines, and heat pump systems that use a condensate vapor as the working fluid; efficiency is measured as work done divided by sensible heat supplied.

Recovery factor: The fraction of total resource that can be extracted for productive uses.

Renewable energy: Energy that is produced using resources that regenerate quickly or are inexhaustible. Wind energy is considered inexhaustible, because although it may blow intermittently it will never stop.

Renewable energy resources: Energy resources that are naturally replenishing but flow limited. They are virtually inexhaustible in duration but limited in the amount of energy that is available per unit of time. Renewable energy resources include biomass, hydro, geothermal, solar, wind, ocean thermal, wave action, and tidal action.

Renewable Portfolio Standard (RPS): A mandate requiring that renewable energy provide a certain percentage of total energy generation or consumption.

Reservoir: A natural underground container of liquids, such as water or steam (or, in the petroleum context, oil or gas).

Residues: Bark and woody materials that are generated in primary wood-using mills when roundwood products are converted to other products. Examples are slabs, edgings, trimmings, sawdust, shavings, veneer cores and clippings, and pulp screenings. Bark residues and wood residues (both coarse and fine materials) are included, but not logging residues.

Resistivity survey: Measurement of the ability of a material to resist or inhibit the flow of an electrical current, measured in ohm-meters. Resistivity is measured by inducing a current between two electrodes and measuring the resulting potential at other electrodes. Resistivity surveys can be used to delineate the boundaries of geothermal fields.

Resource base: All of a given material in the Earth's crust, whether its existence is known or unknown, and regardless of cost considerations.

Ribbon silicon: Single-crystal silicon derived by means of fabricating processes that produce sheets or ribbons of single-crystal silicon. These processes include edge-defined film-fed growth, dendritic web growth, and ribbon-to-ribbon growth.

Roller cone bit: Drill bit with studded rotating cones attached to it that crushes rock.

Rotten tree: A live tree of commercial species that does not contain a saw log now or prospectively primarily because of rot (that is, when rot accounts for more than 50% of the total cull volume).

Rough tree: A live tree of commercial species that does not contain a saw log now or prospectively primarily because of roughness (that is, when sound cull, due to such factors as poor form, splits, or cracks, accounts for more than 50% of the total cull volume), or a live tree of noncommercial species.

Roundwood: Wood cut specifically for use as a fuel.

Roundwood products: Logs and other round timber generated from harvesting trees for industrial or consumer use.

Salvable dead tree: A downed or standing dead tree that is considered currently or potentially merchantable by regional standards.

Salinity: A measure of the quantity or concentration of dissolved salts in water.

Saplings: Live trees 1.0 inch through 4.9 inches in d.b.h.

Secondary wood processing mills: A mill that uses primary wood products in the manufacture of finished wood products, such as cabinets, moldings, and furniture.

Seismic: Pertaining to, of the nature of, or caused by an earthquake or earth vibration, natural or manmade.

Seismicity: Earth movements; also, the frequency, distribution, and intensity of earthquakes.

Seismometer: Electrical device that is used on the surface and within wellbores to measure the magnitude and direction of seismic events.

Self-potential: Self-potential in geothermal systems measures currents induced in the subsurface because of the flow of fluids.

Silicon: A semiconductor material made from silica, purified for photovoltaic applications.

Single-crystal silicon, Czochralski: An extremely pure form of crystalline silicon produced by the Czochralski method of dipping a single crystal seed into a pool of molten silicon under high-vacuum conditions and slowly withdrawing a solidifying single-crystal boule rod of silicon. The boule is sawed into thin wafers and fabricated into single-crystal photovoltaic cells.

Skylight: A window located on the roof of a structure to provide interior building spaces with natural daylight, warmth, and ventilation.

Slim hole: Drill holes that have a nominal inside diameter less than about 6 inches.

Slotted liner: Liner that has slots or holes in it to let fluid pass between the wellbore and surrounding rock.

Sludge: A dense, slushy, liquid-to-semifluid product that accumulates as an end result of an industrial or technological process designed to purify a substance. Industrial sludges are produced from the processing of energy-related raw materials, chemical products, water, mined ores, sewerage, and other natural and manmade products. Sludges can also form from natural processes, such as the runoff produced by rainfall, and can accumulate on the bottom of bogs, streams, lakes, and tidelands.

Smart tracer: Tracer that is useful in determining the flow path between a well injecting fluid into the subsurface and a well producing fluid from an adjacent well; it can also be used to determine temperature along the flow path, the surface area contacted by the tracer, the volume of rock that the tracer interacts with, and the relative velocities of separate phases (gas, oil, and water in petroleum fields; steam and liquid water in geothermal systems).

Solar energy: The radiant energy of the sun which can be converted into other forms of energy, such as heat or electricity.

Solar heat gain coefficient (SHGC): Measures how well a product blocks heat caused by sunlight. The SHGC is expressed as a number between 0 and 1. The lower the SHGC, the less solar heat it transmits.

Solar thermal collector: A device designed to receive solar radiation and convert it into thermal energy. Normally, a solar thermal collector includes a frame, glazing, and an absorber, together with the appropriate insulation. The heat collected by the solar thermal collector may be used immediately or stored for later use.

Solar thermal collector, special: An evacuated-tube collector or a concentrating (focusing) collector; special collectors operate in the temperature range from just above ambient temperature (low concentration for pool heating) to several hundred degrees Fahrenheit (high concentration for air conditioning and specialized industrial processes).

Spent liquor: The liquid residue left after an industrial process; can be a component of waste materials used as fuel.

Spent sulfite liquor: End product of pulp and paper manufacturing processes that contains lignins and has a high moisture content; often reused in recovery boilers. Similar to black liquor.

Spinner survey: The use of a device with a small propeller that spins when fluid passes to measure fluid flow in a wellbore. The device is passed up and down the well, continuously measuring flow to establish where and how much fluid enters or leaves the wellbore at various depths.

Spinning reserve: Generation capacity that is online but unloaded and that can respond within 10 minutes to compensate for unexpected increases in demand or a decrease in supply.

Start-up speed: The wind speed at which a rotor begins to rotate.

Stimulation: A treatment performed to restore or enhance the productivity of a well. Stimulation treatments fall into two main groups: hydraulic fracturing treatments and matrix treatments.

Stress: The forces acting on rock. In the subsurface, the greatest force or stress is generally vertical and caused by the weight of overlying rock.

Structural discontinuity: A discontinuity of the rock fabric that can be a fracture, fault, intrusion, or differing adjacent rock type.

Submersible sump: Pump with both the pumping mechanism and a driving electric motor suspended together at depth in the well.

Subsidence: Sinking of an area of the Earth's crust due to fluid withdrawal and pressure decline.

Syngas: A synthesis gas produced through gasification of biomass; similar to natural gas. It can be cleaned and conditioned to form a feedstock for production of methanol.

TDS (total dissolved solids): Describes the amount of solid materials in water.

Tall oil: The oily mixture of rosin acids, fatty acids, and other materials obtained by acid treatment of the alkaline liquors from the digesting (pulping) of pine wood.

Thermal drawdown: Decline in formation temperature due to geothermal production.

Thermal gradient: The rate of increase in temperature as a function of depth into the Earth's crust.

Thermosiphon system: A solar collector system for water heating in which circulation of the collection fluid through the storage loop is provided solely by the temperature and density difference between the hot and cold fluids.

Thin-film silicon: A technology in which amorphous or polycrystalline material is used to make photovoltaic (PV) cells.

Tiltmeter: Device able to measure extremely small changes in its rotation from horizontal. The tilt measured by an array of tiltmeters placed over a stimulation allows delineation of inflation and fracturing caused by the stimulation.

Tip speed ratio (TSR): The ratio between the wind speed and the speed of the tips of the wind turbine blades.

Tipping fee: A fee for disposal of waste.

Tower: Steel structures that support a turbine assembly; higher towers allow for longer blades and the capture of faster moving air at higher altitudes.

Tracer: A chemical injected into the flow stream of a production or injection well to determine fluid path and velocity.

Transformer: Used to step up or step down AC voltage or AC current.

Transmission line: Structures and conductors that carry bulk supplies of electrical energy from power-generating units.

Transmission system (electric): An interconnected group of electric transmission lines and associated equipment for moving or transferring electric energy in bulk between points of supply and points at which it is transformed for delivery over the distribution system lines to consumers or is delivered to other electric systems.

Transportation sector: An energy-consuming sector that consists of all vehicles whose primary purpose is transporting people or goods from one physical location to another. Included are automobiles; trucks; buses; motorcycles; trains, subways, and other rail vehicles; aircraft; and ships, barges, and other waterborne vehicles. Vehicles whose primary purpose is not transportation (e.g., construction cranes and bulldozers, farming vehicles, warehouse tractors and forklifts) are classified in the sector of their primary use.

Turbine: A machine for generating rotary mechanical power from the energy of a stream of fluid (such as water, steam, or hot gas). Turbines convert the kinetic energy of fluids to mechanical energy through the principles of impulse and reaction, or a mixture of the two.

U-factor: Measures the rate of heat loss or how well a product prevents heat from escaping. It includes the thermal properties of the frame as well as the glazing. The insulating value is indicated by the R-value, which is the inverse of the U-factor. U-factor ratings generally fall between 0.20 and 1.20. The lower the U-factor, the greater a product's resistance to heat flow and the better its insulating value.

Under reamer: A heavy-duty drilling tool for enlarging a borehole significantly beyond the bit diameter. The device is placed about the drill bit; it can be opened during drilling and then closed to be brought back up through smaller diameter holes or casings.

Upwind turbine: A turbine that faces into the wind; it requires a wind vane and yaw drive to maintain proper orientation in relation to the wind.

Useful heat: Heat stored above room temperature (in a solar heating system).

Useful thermal output: The thermal energy made available for use in any industrial or commercial process or used in any heating or cooling application; the total thermal energy made available for processes and applications other than electrical generation.

Vapor dominated: A geothermal reservoir system in which subsurface pressures are controlled by vapor rather than by liquid. Sometimes referred to as a *dry-steam reservoir.*

Viewshed: The scenic characteristics of an area, when referred to as a resource.

Visible light transmittance: The amount of visible light that passes through the glazing material of a window, expressed as a percentage.

Voltage: The measure of electrical potential difference.

Watt (electric): The electrical unit of power; the rate of energy transfer equivalent to 1 ampere of electric current flowing under a pressure of 1 volt at unity power factor.

Watt-hour (Wh): The electrical energy unit of measure equal to 1 watt of power supplied to, or taken from, an electric circuit steadily for 1 hour.

Watt (thermal): A unit of power in the metric system, expressed in terms of energy per second, equal to the work done at a rate of 1 joule per second.

Well log: Logging includes measurements of the diameter of the well and various electrical, mass, and nuclear properties of the rock which can be correlated with physical properties of the rock. The well log is a chart of the measurement relative to depth in the well.

Wind energy: Energy present in wind motion that can be converted to mechanical energy for driving pumps, mills, and electric power generators. Wind pushes against sails, vanes, or blades radiating from a central rotating shaft.

Wind power plant: A group of wind turbines interconnected to a common utility system through a system of transformers, distribution lines, and (usually) one substation. Operation, control, and maintenance functions are often centralized through a network of computerized monitoring systems, supplemented by visual inspection. In Europe, this type of plant is called a *generating station.*

Wind rose: A diagram that indicates the average percentage of time that the wind blows from different directions on a monthly or annual basis.

Wind turbine or windmill: A device for harnessing the kinetic energy of the wind and using it to do work, or generate electricity.

Wind vane: Wind direction measurement device used to send data to the yaw drive.

Windward: Into, or facing the direction of, the wind; opposite of leeward.

Wood/wood waste: Category of biomass energy that includes black liquor, wood/wood waste liquids (red liquor, sludge wood, spent sulfite liquor), and wood/wood waste solids (peat, paper pellets, railroad ties, utility poles, wood/wood waste).

Wood energy: Wood and wood products used as fuel, including roundwood (cord wood), limb wood, wood chips, bark, sawdust, forest residues, charcoal, pulp waste, and spent pulping liquor.

Wood pellets: Sawdust compressed into uniform-diameter pellets to be burned in a heating stove.

Yaw: The rotation of a horizontal axis wind turbine around its tower or vertical axis.

Yaw drive: Motor that keeps an upwind turbine facing into the wind.

Zonal isolation: Various methods to selectively partition portions of the wellbore for stimulation, testing, flow restriction, or other purpose.

Zone: Area within the interior space of a building that is to be cooled, heated, or ventilated. A zone has its own thermostat to control the flow of conditioned air into the space.

Index

D

E

F

R

S

T

U

V

An environmentally friendly book printed and bound in England by www.printondemand-worldwide.com

PEFC Certified
This product is from sustainably managed forests and controlled sources
www.pefc.org
PEFC/16-33-415

MIX
Paper from responsible sources
FSC® C004959
www.fsc.org

This book is made entirely of sustainable materials; FSC paper for the cover and PEFC paper for the text pages.

#0116 - 070114 - C0 - 254/178/18 [20] - CB